Nanopriming Approach to Sustainable Agriculture

Abhishek Singh
Faculty of Biology, Yerevan State University, Armenia

Vishnu D. Rajput
Academy of Biology and Biotechnology, Southern Federal University, Rostov-on-Don, Russia

Karen Ghazaryan
Faculty of Biology, Yerevan State University, Armenia

Santosh Kumar Gupta
National Institute of Plant Genome Research, India

Tatiana Minkina
Academy of Biology and Biotechnology, Southern Federal University, Rostov-on-Don, Russia

IGI Global
PUBLISHER of TIMELY KNOWLEDGE

A volume in the Advances in Environmental Engineering and Green Technologies (AEEGT) Book Series

Published in the United States of America by
IGI Global
Engineering Science Reference (an imprint of IGI Global)
701 E. Chocolate Avenue
Hershey PA, USA 17033
Tel: 717-533-8845
Fax: 717-533-8661
E-mail: cust@igi-global.com
Web site: http://www.igi-global.com

Library of Congress Cataloging-in-Publication Data

Names: Singh, Abhishek, 1994- editor. | Rajput, Vishnu D., editor. | Ghazaryan, Karen, editor. | Gupta, Santosh Kumar, 1978- editor. | Minkna, Tatiana, editor.
Title: Nanopriming approach to sustainable agriculture / edited by: Abhishek Singh, Vishnu D. Rajput, Karen Ghazaryan, Santosh Kumar Gupta, and Tatiana Minkna.
Description: Hershey, PA : Engineering Science Reference, [2023] | Includes bibliographical references and index. | Summary: "Our book deals with the advanced technologies use of nanotechnology, in agriculture for crop production, protection, soil fertility improvement, crop improvement, etc. This book intends to provide an overview of some of the most promising technologies with precision agriculture from an economic point of view. Each chapter has been put together so that it can be read individually should the reader wish to focus on one particular topic. Precision Farming as a farm technology benefits from large-scale advantages due to relatively high investment costs and is primarily adopted on farms with medium to large field areas. However, this edited book deals with the advanced technologies used in agriculture like nanotechnologies with up-to-date information for farmers, students, researchers, and teachers to build new concepts"-- Provided by publisher.
Identifiers: LCCN 2023017045 (print) | LCCN 2023017046 (ebook) | ISBN 9781668472323 (hardcover) | ISBN 9781668472330 (paperback) | ISBN 9781668472347 (ebook)
Subjects: LCSH: Seed technology. | Nanotechnology. | Sustainable agriculture. | Precision farming.
Classification: LCC SB113.2 .N36 2023 (print) | LCC SB113.2 (ebook) | DDC 630.2/086--dc23/eng/20230616
LC record available at https://lccn.loc.gov/2023017045
LC ebook record available at https://lccn.loc.gov/2023017046

This book is published in the IGI Global book series Advances in Environmental Engineering and Green Technologies (AEEGT) (ISSN: 2326-9162; eISSN: 2326-9170)

British Cataloguing in Publication Data
A Cataloguing in Publication record for this book is available from the British Library.

All work contributed to this book is new, previously-unpublished material.
The views expressed in this book are those of the authors, but not necessarily of the publisher.

For electronic access to this publication, please contact: eresources@igi-global.com.

Advances in Environmental Engineering and Green Technologies (AEEGT) Book Series

ISSN:2326-9162
EISSN:2326-9170

MISSION

Growing awareness and an increased focus on environmental issues such as climate change, energy use, and loss of non-renewable resources have brought about a greater need for research that provides potential solutions to these problems. Research in environmental science and engineering continues to play a vital role in uncovering new opportunities for a "green" future.

The **Advances in Environmental Engineering and Green Technologies (AEEGT)** book series is a mouthpiece for research in all aspects of environmental science, earth science, and green initiatives. This series supports the ongoing research in this field through publishing books that discuss topics within environmental engineering or that deal with the interdisciplinary field of green technologies.

COVERAGE

- Air Quality
- Renewable Energy
- Industrial Waste Management and Minimization
- Cleantech
- Biofilters and Biofiltration
- Green Transportation
- Alternative Power Sources
- Green Technology
- Sustainable Communities
- Contaminated Site Remediation

IGI Global is currently accepting manuscripts for publication within this series. To submit a proposal for a volume in this series, please contact our Acquisition Editors at Acquisitions@igi-global.com or visit: http://www.igi-global.com/publish/.

The Advances in Environmental Engineering and Green Technologies (AEEGT) Book Series (ISSN 2326-9162) is published by IGI Global, 701 E. Chocolate Avenue, Hershey, PA 17033-1240, USA, www.igi-global.com. This series is composed of titles available for purchase individually; each title is edited to be contextually exclusive from any other title within the series. For pricing and ordering information please visit http://www.igi-global.com/book-series/advances-environmental-engineering-green-technologies/73679. Postmaster: Send all address changes to above address.

Titles in this Series

For a list of additional titles in this series, please visit:
http://www.igi-global.com/book-series/advances-environmental-engineering-green-technologies/73679

Perspectives on Ecological Degradation and Technological Progress
Veli Yilanci (Faculty of Political Sciences, Canakkale Onsekiz Mart University, Turkey)
Engineering Science Reference • copyright 2023 • 342pp • H/C (ISBN: 9781668467275) • US $225.00 (our price)

Contemporary Developments in Agricultural Cyber-Physical Systems
G.S. Karthick (Department of Software Systems, PSG College of Arts and Science, India)
Engineering Science Reference • copyright 2023 • 320pp • H/C (ISBN: 9781668478790) • US $240.00 (our price)

Nematode-Plant Interactions and Controlling Infection
Waleed Fouad Abobatta (Horticulture Research Institute, Agriculture Research Center, Egypt) and Rehab Yasin Ghareeb (Plant Protection and Biomolecular diagnosis Dep., Arid Lands Cultivation Research Institute, City of Scientific Research and Technological Applications, Egypt)
Engineering Science Reference • copyright 2023 • 300pp • H/C (ISBN: 9781668480830) • US $240.00 (our price)

Perspectives on Global Biodiversity Scenarios and Environmental Services in the 21st Century
Naveen Kumar Chourasia (Government College, Bichhua, India) and Kavita Chahal (Government College, Bichhua, Chhindwara, India)
Engineering Science Reference • copyright 2023 • 300pp • H/C (ISBN: 9781668490341) • US $250.00 (our price)

Handbook of Research on AI-Equipped IoT Applications in High-Tech Agriculture
Alex Khang (Global Research Institute of Technology and Engineering, USA)
Engineering Science Reference • copyright 2023 • 450pp • H/C (ISBN: 9781668492314) • US $335.00 (our price)

For an entire list of titles in this series, please visit:
http://www.igi-global.com/book-series/advances-environmental-engineering-green-technologies/73679

IGI Global
PUBLISHER of TIMELY KNOWLEDGE

701 East Chocolate Avenue, Hershey, PA 17033, USA
Tel: 717-533-8845 x100 • Fax: 717-533-8661
E-Mail: cust@igi-global.com • www.igi-global.com

Table of Contents

Foreword xx

Preface xxi

Acknowledgment xxiii

Section 1
Nanopriming: An Introduction

Chapter 1
An Introduction to Nanopriming for Sustainable Agriculture 1
Shreni Agrawal, Parul Institute of Applied Science, Parul University, India
Richa Das, Parul Institute of Applied Science, Parul University, India
Shivangee Solanki, Parul Institute of Applied Science, Parul University, India
Simran Choudhury, Parul Institute of Applied Science, Parul University, India
Indrani Bhattacharya, Parul Institute of Applied Science, Parul University, India
Pradeep Kumar, MMV, Banaras Hindu University, India
Amit kumar Singh, Indian Institute of Technology, Banaras Hindu University, India
Sunil Kumar Mishra, Indian Institute of Technology, Banaras Hindu University, India
Kavindra Nath Tiwari, MMV, Banaras Hindu University, India

Section 2
Role of Nanopriming for the Sustainability of Agriculture

Chapter 2
Nanopriming Technology for Sustainable Agriculture 21
Sapna Rawat, Delhi University, India

Chapter 3
The Potential of Nano-Based Seed Priming for Sustainable Agriculture............43
Annika Arvind Gupta, B.K. Birla College, India
Arvind Radheyshyam Gupta, Bioxia, India

Section 3
Role of Nanopriming for Crop Production and Protection

Chapter 4
Carbon Nano Tubes (CNTs) as a Tool of Seed Quality Enhancement Using Nanopriming Approach: A Review 90
Manish Kumar Vijay, ICFRE-Tropical Forest Research Institute, Jabalpur, India
Radheshyam Sharma, Jawaharlal Nehru KrishiVishwaVidyalaya, India

Chapter 5
Nanopriming for Crop Management for Sustainable Agriculture: An Overview...110
Muhammad Majeed, University of Gujrat, Pakistan
Murad Muhammad, Chinese Academy of Sciences, China & University of Peshawar, Pakistan
Sehar Nawaz, University of Agriculture, Faisalabad, Pakistan
Tayyaba Naz, University of Agriculture, Faisalabad, Pakistan
Muhammad Mazhar Iqbal, Ayub Agricultural Research Institute, Department of Agriculture, Government of Punjab, Pakistan
Nafeesa Zahid, University of Kotli, Pakistan
Mumtaz Hussain, Arid Zone Research Institute, Pakistan
Allah Nawaz Khan, Chinese Academy of Sciences, China
Ghulam Abbas, University of Agriculture, Pakistan
Allah Bakhsh Gulshan, Ghazi University, Pakistan
Maria Mehboob, University of Poonch, Pakistan

Chapter 6
Nanoparticle-Based Priming-Enhanced Agricultural Practices: Promoting Sustainable Plant Growth and Development...142
Abhishek Singh, Faculty of Biology, Yerevan State University, Armenia
Sapna Rawat, University of Delhi, India
Vishnu D. Rajput, Academy of Biology and Biotechnology, Southern Federal University, Rostov-on-Don, Russia
Karen Ghazaryan, Faculty of Biology, Yerevan State University, Armenia
Tatiana Minkina, Academy of Biology and Biotechnology, Southern Federal University, Rostov-on-Don, Russia
Abdel Rahman Mohammad Al Tawaha, Al Hussein bin Talal University,

Jordan
Athanasios Alexiou, Novel Global Community Educational Foundation, Australia
Karansher Singh Sandhu, Bayer Crop Sciences, USA

Section 4
Nanopriming-Based Plant Breeding and Omics Response for Sustainable Agriculture

Chapter 7
Emerging Frontiers in Plant Breeding Molecular Markers and Nanopriming for Enhanced Crop Improvement 159
Sapna Rawat, Delhi University, India

Chapter 8
Impact of Nanopriming and Omics-Based Applications for Sustainable Agriculture 175
Himani Sharma, Uttaranchal University, India
Atin Kumar, Uttaranchal University, India
Rajat Singh, Uttaranchal University, India
Satendra Kumar, Sardar Vallabhbhai Patel University of Agriculture and Technology, India

Section 5
Nanopriming and Geoinformatics: Combined Approaches for Crop Production and Soil Health Monitoring

Chapter 9
Nanopriming and Geoinformatics: A Combined Approach for Crop Production for Monitoring of Soil Health 193
P. Smriti Rao, Integral University, India
Setu Ratnam, Integral University, India
Ashish David, Sam Higginbottom University of Agricultural Science and Technology, India
Raghunandan Khatana, Krishi Vigyan Kendra, Abusar, India
Akshita Barthwal, R&D Bayer Crop Science, India

Section 6
Nanopriming-Based Biotic and Abiotic Stress Management

Chapter 10
Nanopriming: A Potential Solution for Heavy Metal Toxicity Alleviation in Crops 221

Sachin Kumar, Himachal Pradesh Agriculture University, Palampur, India
Chiranjeev Kumar, Himachal Pradesh Agriculture University, Palampur, India
Shilpa Manhas, Himachal Pradesh Agriculture University, Palampur, India
Ankita Sharma, Punjab Agricultural University, India
Richa Jaswal, Himachal Pradesh Agriculture University, Palampur, India

Chapter 11
Nanopriming Effective Tools for the Field of Horticultural Crops to Copping Up With Biotic and Abiotic Stresses 241
Shipra Singh Parmar, Dr. Y.S. Parmar University of Horticulture and Forestry, India
Divya Pandey, Dr. Y.S. Parmar University of Horticulture and Forestry, India

Chapter 12
Nanopriming-Based Management of Biotic Stresses for Sustainable Agriculture 263
Muhammad Majeed, University of Gujrat, Pakistan
Murad Muhammad, Chinese Academy of Sciences, China & University of Peshawar, Pakistan
Tanveer Hussain, Mirpur University of Science and Technology, Pakistan
Muqarrab Ali, Muhammad Nawaz Shareef University of Agriculture, Pakistan
Tayyaba Naz, Institute of Soil and Environmental Sciences, University of Agriculture, Faisalabad, Pakistan
Sehar Nawaz, University of Agriculture, Faisalabad, Pakistan
Muhammad Mazhar Iqbal, Government of Punjab, Pakistan
Ghulam Abbas, University of Agriculture, Dera Ismail Khan, Pakistan

Chapter 13
Seed Nanopriming: An Innovative Approach for Upregulating Seed Germination and Plant Growth Under Salinity Stress 290
Abhishek Singh, Faculty of Biology, Yerevan State University, Armenia
Shreni Agrawal, Parul Institute of Applied Science, Parul University, India
Vishnu D. Rajput, Academy of Biology and Biotechnology, Southern Federal University, Rostov-on-Don, Russia
Karen Ghazaryan, Faculty of Biology, Yerevan State University,

Armenia
Hasmik S. Movsesyan, Yerevan State University, Armenia
Tatiana Minkina, Academy of Biology and Biotechnology, Southern Federal University, Rostov-on-Don, Russia
Abdel Rahman Mohammad Al Tawaha, Department of Science and Engineering, Al Hussein bin Talal University, Jordan
Athanasios Alexiou, Department of Science and Engineering, Novel Global Community Educational Foundation, Hebersham, Australia
Badal Singh, ICAR-National Bureau of Plant Genetic Resources, Pusa, India
Santosh Kumar Gupta, National Institute of Plant Genome Research, India

Section 7
Future Role of Nanopriming-Based Approaches for Food Security and Sustainable Agriculture

Chapter 14
Nanotechnology for Sustainability and Food Security in Agriculture: A Nanopriming Long Story in Short 315
Omkar Singh, Department of Soil Science, Sardar Vallabhbhai Patel University of Agriculture and Technology, India

Chapter 15
Nanotechnology: A New Innovation in Biofortification and Food Safety 340
Shivangi Singh, Sardar Vallabhbhai Patel University of Agriculture and Technology, India
Omkar Singh, Sardar Vallabhbhai Patel University of Agriculture and Technology, India
Uday Pratap Shahi, Sardar Vallabhbhai Patel University of Agriculture and Technology, India

Chapter 16
Nanopriming Revolutionizing Agribiotechnology for Sustainable Food Security 352
Varsha Gupta, Bureau of Indian Standards, India
Santosh Kumar Gupta, National Institute of Plant Genome Research, India

Compilation of References 363

About the Contributors 442

Index 450

Detailed Table of Contents

Foreword xx

Preface xxi

Acknowledgment xxiii

Section 1
Nanopriming: An Introduction

Chapter 1
An Introduction to Nanopriming for Sustainable Agriculture 1
Shreni Agrawal, Parul Institute of Applied Science, Parul University, India
Richa Das, Parul Institute of Applied Science, Parul University, India
Shivangee Solanki, Parul Institute of Applied Science, Parul University, India
Simran Choudhury, Parul Institute of Applied Science, Parul University, India
Indrani Bhattacharya, Parul Institute of Applied Science, Parul University, India
Pradeep Kumar, MMV, Banaras Hindu University, India
Amit kumar Singh, Indian Institute of Technology, Banaras Hindu University, India
Sunil Kumar Mishra, Indian Institute of Technology, Banaras Hindu University, India
Kavindra Nath Tiwari, MMV, Banaras Hindu University, India

Plant seeds are of paramount importance in the agricultural sector as well as plant life cycle, and are being severely threatened by environmental changes such as biotic and abiotic stress, resource deficiency, climate fluctuation, etc. Cue, nano-priming, a nascent yet innovative approach to seed technology, combining versatility of nanoparticles and engineered nanomaterials with 'seed-priming' to induce adaptive physiological changes, thus, improving seed quality and crop yield. This study reported mechanisms by which nano-priming synchronizes seed germination, breaks seed dormancy, boosts seed vigor, increases tolerance to periodic stress conditions, assists phyto-microbiota to thrive under environmental uncertainties, creates nanopores in shoots along with upregulation of aquaporin genes for better water uptake, and other reported changes at molecular, biochemical level. Withal, this study offers insight on the future of nano-priming when blended with cutting-edge technologies like cold plasma, artificial intelligence, and digital analysis to revolutionize the agrarian sector.

Section 2
Role of Nanopriming for the Sustainability of Agriculture

Chapter 2
Nanopriming Technology for Sustainable Agriculture 21
Sapna Rawat, Delhi University, India

Climate change has had detrimental impacts on the growth and development of crops, leading to reduced crop production and endangering global food security. However, nanotechnology offers a promising solution to address these challenges in agriculture, ushering in the era of "NanoTech-Agril." By employing seed priming with nanoparticles, farmers can mitigate the impacts of climate change such as salinity, drought, and floods, thereby enhancing crop yields. This seed priming method improves seed metabolism and signaling pathways, leading to enhanced germination and overall plant growth throughout their life cycle. Additionally, nano-seed priming activates various metabolic, biochemical, antioxidant, and phytohormone pathways, bolstering plants against abiotic and biotic stresses. Consequently, the reliance on pesticides and fertilizers can be reduced. This review presents a comprehensive overview of the applications of nanoparticles in sustainable agriculture, providing valuable insights into their potential uses.

Chapter 3
The Potential of Nano-Based Seed Priming for Sustainable Agriculture 43
Annika Arvind Gupta, B.K. Birla College, India
Arvind Radheyshyam Gupta, Bioxia, India

Plant seeds are an important input in agriculture; however, environmental stresses can adversely affect seed vitality, development, and crop production during their developmental stages which results in decreased crop production. High salinity, drought, and climate change-resistant seeds can increase crop yield. The "NanoTech-Agril" era is novel in nano-based technology, having solutions to climate change-related problems in agriculture, empowering farmers to produce a greater number of crops while avoiding problems associated with climate change. Nanoparticle-based seed priming can improve seed metabolism and signaling pathways, helping improve seed growth. Nano-priming induces the formation of nanopores in shoots, aiding in water absorption, activates (ROS)/antioxidant mechanisms in seeds, forms hydroxyl radicals to loosen cell walls, and acts as an inducer for starch hydrolysis present chapter aims to provide an overview of nanoparticle-based seed priming for sustainable agriculture.

Section 3
Role of Nanopriming for Crop Production and Protection

Chapter 4
Carbon Nano Tubes (CNTs) as a Tool of Seed Quality Enhancement Using Nanopriming Approach: A Review 90
Manish Kumar Vijay, ICFRE-Tropical Forest Research Institute, Jabalpur, India
Radheshyam Sharma, Jawaharlal Nehru KrishiVishwaVidyalaya, India

Nano seed technology is a new platform with endless possibilities and impacts on applied agriculture and forestry research. The implementation of nanotechnology in seed research is in its infancy and needs more pace to meet the needs of current global demand. Carbon nanotubes (CNTs) are allotropes of carbon and engineered nanomaterials. Based on the number of concentric layers of rolled graphene sheets, they are classified into single-walled CNTs (SWCNTs) and multi-walled CNTs (MWCNTs). CNTs exhibit both beneficial and toxic effects on plant species by altering the morphological and physiological properties of plant cells. In most of the studies, low concentrations of CNTs are considered safe. Therefore, there is plenty of room to use these cutting-edge technologies to alleviate seed dormancy, boost vigor, and solves the problem of slow growth in seed propagated species. This chapter summarizes the results of various studies investigating the effects of carbon nanotubes on seed viability, as well as seedling growth and biomass.

Chapter 5
Nanopriming for Crop Management for Sustainable Agriculture: An Overview...110
Muhammad Majeed, University of Gujrat, Pakistan
Murad Muhammad, Chinese Academy of Sciences, China & University of Peshawar, Pakistan
Sehar Nawaz, University of Agriculture, Faisalabad, Pakistan
Tayyaba Naz, University of Agriculture, Faisalabad, Pakistan
Muhammad Mazhar Iqbal, Ayub Agricultural Research Institute, Department of Agriculture, Government of Punjab, Pakistan
Nafeesa Zahid, University of Kotli, Pakistan
Mumtaz Hussain, Arid Zone Research Institute, Pakistan
Allah Nawaz Khan, Chinese Academy of Sciences, China
Ghulam Abbas, University of Agriculture, Pakistan
Allah Bakhsh Gulshan, Ghazi University, Pakistan

Maria Mehboob, University of Poonch, Pakistan

The purpose of this study is to inform scholars in the aforementioned fields on the state of the art in seed nano-priming methods and the ongoing research that is shaping sustainable agricultural practices for the modern day. As things are, agriculture is hampered by a number of abiotic and variables that reduce yields. These stressors on crops are made worse by a lack of freshwater supplies, changes in the climate, and the inefficiency of currently available agrochemicals. The agricultural sector has recently shown an interest in nanomaterials such carbon tubes, nanofiber, and nanoparticles. Coating the nanoparticle with the target substance before releasing it into the surrounding soil or plant increases absorption of the target substance relative to the non-engineered nanomaterial. In order to strike an evolving equilibrium between agricultural output and environmental sustainability, nanotechnology undoubtedly offers a fresh platform. Nano-priming may also be utilized for targeted biofortification of seeds, which can improve crop yields and food quality.

Chapter 6
Nanoparticle-Based Priming-Enhanced Agricultural Practices: Promoting Sustainable Plant Growth and Development.. 142
Abhishek Singh, Faculty of Biology, Yerevan State University, Armenia
Sapna Rawat, University of Delhi, India
Vishnu D. Rajput, Academy of Biology and Biotechnology, Southern Federal University, Rostov-on-Don, Russia
Karen Ghazaryan, Faculty of Biology, Yerevan State University, Armenia
Tatiana Minkina, Academy of Biology and Biotechnology, Southern Federal University, Rostov-on-Don, Russia
Abdel Rahman Mohammad Al Tawaha, Al Hussein bin Talal University, Jordan
Athanasios Alexiou, Novel Global Community Educational Foundation, Australia
Karansher Singh Sandhu, Bayer Crop Sciences, USA

The challenges of climate change, resource depletion, and biodiversity loss pose significant risks to agriculture. To ensure sustainable food production and tackle these issues, an agricultural revolution is necessary. Nanotechnology offers promising solutions for enhancing agricultural sustainability, and one such approach is seed nano-priming. This technique effectively alters seed metabolism and signaling pathways, influencing not only germination and seedling establishment but also the entire plant life cycle. Extensive research has demonstrated numerous benefits of seed nano-priming, including improved plant growth, increased productivity, and enhanced nutritional quality of food. This chapter provides an overview of recent advancements in nanotechnology for seed nano-priming and explores the challenges and possibilities associated with its implementation in sustainable agriculture.

Section 4
Nanopriming-Based Plant Breeding and Omics Response for Sustainable Agriculture

Chapter 7
Emerging Frontiers in Plant Breeding Molecular Markers and Nanopriming for Enhanced Crop Improvement....159
Sapna Rawat, Delhi University, India

Plant breeding plays a crucial role in improving plant genotypes and ensuring global food security. However, traditional breeding methods have limitations in effectively selecting traits, particularly quantitative traits, and can be time-consuming and unreliable. Breeders have introduced molecular markers into plant breeding practices to address these problems, resulting in the fields of "smart breeding," "molecular breeding," and "marker-assisted breeding." To pinpoint the genetic areas that regulate desired features, a number of DNA-based molecular marker systems, including RFLP, RAPD, AFLP, SSR, and SNP, have been created. These markers eliminate the need for phenotype-based selection by streamlining the identification of target genomic areas based on genotypes.

Chapter 8
Impact of Nanopriming and Omics-Based Applications for Sustainable Agriculture....175
Himani Sharma, Uttaranchal University, India
Atin Kumar, Uttaranchal University, India
Rajat Singh, Uttaranchal University, India
Satendra Kumar, Sardar Vallabhbhai Patel University of Agriculture and Technology, India

The concept of omics technologies in the agriculture sector is gaining importance nowadays due to its diverse applications like genomics which is based on genome sequences, transcriptomic which is based on gene expression and profiling, proteomics which is based on protein content, and metabolomics which involves the study of the metabolism of an organism. A revolutionary transformation in the last decade in the domain of omics has made analytical and scientific procedures cost-effective, easier, and faster and has increased accuracy. With the introduction of technologies involving omics in the field of agriculture, it has now become easier to identify and recognize the presence of different genes and genomic sequences liable for various agronomic traits like genes responsible for increasing yield, and genes for inducing resistance against biotic and abiotic stresses. Utilization of such technologies has resulted in obtaining accurate gene interpretation and identifying gene markers, thus acting as a guiding hand in the development of new crop varieties with some desirable traits.

Section 5
Nanopriming and Geoinformatics: Combined Approaches for Crop Production and Soil Health Monitoring

Chapter 9
Nanopriming and Geoinformatics: A Combined Approach for Crop Production for Monitoring of Soil Health..193
P. Smriti Rao, Integral University, India
Setu Ratnam, Integral University, India
Ashish David, Sam Higginbottom University of Agricultural Science and Technology, India
Raghunandan Khatana, Krishi Vigyan Kendra, Abusar, India
Akshita Barthwal, R&D Bayer Crop Science, India

A soil survey is a component of an improved agricultural research and advisory programme. It provides in-depth knowledge on soils and a full assessment of the region's soil resources. It offers the information needed to organize activities for managing soil and land use. With the help of the soil survey report and field samples, a database of soil characteristics is created. Although the soil survey report is mostly based on a low-intensity survey, it can be used to achieve a number of planning goals. Geographical information systems (GIS), Photogrammetry, remote sensing, global positioning systems (GPS), and other technologies are useful for analysing soil, creating various vegetation indices, creating digital elevation models, determining land use and land cover, and many other tasks. These computational models can then be used by land managers, soil scientists, researchers, and many other professionals to make well-informed suggestions.

Section 6
Nanopriming-Based Biotic and Abiotic Stress Management

Chapter 10
Nanopriming: A Potential Solution for Heavy Metal Toxicity Alleviation in Crops...221
Sachin Kumar, Himachal Pradesh Agriculture University, Palampur, India
Chiranjeev Kumar, Himachal Pradesh Agriculture University, Palampur, India
Shilpa Manhas, Himachal Pradesh Agriculture University, Palampur, India
Ankita Sharma, Punjab Agricultural University, India

Richa Jaswal, Himachal Pradesh Agriculture University, Palampur, India

Sustainable crop production required steady and balanced means of nutrients and other growth factors during the crop growth period. Heavy metal accumulation in plant and soil ecosystems along with its toxicity effects is much more pronounced in crops. Heavy metals interfere with crop germination, plant nutrient uptake, photosynthesis, and response to fertilizer and organic inputs which in turn make crops vulnerable to various adverse conditions. Priming seeds and other plant propagating materials with suitable nano-priming materials may enhance the overall plant photosynthetic rate, enhance vigor, and increase plant biomass with ensuring food security and productivity in a long-term approach. Nano priming with nano-materials has higher efficiency due to its smaller size and large surface area coverage which then perfectly adsorbed on the surface of plant growing materials such as seed and other materials to remove the attachment and concentration of heavy metals in plants.

Chapter 11
Nanopriming Effective Tools for the Field of Horticultural Crops to Copping Up With Biotic and Abiotic Stresses 241
Shipra Singh Parmar, Dr. Y.S. Parmar University of Horticulture and Forestry, India
Divya Pandey, Dr. Y.S. Parmar University of Horticulture and Forestry, India

Horticulture crops are continually impacted by severe biotic and abiotic stresses, which induce various impacts leading to global food scarcity scenarios. In light of this, establishing a balance between horticultural output and food demand necessitates the development of strategies to acclimate plants to stressful environments. In recent years, seed priming has been of significant interest to scientists. It leads to induced changes in seed metabolism promoting successful establishment of crops through accelerated seed germination and enhanced growth of growing seedlings. The application of nanotechnology in the current era holds the promise in combating the environmental stressors through the use of nanomaterials such as nanoseed priming technologies. Seeds are nano primed to hasten germination, which is followed by improved germination, biotic and abiotic stress tolerance, and to increase plant production.

Chapter 12
Nanopriming-Based Management of Biotic Stresses for Sustainable Agriculture 263
Muhammad Majeed, University of Gujrat, Pakistan
Murad Muhammad, Chinese Academy of Sciences, China & University of Peshawar, Pakistan
Tanveer Hussain, Mirpur University of Science and Technology,

Pakistan
Muqarrab Ali, Muhammad Nawaz Shareef University of Agriculture, Pakistan
Tayyaba Naz, Institute of Soil and Environmental Sciences, University of Agriculture, Faisalabad, Pakistan
Sehar Nawaz, University of Agriculture, Faisalabad, Pakistan
Muhammad Mazhar Iqbal, Government of Punjab, Pakistan
Ghulam Abbas, University of Agriculture, Dera Ismail Khan, Pakistan

An increasing number of agricultural uses for nanotechnology has also sparked interest in nano-priming of plants. Plant growth and seed production can be stunted by the physiological, psychological, and molecular alterations that are commonly associated with stressor events. Nano-priming is a novel technique for seed priming that enhances plant resilience in improved germination, growth, and yield in unfavorable environmental conditions. Unfavorable impacts of nano-priming upon seed germination, seedling, and plant growth features were detected despite the fact that the usage of nanoparticles increased environmental stress tolerance of these plants. The most crucial aspects of these plants' germination, morphology, and physiology that influence their establishment, development, and production are also addressed, along with the impact of various nanoparticles. Using nanotechnology for agricultural purposes is to make growing crops easier and more efficient. Nanotechnology is a viable method for sustainability boosting agricultural output, which is essential to achieving this objective.

Chapter 13
Seed Nanopriming: An Innovative Approach for Upregulating Seed Germination and Plant Growth Under Salinity Stress 290
Abhishek Singh, Faculty of Biology, Yerevan State University, Armenia
Shreni Agrawal, Parul Institute of Applied Science, Parul University, India
Vishnu D. Rajput, Academy of Biology and Biotechnology, Southern Federal University, Rostov-on-Don, Russia
Karen Ghazaryan, Faculty of Biology, Yerevan State University, Armenia
Hasmik S. Movsesyan, Yerevan State University, Armenia
Tatiana Minkina, Academy of Biology and Biotechnology, Southern Federal University, Rostov-on-Don, Russia
Abdel Rahman Mohammad Al Tawaha, Department of Science and Engineering, Al Hussein bin Talal University, Jordan
Athanasios Alexiou, Department of Science and Engineering, Novel Global Community Educational Foundation, Hebersham, Australia
Badal Singh, ICAR-National Bureau of Plant Genetic Resources, Pusa, India
Santosh Kumar Gupta, National Institute of Plant Genome Research, India

Soil salinity, one of the main abiotic risks to agricultural products, lowers agricultural yields, and as a result, the world's rising population confronts substantial food challenges. It's one of the world's most pressing abiotic issues, affecting 30% or more of irrigated land. Saline stress is caused by a number of different salt concentrations; however, NaCl is by far the most common salt in the ground. High salt stress is the reason for delay in seed germination, which may be caused by either the osmotic potential of the water or by the effects of certain toxic ions. This chapter covers seed priming and salt resistance approaches that generate biochemical, physiological, and structural processes in saline stress, and the interaction between seed priming and salt resistance. Priming with nanoparticles (NPs) needs more study to reduce the negative effects of salinity stress on a variety of crops grown in extreme conditions.

Section 7
Future Role of Nanopriming-Based Approaches for Food Security and Sustainable Agriculture

Chapter 14
Nanotechnology for Sustainability and Food Security in Agriculture: A Nanopriming Long Story in Short 315
Omkar Singh, Department of Soil Science, Sardar Vallabhbhai Patel University of Agriculture and Technology, India

Sustainable techniques are crucial to ensuring the availability of nourishing food while reducing environmental effects due to the growing global population and increasing demand for agricultural resources. By utilizing nanomaterials and nanodevices to improve crop yield, resource efficiency, and pest management, nanotechnology offers creative solutions. The chapter covers the potential advantages of nanotechnology in agriculture, including the precise pest control abilities of nanopesticides and the controlled-release qualities of nanofertilizers to enhance nutrient delivery and use. Additionally, it emphasizes the use of nanosensors for continuous assessment of crop health and environmental factors, as well as nanoscale imaging methods for early disease detection and treatment. Resource efficiency, precision agriculture, waste reduction, and improved food security are made possible by the integration of nanotechnology into agriculture. However, it also highlights how crucial it is to address the safety, moral, and legal issues raised by the application of nanotechnology in farming.

Chapter 15
Nanotechnology: A New Innovation in Biofortification and Food Safety.........340
Shivangi Singh, Sardar Vallabhbhai Patel University of Agriculture and Technology, India
Omkar Singh, Sardar Vallabhbhai Patel University of Agriculture and Technology, India
Uday Pratap Shahi, Sardar Vallabhbhai Patel University of Agriculture and Technology, India

Nanotechnology brings new prospects to increase food quality and safety through nano-enabled delivery systems, nanosensors, smart packaging and processing assistance. Nanomaterials display unique features at the 1-100 nanometer range that surpass constraints of classic breeding and detection methods. Iron, zinc, and vitamin A nano-encapsulates and nanoparticles can boost crops through tailored nutritional delivery. Nanosensors and smart packaging with antimicrobial nanoparticle coatings offer rapid, ultrasensitive pathogen and toxin detection while enhancing food stability. Nano-filters utilizing photocatalysts remove pollutants from agricultural water. However, the putative hazards from nanoparticle toxicity and accumulation require continuous investigation. Regulations and public acceptance factors are also crucial for sustainable progress. If developed properly, nanotechnologies can considerably boost food security through augmented micronutrition, antibacterial capabilities, and safety monitoring from production to consumption.

Chapter 16
Nanopriming Revolutionizing Agribiotechnology for Sustainable Food Security 352
Varsha Gupta, Bureau of Indian Standards, India
Santosh Kumar Gupta, National Institute of Plant Genome Research, India

The most promising branch of science and technology is nanotechnology. This area is constantly developing new tools and ideas for the development of sustainable agriculture. Nanoparticles (NPs) are used in agriculture due to their distinct physicochemical characteristics that revolutionize agriculture farming and global food security. One of the innovative seed-priming techniques is nano-priming, which helps to improve seed germination, seed growth, and yield by providing resistance to various stresses in crop plants. Nano-priming is a considerably more effective method compared to all other seed-priming methods. The interaction of nanomaterials with soil components is strongly determined in terms of soil quality and plant growth. NPs would be applied to improve the quality and reduce post-harvest loss of agricultural products by extending their shelf life, particularly for fruits and vegetables. This chapter assesses the available literature on nanotechnology and comprises a collection of information, opinions, and perspectives on a single platform.

Compilation of References 363

About the Contributors 442

Index 450

Foreword

This edited book brings together the expertise and insights of renowned researchers and experts in the field of sustainable agriculture, focusing on the innovative approach of nanopriming. The book provides comprehensive exploration techniques in improving crop of the application of nanopriming productivity, nutrient uptake efficiency, and stress tolerance in plants. The chapters in the book cover various aspects related to nanopriming, including its underlying mechanisms, methodology, optimization strategies, practical applications, and future directions. Each chapter is authored by established scientists who have made significant contributions to the field. The book showcases the latest advancements in nanopriming and its potential implications for sustainable agricultural practices. It addresses the pressing challenges faced by the agriculture industry, such as increasing food demand, limited resources, and environmental concerns. The meticulous editing process has ensured that the chapters are well-structured, coherent, and provide valuable insights to readers. The content is supported by thorough research, data analysis, and evidence-based discussions.

I believe that *Nanopriming Approach to Sustainable Agriculture* will addition to the existing literature in the field. It offers a comprehensive be a valuable overview of nanopriming techniques, their benefits, and their potential applications, making it an essential resource for researchers, agronomists, policymakers, and professionals involved in sustainable agriculture.

Narendra Pratap Singh
Vice Chancellor

Preface

The challenges posed by climate change have significantly impacted global food security, with crop yields suffering from both abiotic and biotic stresses. The book *Nanopriming Approach to Sustainable Agriculture* offers a promising solution to this problem by introducing a sustainable technology that ensures the growth and development of healthy plants even in adverse conditions. By using nanoparticles to enhance seed germination and growth, nanopriming establishes pre-resistance against diseases and stresses, thereby reducing the need for pesticides and fertilizers. This reduction in the use of harmful chemicals not only benefits the environment but also increases the income of farmers worldwide.

The book is divided into seven sections having sixteen chapters with the aim to provide a comprehensive understanding of emerging knowledge and technology. Section One, "Nanopriming: An Introduction," covers the introduction of nanopriming. Section Two, "Role of Nanopriming for the Sustainability of Agriculture," contains two chapters that explore the application of nanopriming for sustainable agriculture. Section Three, "Role of Nanopriming for Crop Production and Protection," comprises three chapters that provide information on nanopriming role in crop production and protection applications. Section Four, "Nanopriming Based Plant Breeding and Omics Response For Sustainable Agriculture," includes two chapters discussing the roles of nanopriming at the omics level in plant breeding and crop science for sustainable agriculture. Section Five consists of one chapter on "Nanopriming and Geoinformatics: Combined Approaches for Crop Production and Soil Health Monitoring," which discusses nutrient management in crop production and soil health. Section Six, "Nanopriming Based Biotic and Abiotic Stress Management," has four chapters that mainly focus on cutting-edge nanopriming methodologies for managing biotic and abiotic stresses in the context of climate change. Lastly, Section Seven, "Future Role of Nanopriming Based Approaches for Food Security and Sustainable Agriculture," has three chapters that delve into the nanopriming strategy for crop improvement, biofortification, and food security, which are essential for farmers' welfare.

Undoubtedly, this new book is an important addition to the fields of crop science, food production, and human health.

Abhishek Singh
Faculty of Biology, Yerevan State University, Armenia

Vishnu D. Rajput
Academy of Biology and Biotechnology, Southern Federal University, Rostov-on-Don, Russia

Karen Ghazaryan
Faculty of Biology, Yerevan State University, Armenia

Santosh Kumar Gupta
National Institute of Plant Genome Research, India

Tatiana Minkina
Academy of Biology and Biotechnology, Southern Federal University, Rostov-on-Don, Russia

Acknowledgment

We would like to thank the Science Committee of RA (Republic of Armenia) and the Ministry of Science and Higher Education of the Russian Federation for their support and funding, grant numbers 21AG-4C075 and FENW-2023-0008, respectively. Their contributions have been essential in making this edited book, "Nanopriming Approach to Sustainable Agriculture," a reality. We are grateful for their support, which has enabled us to share valuable insights and knowledge in the field of sustainable agriculture and nanopriming.

Section 1

Nanopriming: An Introduction

Chapter 1
An Introduction to Nanopriming for Sustainable Agriculture

Shreni Agrawal
Parul Institute of Applied Science, Parul University, India

Richa Das
Parul Institute of Applied Science, Parul University, India

Shivangee Solanki
https://orcid.org/0009-0003-9194-5690
Parul Institute of Applied Science, Parul University, India

Simran Choudhury
https://orcid.org/0009-0000-4572-3200
Parul Institute of Applied Science, Parul University, India

Indrani Bhattacharya
Parul Institute of Applied Science, Parul University, India

Pradeep Kumar
MMV, Banaras Hindu University, India

Amit kumar Singh
Indian Institute of Technology, Banaras Hindu University, India

Sunil Kumar Mishra
Indian Institute of Technology, Banaras Hindu University, India

Kavindra Nath Tiwari
MMV, Banaras Hindu University, India

ABSTRACT

Plant seeds are of paramount importance in the agricultural sector as well as plant life cycle, and are being severely threatened by environmental changes such as biotic and abiotic stress, resource deficiency, climate fluctuation, etc. Cue, nano-priming, a nascent yet innovative approach to seed technology, combining versatility of nanoparticles and engineered nanomaterials with 'seed-priming' to induce adaptive physiological changes, thus, improving seed quality and crop yield. This

DOI: 10.4018/978-1-6684-7232-3.ch001

study reported mechanisms by which nano-priming synchronizes seed germination, breaks seed dormancy, boosts seed vigor, increases tolerance to periodic stress conditions, assists phyto-microbiota to thrive under environmental uncertainties, creates nanopores in shoots along with upregulation of aquaporin genes for better water uptake, and other reported changes at molecular, biochemical level. Withal, this study offers insight on the future of nano-priming when blended with cutting-edge technologies like cold plasma, artificial intelligence, and digital analysis to revolutionize the agrarian sector.

INTRODUCTION

The world population has been estimated to reach 9.6 billion by 2050, to catch up with, the agricultural production must also rise up by 70-100% (Shelar A. et al., 2021). Utilization of global water to support food production and agronomy has been estimated to rise from 70% to an alarming 83%. Additionally, modern agricultural practices have resulted in extensive use of agrochemicals (fertilizers, pesticides), an annual consumption of 2.5 million tones, large portion of which is lost to evaporation, run-off into bodies of water, or cause more harm than intended good. To put in figures, 70% of crop yield is lost to major abiotic stressors (drought, salinity, temperature extremes, nutrient deficiency, metal toxicity). According to reports, 70% of heavy metals with their amalgams cause significant quality loss at early developmental stages of plants (Salam A. et al., 2022; Imtiaz H., 2023).

Several strategic techniques have been implemented to boost crop quality and yield. Seed priming is one such inexpensive, sensitization technique which involves pre-treating seeds, causing them to imbibe a small amount of water which triggers essential metabolism in seeds (pre-germination) without protrusion of radical, followed by rapid desiccation. Deliberate exposure to minor drought like condition for a short, predetermined time interval generates primary stress memory. Hence, upon subsequent exposure, primed seeds acclimatize faster than those left un-primed. Cross-tolerance is an alternate term for this pre-programming of resilience against stressful conditions prior to naturally occurring recurring exposures. Seeds are primed to promote vital metabolic processes that typically occur during the first few days of germination.

Traditionally, reduction of seed dormancy, rapidity of germination, increment in stress tolerance and micronutrient concentration was achieved by soaking seeds in salt solution (halo-priming), water (hydro-priming), beneficial microbial solution (bio-priming), solid carrier (matrix-priming) or phytohormone solution (hormonal priming). But, with the advent of engineered nanoparticles in almost all scientific domains, a new paradigm in seed technology was born—*nano-priming* (Nile S.

H. et al., 2022; Rahman M. S. et al., 2020). The cost effectivity, usage in low, environmentally realistic dosages, simplicity of interaction, uptake, translocation and accumulation of NPs into plants while curtailing treatment with harsh chemicals or salts has intrigued researchers and scientists to uncover full potential of nano-priming. Reports have elucidated the direct, positive effects of NPs on production of secondary plant metabolites namely, terpenoids, alkaloids, phenylpropanoids, glucosinolates and more as defense compounds against environmental stressors. Alongside stimulating stress resistant genes, nano-priming has demonstrated possibility of achieving seed biofortification (nano-biofortificatio) and protection as nanoparticles possess antimicrobial properties, however, the notion is under development (Nile S. H. et al., 2022; Salam A. et al., 2022).

SEED NANO-PRIMING: PRE-REQUISITES

Seeds are direct reflection of successful plant establishment and crop quality, and are extremely susceptible to fluctuations of the external environment. To circumvent this inconsistency, seed nano-priming was first effectuated in 2009 by Khodakovskaya using nanocarbon to ensure seeds of only prime quality, able to withstand adversities, reached the field. Although, the origin of this technology dates back to 371 B.C, when Theophrastus first hydro-primed cucumber seeds, it catalyzed the emergence of newer and better seed priming methods of the present. Following section summarizes rudimentary information concerning seed nano-priming and briefings on mechanistic action of NPs (Shelar A. et al., 2021). Furthermore, fullerene, a carbon-based nanomaterial, was the first nanomaterial to be manufactured in 1985 (Szollosi R. et al., 2020).

Nano-Priming Agents

Nanotechnology is the recent trend in almost all disciplines of science, so is the expansive applications of nanomaterials. Owning to their special attributes and widespread manufacture at minimum production cost (for instance, the annual market production of zinc oxide nanoparticles ranges from 550 to 5550 tons), they have been adopted as smart 'nano-catalysts' to accomplish astounding aims such as agricultural sustainability. Due to their eco-friendliness and the advent of 'green NPs', the Indian government formulated the *Nano Mission* to promote the implementation of nanomaterials by the private sector (Szollosi R. et al., 2020).

Materials with at least two dimensions ranging between 10-100 nm are recognized as nanomaterials (Nile S. H. et al., 2022; Shelar A. et al., 2021; Salam A. et al., 2022; Szollosi R. et al., 2020). Due to their unique properties such as nano-size, large surface area, magnetism, high surface energy (in view of high surface to mass ratio), quantum confinement, precise delivery of active bio-compounds, sorption potential, etc., nanomaterials have long been favored as debugging tools for contemporary challenges. Conforming to nano-structural diversity found in nanomaterials, they are classified as 0D NPs (zero dimensions) whose mobility is entirely restricted (quantum dots), 1D NPs possessing the ability to move along one particular direction (quantum wires, fibers, nanotubes) and 3D NPs with absolute freedom of movement (Szollosi R. et al., 2020). Depending upon their source, mode of action and structural features, they are grouped as nano-emulsions and nanoparticles. Seed nano-priming technology makes considerable use of organic (fullerenes, fullerols, carbon nanotubes, carbon nanohorns, dendrimers, liposomes) as well as inorganic (metals, metal oxides, quantum dots) nanoparticles. In practice, high percentage of NPs are derived either from metals (Zn, Ag, Co, Cu, Cd, etc.) or metalic oxides (TiO_2, ZnO, SiO_2, MnO, Fe_3O_4, MnO_3, CeO_2, etc.). ENMs have recently grown in popularity because they provide an open possibility of manufacturing nanomaterials in the presence of biological agents (green NPs), so as to make them environmentally benign and lower their hazardous side effects (Salam A. et al., 2022; Imitaz H. et al., 2023, Szollosi R. et al., 2020).

Underlying Mechanisms of Nano-Priming

Standard (unprimed) seed germination occurs through three sequential developmental phases, as illustrated in *Fig.1*. *Phase I* or *imbibition* wherein absorption of water by dry seeds is driven by low water potential. Water uptake is facilitated through apoplastic pathway (cell wall) which is then employed for mitochondrial repair, protein synthesis from available DNA, RNA etc. *Phase II* is marked by sudden drop of imbibition upon reaching a dynamic equilibrium. Reactivated metabolism results in the synthesis of mRNA and proteins, the growth of the embryo, etc. This phase is analogous to *lag phase* observed in growth curve of microorganisms. With the onset of *Phase III*, seeds initiate water uptake, mobilize stored food reserves in embryo and use it for radical protrusion and cell elongation. Phase I and III are indistinguishable for both, standard (unprimed) and primed seeds. However, prolonged hydration with highly regulated water uptake to provoke pre-germinative metabolism characterizes *phase II* in germination of primed seeds (Imtiaz H. et al., 2023).

Nano-priming entails soaking dried seeds in suspensions of nanomaterials or nano-formulations for a predetermined amount of time. Nanoparticles in the suspension may or may not be adsorbed by the seeds, those which do create a coat, the process known as *seed dressing*. To avoid microbial spoilage during storage, so formed nano-coat maybe be supplemented with bactericides or fungicides. After soaking, seeds are subjected to desiccation, this removal of water warrants longevity

Figure 1. A graphical comparison of germination stages observed in primed and unprimed seeds

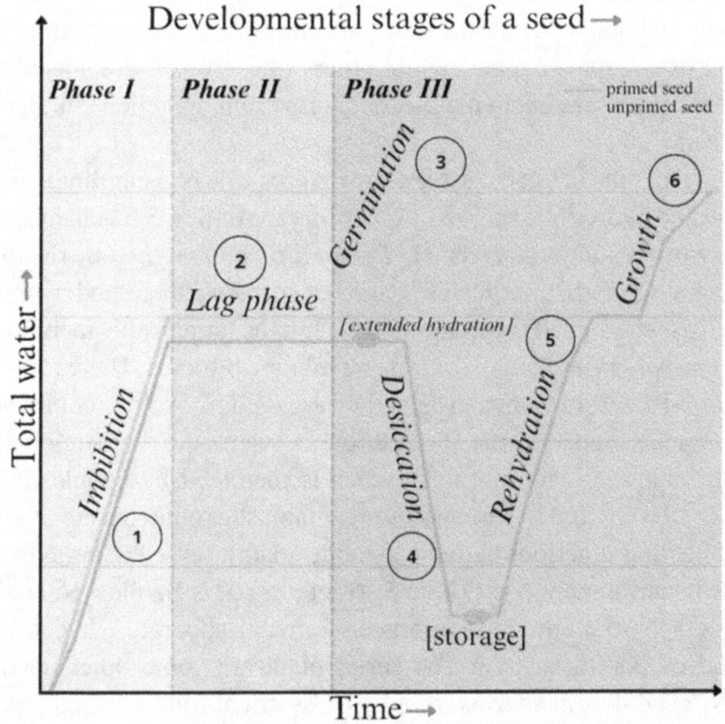

of treatment and quality of seeds when stored, and creation of stress memory imprints. When seeds are re-hydrated (at the time of planting) seeds demonstrate an array of cellular processes like synthesis of nucleic acids and proteins *de novo*, ATP synthesis, accumulation of phospholipids and sterols, activation of DNA, repair mechanism, antioxidant system, loosening of cell wall, endosperm softening, accelerated hydrolysis of starch and more which contribute to enhanced germination, tolerance and growth (Imtiaz H. et al., 2023).

In the natural world, a balanced abscisic acid to gibberellic acid (ABA/GA) ratio regulates the amount of water seeds absorb during germination. Numerous investigations have shown that abscisic acid tightly controls the aquaporin family's *PIP2, NIP1, TIP3,* and *TIP4* genes. Auxins and abscisic acid function in a coordinated manner to enhance nanoparticle adsorption and subsequent transport into the endosperm (Nile S. H. et al., 2022; Imtiaz H. et al., 2023; Kandhol N. et al., 2022).

Route of Entry

When adsorbed onto the seed coat, priming agents (nanoparticles) are merely external agents. Some nanoparticles are internalized through pores created as a result of upregulation of aquaporin genes, while others like Zn-NPs are adequately sturdy to penetrate through sclerenchymatous outer covering and elicit cellular responses inside the seed.

In order to enter the target cells, NPs have to cross the phospholipid bilayer rich in hydrophobic and hydrophilic groups. According to studies, 3 mechanisms carry out general entry of NPs into the cells: (1) *Direct diffusion method* by the dint of their minute size and other distinct features like net surface charge and hydrophobicity, (2) *Endocytosis* wherein NPs are engulfed within the target cell's membrane and (3) *Facilitated transport* via channels or transmembrane proteins. There is no downward movement of NPs within plants, instead they are predominantly mobilized from the root system and ascended to the shoot through xylem. Abovementioned transport of NPs from plant cells to vascular tissues is mediated by apoplastic route (via neighboring cell walls and xylem vessels) or symplastic route (through the cytoplasm and cell connecting junctions called plasmodesmata), both leading NPs to the pith and eventually aerial plant parts (Nile S. H. et al., 2022; Szollosi R. et al., 2020).

Ionic channels, the cuticle membrane, stomatal apparatus, hydathodes lining the leaf, and casparian strips in the aerial plant are some other potential entry routes for NPs, while rhizoderms, root tips, absorbent root hairs, etc. are all found in the root system. Augmenting NP solution with organic additives including citric acid, fulvic acid, humic acid or polymers can presumably stabilize their entry and avoide precipitation. An analysis demonstrated how microorganisms inhabiting the rhizoshpere enhance uptake of NPs by plants (Nile S. H. et al., 2022).

Biochemical and Molecular Responses in Nano-Primed Seeds

Bibliometric statistic reveals a dearth of studies elucidating the specific molecular changes, modifications to signaling pathways and phytohormone crosstalk in nano-primed seeds. Although nanoparticles play a diverse role for and against numerous plant factors, major part of this diversity is yet to be discovered and explored. It must be noted that all physio-biochemical alterations detected as a consequence of nano-priming cannot be standardized and vary depending upon the target species, nano-priming agents or stage at which priming occurs. However, we have attempted to condense general yet major response patterns to ubiquitously used NPs in the following section and demonstrated in *Fig.2*.

Abiotic and biotic stress mitigation: Downstream signaling cascades of the seed are altered by internalization of nanoparticles, prompting reactive oxygen and nitrogen species (ROS, RNS) to agglomerate (mimicking natural salinity stress situation). ROS are prime players in eliciting cellular defenses mechanisms and reducing susceptibility against stress conditions, although optimization of ROS levels vary with target specie and. While imparting stress tolerance, increasing ROS concentration poses a threat of apoptotic death to cells. NPs counterpoise such damages by amplifying detoxification and enzymatic activities of guaiacol-peroxidase, SOD (superoxide dismutase), CAT (catalase), APX (ascorbate peroxidase), POD (peroxidase) and non-enzymatic, lipophilic antioxidants (ascorbic acid, TPC or total phenolic content, TFC or total flavonoid content). SOD particularly scavenges oxygen free radicals and converts them into hydrogen peroxidase which is further acted upon by CAT, releasing water. In a study concerning PNC priming (polyacrylic acid coated cerium oxide nanoparticles), upregulation of POD and GST (glutathione-S-transferase), downregulation of PRX (peroxiredoxins) was observed (Shelar A. et al., 2021). When primed with CS (chitosan), green tea polyphenols (GTPs) or catechins–class of flavonoids–contribute their exceptional antioxidant activity under a well-established antioxidant potential by acting as direct antioxidants, triggering pro-oxidant responses and ensuring consistent, real-time scavenging of free radicals. ROS or RNS react with polyphenols to form rather stable phenol-oxygen radicals. Additionally, NPs are implicated in diminishing MDA (malondialdehyde) activity by 65-75% as they are prominent markers of afflictions caused by stress (Farooq T. et al., 2022). Certain species demonstrated high seedling germination and vigor in the face of abiotic stressors by heightened expression of metallothionine genes (*NnMT2a, NnMT2b* genes in Arabidopsis, *MT1* and *MT4* genes in tomato), genes encoding for PAL (phenylalanine ammonia-lyase) and CHS (chalcone synthase) and RACK1 (receptor for activated C kinase 1 is a key player in the ROS signaling cascade) (Nile S. H. et al., 2022; Shelar A. et al., 2021; Salam A. et al., 2022). Salam A. et al registered an upliftment in heat stock as LEA (late embryogenesis abundant) proteins in the face of drought (Salam A. et al., 2022; Imtiaz H. et al., 2023).

Unique NPs with antibacterial properties, such CGNPs (chitosan guar nanoparticles), serve as bioprotectants. Due to their diminutive size and need for a concentration of the same magnitude, they are widely used for the precise delivery of biological substances to specific locations within the target. Maize crops are successfully protected against fungal pathogens and have their nutritional quality boosted by Zn-chitosan foliar spraying. As was previously mentioned, seed dressing with NPs significantly reduces damage from both pesticides (carbofuran, thiran) and microbial pests. For instance, cinnamaldehyde-encapsulated MSNP (mesoporous silica NP) supplemented to innate antimicrobial properties of NPs. This strategy curbs excessive application of lethal pesticidal and herbicidal agrochemicals, used at the expense of environmental safety (Shelar A. et al., 2021; Imtiaz H. et al., 2023).

Growth and germination: Signaling factors associated with ABA and GA (one of them being *AGL62* transcription factor) elevate α-amylase (verified by protruding action of amylase into and from within the starch granules) activity at transcriptional level, increasing reserved starch degradation rates thus, leading to high soluble sugar concentration inside the endosperm. Consequently, the osmotic potential and water potential within the seed are reduced, creating a potential difference between external and internal environment. This gradient propels the passage of water into the seed thus, improving emergence by breaking seed dormancy, and aids in bolstering plant strength (Nile S. H. et al., 2022). A dynamic ABA/GA ratio is achieved and maintained by upregulation of *ABA 8'-hydroxylase* and *GA3 oxidase 1* genes at transcriptome level Imtiaz H. et al., 2023; Kandhol N. et al., 2022). Other proteins, namely, glyceraldehyde-3-phosphate dehydrogenase 2, UDP-glucose-6-dehydrogenase are stimulated while globulin and vicilin-like storage proteins are inhibited. Nucleic acid and protein synthesis, metabolic repair (particularly, BER or base excision repair, NER or nucleotide excision repair or both) supplemented with ROS overproduction leads to rapid germination, high germination energy and seed vigor in phase I of nano-primed seeds. Statistically, 8-10% increment of germination index was observed. In later developmental stages, advancement of cell cycle from G1 to G2 phase is halted to intensify the effects of nano-priming. NPs mediated induction of mitogen activated protein kinase (MAPK) cascade and Ca^{2+} accumulation in cytoplasm result in rapid plant growth (Nile S. H. et al., 2022; Salam A. et al., 2022; Imtiaz H. et al., 2023; Farooq T. et al., 2022).

Seed dormancy: Although seed dormancy is a natural, precautionary phenomenon to prevent seeds from sprouting prematurely or off-season, it also inhibits seeds growth under normal conditions. According to metabolic response patterns of the seed, there are three main types of dormancies: (1) *Primary dormancy* imparted by germination suppressing compounds involving cis-ABA, lucopyranosyl ester, benzoic acid, salicylic acid, chlorogenic acid, coumarin, etc.; (2) *Physical dormancy* wherein the rigid, impermeable seed coat prevents water and gaseous exchange crucial to germination and (3) *Internal dormancy* driven by hormonal homeostasis within seeds featuring cytokinins, ABA, IAA (indole-3-acetic acid) and gaseous ethylene. NPs carefully modulate the ABA/GA balance to control rate of water absorption, which ultimately improves the SVI (seed vigor index) and releases dormancy (Nile S. H. et al., 2022).

It has been discovered that ROS are interlinked to breaking seed dormancy as the ROS buildup catalyzes cleavage of hydrolytic polysaccharide bonds in cell wall of endosperm thus, loosening it or drilling pores into the seed coat (Nile S. H. et al., 2022). Szollosi R. et al. noted that seed dormancy was positively impacted by reconfiguration of cell wall lipidome (increase in amount of C18:3 fatty acids) (Szőllősi R. et al., 2020).

Metabolite synthesis: There exists fine tuning between ROS concentration within cells and other metabolic activities, for instance, ROS act as direct or indirect signaling cues for the production of primary as well as secondary metabolites. Evidences of ROS inducing brassinosteroids (BRs), nitric oxide (NO), ethylene (ET), salicylic acid (SA) and jasmonic acid (JA) have been reported, which are messengers of regulation of secondary metabolism. Yet another induction junction of secondary metabolism in plants–lignin-phenyl propanoid pathway is upregulated by ROS (Nile S. H. et al., 2022).

Photosynthesis: It is an autotrophic phenomenon of utmost vitality for survival and is directly responsible for gaining seed biomass (dry weight, fresh weight). Higher photosynthetic quantum yield in primed seeds has been evinced by large NAR (net assimilation rate) values (Imtiaz H. et al., 2023). Efficient nutrient uptake due to NPs and their capability of attaching to, entrapping and absorbing, particularly Fe, has been discovered to have a correlation with stimulated biosynthesis of chlorophylls, redox reactions, etc. Nature-derived Fe or Fe_2O_3 priming increases its concentration in the seed, which results in higher production ALA (theta-aminolevulinic acid), a precursor to chlorophyll. Additionally, Si priming promotes enhanced phytopigment production and the recovery of leaf relative water content (LRWC). Priming with TiO_2 stimulated RuBisCO (ribulose-1,5-bisphosphate carboxylase/oxygenase) activity—nature's most abundant, primary CO_2 fixer. As a consequence, PSI and PSII (photosystem) functionality enhanced, hydrolysis of water and evolution of O_2 into the ETS (electron transport system) was increased. High concentration of O_2 lead to increased water exchange via stomata. With amplified water conductance, respiration and photosynthesis increase as well (Nile S. H. et al., 2022; Salam A. et al., 2022; Imtiaz H. et al., 2023). A study examining CS-GTE (chitosan-green tea extract) priming documented a surge in phytopigments associated with photosynthesis. Chlorophyll a content increased by an approximate 11-17%, chlorophyll b by a massive 3-38%, total chlorophyll content rose by 8-21% and carotenoids by 15-18% under typical circumstances. A similar trend was observed in response to extrinsic stress wherein *chl a, chl b* and total *chl* content increased by 4-6%, 34-36% and 7-14% respectively. On the contrary, synthesis of lycopene and tannins experienced a drop, and increased by a slight margin only while encountering unfavorable conditions (Farooq T. et al., 2022).

Storage and viability: As discussed earlier, ROS signals stimulate production of cellular messengers like BRs which are responsible for regulation of viability during storage. Contradicting outcomes from tests produce no clarity regarding mechanism of retaining viability thus, demanding more attention and extensive research. However, it has been affirmed that *BEN1*, *DWF1* and *EXO* genes associated with BR synthesis pathway along with *EXPA1, EXPA2, DUF642, TRG1/XLY1* genes of cell wall modification partially contribute to seed viability (Farooq T. et al., 2022).

Microbiota of the rhizosphere: Plant's underground ecosystem surrounding the roots, also known as rhizosphere, is densely populated with mutualistic microorganisms collectively known as phytomicrobiota. The two entities, through intricate communication interfaces, complement each other's metabolic processes and affect the plant favourably as a whole. They achieve so by releasing minuscule amounts of phytohormones or plant growth regulators offering adequate protection again stresses and supporting growth thus, making them natural reinforces of sustainable agriculture. In accordance with certain studies, application of a *'Phytoguard'* consortium made up of NPs and PGPRs (plant growth promoting rhizobacteria) significantly benifit the plant. Their synergistic action reduced MDA content while boosted photosynthesis and antioxidant defense mechanism. Additionally, bacterial population in the rhizosphere flourished immensely as a result of nano-silica treatment. Close inspections also revealed an increase in the rhizosphere zone, bacterial growth zone, root extension, and several soil enzymes like alkaline phosphatases, acid phosphatases, and phytases (Salam A. et al., 2022).

NANO-PRIMING: POTENTIAL EFFECTS ON VARIOUS PLANT ASPECTS

An unfathomably wide suite of nanoparticles has been discovered to have the potential of imparting desirable properties like rapidity of growth, resistance to damaging, external forces, nutrient richness in seeds. Following *Table 1* partially summarizes widely used nanoparticles employed as priming agents and their corresponding effects on functional aspects of plants as well as gene networks associated to said functions.

CURRENT DEVELOPMENTAL STATUS

Even though nano-priming was first effectuated in 2009 (Shelar A. et al., 2021), and more than a decade has passed, yet this technique is still in its nascent phase and much needs to be done in order for it to be as feasible as many of its counterparts. Although, nanoparticle treatment has been efficaciously used in numerous fields (such as nanomedicine) for over a decade, and has shown its rising influence in the agrarian sector, its true potential in sustainable agriculture and food safety is yet to be unleashed. This is primarily due to the lack of knowledge about a myriad of multiplex interactions between the seed's internal system and nano-priming agents.

Till date a number of seeds of various species have been nano-primed in order to improve their attributes, such as drought resistance, better delivery and targeting of fertilizers and pesticides, better seed germination, etc. But there are dominant

Figure 2. An illustration of nanoparticles targeting various signaling pathways and biomolecules to boost overall seed and plant performance

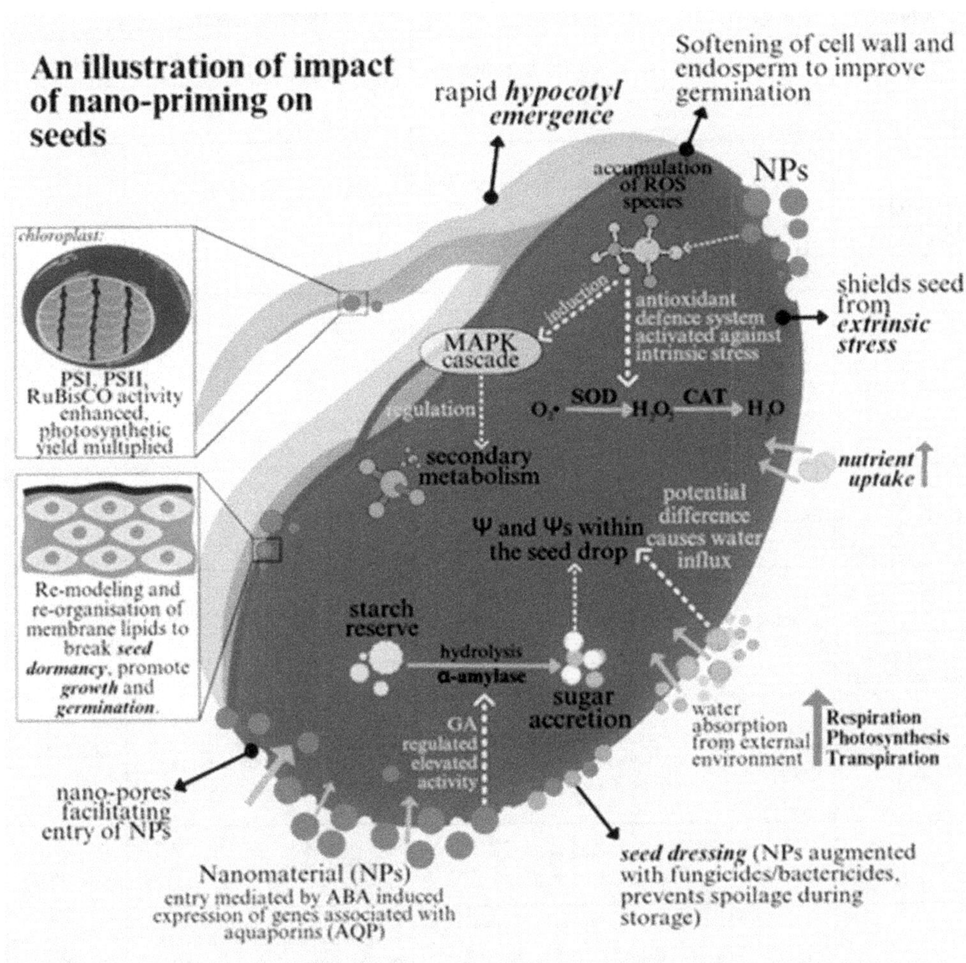

questions regarding the existence of certain unidentified factors in the germination of nano-primed seeds; factors concerning the practical field, such as interference due to interaction of different biomolecules present in the seed. Nonetheless, this technique has recently been in the limelight with lucrative reports of nano-primed seeds (both dicot and monocot) having reinforced seed germination potential (Chandrasekaran U. et al., 2020). Currently, nano-priming is being extensively researched in order to find ways of producing crops that are much suited to the changing times. These enhancements include grains with better yield, better dormancy release, resilience to various environmental stress, better resource utilization, etc.

Table 1. Common and emerging NPs employed to prime seeds and their effects

NPs used for priming	Target species	Physiological effects (improved attributes)	Molecular targets	References
MWCNTs (multiwalled carbon nanotubes)	Tobacco seeds	Water uptake, cell wall loosening, rapid growth due to increased cell division	Active expression of cell wall extension gene (*NtLRX1*), cell division gene (*CycB*), aquaporin gene (*NtPIP1*)	(Nile S. H. et al., 2022; Imtiaz H. et al., 2023)
MWCNT—COOH	*A. Viridis* L. (green alder seeds) and forest species	90% improvement in germination and cell wall lipidome (enrichment with C18:3 fatty acids resulted in higher SVI)	N/A	(Nile S. H. et al., 2022)
CNTs (carbon nanotubes)	Upland boreal (forest) species	Regulation of lipid metabolism, lipid remodeling (phosphatidylinositol, phosphatidylcholine, phosphatidic acid, phosphatidylglycerol, phosphatidylserine) to break seed dormancy, increases germination potential	N/A	(Imtiaz H. et al., 2023)
CNTs	Broccoli	Electrostatic reactions balanced in cell membrane, emergence of new lipid domains/rafts in phospholipid bilayer	N/A	(Imtiaz H. et al., 2023)
PNCs (polyacrylic acid coated cerium oxide nanoparticles)	Cotton seeds	ROS and RNS accretion (under salinity stress), elevated Mg^{2+} in roots (post-germination) for effective plant establishment	Upregulation of 13 genes associated with ROS production and 10 genes involved in ion homeostasis, genes of terpene synthesis (*CAD1, TPS*)	(Shelar A. et al., 2021; Imtiaz H. et al., 2023)
CS (chitosan)	Omycete species	Germination, seed vigor, immunity against Downy mildew causing *Sclerospora graminicola*	Increased expression of genes of alkaline peroxidase, PAL (phenylalanine ammonia lyase), genes of salicylic acid pathway (*PRI, PR5*)	(Shelar A. et al., 2021)
CuO	*B. rapa* spp.	Flavonoid, phenolic content as response to a/biotic stressors	Upregulation of genes liked to flavonoid, phenolic compounds (*PAP1, ANS, PAL, FLS*)	(Nile S. H. et al., 2022)
AgNPs	Chinese cabbage	Glucosinolate (GSL) synthesis to confer defence against stress	Heightened regulation of genes of GSL biosynthesis (*BrMYB28, BrMYB29, BrMYB34, BrMYB51, St5C, SUR1*)	(Nile S. H. et al., 2022)
ZnO	Barley seeds	Stress resistance by upregulating expression of microRNAs	Active expression of *miR156*, *miR159* genes	(Salam A. et al., 2022)
ZnO	*Brassica napus*	Seed germination, seed imbibition, high production of linoleic acid and linolenic acid under salt stress	Upregulation of *BnCAM, BnPER* genes	(El-Badri et al., 2021)
nZVI (low zero valent iron)	*Oryza sativa*	Modulation of hydrolase enzyme activity, mobilization of endosperm food reserves to promote photosynthesis; eventually increase germination	Increased expression of genes related to mobilization and enzyme activity (*OsGAMYB, OsGA30x2*)	(Nile S. H. et al., 2022; Salam A. et al., 2022)
PHF (polyhydroxy fullerene or fullerol)	Seeds under drought condition	Elevated ABA content to enhance AQP expression for improved water absorption	Downregulation of genes involved in ABA catabolism (*CYP707 A3*)	(Szoľlosi R. et al., 2020)

In addition to being used for better seed germination, nano-priming is significantly being done in order to develop better abiotic stress tolerance in nano-primed seeds, compared to the untreated ones. This was recently done in seeds of tomato, cucumber, and wheat, etc. Also, as the world keeps progressing, there have been significant changes in the climatic conditions over the decades. This has resulted in many unpredictable drought-like situations in many places across the globe, due to reduced and irregular rainfall, and also there is a lack of proper irrigation facilities and reduced groundwater, adding to the troubles in the agrarian sector. With the groundwater levels falling over time, there is a significant reduction in this precious resource, thus, it calls for the use of seeds that are drought-tolerant, which would help the farmers in drought prone areas to ensure better yield and minimize crop loss.

Yet, this is just the tip of a huge iceberg, and much work needs to be done in unlocking its true potential. Although, nanomaterials have been in use in the biomedical and industrial sectors for quite some time, their use in the field of agriculture is still a very new approach.

SIGNIFICANCE OF NANO-PRIMING IN SUSTAINABLE AGRICULTURE

As discussed above, nano-priming has emerged as a significant technique in sustainable agriculture, but also it has shown promising results in combatting issues related to food security. With the world facing severe food security issues (nearly one in three people in the world, i.e., 2.37 billion people, did not have access to adequate food in 2020, which is an increase of approximately 320 million from 2019), use of seeds that can result in crops with better yield and better tolerance to various abiotic stresses is of utmost importance. Also, there exist evidences of nano-priming aiding in enhanced germination and starch metabolism of aged rice seeds (using photosynthesized silver nanoparticles); this would help in the utilization of seeds that are aged and might be on the verge of losing their viability. This would prevent such seeds from being wasted, and a higher and much more efficient utilization of the existing resources, thus also improving the economy of the food industry.

Apart from obtaining better dormancy release and better germination results (as seen in nano-priming of maize seeds using phytochemicals-capped gold nanoparticles (Mahakham W. et al., 2016), another extensively experimented area is the use of nano-priming to produce drought resistant crops. Some significant examples of nano-priming to improve drought stress alleviation include, nano-priming of wheat plants by silicon oxide (Rai-Kalal et al., 2021), nano-priming *Alnus subcordata* with multi walled carbon nanotubes (Rahimi D. et al., 2016), nano-priming of wheat by zinc oxide (due to prevention of chlorophyll degradation during drought stress; also, it increases seedling vigor by H_2O_2 signaling networks for better seed germination) (Rai-Kalal et al., 2021), nano-priming of rice by zinc oxide nanoparticles (Sharma D. et al., 2021), etc. Nonetheless, nano-priming is also seen to be an approach toward mitigating salt/salinity stress in various crops, as is seen in nano-priming of *Capsicum annum* using manganese nanoparticles (Yuqing Y. et al., 2020), nano-priming of rapeseeds using CeO_2 nanoparticles (Khan M. N. et al., 2022), nano-priming of bitter melon plants using selenium-chitosan nanoparticles (Sheikhalipour M. et al., 2023), etc.

Another significance of nano-priming approach lies in obtaining crops with better agronomic profiles, which can consist of better crop yield (increased production per capita), better suitability to the changing climatic conditions, and improved tolerance towards abiotic stresses (such as water stress, salinity, salt toxicity, heavy metal toxicity etc.), finer metabolism (in terms of phytopigment content, activity of enzymes, nitrogen and phosphorus bio-cycles, TAA or total antioxidant activity, DPPH or 2,2-diphenyl-1-picrylhydrazyl content, etc.), improved morphology (wider leaves, firmer stems, healthier roots) (Nile S. H. et al., 2022). This, in turn, would lead to efficient utilization of the limited resources and hence, a way forward towards sustainable agricultural practices.

CRITICAL STUDY: CHALLENGES AND FUTURE INSIGHTS

Researchers and scientists are unable to fully exploit the potential of seed nano-priming due to a persistent lack of understanding regarding the interface between nanoparticles and plant ontology. Also, there are speculations amongst researchers that there might exist certain unidentified factors that influence this technique, positively or negatively. A thorough understanding can allow us to coax nano-priming into being much more potent than it is today. Nanoparticles function in a strictly concentration/dose dependent fashion, raising toxicity (phototoxicity, genotoxicity) risks for seeds (negative effects observed in commonly consumed crop plants such as tomato, wheat, etc.) as well as environmental safety and health concerns for surrounding life forms including humans (Nile S. H. et al., 2022). Premature seeds may suffer irreparable damage or lose viability if they are dried to their initial moisture level more quickly than mature seeds. Furthermore, repeated priming may make seeds resistant to priming thus, jeopardizing intended effects of sensitization. Revitalization of such seeds demands more time and resources. At different stages of the procedure, there may exist minor possibilities of fungal or bacterial contamination therefore, nano-priming cannot provide absolute reproducibility. Priming may cause seeds to become unstable and require scrutinized maintenance and storage. As expensive and automated the procedure is, it fails to generate employment (Shelar A. et al., 2021).

Overtime, conventional priming techniques witnessed drastic modifications and leaps of advancement paving way for the present-day nano-priming method. In an effort to increase the output of nano-priming while coping up with the market's supply and demand for high-quality food, many technologies have been tested integrated with it. For instance, combining cold plasma technology with nano-priming to develop an eco-agricultural, quick, risk-free method. Cold plasma technology has provided remarkable results during testing–actively enhancing germination, flowering, nutritional composition, plant resistance to stress by increasing POD activity, even inactivating plant pathogens. It induces the expression of PAL (phenylalanine ammonia lyase) and USPs (universal stress proteins) at the molecular level. Surprisingly, applying cold plasma technology to seeds that have been nano-primed has broken a significant constrain by reducing the toxicity of NNPs. With advanced technologies permeating the research sector, seed profile assessment is now being carried out by high processing instruments such as FT-NIR spectroscopy and X-ray imaging. Manual calculation of germination indices, seed vigour indices, and effectivity scores limits the scope of testing and introduces the possibility of human error. 'Automated seed diagnostics and phenotypic analysis' has been recently developed to address these restrictions. An amalgamation of seed nano-priming, machine learning, and artificial intelligence to generate rather accurate predictions about primed seeds. In a study, seeds of Thai rice varieties were digitally assessed for germination state and viability with the help of an artificial neural nexus fed with specific algorithms based on evaluations confered by FT-NIR and X-ray imaging. With the aid of such an evaluation method, standardization of responses with enormous cultivars that would not have been manually conceivable is made possible and simultaneously increasing confidence in the outcomes (Shelar A. et al., 2021).

CONCLUSION

At this stage when researches and developments are still ongoing in this field, it would be challenging to draw any firm conclusions, but nonetheless, it can be concluded that nano-priming is a propitious candidate than all its counterparts in seed priming. Additionally, it has shown lower hazard potentials and greater efficacy in achieving its goals. As it has been discussed, nano-priming can emerge as a clever approach in sustainable agriculture, if thorough and extensive work continues, most importantly in understanding the interaction of nanomaterials with the various biomolecules and their influence on signaling pathways and plant ontology. Decoding this information may give an insight into the utilization of nanomaterials in plant disease prevention and control, enhance their resistance to environmental stressors, and ultimately shape a modern-era agrotechnology. This would lead to an improved food security across the world and make significant amendments to the agrarian sector, food industry, and world economy.

As this approach has the untapped prospective of making a significant impact on environmental economy, a far more "greener approach" would be the synthesis of nanomaterials for nano-priming using food and plant wastes, or some other plants themselves. An important application of such a method was seen in the use of zero valent iron synthesized from pomegranate peel waste, to nano-prime Oryza sativa (Guha T. et al., 2021), and yet another significant example being the use of phytochemical-capped gold nanoparticles, synthesized from galanga plant rhizome extract, to nano-prime maize seeds for better germination (Mahakham W. et al., 2016). Therefore, the constant pursuance of a greener alternatives would help in consistently achieving a dynamic equanimity between agricultural productivity and environmental sustainability.

REFERENCES

Chandrasekaran, U., Luo, X., Wang, Q., & Shu, K. (2020). Are there unidentified factors involved in the germination of nanoprimed seeds? *Frontiers in Plant Science*, *11*, 832. doi:10.3389/fpls.2020.00832 PMID:32587599

Chokheli, V. A., Shmaraeva, A. N., Kornienko, I. V., Stepanenko, V. V., Rajput, V. D., Singh, A., Kumari, A., Dmitriev, P. A., Azarov, A. S., Minkina, T. M., & Varduni, T. V. (2022). Assessment of genetic diversity and DNA fingerprinting of rare species of the genus Crambe (Brassicaceae). *Crop & Pasture Science*. Advance online publication. doi:10.1071/CP22027

El-Badri, A. M., Batool, M., Mohamed, I. A., Khatab, A., Sherif, A., Wang, Z., Salah, A., Nishawy, E., Ayaad, M., Kuai, J., Wang, B., & Zhou, G. (2021). Modulation of salinity impact on early seedling stage via nano-priming application of zinc oxide on rapeseed (Brassica napus L.). *Plant Physiology and Biochemistry*, *166*, 376–392. doi:10.1016/j.plaphy.2021.05.040 PMID:34153882

Farooq, T., Akram, M. N., Hameed, A., Ahmed, T., & Hameed, A. (2022). Nanopriming-mediated memory imprints reduce salt toxicity in wheat seedlings by modulating physiobiochemical attributes. *BMC Plant Biology*, *22*(1), 540. doi:10.118612870-022-03912-2 PMID:36414951

Guha, T., Gopal, G., Das, H., Mukherjee, A., & Kundu, R. (2021). Nanopriming with zero-valent iron synthesized using pomegranate peel waste: A "green" approach for yield enhancement in Oryza sativa L. cv. Gonindobhog. *Plant Physiology and Biochemistry*, *163*, 261–275. doi:10.1016/j.plaphy.2021.04.006 PMID:33872831

Imtiaz, H., Shiraz, M., Mir, A. R., Siddiqui, H., & Hayat, S. (2023). Nano-Priming Techniques for Plant Physio-Biochemistry and Stress Tolerance. *Journal of Plant Growth Regulation*, 1–21. doi:10.100700344-023-10981-6

Kandhol, N., Singh, V. P., Ramawat, N., Prasad, R., Chauhan, D. K., Sharma, S., Grillo, R., Sahi, S., Peralta-Videa, J., & Tripathi, D. K. (2022). Nano-priming: Impression on the beginner of plant life. *Plant Stress*, *5*, 100091. doi:10.1016/j.stress.2022.100091

Khan, M. N., Li, Y., Fu, C., Hu, J., Chen, L., Yan, J., Khan, Z., Wu, H., & Li, Z. (2022). CeO2 nanoparticles seed priming increases salicylic acid level and ROS scavenging ability to improve rapeseed salt tolerance. *Global Challenges (Hoboken, NJ)*, *6*(7), 2200025. doi:10.1002/gch2.202200025 PMID:35860396

Kumar, P., Singh, A., Rajput, V. D., Yadav, A. K. S., Kumar, P., Singh, A. K., & Minkina, T. (2022). Role of artificial intelligence, sensor technology, big data in agriculture: next-generation farming. In *Bioinformatics in Agriculture* (pp. 625–639). Academic Press. doi:10.1016/B978-0-323-89778-5.00035-0

Mahakham, W., Theerakulpisut, P., Maensiri, S., Phumying, S., & Sarmah, A. K. (2016). Environmentally benign synthesis of phytochemicals-capped gold nanoparticles as nanopriming agent for promoting maize seed germination. *The Science of the Total Environment*, *573*, 1089–1102. doi:10.1016/j.scitotenv.2016.08.120 PMID:27639594

Nile, S. H., Thiruvengadam, M., Wang, Y., Samynathan, R., Shariati, M. A., Rebezov, M., Nile, A., Sun, M., Venkidasamy, B., Xiao, J., & Kai, G. (2022). Nano-priming as emerging seed priming technology for sustainable agriculture—Recent developments and future perspectives. *Journal of Nanobiotechnology*, *20*(1), 1–31. doi:10.118612951-022-01423-8 PMID:35659295

Rahimi, D., Kartoolinejad, D., Nourmohammadi, K., & Naghdi, R. (2016). Increasing drought resistance of Alnus subcordata CA Mey. seeds using a nano priming technique with multi-walled carbon nanotubes. *Journal of Forest Science*, *62*(6), 269–278. doi:10.17221/15/2016-JFS

Rai-Kalal, P., Tomar, R. S., & Jajoo, A. (2021). H2O2 signaling regulates seed germination in ZnO nanoprimed wheat (Triticum aestivum L.) seeds for improving plant performance under drought stress. *Environmental and Experimental Botany*, *189*, 104561. doi:10.1016/j.envexpbot.2021.104561

Rai-Kalal, P., Tomar, R. S., & Jajoo, A. (2021). Seed nanopriming by silicon oxide improves drought stress alleviation potential in wheat plants. *Functional Plant Biology*, *48*(9), 905–915. doi:10.1071/FP21079 PMID:34366002

Rhaman, M. S., Rauf, F., Tania, S. S., & Khatun, M. (2020). Seed priming methods: Application in field crops and future perspectives. *Asian Journal of Research in Crop Science*, *5*(2), 8–19. doi:10.9734/ajrcs/2020/v5i230091

Salam, A., Afridi, M. S., Javed, M. A., Saleem, A., Hafeez, A., Khan, A. R., Zeeshan, M., Ali, B., Azhar, W., Sumaira, Ulhassan, Z., & Gan, Y. (2022). Nano-priming against abiotic stress: A way forward towards sustainable agriculture. *Sustainability (Basel)*, *14*(22), 14880. doi:10.3390u142214880

Sharma, D., Afzal, S., & Singh, N. K. (2021). Nanopriming with phytosynthesized zinc oxide nanoparticles for promoting germination and starch metabolism in rice seeds. *Journal of Biotechnology*, *336*, 64–75. doi:10.1016/j.jbiotec.2021.06.014 PMID:34116127

Sheikhalipour, M., Mohammadi, S. A., Esmaielpour, B., Spanos, A., Mahmoudi, R., Mahdavinia, G. R., Milani, M. H., Kahnamoei, A., Nouraein, M., Antoniou, C., Kulak, M., Gohari, G., & Fotopoulos, V. (2023). Seedling nanopriming with selenium-chitosan nanoparticles mitigates the adverse effects of salt stress by inducing multiple defence pathways in bitter melon plants. *International Journal of Biological Macromolecules*, *242*, 124923. doi:10.1016/j.ijbiomac.2023.124923 PMID:37211072

Shelar, A., Singh, A. V., Maharjan, R. S., Laux, P., Luch, A., Gemmati, D., Tisato, V., Singh, S. P., Santilli, M. F., Shelar, A., Chaskar, M., & Patil, R. (2021). Sustainable agriculture through multidisciplinary seed nanopriming: Prospects of opportunities and challenges. *Cells*, *10*(9), 2428. doi:10.3390/cells10092428 PMID:34572078

Singh, A., Rajput, V. D., Rawat, S., Sharma, R., Singh, A. K., Kumar, P., & Singh, S. (2022). Geoinformatics and Nanotechnological Approaches for Coping Up Abiotic and Biotic Stress in Crop Plants. *Sustainable Agriculture Systems and Technologies*, 337-359.

Singh, A., Sengar, R. S., Rajput, V. D., Minkina, T., & Singh, R. K. (2022). Zinc oxide nanoparticles improve salt tolerance in rice seedlings by improving physiological and biochemical indices. *Agriculture*, *12*(7), 1014. doi:10.3390/agriculture12071014

Singh, A., Sengar, R. S., Shahi, U. P., Rajput, V. D., Minkina, T., & Ghazaryan, K. A. (2023). Prominent Effects of Zinc Oxide Nanoparticles on Roots of Rice (Oryza sativa L.) Grown under Salinity Stress. *Stresses*, *3*(1), 33–46. doi:10.3390tresses3010004

Singh, A., Singh, A. K., Rawat, S., Pal, N., Rajput, V. D., Minkina, T., Sharma, R., Singh, N. P., & Tripathi, J. N. (2022). Satellite-Based Quantification of Methane Emissions from Wetlands and Rice Paddies Ecosystems in North and Northeast India. *Hydrobiology*, *1*(3), 317–330. doi:10.3390/hydrobiology1030023

Szőllősi, R., Molnár, Á., Kondak, S., & Kolbert, Z. (2020). Dual effect of nanomaterials on germination and seedling growth: Stimulation vs. phytotoxicity. *Plants*, *9*(12), 1745. doi:10.3390/plants9121745 PMID:33321844

Ye, Y., Cota-Ruiz, K., Hernandez-Viezcas, J. A., Valdes, C., Medina-Velo, I. A., Turley, R. S., Peralta-Videa, J. R., & Gardea-Torresdey, J. L. (2020). Manganese nanoparticles control salinity-modulated molecular responses in Capsicum annuum L. through priming: A sustainable approach for agriculture. *ACS Sustainable Chemistry & Engineering*, *8*(3), 1427–1436. doi:10.1021/acssuschemeng.9b05615

Section 2

Role of Nanopriming for the Sustainability of Agriculture

Chapter 2
Nanopriming Technology for Sustainable Agriculture

Sapna Rawat
Delhi University, India

ABSTRACT

Climate change has had detrimental impacts on the growth and development of crops, leading to reduced crop production and endangering global food security. However, nanotechnology offers a promising solution to address these challenges in agriculture, ushering in the era of "NanoTech-Agril." By employing seed priming with nanoparticles, farmers can mitigate the impacts of climate change such as salinity, drought, and floods, thereby enhancing crop yields. This seed priming method improves seed metabolism and signaling pathways, leading to enhanced germination and overall plant growth throughout their life cycle. Additionally, nano-seed priming activates various metabolic, biochemical, antioxidant, and phytohormone pathways, bolstering plants against abiotic and biotic stresses. Consequently, the reliance on pesticides and fertilizers can be reduced. This review presents a comprehensive overview of the applications of nanoparticles in sustainable agriculture, providing valuable insights into their potential uses.

INTRODUCTION

A few of the current issues the agricultural sector is dealing with include crop yield losses brought on by pests, the depletion of natural resources, and the effects of climate change (De La Torre-Roche et al., 2020; Kah et al., 2019). Pesticides and fertilizers are used extensively in conventional farming practices, which are harmful to the environment (Rajput et al., 2018). With the world's population expected to

DOI: 10.4018/978-1-6684-7232-3.ch002

reach 9–10 billion by 2050, there is an immediate need to increase food production by 25–70% from its current levels to meet the rising demand (Scott et al., 2018). Therefore, the implementation of cutting-edge agricultural technologies is essential for sustaining agricultural systems and increasing agricultural output (Fraceto et al., 2016; Panpatte et al, 2016). In ecosystems like rangelands and agricultural lands, seed germination is particularly important for maintaining plant species (Manjaiah et al., 2018). It is the key first phase of a plant's life cycle. During the germination stage, a plant's capacity to withstand challenges like drought and salt stress is crucial since it decides whether or not the plant will survive to the following developmental stage (Akter et al., 2018). The rate and uniformity of seedling development determines a plant's resistance to environmental challenges such drought, salinity, temperature, and moisture (Rajput et al., 2015; Friedrichs et al., 2019; Maity et al., 2018; Zahedifar and Zohrabi, 2016). Protecting crops may be aided by the use of nanomaterials, particularly nanoparticles, in agriculture (Scott et al., 2018). Many farms are researching the potential uses of nanoparticles in agriculture, particularly for pest management, plant nutrition, and greener manufacturing processes (Chau et al., 2019; Acharya et al., 2020; Rajput et al., 2021c). The effects of nanoparticles on seeds and plants have been studied recently and have been found to be varied. Some have been found to stimulate plant growth by altering metabolic processes and vigor in seeds and young plants (Acharya et al., 2020; Pérez-de-Luque, 2017; Rajput et al., 2020a; Abbasi Khalaki et al., 2020). Some nanoparticles, however, have been shown to have unfavorable effects, such as phytotoxicity or the suppression of germination. Nanoparticle concentration, size, and zeta potential all play important roles in controlling biological reactions and uptake and movement within plants (Abbasi Khalaki et al., 2020; Acharya et al., 2019; Singh et al., 2021). The ability of nanoparticles to cross biological barriers is correlated with their size and surface charge, and smaller nanoparticles are more efficiently absorbed by leaves and roots (Bombo et al., 2019; Hu et al., 2020; Palocci et al., 2017; Rajput et al., 2022; Valletta et al., 2014; Avellan et al., 2017; Spielman-Sun et al., 2019). Nanopriming, the process of applying nanoparticles to seeds, has the potential to improve plant health, germination and synchronization, crop resistance to biotic and abiotic stressors, and seed protection while stored.

NANOPARTICLES AND NANO-ENABLED PRODUCTS

Nanotechnology has emerged as a promising solution for addressing numerous challenges in agriculture (Singh et al., 2022). According to Rajput et al., (2021b; Singh et al., 2022) and others, nanoparticles have special features that are a result of their small size (1–100 nm) and vast surface area. This contributes to their major physical and chemical importance. These advantageous characteristics have facilitated the development of various nano-enabled products, including nanofertilizers, nanoinsecticides, and nanopesticides, aiming for sustainable agriculture. Nanoparticles exhibit a wide range of chemical compositions, which can be classified into four main categories: carbon-based, metal-oxide-based, bio-organic-based, and composite-based (Lowry et al., 2019). These categories encompass various types of nanoparticles, each with distinct chemical compositions and properties. Another classification for nanoparticles is based on their composition, distinguishing between organic and inorganic nanoparticles. Biodegradability is a characteristic of organic nanoparticles like polymeric nanoparticles, liposomes, carbon-based nanomaterials, lipid-based nanocarriers, and solid lipid nanoparticles. In contrast, inorganic materials such as metals and metal oxides like zinc oxide and silver oxide make up inorganic nanoparticles. This classification scheme allows for a comprehensive understanding of the diverse types of nanoparticles and their composition in the field of nanotechnology. Silver nanoparticles (Ag-NPs) are particularly prevalent among synthesized nanoparticles, comprising over 25% of diverse consumer products (Scott et al., 2018; Rajput et al., 2021b). Ag-NPs find extensive application in antifungal, antiviral, and antibacterial medicines.

Every year, advancements in state-of-the-art technology lead to the generation of new types of nanoparticles (NPs) with diverse applications across various industries. The top-down method and the bottom-up method are two frequently used techniques for NP synthesis (Fraceto et al., 2016; Panpatte et al., 2016). Additionally, three alternative approaches—physical, chemical, and biological—can be used to synthesise NPs. Utilising plant biomass or extract and a metal salt solution, green NPs for seed priming are created, which reduces and stabilises the metal ions present in the plant (Mohamad et al., 2014). Previous research has shown that live lucerne plants grown in an environment rich in AuCl4 can produce gold nanoparticles (Camerel et al., 2002). The ability of *Medicago sativa* and *Brassica juncea* plants to accumulate gold NPs from aqueous $KAuCl_4$ solutions has also been demonstrated in related investigations (Bali and Harris). Most of the research in recent years has focused on synthesizing NPs using the plant's inactive components, such as powders or extracts (Mohamad et al., 2014). Metallic nanoparticles can be synthesized using different components of plants, such as leaves, stems, flowers, fruits, roots, seed coats, seeds, and latex (Mohamad et al., 2014). The process of nanoparticle synthesis typically entails mixing plant biomass or extract with a solution containing metal salts at specific temperature and pH conditions. A noticeable colour shift in the solution signals the production of nanoparticles (Mohamad et al., 2014). Utilizing methods such the Soxhlet device, cold treatment, and hot treatment, plant extracts are produced for use in NP synthesis. Due to its simplicity in scaling up, downstream processing, environmental friendliness, non-toxicity, biocompatibility, and renewability (Camerel et al., 2002; Dikshit et al., 2021; Mittal et al., 2013; Mohamad et al., 2014), this technique of NP synthesis is favourable. Numerous biological uses exist for these biocompatible nanoparticles. Plant extract is added to a metal precursor solution that contains metal salts in order to produce metal nanoparticles (NPs). Commonly used metal precursor solutions for the production of NPs include $AgNO_3$, $HAuCl_4$, $PdCl_2$, H2PtCl$_6$, $Cu(NO_3)$ 23H_2O, $FeCl_3$6H_2O, Na_2SeO_3, and ($NiNO_3$)26H_2O (Dikshit et al., 2021). The synthesis process of metallic nanoparticles using plant extracts can be divided into three main phases. In the initial phase, metal ions (M+ or M2+) are reduced to metal atoms (M0) through a reduction process, followed by the nucleation of these reduced metal atoms (Dikshit et al., 2021). The second phase involves the aggregation of small nanoparticles that are in close proximity, leading to the formation of larger particles and an increase in thermodynamic stability. Finally, in the last phase, the nanoparticles attain their final shape, completing the synthesis process (Si and Mandal (2007); Makarov et al., (2014)). Plant extracts contain a variety of useful biomolecules that help reduce and stabilise metal ions in solution, though it can be difficult to pinpoint the precise molecules involved because the plant extract also contains a wide variety of phytochemicals.

GERMINATION AND PRINCIPLES OF SEED PRIMING

Seed priming is an important practice in agriculture that aims to improve seed germination and subsequent plant development (Carrillo-Reche et al., 2018). It involves pre-conditioning seeds before planting using various techniques such as water-based methods and physical treatments. Water-based methods include soaking the seeds in water and then drying them, while physical treatments may involve UV priming (Lemmens et al., 2019). The goal of seed priming is to activate metabolic pathways associated with pre-germination phases, particularly phases I and II, by facilitating water absorption without triggering radicle emergence. At the molecular and cellular levels, seed metabolism undergoes changes, resulting in increased mobilization of reserves, transcriptomic reprogramming, cell wall modification, enhanced translation, and post-translational modifications (Carrillo-Reche et al., 2018). Priming also induces antioxidant mechanisms, heat shock proteins, and stress-related responses, which confer cross-resistance to additional stressors. Rapid germination reduces the exposure time of germinating seeds to unfavorable soil conditions. Overall, seed priming enhances seedling vigor, stress tolerance, synchronized germination, and ultimately improves food quality and crop yield (Carrillo-Reche et al., 2018; Lemmens et al., 2019). There are various methods available for seed priming. Hydro-priming or hydro-thermopriming involves hydrating the seeds through water treatment for a specific duration, usually 7-14 hours, to facilitate phase II of germination. Temperature manipulation, such as cold or hot treatments, can be combined with this approach (Noorhosseini et al., 2017). Osmo-priming utilizes low water potential solutions (around 10-20%) to manage hydration and induce abiotic stress-mediated changes in seed metabolism (Carrillo-Reche et al., 2018). Additionally, microorganism-based priming is another pre-sowing treatment option (Lemmens et al., 2019).

In summary, seed priming is an effective technique that can positively influence seed germination and subsequent plant growth. By employing various methods, seed priming enhances seedling vigor, stress tolerance, and overall crop yield, leading to improved agricultural productivity.

Seed nano-priming is an emerging technique that harnesses the power of nanomaterials, specifically nanoparticles, to enhance seed priming effects (see Table 1). It is important to differentiate between traditional seed priming and seed nano-priming. While traditional seed priming involves soaking seeds in water or nutrient solutions to form a coating with beneficial substances, seed nano-priming takes it a step further by using suspensions and nanoformulations that allow nanoparticles to be absorbed or create a protective coating on the seed surface (Acharya et al., 2019). This approach can also incorporate fungicides or bactericides into the seed coating to protect against diseases (Gross et al., 2020).

Early research by Khodakovskaya et al., (2009) demonstrated the influence of nanomaterials on seed germination. They observed that tomato seeds absorbed carbon

Table 1. Use of nanomaterial for seed priming and coating

Nanoparticle	Characteristics	Crops	Application	References
Fe-NPs	Particle size <50 nm	*Sorghum bicolor*	Improved *S. bicolor*'s germination, seedling development, and tolerance to salinity.	(Maswada et al., 2018)
Biogenic Fe-NPs	Particle size of 19–30 nm	*Citrullus lanatus*	Manipulation of antioxidant capacity and defense-related hormones in seedlings of *C. lanatus*.	(Kasote et al., 2019)
Biogenic Ag-NPs	Particle size of 6-26 nm	*Oryza sativa*	Aged *O. sativa* seeds exhibit improved germination and starch metabolism.	(Mahakham et al., 2017)
ZnO-NPs and Fe_2O_4-NPs	ZnO-NPs 20–30 nm Fe_2O_4-NPs 50-100 nm	*Triticum aestivum*	Increase in plant growth, a decrease in oxidative stress, and a drop in cadmium levels in *T. aestivum*.	(Rizwan et al., 2019)
Si-NPs	Si-NPs 90nm	*T. aestivum*	Reduced oxidative stress, increased biomass, and increased yield in *T. aestivum* grains, as well as decreased cadmium levels.	(Hussain et al., 2019)
Au-NPs	Au-NPs 10-30nm	*Zea mays L.*	Facilitating seed germination in *Z. mays*.	(Mahakham et al., 2016)
Nano-pyrite (FeS_2)	FeS_2-NPs10-30nm	*O. sativa*	Utilization of NPK fertilizer *O. sativa* production	(Das et al., 2018)

nanotubes, resulting in improved water intake and increased flower production in tomato plants. Carbon nanotubes were found to impact the expression of water channel proteins and alter seed metabolism in crops like barley, soybean, and maize (Villagarcia et al., 2012). Different types of nanomaterials, including metallic, biogenic metallic, and polymeric nanoparticles, have shown promise in seed nano-priming by affecting gene expression, metabolic processes, hormone production, and promoting accelerated shoot and root development (Siddaiah et al., 2018). Additionally, seed nano-priming has been reported to enhance plant resistance against pests and various environmental stresses by increasing antioxidant activity and activating enzymatic defense systems (Itroutwar et al., 2019; Siddaiah et al., 2018).

In summary, seed nano-priming is an innovative technique that utilizes nanoparticles to enhance seed priming effects. By modifying gene expression, metabolic processes, and hormone regulation in seeds, it offers potential benefits such as improved germination, accelerated plant growth, and enhanced resistance against biotic and abiotic stresses. Ongoing research aims to further explore the efficacy, safety, and environmental implications of seed nano-priming in agricultural practices.

APPLICATION OF DIFFERENT NANOPARTICLES IN SEED PRIMING

Silver Nanoparticles (Ag-NPs)

Silver nanoparticles (Ag-NPs) have emerged as a versatile nanomaterial with a wide range of applications, including agriculture and gardening. They possess antimicrobial properties and have been extensively studied for their potential in various fields (Awasthi et al., 2017). Ag-NPs exhibit unique optical, electrical, and magnetic characteristics, making them suitable for diverse applications such as cryogenic superconducting components, electronic materials, biosensor materials, cosmetics, composite fibers, ethylene regulation, and antimicrobial treatments in plants (Awasthi et al., 2017). In the realm of agriculture, Ag-NPs have shown promise in protecting seeds from bacterial and fungal attacks, making them valuable in different agricultural practices (Parveen and Rao., 2015). These nanoparticles offer a means to mitigate the harmful effects of pathogens on seeds, thereby enhancing seed quality and viability.

It is important to note that the use of Ag-NPs in agriculture is an area of ongoing research, and further studies are needed to fully understand their efficacy, safety, and potential environmental impacts. Nonetheless, their antimicrobial properties and potential applications make Ag-NPs an intriguing prospect for improving seed health and overall agricultural productivity.

Silica Nanoparticles (Si-NPs)

Silicon (Si) is an abundant element in the Earth's crust and has diverse applications in various fields. Nanosilica, derived from Si, finds utility in research, technology, biology, and electronics. In the realm of plant science, Si has been implicated in aiding plants in coping with salt stress, among other benefits (Wang et al., 2011). It has been observed that Si enhances light absorption in leaves, improves photosynthetic activity, boosts plant resistance to both biotic and abiotic stressors, prolongs organ longevity, reduces evapotranspiration, and enhances leaf mechanical strength. The presence of Si in the epidermal tissue of plant secretory organs is noteworthy (Rastogi et al., 2019). The application of silicon fertilizers has shown promising results in improving crop quantity and quality across diverse plants, climates, and soil conditions. Silicon nanoparticles (NPs) possess unique physiological characteristics that enable them to exert influence on plant metabolism (Rastogi et al., 2019). Moreover, silica nanoparticles (NPs) have been found to facilitate the transport of DNA and other substances into animal and plant cells (Wang et al., 2011). The use of nanosilica has demonstrated positive effects on germination traits, including accelerated germination rate, increased dry weight, and enhanced radicle height in plants (Rastogi et al., 2019; Wang et al., 2011). It is important to note that further research is needed to fully comprehend the mechanisms by which Si and Si nanoparticles exert their effects on plants. Nonetheless, the application of Si-based products, including nanosilica, holds promise in improving various aspects of plant growth and development.

Copper Nanoparticles (Cu-NPs)

Copper (Cu) is an essential element for plant growth and biochemical reactions, including photosynthesis and enzyme activity (Leng et al., 2015). However, excessive copper levels can be toxic, while copper deficiency leads to curled leaves (Leng et al., 2015). The release of copper nanoparticles (Cu-NPs) has raised concerns about potential harm to human health and ecosystems (Chen et al., 2012). Studies have shown that copper oxide (CuO) NPs can be harmful to aquatic organisms (Chen et al., 2012). However, lower concentrations of CuO have been found beneficial for *Vigna radiata* (L.) seedling development and germination (Singh et al., 2017). Further research is needed to fully understand the effects of Cu-NPs on plants (Leng et al., 2015). Proper investigation is necessary to ensure safe and effective use of Cu-NPs in agriculture.

Iron Nanoparticles (Fe-NPs)

Iron (Fe) is an abundant and non-toxic element that serves as a micronutrient for various organisms, including plants (Li et al., 2006). It plays a crucial role in essential biological processes like chlorophyll formation, photosynthesis, and respiration (Najafi Disfani et al., 2017). Iron oxide nanoparticles (FeO-NPs), specifically iron oxide (FeO) NPs, are widely used in environmental applications and wastewater restoration (Fu et al., 2014; Najafi Disfani et al., 2017). These FeO-NPs have a significant impact on seed germination, plant development, and crop yield enhancement. Higher concentrations of FeO-NPs can alleviate iron deficiency by facilitating chlorophyll synthesis and increasing chlorophyll levels (Ghafariyan et al., 2013). FeO-NPs play a crucial role in advancing agriculture and related fields due to their influence on plant growth and development (Li et al., 2006). They are also employed as nano-fertilizers to improve iron availability, resulting in increased plant height, biomass, root length, and modulation of hormone activity and antioxidant enzymes (Ghafariyan et al., 2013). These nano-fertilizers hold promise for enhancing agricultural practices and promoting plant growth.

Zinc Oxide Nanoparticles (ZnO-NPs)

Zinc (Zn) is an essential element for plant growth, contributing to chlorophyll synthesis and carbohydrate production. It also aids in reducing the uptake of harmful heavy metals and mitigating their toxic effects in plants. Nanoparticles have found diverse applications in industries such as biosensors, healthcare, and household products. In agriculture, zinc nanoparticles are widely utilized due to their significant impact on plant physiology and anatomy (Awasthi et al., 2017). Zinc oxide (ZnO) plays a crucial role in regulating phytohormone metabolism and various enzyme functions, including superoxide dismutase and dehydrogenases (Rajput et al., 2021a). ZnO nanoparticles possess favorable characteristics such as a wide surface area, photodegradability, low toxicity, extended lifespan, and high pore volume, making them suitable for various applications such as catalysts, polymer additives, chemical absorbents, and antibacterial agents. Recent research has focused on studying the effects of ZnO nanoparticles on seed germination in different plant species (Awasthi et al., 2017; Gaafar et al., 2020; Sharma et al., 2021). However, caution must be exercised when using ZnO nanoparticles as nano-fertilizers, as they have been observed to have detrimental effects on plants due to the biological activity of metal-based nanoparticles (Itroutwar et al., 2020). Some studies have shown that nano-ZnO can be toxic to plant root development. Therefore, the application of ZnO nanoparticles in agriculture requires careful consideration, and further research is needed to comprehensively understand their impacts on plant growth and development.

IMPACT OF NANO-PRIMING IN ABIOTIC STRESS

Crop yields can be significantly reduced due to pollution and soil salinity, leading to decreased harvest quantities (Maswada et al., 2018). High salt levels in the soil directly impact plant metabolism, resulting in nutrient and water deficiencies that hinder plant growth (Abdel Latef et al., 2017). The presence of heavy metals in the soil can originate from both human activities and natural sources. Anthropogenic sources include waste disposal, burning, sludge and sewage discharge, industrial processes, and the use of fertilizers. On the other hand, natural or geogenic causes of heavy metal accumulation may stem from local geological events or atmospheric deposition (Qayyum et al., 2017). These factors collectively contribute to soil contamination with heavy metals, posing risks to plant growth and the overall well-being of ecosystems.

Seed priming with nanoparticles, such as manganese, zinc, iron, and copper, has shown promising results in enhancing seed germination and plant growth under various stress conditions. For instance, manganese nanoparticles have been found to improve germination and root elongation of Jalapeno pepper seeds under high salinity conditions (Ye et al., 2020). Zinc nanoparticles have been effective in enhancing salt tolerance and promoting growth in lupin seeds (Fig. 1), while iron nanoparticles

Figure 1. Diagrammatic representation illustrating the application of seed priming with nanoparticles to alleviate abiotic stress caused by salinity during the germination stage of rice plants
(Singh A et al., 2022)

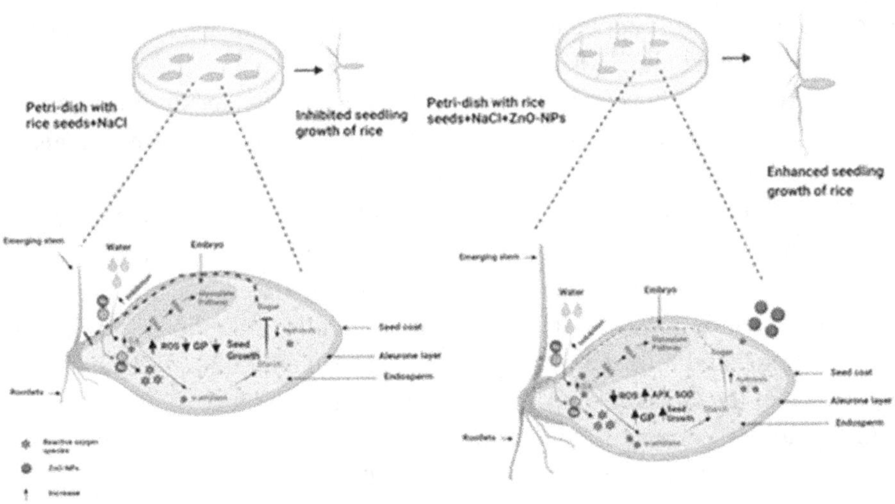

have shown positive effects on sorghum seeds in salty conditions (Maswada et al., 2018). Seed priming with zinc and iron nanoparticles has also been demonstrated to reduce cadmium accumulation in wheat grains (Rizwan et al., 2019). Additionally, silicon nanoparticles used in seed priming have shown benefits in reducing cadmium absorption and enhancing plant growth in cadmium-contaminated soil (Rajput et al., 2020a). Biogenic iron nanoparticles derived from onion extracts have been utilized for seed priming, resulting in stress tolerance and increased levels of jasmonic acid in watermelon seedlings (Kasote et al., 2019). Furthermore, seed priming with copper nanoparticles has improved drought resilience in maize by reducing oxidative stress and enhancing leaf pigments (Nguyen et al., 2020).

These findings highlight the potential of seed nano-priming as a strategy to enhance stress tolerance and promote plant growth under adverse environmental conditions. By regulating plant metabolism and promoting the synthesis of essential compounds, seed nano-priming holds promise in mitigating the negative effects of climate change, heavy metal contamination, drought, and nutritional deficiencies. Implementing seed nano-priming techniques can contribute to reducing yield losses caused by both human activities and natural environmental factors.

CONCERNS

It is indeed crucial to exercise caution and implement responsible regulations when using nanomaterials for seed coating and priming, not only in agriculture but also in other industries. Understanding the ecological impacts and potential environmental toxicity of nanoparticles is essential for their safe and sustainable use. Adhering to legal guidelines and conducting thorough research are necessary for the commercial production, agricultural applications, and disposal of industrial waste related to nanoparticles (Kah et al., 2018).

Nanoparticles can have significant effects on various ecosystems associated with agricultural practices, and it is important to comprehend their mechanisms of action to develop nanomaterials that are safe for the soil and the environment (Lowry et al., 2019). When applying seed priming, careful consideration should be given to various factors, including nanoparticle size, quantity, and treatment duration. Inappropriate priming conditions can lead to adverse effects such as inhibition of germination, metabolic and structural changes in cells, alterations in microbiota-root interactions, and reduced plant growth (Rahman et al., 2020). Understanding the influence of the chemical and physical properties of nanoparticles on seeds and related organisms is crucial for the development of efficient nanomaterials with minimal adverse consequences (Camara et al., 2019).

Different types of nanoparticles, such as polymeric, metallic, and biogenic metallic nanoparticles, exhibit distinct physical, chemical, and biological properties. Seed nano-priming can be customized for specific purposes, such as enhancing plant tolerance to stress, biofortification, or seed protection. Compared to soil and foliar applications, seed treatments reduce nanoparticle exposure, and lower quantities of nanoparticles are required, which can be regulated by manufacturers to prevent excessive release into the environment. Although further research is needed to better understand the behavior and interactions of different nanoparticles with plant growth, it is likely that the amount of nanoparticles remaining in the plant is negligible, very small, or even nonexistent. Overall, responsible use, thorough research, and adherence to regulations are necessary to ensure the safe and sustainable application of nanomaterials in seed priming and other related practices.

CONCLUSION

The field of nanotechnology indeed holds immense potential for revolutionizing agriculture, and seed nano-priming is a promising application that can contribute to enhancing sustainability in farming practices. By improving plant growth and protecting plants from stress conditions, seed nano-priming can lead to increased food quality and quantity. This innovative technique has the ability to replace conventional methods, such as excessive agrochemical usage, with more sustainable agricultural practices.

The adoption of nanotechnology in seed treatment can mitigate environmental damage caused by traditional approaches and provide farmers and consumers with safer and more reliable agricultural products. However, it is important to carefully consider various factors during the development and implementation of nanotechnological materials in the field. This includes determining optimal priming conditions for seeds, scaling up production, and assessing potential adverse impacts on plants and associated organisms. Thorough evaluation of these concerns throughout the manufacturing process and field application of nanotechnological materials is crucial.

By addressing these considerations, the agricultural industry can embrace safer practices that reduce reliance on pesticide applications and mitigate their negative effects on the environment, farmers, and consumers. Nanotechnology has the potential to contribute to sustainable agriculture and foster a more environmentally friendly and resilient food production system.

ACKNOWLEDGMENTS

The author expresses their gratitude to Dr. Abhishek Singh and Dr. Vishnu Rajput for their valuable assistance during the preparation of this manuscript.

REFERENCES

Abbasi Khalaki, M., Moameri, M., Asgari Lajayer, B., & Astatkie, T. (2020). Influence of Nano-Priming on Seed Germination and Plant Growth of Forage and Medicinal Plants. *Plant Growth Regulation*, *931*(93), 13–28. doi:10.100710725-020-00670-9

Abdel Latef, A. A. H., Abu Alhmad, M. F., & Abdelfattah, K. E. (2017). The Possible Roles of Priming with ZnO Nanoparticles in Mitigation of Salinity Stress in Lupine (*Lupinus termis*) Plants. *Journal of Plant Growth Regulation*, *36*(1), 60–70. doi:10.100700344-016-9618-x

Acharya, P., Jayaprakasha, G. K., Crosby, K. M., Jifon, J. L., & Patil, B. S. (2019). Green-Synthesized Nanoparticles Enhanced Seedling Growth, Yield, and Quality of Onion (*Allium cepa* L.). *ACS Sustainable Chemistry & Engineering*, *7*(17), 14580–14590. doi:10.1021/acssuschemeng.9b02180

Acharya, P., Jayaprakasha, G. K., Crosby, K. M., Jifon, J. L., & Patil, B. S. (2020). Nanoparticle-Mediated Seed Priming Improves Germination, Growth, Yield, and Quality of Watermelons (*Citrullus lanatus*) at multi-locations in Texas. *Scientific Reports*, *101*(10), 1–16. doi:10.103841598-020-61696-7 PMID:32193449

Akter, L., Fakir, O. A., Alam, M. K., Islam, M. U., Chakraborti, P., Alam, M. J., Rashid, M. H., Begum, M., & Kader, M. A. (2018). Amelioration of Salinity Stress in Maize Seed Germination and Seedling Growth Attributes through Seed Priming. *Open J. Soil Sci*, *08*(5), 137–146. doi:10.4236/ojss.2018.85011

Avellan, A., Schwab, F., Masion, A., Chaurand, P., Borschneck, D., Vidal, V., Rose, J., Santaella, C., & Levard, C. (2017). Nanoparticle Uptake in Plants: Gold Nanomaterial Localized in Roots of Arabidopsis thaliana by X-ray Computed Nanotomography and Hyperspectral Imaging. *Environmental Science & Technology*, *51*(15), 8682–8691. doi:10.1021/acs.est.7b01133 PMID:28686423

Awasthi, A., Bansal, S., Jangir, L. K., Awasthi, G., Awasthi, K. K., & Awasthi, K. (2017). Effect of ZnO Nanoparticles on Germination of *Triticum aestivum* Seeds. *Macromolecular Symposia*, *376*(1), 1700043. doi:10.1002/masy.201700043

Bailly, C. (2019). The signalling role of ROS in the regulation of seed germination and dormancy. *The Biochemical Journal, 476*(20), 3019–3032. doi:10.1042/BCJ20190159 PMID:31657442

Bombo, A. B., Pereira, A. E. S., Lusa, M. G., De Medeiros Oliveira, E., De Oliveira, J. L., Campos, E. V. R., De Jesus, M. B., Oliveira, H. C., Fraceto, L. F., & Mayer, J. L. S. (2019). A Mechanistic View of Interactions of a Nanoherbicide with Target Organism. *Journal of Agricultural and Food Chemistry, 67*(16), 4453–4462. doi:10.1021/acs.jafc.9b00806 PMID:30933503

Camara, M. C., Campos, E. V. R., Monteiro, R. A., Do Espirito Santo Pereira, A., De Freitas Proença, P. L., & Fraceto, L. F. (2019). Development of stimuli-responsive nano-based pesticides: Emerging opportunities for agriculture. *Journal ofNanobiotechnology, 17*(1), 1–19. doi:10.118612951-019-0533-8 PMID:31542052

Camerel, F., Gabriel, J. C. P., Batail, P., Davidson, P., Lemaire, B., Schmutz, M., Gulik-Krzywicki, T., & Bourgaux, C. (2002). Original Single Walled Nanotubules Based on Weakly Interacting Covalent Mineral Polymers, 1∞[Nb2PS10-] in N-Methylformamide. *Nano Letters, 2*(4), 403–407. doi:10.1021/nl0100901

Carrillo-Reche, J., Vallejo-Marín, M., & Quilliam, R. S. (2018). Quantifying the potential of 'on-farm' seed priming to increase crop performance in developing countries. A meta-analysis. *Agronomy for Sustainable Development, 38*(6), 1–14. doi:10.100713593-018-0536-0

Chau, N. H., Doan, Q. H., Chu, T. H., Nguyen, T. T., Dao Trong, H., & Ngo, Q. B. (2019). Effects of Different Nanoscale Microelement-Containing Formulations for Presowing Seed Treatment on Growth of Soybean Seedlings. *Journal of Chemistry, 2019*, 1–8. doi:10.1155/2019/8060316

Chen, Y., Wang, D., Zhu, X., Zheng, X., & Feng, L. (2012). Long-term effects of copper nanoparticles on wastewater biological nutrient removal and N_2O generation in the activated sludge process. *Environmental Science & Technology, 46*(22), 12452–12458. doi:10.1021/es302646q PMID:23110389

Chokheli, V. A., Shmaraeva, A. N., Kornienko, I. V., Stepanenko, V. V., Rajput, V. D., Singh, A., Kumari, A., Dmitriev, P. A., Azarov, A. S., Minkina, T. M., & Varduni, T. V. (2022). Assessment of genetic diversity and DNA fingerprinting of rare species of the genus Crambe (Brassicaceae). *Crop & Pasture Science*. Advance online publication. doi:10.1071/CP22027

Das, C. K., Jangir, H., Kumar, J., Verma, S., Mahapatra, S. S., Philip, D., Srivastava, G., & Das, M. (2018). Nano-pyrite seed dressing: A sustainable design for NPK equivalent rice production. Nanotechnol. *Environ. Eng*, *31*(3), 1–14. doi:10.100741204-018-0043-1

De La Torre-Roche, R., Cantu, J., Tamez, C., Zuverza-Mena, N., Hamdi, H., Adisa, I. O., Elmer, W., Gardea-Torresdey, J., & White, J. C. (2020). Seed Biofortification by Engineered Nanomaterials: A Pathway to Alleviate Malnutrition? *Journal of Agricultural and Food Chemistry*, *68*(44), 12189–12202. doi:10.1021/acs.jafc.0c04881 PMID:33085897

Dietz, K. J., & Herth, S. (2011). Plant nanotoxicology. *Trends in Plant Science*, *16*(11), 582–589. doi:10.1016/j.tplants.2011.08.003 PMID:21906987

Dikshit, P. K., Kumar, J., Das, A. K., Sadhu, S., Sharma, S., Singh, S., Gupta, P. K., & Kim, B. S. (2021). Green Synthesis of Metallic Nanoparticles: Applications and Limitations. *Catalysts*, *11*(8), 902. doi:10.3390/catal11080902

Fraceto, L. F., Grillo, R., de Medeiros, G. A., Scognamiglio, V., Rea, G., & Bartolucci, C. (2016). Nanotechnology in agriculture: Which innovation potential does it have? *Frontiers in Environmental Science*, *4*, 20. doi:10.3389/fenvs.2016.00020

Friedrichs, S., Takasu, Y., Kearns, P., Dagallier, B., Oshima, R., Schofield, J., & Moreddu, C. (2019). An overview of regulatory approaches to genome editing in agriculture. *Biotechnol. Res. Innov.*, *3*(2), 208–220. doi:10.1016/j.biori.2019.07.001

Gaafar, R. M., Diab, R. H., Halawa, M. L., El-Shanshory, A. R., El-Shaer, A., & Hamouda, M. M. (2020). Role of Zinc Oxide Nanoparticles in Ameliorating Salt Tolerance in Soybean. *Egyptian Journal of Botany*, *60*(0), 733–747. doi:10.21608/ejbo.2020.26415.1475

Ghafariyan, M. H., Malakouti, M. J., Dadpour, M. R., Stroeve, P., & Mahmoudi, M. (2013). Effects of Magnetite Nanoparticles on Soybean Chlorophyll. *Environmental Science & Technology*, *47*, 10645–10652. doi:10.1021/es402249b PMID:23951999

Gross, M. S., Bean, T. G., Hladik, M. L., Rattner, B. A., & Kuivila, K. M. (2020). Uptake, Metabolism, and Elimination of Fungicides from Coated Wheat Seeds in Japanese Quail (*Coturnix japonica*). *Journal of Agricultural and Food Chemistry*, *68*(6), 1514–1524. doi:10.1021/acs.jafc.9b05668 PMID:31977218

Hu, P., An, J., Faulkner, M. M., Wu, H., Li, Z., Tian, X., & Giraldo, J. P. (2020). Nanoparticle Charge and Size Control Foliar Delivery Efficiency to Plant Cells and Organelles. *ACS Nano*, *14*(7), 7970–7986. doi:10.1021/acsnano.9b09178 PMID:32628442

Hussain, A., Rizwan, M., Ali, Q., & Ali, S. (2019). Seed priming with silicon nanoparticles improved the biomass and yield while reduced the oxidative stress and cadmium concentration in wheat grains. *Environmental Science and Pollution Research International*, *26*(26), 7579–7588. doi:10.100711356-019-04210-5 PMID:30661166

Itroutwar, P. D., Govindaraju, K., Tamilselvan, S., Kannan, M., Raja, K., & Subramanian, K. S. (2019). Seaweed-Based Biogenic ZnO Nanoparticles for Improving Agro-morphological Characteristics of Rice (*Oryza sativa* L.). *Journal of Plant Growth Regulation*, *39*(2), 717–728. doi:10.100700344-019-10012-3

Itroutwar, P. D., Kasivelu, G., Raguraman, V., Malaichamy, K., & Sevathapandian, S. K. (2020). Effects of biogenic zinc oxide nanoparticles on seed germination and seedling vigor of maize (*Zea mays*). *Biocatalysis and Agricultural Biotechnology*, *29*, 101778. doi:10.1016/j.bcab.2020.101778

Kah, M., Kookana, R. S., Gogos, A., & Bucheli, T. D. (2018). A Critical Evaluation of Nanopesticides and Nanofertilizers Against Their Conventional Analogues. *Nature Nanotechnology*, *13*(8), 677–684. doi:10.103841565-018-0131-1 PMID:29736032

Kah, M., Tufenkji, N., & White, J. C. (2019). Nano-Enabled Strategies to Enhance Crop Nutrition and Protection. *Nature Nanotechnology*, *14*(6), 532–540. doi:10.103841565-019-0439-5 PMID:31168071

Kasote, D. M., Lee, J. H. J., Jayaprakasha, G. K., & Patil, B. S. (2019). Seed Priming with Iron Oxide Nanoparticles Modulate Antioxidant Potential and Defense-Linked Hormones in Watermelon Seedlings. *ACS Sustainable Chemistry & Engineering*, *7*(5), 5142–5151. doi:10.1021/acssuschemeng.8b06013

Khan, M. N., Mobin, M., Abbas, Z. K., AlMutairi, K. A., & Siddiqui, Z. H. (2017). Role of Nanomaterials in Plants under Challenging Environments. *Plant Physiology and Biochemistry*, *110*, 194–209. doi:10.1016/j.plaphy.2016.05.038 PMID:27269705

Khodakovskaya, M., Dervishi, E., Mahmood, M., Xu, Y., Li, Z., Watanabe, F., & Biris, A. S. (2009). Carbon Nanotubes are Able to Penetrate Plant Seed Coat and Dramatically Affect Seed Germination and Plant Growth. *ACS Nano*, *3*(10), 3221–3227. doi:10.1021/nn900887m PMID:19772305

Kumar, P., Singh, A., Rajput, V. D., Yadav, A. K. S., Kumar, P., Singh, A. K., & Minkina, T. (2022). Role of artificial intelligence, sensor technology, big data in agriculture: next-generation farming. *Bioinformatics in Agriculture*, 625–639. doi:10.1016/B978-0-323-89778-5.00035-0

Lemmens, E., Deleu, L. J., De Brier, N., De Man, W. L., De Proft, M., Prinsen, E., & Delcour, J. A. (2019). The Impact of Hydro-Priming and Osmo-Priming on Seedling Characteristics, Plant Hormone Concentrations, Activity of Selected Hydrolytic Enzymes, and Cell Wall and Phytate Hydrolysis in Sprouted Wheat (*Triticum aestivum* L.). *ACS Omega*, *4*(26), 22089–22100. doi:10.1021/acsomega.9b03210 PMID:31891089

Leng, X., Jia, H., Sun, X., Shangguan, L., Mu, Q., Wang, B., & Fang, J. (2015). Comparative Transcriptome Analysis of Grapevine in Response to Copper Stress. *Scientific Reports*, *5*(1), 17749. doi:10.1038rep17749 PMID:26673527

Li, L., Fan, M., Brown, R. C., Van Leeuwen, J., Wang, J., Wang, W., Song, Y., & Zhang, P. (2006). Synthesis, Properties, and Environmental Applications of Nanoscale Iron-Based Materials: A Review. *Critical Reviews in Environmental Science and Technology*, *36*(5), 405–431. doi:10.1080/10643380600620387

Lowry, G. V., Avellan, A., & Gilbertson, L. M. (2019). Opportunities and Challenges For Nanotechnology in the Agri-Tech Revolution. *Nature Nanotechnology*, *14*(6), 517–522. doi:10.103841565-019-0461-7 PMID:31168073

Mahakham, W., Sarmah, A. K., Maensiri, S., & Theerakulpisut, P. (2017). Nanopriming Technology For Enhancing Germination And Starch Metabolism Of Aged Rice Seeds Using Phytosynthesized Silver Nanoparticles. *Scientific Reports*, *7*(1), 1–21. doi:10.103841598-017-08669-5 PMID:28811584

Mahakham, W., Theerakulpisut, P., Maensiri, S., Phumying, S., & Sarmah, A. K. (2016). Environmentally Benign Synthesis Of Phytochemicals-Capped Gold Nanoparticles As Nanopriming Agent For Promoting Maize Seed Germination. *The Science of the Total Environment*, *573*, 1089–1102. doi:10.1016/j.scitotenv.2016.08.120 PMID:27639594

Maity, A., Natarajan, N., Pastor, M., Vijay, D., Gupta, C. K., & Wasnik, V. K. (2018). Nanoparticles influence seed germination traits and seed pathogen infection rate in forage sorghum (*Sorghum bicolour*) and cowpea (*Vigna unguiculata*). *Indian Journal of Experimental Biology*, *56*, 363–372.

Makarov, V. V., Love, A. J., Sinitsyna, O. V., Makarova, S. S., Yaminsky, I. V., Taliansky, M. E., & Kalinina, N. O. (2014). "Green" Nanotechnologies: Synthesis of Metal Nanoparticles Using Plants. *Acta Naturae (English Ed.)*, *6*(1), 35–44. doi:10.32607/20758251-2014-6-1-35-44 PMID:24772325

Malik, A., Mor, V. S., Tokas, J., Punia, H., Malik, S., Malik, K., Sangwan, S., Tomar, S., Singh, P., Singh, N., Himangini, Vikram, Nidhi, Singh, G., Vikram, Kumar, V., Sandhya, & Karwasra, A. (2020). Biostimulant-Treated Seedlings under Sustainable Agriculture: A Global Perspective Facing Climate Change. *Agronomy (Basel)*, *11*(1), 14. doi:10.3390/agronomy11010014

Manjaiah, K.M., Mukhopadhyay, R., Paul, R., Datta, S.C., Kumararaja, P., Sarkar, B. (2018). Clay Minerals And Zeolites For Environmentally Sustainable Agriculture. Modif. Clay Zeolite Nanocomposite Mater. *Environ. Pharm. Appl.*, 309–329. doi:10.1016/B978-0-12-814617-0.00008-6

Voloshina, M. (2020). Monitoring Soil Salinity and Recent Advances in Mechanism of Salinity Tolerance in Plants. *Biogeosystem Tech.*, *7*, 66–87. https://doi.org/http://dxdoi.org/10.13187/bgt.2020.2.66

Maswada, H. F., Djanaguiraman, M., & Prasad, P. V. V. (2018). Seed Treatment With Nano-Iron (III) Oxide Enhances Germination, Seeding Growth And Salinity Tolerance Of Sorghum. *Journal Agronomy & Crop Science*, *204*(6), 577–587. doi:10.1111/jac.12280

Mittal, A. K., Chisti, Y., & Banerjee, U. C. (2013). Synthesis Of Metallic Nanoparticles Using Plant Extracts. *Biotechnology Advances*, *31*(2), 346–356. doi:10.1016/j.biotechadv.2013.01.003 PMID:23318667

Mohamad, N. A. N., Arham, N. A., Jai, J., & Hadi, A. (2014). Plant Extract as Reducing Agent in Synthesis of Metallic Nanoparticles: A Review. *Advanced Materials Research*, *832*, 350–355. . doi:10.4028/www.scientific.net/AMR.832.350

Montanha, G. S., Rodrigues, E. S., Marques, J. P. R., de Almeida, E., Colzato, M., & Pereira de Carvalho, H. W. (2020). Zinc Nanocoated Seeds: An Alternative To Boost Soybean Seed Germination And Seedling Development. *SN Applied Sciences*, *2*(5), 1–11. doi:10.100742452-020-2630-6

Najafi Disfani, M., Mikhak, A., Kassaee, M. Z., & Maghari, A. (2017). Effects Of Nano Fe/SiO_2 Fertilizers On Germination And Growth Of Barley And Maize. *Archives of Agronomy and Soil Science*, *63*(6), 817–826. doi:10.1080/03650340.2016.1239016

Nonogaki, H. (2014). Seed Dormancy And Germination-Emerging Mechanisms And New Hypotheses. *Frontiers in Plant Science*, *5*, 233. doi:10.3389/fpls.2014.00233 PMID:24904627

Nonogaki, H., Bassel, G. W., & Bewley, J. D. (2010). Germination—Still a mystery. *Plant Science*, *179*(6), 574–581. doi:10.1016/j.plantsci.2010.02.010

Noorhosseini, S. A., Jokar, N. K., & Damalas, C. A. (2017). Improving Seed Germination and Early Growth of Garden Cress (*Lepidium sativum*) and Basil (*Ocimum basilicum*) with Hydro-priming. *Journal of Plant Growth Regulation*, *37*(1), 323–334. doi:10.100700344-017-9728-0

Palocci, C., Valletta, A., Chronopoulou, L., Donati, L., Bramosanti, M., Brasili, E., Baldan, B., & Pasqua, G. (2017). Endocytic Pathways Involved In PLGA Nanoparticle Uptake By Grapevine Cells And Role Of Cell Wall And Membrane In Size Selection. *Plant Cell Reports*, *36*(12), 1917–1928. doi:10.100700299-017-2206-0 PMID:28913707

Panpatte, D. G., Jhala, Y. G., Shelat, H. N., & Vyas, R. V. (2016). *Nanoparticles: The Next Generation Technology for Sustainable Agriculture. Microb. Inoculants Sustain. Agric. Product* (Vol. 2). Funct. Appl. doi:10.1007/978-81-322-2644-4_18

Parveen, A., & Rao, S. (2015). Effect of Nanosilver on Seed Germination and Seedling Growth in *Pennisetum glaucum*. *Journal of Cluster Science*, *26*(3), 693–701. doi:10.100710876-014-0728-y

Pérez-de-Luque, A. (2017). Interaction Of Nanomaterials With Plants: What Do We Need For Real Applications In Agriculture? *Frontiers in Environmental Science*, *5*, 12. doi:10.3389/fenvs.2017.00012

Pirzada, T., de Farias, B. V., Mathew, R., Guenther, R. H., Byrd, M. V., Sit, T. L., Pal, L., Opperman, C. H., & Khan, S. A. (2020). Recent Advances In Biodegradable Matrices For Active Ingredient Release In Crop Protection: Towards Attaining Sustainability In Agriculture. *Current Opinion in Colloid & Interface Science*, *48*, 121–136. doi:10.1016/j.cocis.2020.05.002 PMID:33013179

Qayyum, M. F., Rehman, M. Z., Ali, S., Rizwan, M., Naeem, A., Maqsood, M. A., Khalid, H., Rinklebe, J., & Ok, Y. S. (2017). Residual Effects Of Monoammonium Phosphate, Gypsum And Elemental Sulfur On Cadmium Phytoavailability And Translocation From Soil To Wheat In An Effluent Irrigated Field. *Chemosphere*, *174*, 515–523. doi:10.1016/j.chemosphere.2017.02.006 PMID:28189896

Rahman, M. S., Chakraborty, A., Mazumdar, S., Nandi, N. C., Bhuiyan, M. N. I., Alauddin, S. M., Khan, I. A., & Hossain, M. J. (2020). Effects Of Poly(Vinylpyrrolidone) Protected Platinum Nanoparticles On Seed Germination And Growth Performance Of *Pisum sativum*. *Nano-Structures & Nano-Objects*, *21*, 100408. doi:10.1016/j.nanoso.2019.100408

Rajput, V., Minkina, T., Fedorenko, A., Sushkova, S., Mandzhieva, S., Lysenko, V., Duplii, N., Fedorenko, G., Dvadnenko, K., & Ghazaryan, K. (2018). Toxicity Of Copper Oxide Nanoparticles On Spring Barley (*Hordeum sativum distichum*). *The Science of the Total Environment*, *645*, 1103–1113. doi:10.1016/j.scitotenv.2018.07.211 PMID:30248835

Rajput, V., Minkina, T., Mazarji, M., Shende, S., Sushkova, S., Mandzhieva, S., Burachevskaya, M., Chaplygin, V., Singh, A., & Jatav, H. (2020). Accumulation Of Nanoparticles In The Soil-Plant Systems And Their Effects On Human Health. *Annals of Agricultural Science*, *65*(2), 137–143. doi:10.1016/j.aoas.2020.08.001

Rajput, V., Minkina, T., Sushkova, S., Behal, A., Maksimov, A., Blicharska, E., Ghazaryan, K., Movsesyan, H., Barsova, N. (2020). ZnO and CuO Nanoparticles: A Threat To Soil Organisms, Plants, And Human Health. *Environmental Geochemistry and Health*, 42(1), 147–158. doi:10.1007/s10653-019-00317-3

Rajput, V.D., Chen, Y., Ayup, M. (2015). Effects Of High Salinity On Physiological And Anatomical Indices In The Early Stages Of *Populus euphratica* Growth. *Russian Journal of Plant Physiology, 62*(2), 229–236. (https://doi.org/) doi:10.1134/S1021443715020168

Rajput, V.D., Minkina, T., Fedorenko, A., Chernikova, N., Hassan, T., Mandzhieva, S., Sushkova, S., Lysenko, V., Soldatov, M.A., Burachevskaya, M. (2021). Effects Of Zinc Oxide Nanoparticles On Physiological And Anatomical Indices In Spring Barley Tissues. *Nanomaterials, 11*(7), 1722. (https://doi.org/) doi:10.3390/nano11071722

Rajput, V.D., Singh, A., Minkina, T., Rawat, S., Mandzhieva, S., Sushkova, S., Shuvaeva, V., Nazarenko, O., Rajput, P., Komariah, Verma, K.K., Singh, A.K., Rao, M., & Upadhyay, S.K. (2021). Nano-Enabled Products: Challenges and Opportunities For Sustainable Agriculture. *Plants, 10*. (https://doi.org/) doi:10.3390/plants10122727

Rajput, V.D., Singh, A., Minkina, T.M., Shende, S.S., Kumar, P., Verma, K.K., Bauer, T., Gorobtsova, O., Deneva, S., Sindireva, A. (2021). Potential Applications of Nanobiotechnology in Plant Nutrition and Protection for Sustainable Agriculture. *Nanotechnology in Plant Growth Promotion and Protection,* 79–92. (https://doi.org/) doi:10.1002/9781119745884.ch5

Rajput, V. D., Tatiana, M., Kumari, A., Shende, S. S., Ranjan, A., Faizan, M., & Barakvov, A. (2022). A Review On Nanobioremediation Approaches For Restoration Of Contaminated Soil. *Eurasian Journal of Soil Science*, *11*(1), 43–60. doi:10.18393/ejss.990605

Rastogi, A., Tripathi, D.K., Yadav, S., Chauhan, D.K., Živčák, M., Ghorbanpour, M., El-Sheery, N.I., Brestic, M. (2019). Application of Silicon Nanoparticles in Agriculture. *3 Biotech, 9*. (https://doi.org/) doi:10.1007/s13205-019-1626-7

Rizwan, M., Ali, S., Ali, B., Adrees, M., Arshad, M., Hussain, A., Zia ur Rehman, M., & Waris, A.A. (2019). Zinc And Iron Oxide Nanoparticles Improved The Plant Growth And Reduced The Oxidative Stress And Cadmium Concentration In Wheat. *Chemosphere, 214*, 269–277. (https://doi.org/) doi:10.1016/j.chemosphere.2018.09.120

Saddiq, M.S., Iqbal, S., Afzal, I., Ibrahim, A.M.H., Bakhtavar, M.A., Hafeez, M.B., & Jahanzaib, M.M. (2019). *Mitigation Of Salinity Stress In Wheat (Triticum aestivum L.) Seedlings through Physiological Seed Enhancements*. Taylor & Francis. (https://doi.org/) doi:10.1080/01904167.2019.1609509

Scott, N.R., Chen, H., Cui, H. (2018). Nanotechnology Applications And Implications Of Agrochemicals Toward Sustainable Agriculture And Food Systems. *Journal of Agricultural and Food Chemistry, 66*, 6451–6456. (https://doi.org/) doi:10.1021/acs.jafc.8b00964

Sharma, D., Afzal, S., Singh, N.K. (2021). Nanopriming With Phytosynthesized Zinc Oxide Nanoparticles For Promoting Germination And Starch Metabolism In Rice Seeds. *Journal of Biotechnology, 336*, 64–75. (https://doi.org/) doi:10.1016/j.jbiotec.2021.06.014

Si, S., Mandal, T.K. (2007). Tryptophan-Based Peptides To Synthesize Gold And Silver Nanoparticles: A Mechanistic And Kinetic Study. *Chemistry - A European Journal, 13*, 3160–3168. (https://doi.org/) doi:10.1002/chem.200601492

Siddaiah, C. N., Prasanth, K. V. H., Satyanarayana, N. R., Mudili, V., Gupta, V. K., Kalagatur, N. K., Satyavati, T., Dai, X. F., Chen, J. Y., Mocan, A., Singh, B. P., & Srivastava, R. K. (2018). Chitosan Nanoparticles Having Higher Degree Of Acetylation Induce Resistance Against Pearl Millet Downy Mildew Through Nitric Oxide Generation. *Scientific Reports, 81*(8), 1–14. doi:10.103841598-017-19016-z PMID:29410438

Singh, A., Rajput, V., Singh, A.K., Sengar, R.S., Singh, R.K., & Minkina, T. (2021). Transformation Techniques And Their Role In Crop Improvements: A Global Scenario Of Gm Crops. *Policy Issues in Genetically Modified Crops*, 515–542. (https://doi.org/) doi:10.1016/B978-0-12-820780-2.00023-6

Singh, A., Rajput, V.D., Rawat, S., Sharma, R., Singh, Anil Kumar, Kumar, P., Singh, Awani Kumar, Minkina, T., Singh, R.P., & Singh, S. (2022). Geoinformatics and Nanotechnological Approaches for Coping Up Abiotic and Biotic Stress in Crop Plants. *Sustainable Agriculture Systems and Technology,* 337–359. (https://doi.org/) doi:10.1002/9781119808565.ch17

Singh, A., Rajput, V. D., Rawat, S., Sharma, R., Singh, A. K., Kumar, P., Singh, A. K., Minkina, T., Singh, R. P., & Singh, S. (2022, March 4). Geoinformatics and Nanotechnological Approaches for Coping Up Abiotic and Biotic Stress in Crop Plants. *Sustainable Agriculture Systems and Technologies,* 337–359. doi:10.1002/9781119808565.ch17

Singh, A., Sengar, R. S., Rajput, V. D., Minkina, T., & Singh, R. K. (2022, July 13). Zinc Oxide Nanoparticles Improve Salt Tolerance in Rice Seedlings by Improving Physiological and Biochemical Indices. *Agriculture, 12*(7), 1014. doi:10.3390/agriculture12071014

Singh, A., Sengar, R. S., Shahi, U. P., Rajput, V. D., Minkina, T., & Ghazaryan, K. A. (2022). Prominent Effects of Zinc Oxide Nanoparticles on Roots of Rice (*Oryza sativa* L.) Grown under Salinity Stress. *Stresses, 3*(1), 33–46. doi:10.3390tresses3010004

Singh, A., Singh, A. K., Rawat, S., Pal, N., Rajput, V. D., Minkina, T., Sharma, R., Singh, N. P., & Tripathi, J. N. (2022). Satellite-Based Quantification of Methane Emissions from Wetlands and Rice Paddies Ecosystems in North and Northeast India. *Hydrobiology, 1*(3), 317–330. doi:10.3390/hydrobiology1030023

Singh, A., Singh, N.B., Hussain, I., Singh, H., Yadav, V. (2017). Synthesis And Characterization Of Copper Oxide Nanoparticles And Its Impact On Germination Of *Vigna radiata* (L.) R. Wilczek. *Trends in Phytochemical Research, 4*(2), 66–73. https://doi.org/ doi:10.22271/tpr.2017.v4.i2.034

Singh, A., Singh Sengar, R., Singh, R., Shahi, U. P., Yadav, M. K., Vaishali, Gangwar, L. K., & Rajput, V. D. (2022). Effects of zinc oxide nanoparticles for promoting seed germination of rice (*Oryza sativa* L.) under salinity stress. Ecology. *Environmental Conservation, 28*, 254–259. doi:10.53550/EEC.2022.v28i07s.042

Spielman-Sun, E., Avellan, A., Bland, G.D., Tappero, R.V., Acerbo, A.S., Unrine, J.M., Giraldo, J.P., Lowry, G.V. (2019). Nanoparticle Surface Charge Influences Translocation And Leaf Distribution In Vascular Plants With Contrasting Anatomy. *Environmental Science: Nano,* 6, 2508–2519. (https://doi.org/) doi:10.1039/C9EN00626E

Sytar, O., Kumari, P., Yadav, S., Brestic, M., Rastogi, A. (2018). Phytohormone Priming: Regulator For Heavy Metal Stress In Plants. *Journal of Plant Growth Regulation*, *38*(2), 739–752. (https://doi.org/) doi:10.1007/s00344-018-9886-8

Valletta, A., Chronopoulou, L., Palocci, C., Baldan, B., Donati, L., Pasqua, G. (2014). Poly(Lactic-Co-Glycolic) Acid Nanoparticles Uptake By *Vitis Vinifera* And Grapevine-Pathogenic Fungi. *Journal of Nanoparticle Research*, *16*(12), 2744. (https://doi.org/) doi:10.1007/s11051-014-2744-0

Van Nguyen, D., Nguyen, H. M., Le, N. T., Nguyen, K. H., Le, H. M., Nguyen, A. T., Dinh, N. T. T., Hoang, S. A., & Van Ha, C. (2020). Copper Nanoparticle Application Enhances Plant Growth And Grain Yield In Maize Under Drought Stress Conditions. bioRxiv 2020.02.24.963132. doi:10.1101/2020.02.24.963132

Villagarcia, H., Dervishi, E., De Silva, K., Biris, A. S., & Khodakovskaya, M. V. (2012). Surface Chemistry Of Carbon Nanotubes Impacts The Growth And Expression Of Water Channel Protein In Tomato Plants. *Small*, *8*(15), 2328–2334. doi:10.1002mll.201102661 PMID:22514121

Wang, H., Kou, X., Pei, Z., Xiao, J. Q., Shan, X., & Xing, B. (2011). Physiological Effects Of Magnetite (Fe_3O_4) Nanoparticles On Perennial Ryegrass (*Lolium perenne* L.) And Pumpkin (*Cucurbita mixta*) Plants. *Nanotoxicology*, *5*(1), 30–42. doi:10.3109/17435390.2010.489206 PMID:21417686

Wu, M., Wu, J., & Gan, Y. (2020). The New Insight Of Auxin Functions: Transition From Seed Dormancy To Germination And Floral Opening In Plants. *Plant Growth Regulation*, *912*(91), 169–174. doi:10.100710725-020-00608-1

Ye, Y., Cota-Ruiz, K., Hernández-Viezcas, J. A., Valdés, C., Medina-Velo, I. A., Turley, R. S., Peralta-Videa, J. R., & Gardea-Torresdey, J. L. (2020). Manganese Nanoparticles Control Salinity-Modulated Molecular Responses In *Capsicum annuum* L. Through Priming: A Sustainable Approach For Agriculture. *ACS Sustainable Chemistry & Engineering*, *8*(3), 1427–1436. doi:10.1021/acssuschemeng.9b05615

Zahedifar, M., & Zohrabi, S. (2016). Germination And Seedling Characteristics Of Drought-Stressed Corn Seed As Influenced By Seed Priming With Potassium Nano-Chelate And Sulfate Fertilizers. *Acta Agriculturae Slovenica*, *107*(1), 113–128. doi:10.14720/aas.2016.107.1.12

Chapter 3
The Potential of Nano-Based Seed Priming for Sustainable Agriculture

Annika Arvind Gupta
https://orcid.org/0000-0001-9956-476X
B.K. Birla College, India

Arvind Radheyshyam Gupta
Bioxia, India

ABSTRACT

Plant seeds are an important input in agriculture; however, environmental stresses can adversely affect seed vitality, development, and crop production during their developmental stages which results in decreased crop production. High salinity, drought, and climate change-resistant seeds can increase crop yield. The "NanoTech-Agril" era is novel in nano-based technology, having solutions to climate change-related problems in agriculture, empowering farmers to produce a greater number of crops while avoiding problems associated with climate change. Nanoparticle-based seed priming can improve seed metabolism and signaling pathways, helping improve seed growth. Nano-priming induces the formation of nanopores in shoots, aiding in water absorption, activates (ROS)/antioxidant mechanisms in seeds, forms hydroxyl radicals to loosen cell walls, and acts as an inducer for starch hydrolysis present chapter aims to provide an overview of nanoparticle-based seed priming for sustainable agriculture.

INTRODUCTION

"Zero hunger" is Goal 2 of the Sustainable Development Goals (SDGs), which was established in 2015 by the UN (17 SDG). The official declaration reads, "End hunger,

DOI: 10.4018/978-1-6684-7232-3.ch003

establish food security, improve nutrition, and promote sustainable agriculture." (2015 Sustainable Development Agenda for 2030); 2022 SDG indicators- Sustainable agriculture, nutrition, food security, and rural transformation are all intricately intertwined in SDG 2 (High-Level Political Forum on Sustainable Development, 2017). The UN estimates that 690 million people or somewhat less than 10% of the globe's people, are starving, (World Population Clock, 2022)_

Pest-related production deficits, depletion of natural reserves, and the effects of global climate shift are just a few of the issues facing agriculture (Kah et al., 2019; De La Torre- Roche et al., 2020). Additional concern is that traditional agricultural practices necessitate the continual application of pesticides and fertilizers, both of which are harmful to the ecosystem (Rajput et al., 2018). By 2050, the Earth's populace is expected to reach 9 to 10 billion people, implying that food productivity may need to increase by 25 to 70 percent from its current levels (Scott et al., 2018). Modern agricultural technology is required to ensure sustainability and boost yield (Panpatte et al., 2016; Fraceto et al., 2016). The existence and preservation of plant species depend on good sprouting, especially in particular environments like rangeland and farming land (Manjaiah et al., 2018). A plant's life begins with the sprouting of seeds. Drought and salt stress are also particularly dangerous during the germination stage. A plant will proceed to the subsequent growth phase if it passes these tests (Akter et al., 2018). Plant development and productivity are controlled by the rate and uniformity of seedlings, which prevent buds from being damaged by unfavourable ecological variables (Rajput et al., 2015). Seed growth and the successive development of seedlings are influenced by a variety of factors, including drought, salt, temperature, moisture, and others. (Friedrichs et al., 2019; Zahedifar and Zohrabi, 2016; Maity et al., 2018).

Since nanotechnology has uses in a variety of fields, including catalysis, materials, energy, health, and pharmaceuticals, nanotechnology has earned a terrific deal of interest in current years. Nanoparticles with a surface area of one to one hundred nanometers (nm) or larger may have applications in agriculture, engineering, and medicine. Using a variety of methods, researchers have attempted to synthesize nanoparticles extensively, including chemical, biological, and physical techniques (Ghidan et al., 2017a). These methods have a number of drawbacks because they are difficult to scale up, require a large amount of surfactants, and require nanoparticle separation and refinement from the microemulsion (surfactant, oil, co-surfactant, and aqueous phase). Green techniques for making nanoparticles from plant extracts are good because they are easy to use, don't hurt the environment, and take little time to do. According to Ghidan et al., (2017b) nanomaterials produced through environmentally friendly and green processes have the potential to enhance agriculture by enhancing fertilizers, plant growth regulators, and insecticides. Additionally, they lessen the environmental harm caused by toxic compounds. As a result, this technology encourages a reduction in contaminants in the environment (Huang et al., 2015), and the numerous uses of nanotechnology in fields like agriculture, the environment, and health have recently piqued interest (Kah and Hofmann, 2014). Small nanoparticles, in particular, are appealing for tackling challenges that are beyond the capabilities of physical, chemical, and biological methods of pest control.

Nanotechnology recommends innovative agrochemical mediators and delivery systems to enhance harvest output while diminishing pesticide usage. A rise in agricultural output is possible with nanotechnology by using:

- Nanoformulations of agrochemicals for utilizing pesticides and fertilisers for yield enhancement;
- Nanosensors in crop security for identifying infections and agrochemical deposits;
- Nanodevices for genetically modifying of plant;
- Plant infection detection;
- Animal wellbeing and reproduction, poultry production; and
- Postharvest managing.

Nanomaterials are defined by their chemical, physical, and mechanical characteristics. Figure 1 displays several nanoparticle formations processes.

Crop security can be aided by nanomaterials, specifically nanoparticles (Scott et al., 2018). This is a substantial field of research that has intrigued a numerous agrarian business, leading to the utilization of nanoparticles in preparations (Chau et al., 2019). Crop nourishment, insect management, and ecological friendly manufacturing processes can all profit from the usage of nanopesticides and nanofertilizers (Rajput et al., 2021c; Acharya et al., 2020). According to a recent study, nanoparticles have varying effects on plants and seeds (Pérez-de-Luque, 2017; Acharya et al., 2020). According to Rajput et al. (2020a) nanoparticles may have negative effects like phytotoxicity or germination inhibition. On the other hand, some nanoparticles enhance cellular signalling pathways, increasing the metabolic rate of seeds, the vigour of seedlings, and crop growth (Abbasi Khalaki et al., 2020). The numerous uses of nanoparticles in farming are depicted in Figure 2.

The nanoparticles' physical and chemical properties, such as biological responses are ultimately determined by zeta potential, size, and intensity (Acharya et al., 2019; Khalaki Abbasi et al., 2020; Singhet al., 2021). When it comes to the transport and absorption of nanoparticles in crops, these characteristics are crucial. For instance, smaller nanoparticles are better at overcoming biological obstacles (Hu et al., 2020; Bombo et al., 2019; Rajput et al., 2022). The external nanoparticle's charge is also crucial. The leaves may transport nanoparticles with either a negative or a positive charge to the roots. Though, the roots quickly absorb only nanoparticles which

Figure 1. Various methods (physical, chemical, and biological) for nanomaterials production

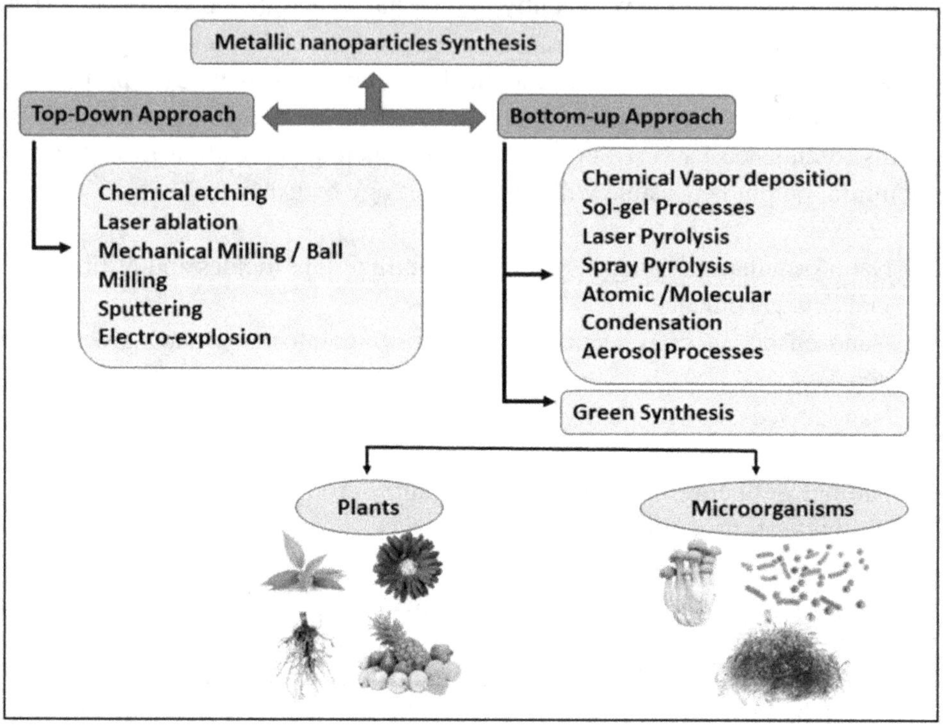

are having a negative charged. Positive charges that prevent plants from absorbing them lead to the formation of mucilage (Spielman-Sun et al., 2019; Avellan et al., 2017). Nanopriming can be sprayed on seeds to preserve them while they are stored, speed up germination and synchronization, improve plant development, make crops more resistant to abiotic and biotic stress, and make pesticides and fertilizers less necessary (Malik et al., 2020). According to new research, seed nanopriming may be able to activate a number of genes during germination, including those involved in plant stress tolerance. The use of nanomaterials in seed priming is a new area of study (Ye et al., 2020; Hussain et al., 2019). Seed nano-priming can also be used to preserve seeds because many nanoparticles have antibacterial properties and may carry compounds that kill bacteria. In addition, nano-priming can be used to concentrate seed biofortification in order to enhance the quantity and quality of food (Hussain et al., 2019; Pirzada et al., 2020) (Figure 3).

Figure 2. Application of nanomaterials in agriculture

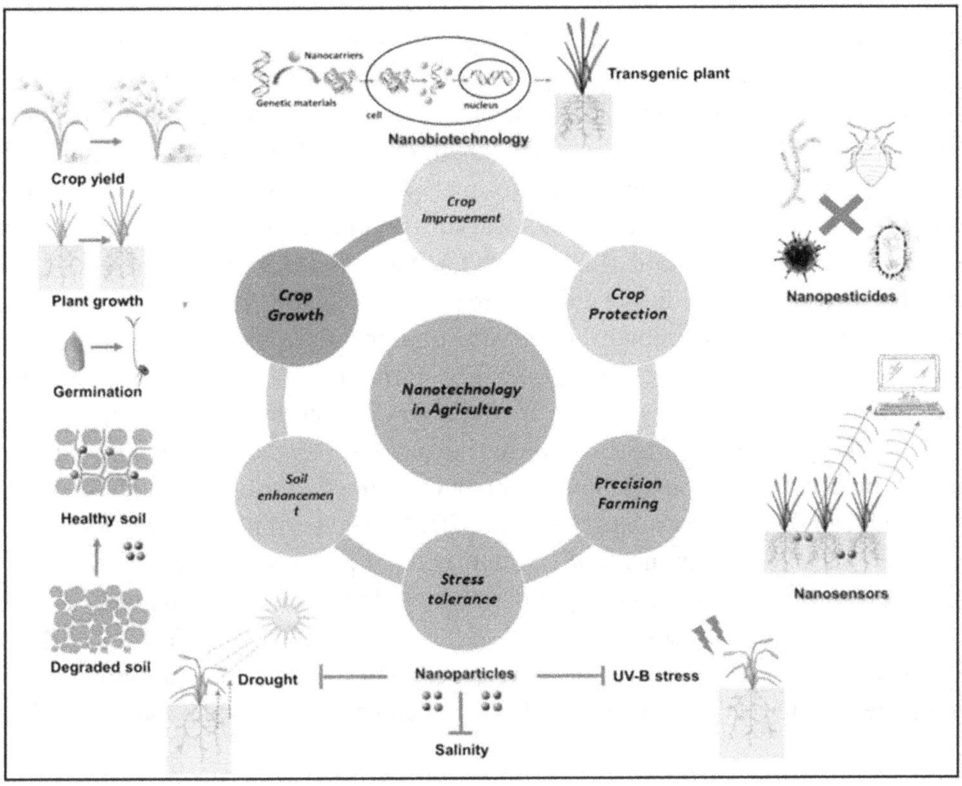

Figure 3. Nanomaterials used for seed priming

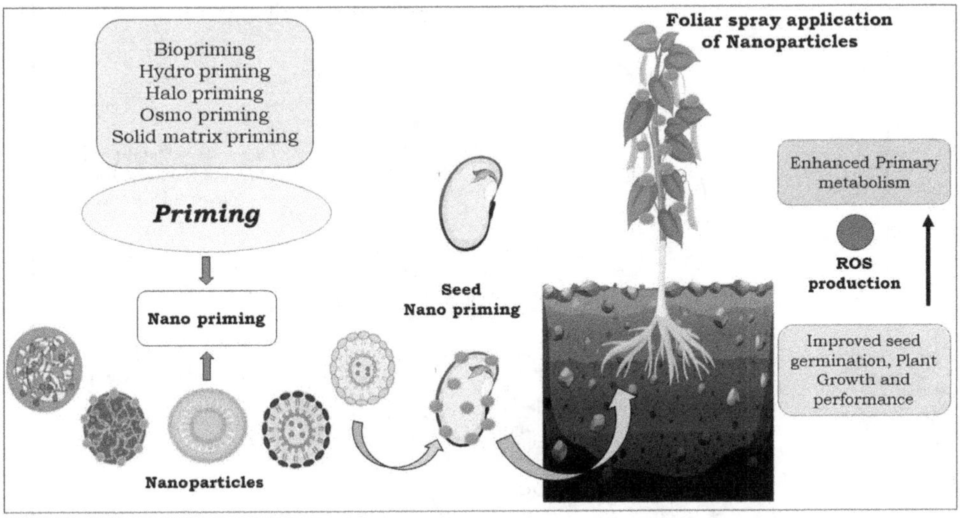

GERMINATION AND PRINCIPLES OF SEED PRIMING

According to Abbasi Khalaki et al., (2020) germination is a vital stage in the growth of plants and agricultural crops. The rapid development of seedlings results in rapid leaf and root elongation, which facilitates the intake of nutrients, production of biomass, and transfer via respiration flow (Mahakham et al., 2016). Sluggish germination puts the initial seedling, one of the extremely vulnerable stages of the crop's life cycle, at risk for a wide range of ecological stresses or illnesses, developing in lower yields and diminished vitality for farmers (Acharya et al., 2019).

The seeds germination occurs in three stages (Nonogaki et al., 2010). Seed basal metabolism, mitochondrial activity, transcription, and protein production all begin with rapid water absorption. The formation of enzymes like phytase endoxylanase and amylases, which are necessary for the mobilization of reserves and the embryo's development, occurs during phase II (lag phase or activation). The radicle protuberance is the result of the embryo's growth during the III phase, when the seeds rapidly absorb water (Nonogaki, 2014). The germination or inactivity of seeds is controlled by auxins. ROS regulate cellular processes associated with seed germination by controlling gene regulation and phytohormone signalling, as well as gibberellin, abscisic acid, ethylene, and auxin homeostasis (Wu et al., 2020). However, significant oxidative damage occurs when ROS concentrations are too high, preventing seed germination. ROS levels must be spatiotemporally regulated

Figure 4. Effects of seed priming

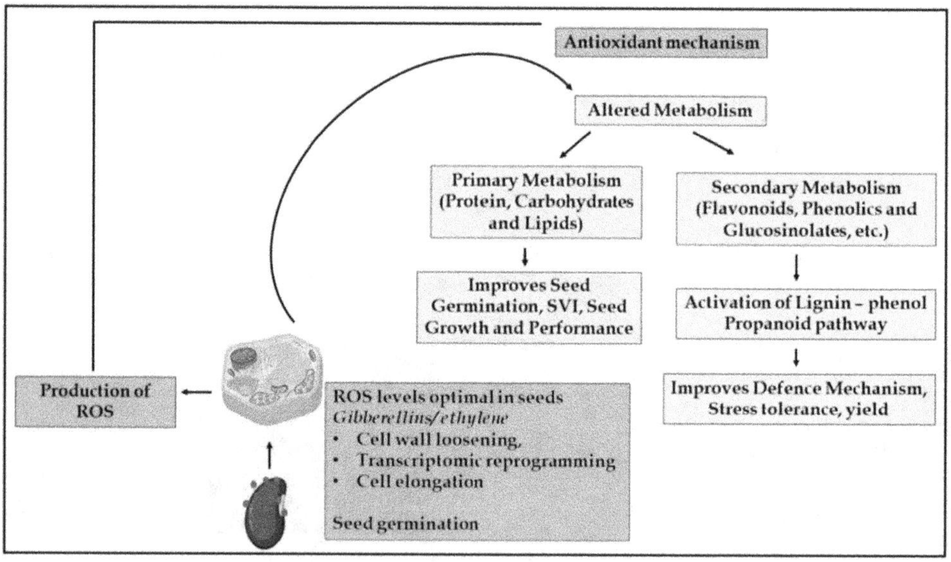

in order to be included in the so-called oxidative window, which enables optimal germination completion (Bailly, 2019). (Figure 4).

Preparing seeds prior to planting is an old-fashioned agricultural practice that aims to improve seed sprouting and plant development (Carrillo-Reche et al., 2018). It is usually a water-based technique in which the seeds are immersed in water and then dried (Lemmens et al.,2019), or it can be done physically by using UV priming. Without causing radicle emergence, pre-germination metabolic processes (phases I and II) should be activated by water absorption alone. Seed metabolism is altered at the cellular and molecular level by this process, which improves reverse mobilisation capability, reprogrammes the transcriptome, loosens cell walls, encourages translation, and alters post-translational processes. During absorption, it causes a physiological condition that improves primed seed germination and viability (Carrillo-Reche et al., 2018). Antioxidant processes, heat shock proteins, and other stress-related responses are triggered by wetting and drying, resulting in cross resistance to supplementary stressors. Additionally, seeds that germinate quickly spend less time in unfavorable soil conditions. As a result, priming of seeds has been used to improve the strength of seedlings, make plants more adaptable to harsh environments, and boost and manage germination, thereby improving food quality and increasing production (Carrillo-Reche et al., 2018; Lemmens and other, 2019).

Preparing seeds prior to planting is a common agricultural practice known as seed priming (Figure 5). This practice encourages seed germination and plant establishment (Acharya et al., 2020; Sundaria et al., 2019; Abdel Latef et al., 2017; Savassa et al.,2018). Although certain physical methods, such as ultraviolet radiation (UV) priming, are utilized, it is typically a water-based process in which the seeds are submerged in water prior to drying (Duran et al., 2017). To prevent radicle emergence, fluid intake must be sufficient to stimulate the metabolic processes that occur prior to germination (phases I and II). This process alters the metabolic activity of seeds at the cellular and molecular levels (for example, transcriptomic reprogramming, enhanced reserve mobilization capacity, cell wall softening, amplified capacity for protein synthesis, and post-translational improvements), resulting in a specific physiological condition that accelerates or even enhances seed primed germination and vigour on a novel absorption (Acharya et al., 2020; Sundaria et al., 2019; Abdel Latef et al., 2017; Savassa et al., 2018). Stress-related reactions, like antioxidant procedures and heat shock proteins, can be induced by soaking and drying, both of which can enhance resistance to other stressors.

Additionally, rapid sprouting reduces the susceptibility of germination seeds to adverse soil conditions. So, seed priming has been used to make crops more tolerant of abiotic and biotic stressors, increase yield, and increase food value (Sanzari et al., 2019), increase seedling vigor, accelerate and synchronize germination (Lowry et al., 2019; Savassa et al., 2018). Seed priming can be carried out in various ways,

Figure 5. A timeline of the progress of seed priming technology toward nanopriming to enhance seed germination

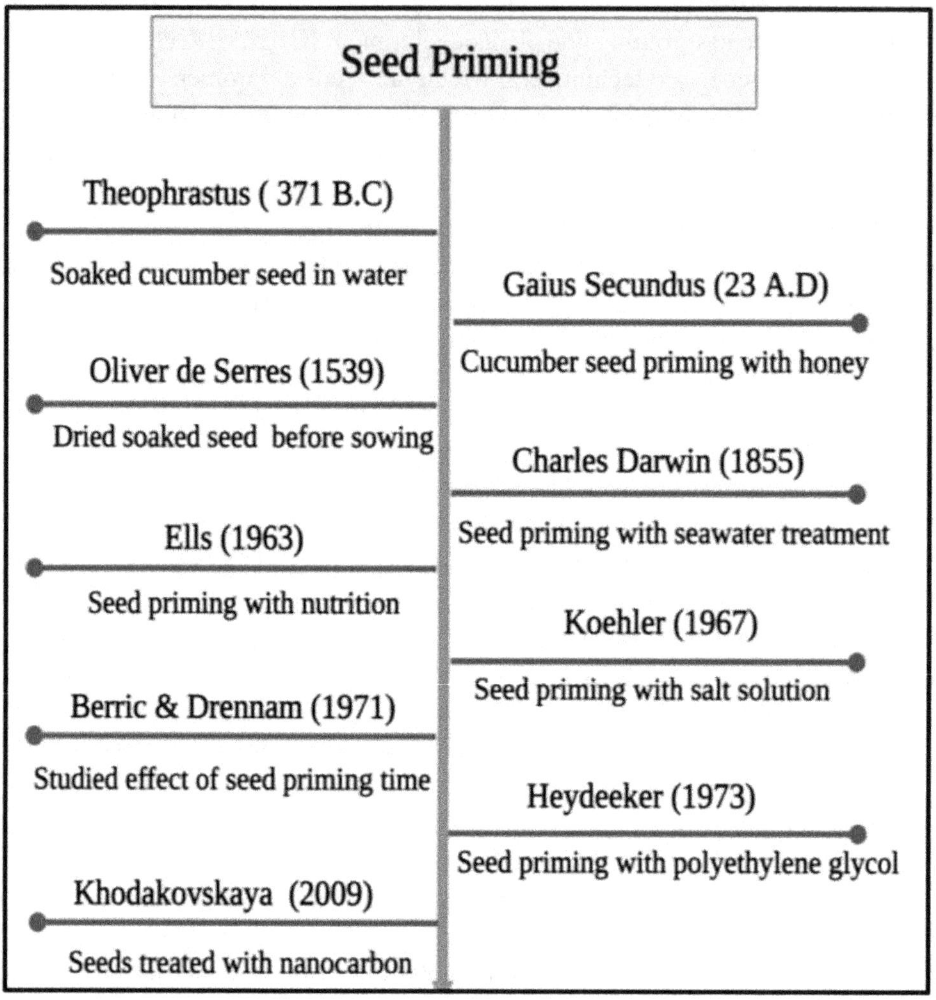

such as hydro-priming or hydro-thermopriming, in which seeds are submerged in water for seven to fourteen hours to maintain their hydration and facilitate phase II germination (Carrillo-Reche et al., 2018). According to Noorhosseini et al. (2017), this method can take advantage of both cold and hot temperature changes. Solutions with a low water potential are used in osmo-priming to alter seed metabolism through abiotic stress and control hydration (about 10% to 20%). Additional salt-based pre-sowing treatments include halopriming, hormone priming, and bio priming (Ye et al., 2020), regulators of plant growth (Li et al., 2019), and bacteria (Duran et al., 2017). A new technique for priming seeds that makes use of nanomaterials, particularly nanoparticles, is called seed nano-priming (Table 1).

The main difference between seed nano priming and seed priming is the process of hydropriming the seed with water or other solutions, where the chemicals are adsorbed on the seed and a seed coating is made of these nutrients, biopolymers, or hormones. Suspensions and nanoformulations are used in seed nano-priming; the seeds may or may not absorb the nanoparticles (Acharya et al., 2019). Even after being taken up, the majority of nanoparticles remain coated on the surface of the seed (Acharya et al., 2019; Montanha et al., 2020). Whether in the field or in storage, these seed covers contain fungicides or bactericides to guard against illness (Gross et al., 2020).

Khodakovskaya and others, (2009), demonstrated that nanomaterials can affect seed germination. These investigators discovered that carbon nanotubes are absorbed on tomato seeds, despite not priming the seeds. Tomato plants were able to absorb twice as much water thanks to these nanotubes. Furthermore, it has been established (Villagarcia et al., 2012) that carbon nanotubes alter the metabolism of barley, soybean, and maize seeds and boost the gene expression of a number of water channel proteins. Seed nano-priming has been demonstrated to be effective with a variety of nanomaterials, including metallic, biogenic metallic, and polymeric nanoparticles (Siddaiah et al., 2018). The change in gene expression that occurs as a result can have an impact on metabolic processes like hormone production and accelerate shoot and root growth. The antioxidant and enzymatic activities of the defense system increase after seed priming, making plants extremely resistant to bugs and other abiotic and biotic stimuli in the field (Itroutwar et al., 2019; Siddiah et al., 2018).

Carbon nanotubes also affect seed metabolism in plants like barley, maize, soybean, and tomato by increasing gene expression for a variety of water channel proteins, according to other studies (Müller et al., 2009; Gomes and Garcia, 2013). Among the nanomaterials that have demonstrated promise for seed nano-priming are polymeric nanoparticles, metallics, and biogenic metallics (Azimi et al., 2021; Kasote et al., 2019; Ernest et al., 2012), which followed by improving metabolic and morphological properties and encouraging plant growth. This mechanism has the potential to accelerate root and shoot growth by altering the expression of genes that affect metabolic processes like the synthesis of phytohormones. Seed nanopriming, according to researchers, alters the function of the defense system, increases antioxidant concentrations and enzyme activity, and makes plants more resistant to pests and other abiotic and biotic challenges in the field (Gohari et al.,2021; Prasad et al., 2014; Maswada et al., 2018).

Table 1. Nanoparticles used for seed priming

Nanomaterials	Size	Seed Species	Effect	Reference
Gibberellic acid loaded in Chitosan	450 ± 10 nm	Tomato (*Solanum lycopersicum* var. cerasiforme)	increased biomass with improved seed vitality and plant morphology	Pereira et al., 2019
Copper-containing chitosan nanoparticles	374.3 ± 8.2 nm	Maize seeds (*Zea mays* L.)	Strengthened seeds and seedlings	Saharan et al., 2016
Zinc-containing chitosan nanoparticles	387.7 ±4 nm	Maize seeds (*Zea mays* L.)	Biotic resistance, Enhanced vitality of seeds and seedlings	Choudhary et al., 2018
Multi-walled carbon nanotubes	13–14 nm	Wheat (*Triticum aestivum* L.)	Increased plant morphology and vitality in the seeds	Joshi et al., 2018
Iron nanoparticles	~80 nm	Wheat (*Triticum aestivum* L.)	Increased plant morphology and vitality in the seeds	Sundaria et al., 2019
Molybdenum and Cobalt oxide nanoparticles	60–80 nm	Soybean seeds (*Glycine max* (L). Merr.)	increased biomass in conjunction with improved seed vitality and plant morphology	Chau et al., 2019
Ligin nanoparticles loaded with gibberellic acid	200–250 nm	Arugula (*Eruca visicaria* (L.) Cav. subsp. sativa), tomato (*Solanum lycopersicum* L. cv. Ciliegino), and chickpea (*Cicer arietinum* L.)	Strengthened seeds and seedlings	Falsini et al., 2019
Silver nanoparticles	141.3 ± 0.78 nm	Watermelon (*Citrullus lanatus* (Thunb.) Matsum. and Nakai)	Increased plant morphology and vitality in the seeds	Acharya et al., 2020
Manganese (III) Oxide nanoparticles	50 nm	Jalapeño (*Capsicum annuum* L.)	Enhanced resistance to salinity and antioxidant enzymes	Ye et al., 2020
Zinc nanoparticles	21.3 nm	Lupin (*Lupinis termis* L.).	Enhanced Biochemical activity and resistance to salinity	Abdel Latef et a., 2017
	20 nm	Common bean (*Phaseolus vulgaris* L.)	Boosted biomass	Savassa et al., 2017
Copper nanoparticles	25, 40, and 80 nm	Common bean (*Phaseolus vulgaris* L.)	Improved seed vitality and biomass	Duran et al., 2017
Iron (II) sulfide aqua Nanoparticles	6–20 nm	Rice (*Oryza sativa* L.)	Increased seed vitality and disease endurance	Ahuja et al., 2019
Iron oxide nanoparticles	<50 nm	Sorghum (*Sorghum bicolor* (L.) Moench)	Improved biochemical activity and biomass, enhanced content of water in leaves	Maswada et al., 2018
Chitosan/ tripoly phosphate nanoparticles	259.4 ± 4.7 nm	Wheat (*Triticum aestivum* L.)	Improved plant morphology and increased expression of the growth regulator	Li et al., 2019
Silver nanoparticles	6–26 nm	Rice seeds (*Oriza sativa* L. cv. KDML 105)	Enhanced vigor of seeds and seedlings, expression of the aquaporin gene upregulated	Mahakham et al., 2017
Zinc, titanium and Silver	ZnO, TiO2, A g 35–40, 100, 85 nm, resp	Chilli (*Capsicum annuum* L.)	Enhanced disease resistance and increased seed vitality	Gnanasekaran et al., 2019
Iron nanoparticles	19–30 nm	Watermelon (*Citrullus lanatus* (Thunb.) Matsum and Nakay varieties).	Reduced phytotoxicity and enhanced plant morphology	Kasote et al., 2019
Silicon nanoparticles	90 nm	Wheat (*Triticum aestivum* L.)	Decreased cadmium absorption, increased biomass and biochemical activity	Hussain et al., 2019

NANOCARRIERS SYSTEMS AND ACTIVE NANOPARTICLES

Nanoparticle applications will be divided into two categories: (i) active nanoparticles and (ii) nanocarrier systems with sustained release. Active nanoparticles can have a biological influence by acting as a stimulant, an anti-pathogen, or both (Azimi et al., 2021; Ernest et al., 2012; Mission et al., 2015). Sustained release nanocarriers are systems in which a nanoparticle (whether inactive or active) is loaded with an active component (synthetic or biological) and provides a long-term delivery of this material over time (Khodakovskaya et al., 2009; Abbasi Khalaki et al., 2021) (Figure 6).

Active nanoparticles that can be synthesized chemically or biologically include metallic nanoparticles (Azimi et al., 2021; Prasad et al.,2017). Since many of the polymers used in the production of polymeric nanoparticles exhibit biological activity (Naguib Abdalla, 2019; Kamyab and Hojjat, 2017). These systems can be utilized for sustained release after being loaded with pesticides, fertilizers, biological agents, and even other nanoparticles. It has been demonstrated that these compounds' prolonged release increases biological activity while reducing adverse effects. Nanocarrier systems can be used to improve the biological activity

Figure 6. Priming of Seed using nanomaterials. The nanoparticles are categorized as either (i) Systems that can encourage plant growth and development include active nanoparticles. or (ii) nanocarrier systems, which can be inert or contain active compounds and provide a prolonged release.

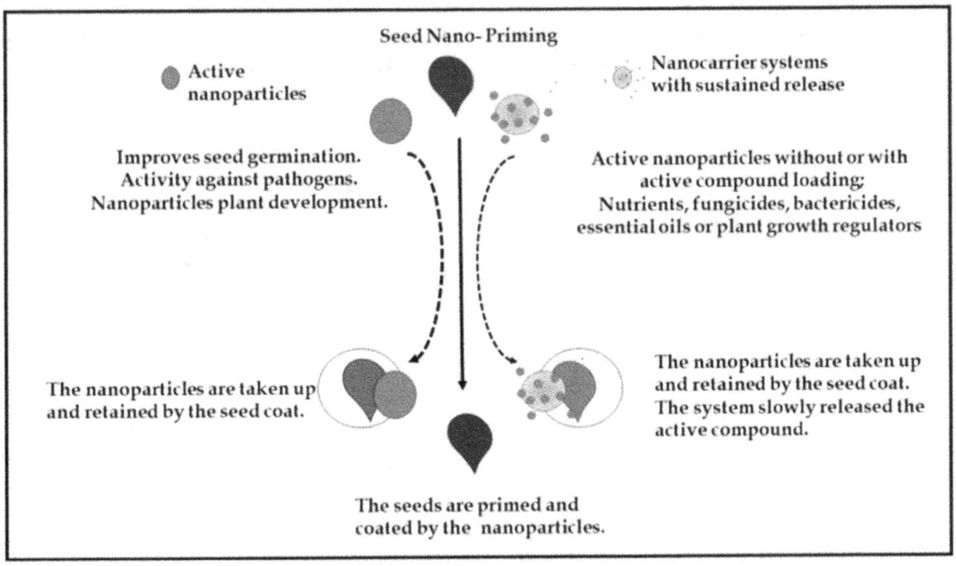

of many seed priming products (Vejan et al., 2018; Abedi et al., 2020; Singh et al., 2021a) (Figure 7).

FUNCTIONS OF VARIOUS NANOPARTICLES IN SEED PRIMING

Silver Nanoparticles (Ag-NPs)

Ag-NPs are currently one of the nanoparticles that are utilized the most frequently, following carbon nanotubes. The antimicrobial properties of these NPs are widely used in textiles, detergents, and polymers, among other industries (Awasthi et al., 2017). Because of their favoured magnetic, optical, and electrical capabilities, Ag-NPs are highly prized. Cryogenic superconducting components, electronic materials, ethylene testing, cosmetics, biosensor materials, composite fibres, and plant antimicrobial applications are just a few examples of their potential applications. Due to their ability to shield seeds from bacterial and fungal attack, Ag-NPs may also be beneficial in horticulture and other agricultural pursuits (Parveen and Rao, 2015).

Figure 7. The NMs' versatile catalytic function selectively promotes enzymatic and biochemical pathways that encourage root and shoot growth.

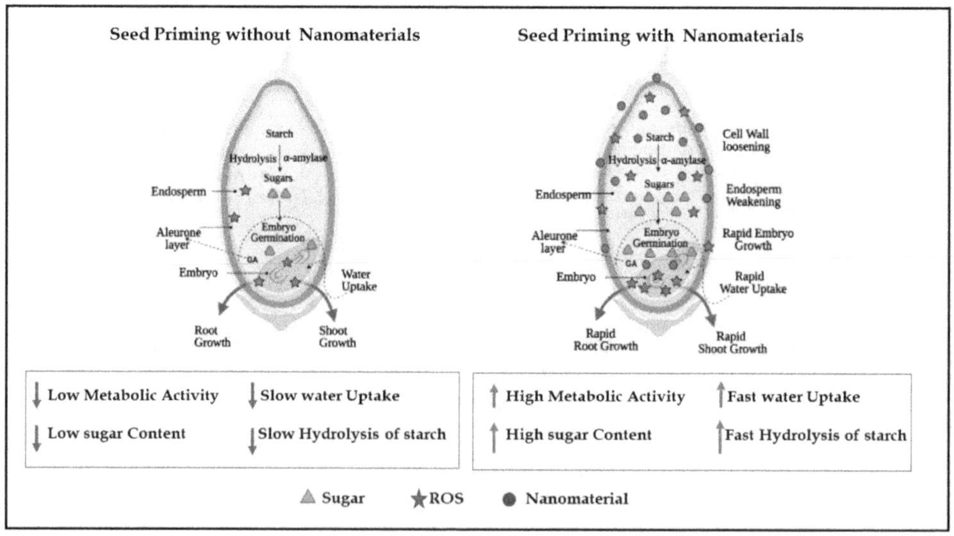

Silica Nanoparticles (Si-NPs)

According to Dietz and Herth (2011), About 31% of the crust is composed of silicon (Si), the second most ample element on Earth after oxygen. Nanosilica is a significant chemical with numerous applications in electronics, biology, and science and technology. Silicon, according to Wang et al., (2011) may assist plants in coping with salt stress. Leaf mechanical strength, evapotranspiration reduction, plant tolerance to biotic and abiotic stress, and light absorption are all enhanced by Si. Plant secretory structures contain Si in the epidermal tissue (Rastogi et al., 2019). In a variety of climates and soil types, it has been demonstrated that silicon fertilizers increase crop quantity and quality in various plants. Due to their distinct physiological properties, silicon nanoparticles can influence plant metabolism (Rastogi et al., 2019). Additionally, according to Wang et al., (2011) silica nanoparticles may be able to transfer DNA and other substances into plant and animal cells. Nanosilica enhances germination properties like germination speed, dry weight, and plant radicle height (Rastogi et al., 2019; Wang et al., 2011).

Copper Nanoparticles (Cu-NPs)

Copper (Cu) is a necessary component for the healthy development and response of photosynthetic cells. It participates in the exchange of hydrocarbons and proteins. Polyphenol and ascorbic acid oxidases, among other oxidizing enzymes, contain copper (Leng et al., 2015). Cu is essemtial for plant metabolism and growth.. Curled leaves indicate a lack of copper, but an excess can be poisonous (Leng et al., 2015). Numerous products emit Cu-NPs, which have the potential to have an effect on people and natural structures (Chen et al., 2012). Copper oxide (CuO) nanoparticles are toxic to marine animals like crustaceans, zebrafish, algae, and protozoa, according to recent research (Chen et al., 2012). On the other hand, it was found that Vigna radiata (L.) R. Wiclczek seedling growth, metabolic rate, and sprouting were aided by lower CuO concentrations (Singh et al., 2017). However, because the effect of Cu-NPs on plants has not been fully investigated, some previous findings are questionable (Leng et al.,2015).

Iron Nanoparticles (Fe-NPs)

According to Li et al., (2006) iron (Fe), the fourth most ample element in the earth's crust, is regarded as non-toxic. and all organisms require very small amounts of it. It is necessary for respiration, the photosynthetic process, and the formation of chlorophyll. One of the many nanoparticles utilized in environmental and wastewater restoration projects is Fe-NP (Najafi Disfani et al., 2017). Nanoparticles of iron oxide

(FeO) play a crucial role in the sprouting, growth, and output of plants. According to Ghafariyan et al., (2013) increasing FeO-NPs improves iron deficiency while simultaneously raising chlorophyll a and b levels because iron is the most ample element of chlorophyll. As a result, the development of agriculture and other fields is significantly impacted by iron oxide nanoparticles (Li et al., 2006). In addition, FeO-NPs are utilized as nanofertilizers to alter the activity of hormones and antioxidant enzymes, increase plant height, biomass, and root length, and increase iron availability to plants.

Zinc Oxide Nanoparticles (ZnO-NPs)

Due to its role in the production of chlorophyll and carbohydrates, zinc is essential for crop growth. Plants that have higher zinc levels are less likely to absorb harmful heavy metals and, as a result, have fewer toxic effects. Biosensors, electrodes, health care, and consumer goods are just a few of the many applications for nanoparticles. Due to their significant roles in the physiological responses and architecture of plants, zinc nanoparticles are frequently used in agriculture (Awasthi et al., 2017). Numerous enzyme activities, including superoxide dismutase and dehydrogenases, depend on ZnO for regulation (Rajput et al., 2021a). Due to their photo degradability, large surface area, low cytotoxicity, long lifespan, and large pore volume, zinc oxide nanoparticles can be utilized as catalysts, chemical absorbents, polymer additives, and antimicrobials. Numerous plant species' seed germination has been the subject of recent research (Awasthi et al., 2017; Gaafar et al., 2020; Sharma et al., 2021). However, it has been shown that using ZnO-NPs as a nano-fertilizer shows adverse effects in plants due to the biological activity of metal-based NPs (Itroutwar et al., 2020). Various studies have shown that the noxious nature of nano-ZnO stops plant roots from getting longer. Two seedling genotypes of rice (*Oryza sativa* L.) were studied to see how ZnO-NPs induced salt tolerance. According to Singh et al.'s (2022), findings, the addition of ZnO-NPs enhanced the cultivation of undocumented landrace (Kargi) and basmati (CSR 30) genotypes of O. sativa in salinity-affected areas and mitigated salinity stress in *O. sativa*. Singh et al. (2023) investigated how zinc oxide nanoparticles (ZnO-NPs) and salinity-induced morphogenic responses affected rice (*Oryza sativa* L.) root characteristics, growth, MDA content, antioxidant enzymatic activity, and root ion accumulation. that O. sativa can thrive in salt-stress conditions when treated with ZnO-NPs, making it possible to grow the plant in harsh environments.

Sustained Release Nanocarrier Systems

Lipids, polysaccharides, and proteins can all be used to build biopolymeric nanoparticle systems. Biodegradable and biocompatible nanoparticles can be tailored to respond to a range of environmental factors as a consequence of these systems (Marthandan et al., 2020; Abbasi Khalaki et al., 2021). Anti-fungals (Nguyen et al., 2020), oil extracts (Falk et al., 2018), plant growth regulators (Dantas de Medeiros et al., 2020), and fertilizers (Singh et al., 2020). 2020) might all make it into these systems. Seed coating polymeric nanoparticles typically have a size greater than 100 nm (Kasote et al., 2019), which makes it possible to prolong the discharge of active compounds that are used to control plant biochemistry or fight pathogens (Babajani et al.,2018). Alginate (Kah et al., 2018) and other biopolymer-based nanocarrier systems viz, cellulose (Tanabata et al., 2019), zein (Abedi et al., 2012), 2020), synthetic biopolymers (polylactic acid, poly(lactic-co-glycolic acid), and poly-epsilon-caprolactone) (Abbasi Khalaki et al., 2020), as well as lipid nanoparticles, have the potential to be utilized in foliar spray (Komyshev. et al., 2017)

Alternative biopolymer that has been used to construct nanocarrier systems for field crops is chitosan. This polymer helps plants by stimulating protective mechanisms and is fungistatic (Marmiroli et al., 2016). According to Hojjat and Kamyab (2017), a high concentration of this biopolymer slows down germination. This is due to the fact that oligosaccharides and polysaccharides can cause physiologic reactions in plants, with large amounts causing cell death and small amounts promoting (An et al., 2020). Biopolymeric nanoparticles and other metal micronutrient carrier systems can be used to enable the gradual discharge of these compounds. Saharan and others (2016) demonstrated that copper-containing chitosan nanoparticles promoted -amylase and protease activity, mobilization of reserves, and improved seedling growth when used to stimulate maize seedlings. In contrast, bulk copper therapy inhibited growth. Similarly, zinc-loaded chitosan nanoparticles stimulated the sprouting of maize seeds by increasing hydrolytic enzyme activity.

From seed germination to production, plant growth regulator nanocarrier systems have demonstrated their ability to prime seedlings (Kasote et al., 2019). Falsini and others 2019) used gibberellic acid-containing lignin nanoparticles to coat tomato and arugula seeds. The outcomes varied based on when the seedlings were sown following the process. The formulations increased sprouting, stem and root length, and plant biomass in arugula seeds. The effects of tomato seeds varied depending on the variable and the quantity used, but they were similar. The ability of active nanoparticles or nanocarrier systems to increase seed germination may be accounted for by the fact that nanoparticles can enter the outer layer at optimal concentrations, increasing the number of pores in the coat, which improves water absorption and oxygen transport (Sathiyabama et al., 2020).

From seed germination to production, plant growth regulator nanocarrier systems have demonstrated their ability to prime seedlings (Kasote et al., 2019). Falsini and others (2019) used gibberellic acid-containing lignin nanoparticles to coat tomato and arugula seeds. The outcomes varied based on when the seedlings were sown following the process. The formulations increased sprouting, stem and root length, and plant biomass in arugula seeds. The effects of tomato seeds were similar and varied depending on the variable and the quantity used. The fact that nanoparticles can enter the outer layer at optimal concentrations, increasing the number of pores in the coat and thus improving water absorption and oxygen transport may account for the ability of active nanoparticles or nanocarrier systems to increase seed germination (Sathiyabama et al., 2020).

SEED NANO-PRIMING AND IMPACTS ON CROP METABOLISM UNDER BIOTIC AND ABIOTIC STRAINS

Both abiotic and biotic forces restrict output and result in losses for the economy. Examples of abiotic stressors include flooding, drought, extreme temperatures, salinity, and soils deficient in nutrients. Insects, weeds, and bacteria that compete for nutrients are bifid stressors (Paparella et al., 2015; Khodakovskaya et al., 2008). The seed's condition and tolerance to trauma can be enhanced through the use of nano-priming in cultivation. The innate immune system can be strengthened, hormone synthesis can be altered, and crops can become more tolerant of diseases or abiotic stress thanks to nanoparticles' ability to precisely combat pathogens (Gilbertson et al.,2020).

By working in multiple metabolic pathways, increasing the quantity of effective gibberellins, and mobilizing storage proteins, the absorption of nanoparticles beneath the seed layer can stimulate ROS generation (Mahakham et al., 2017). Additionally, the nanoparticle's effect on the intake of water by seeds may cause sufficient strain to initiate germination, thereby increasing enzyme activity in stages I and II (Xu et al., 2020). It has been demonstrated that seedling nano-priming increases sprouting because these systems are able to maintain ROS levels within the acceptable ranges indicated by the oxidative frame that encourages seed sprouting (Mahakham et al., 2017) (Figure 7).

By improving the action of enzymes like superoxide dismutase, catalases, and guaiacolperoxidase, nanoparticles can decrease the levels of reactive oxygen species (ROS) in seedlings under stressful conditions. This results in less damage to the cells of the seeds. Long-term storage of seeds at low temperatures leads to ageing, which can dramatically reduce germination rates (Lowry et al., 2014). The sprouting index decreases as a result of metabolic adverse reactions caused by the seed cells engaging in activities that increase ROS generation and decrease antioxidant concentrations over time. It has been demonstrated that the application of biogenic metallic nanoparticles can maximize ROS absorption, resulting in improved germination of aged seeds. A wide variety of natural reductants can be used to treat biologic nanoparticles, which reduce ROS levels in seedlings (Sathiyabama et al., 2020).

Biotic Stress

Seed security is necessary at a number of points throughout the farming process. In the field, it is necessary to protect against fungi, insects, and bacteria that could harm the seeds of the plant and slow down sprouting or seedling growth. Seeds carry germs from the field during harvest and storage. These may result in the food becoming unfit for human consumption or shortening the product's shelf life due to the accumulation of microbial metabolites (Gilbertson et al., 2020). To lessen the harm done by fungi during plant establishment, fungicides and neonicotinoids can be used to treat seeds in the field (Schmitt et al., 2013). In comparison to foliar treatments, this method has demonstrated tremendous potential (Matyjaszczyk et al., 2017; Farias et al., 2015). Seeds can be treated with pesticides, which, in comparison to foliar treatments, can decrease the amount of chemicals used in the field and the amount of agricultural machines used on the land, thereby preventing soil compaction. The strategy can be evaluated at a low cost as a result.

During the seeding process, fungicides may be sprayed into the soil, causing environmental damage. Carbofuran and thiran, for instance, are particularly toxic pesticides that are also stable in the ecosystem (Chau et al., 2019), so animals that eat treated seeds, like granivorous birds, might get sick or die from them (Gross et al., 2015). Since they are materials that are both reactive and physiochemically active, nanoparticles prefer to hetero assemble in ambient fluids. This may result in an efficient seed treatment that prevents these contaminants from entering the environment (Lowry et al., 2012). However, eco-friendly options may be available by utilizing nanoparticles for seed priming and coating. The bioefficacy of active ingredients in small doses can be enhanced by many of these systems, which are recyclable and harmless (Sathiyabama et al., 2020; Bravo Cadena et al., 2018).

The effectiveness of numerous anti-pathogen methods has been demonstrated. Ahuja and colleagues (2019) shown that iron (II) sulfide nanoparticles inhibited the fungus *Fusarium verticillioides* in rice seedlings more effectively than the fungicide carbendazim. Silver nanoparticle-resistant pathogens like *Aspergillus niger, Aspergillus flavus, Aspergillus fumigatus*, and *Colletrotrichum capsica* hold significant promise for seed preservation.

Copper-containing metallic nanoparticle systems (Nguyen et al., 2020), iron (Ahuja et al., 2019), silica (Bravo Cadena et al., 2018), silver, and gold (Mahakham et al., 2017), can be used to preserve seeds as well as prime seeds. The use of chitosan and guar nanoparticles to prime rice seedlings by Sathiyabama and Muthukumar (2020) caused faster plant development, higher levels of chlorophyll and carotenoids, and 71% anti-fungal activity against the rice pathogen *Pyricularia grisea*. In pearl millet, Nandhini and colleagues (2019) discovered that treatment with foliar zinc nanoparticles and seed nano-priming reduced downy mildew caused by *Sclerospora graminicola*. Dileep Kumar and others (2020) claim that by including silver, zinc, and titanium nanoparticles in chili seeds, *Aspergillus flavus*, *Aspergillus niger, Aspergillus fumigatus*, and *Colletrotrichum capsica* may be less likely to be infected with pathogens.

Millet seedlings treated with chitosan nanoparticles, as shown by Siddaiah and others (2018), had a different innate immune system and were more resistant to infections. Plant resistance responses to biotrophic infections are linked to an increase in the proteins expression in the pathway of salicylic acid, which was induced by the treatment. Choudhary and colleagues (2019) discovered that treating maize seeds with chitosan nanoparticles having zinc increased the antioxidant enzyme activity and lignin concentrations of the plants, making them more resistant to disease. According to Bravo Cadena and colleagues (2018), spraying pea seeds with silica nanoparticles infused with cinnamon essential oil acted as seed primers but only marginally killed *Pseudomonas syringae* pv. pisi. By a factor of ninety thousand, the cinnamon essential oil performed better than the unencapsulated oil. The seeds quickly developed. The use of essential oils with antifungal or antibacterial properties as a substitute for chemical fungicides or bactericides in seed priming may be an environmentally friendly option (Kumar et al., 2020). Grain conservation methods include regulating temperature and moisture to prevent the growth of microorganisms and using chemical agents, but customers who prefer natural products may view the latter negatively (Schmitt et al., 2020). In order to accomplish this, anti-pest biopolymers or essential oils can be used to produce agrochemicals in a way that is friendly to the environment and frees them from problems caused by residual pollution.

Abiotic Stress

According to Maswada et al., (2018) soil salinity and pollution can significantly lower harvest yields. Salinity slows plant development due to a lack of water and nutrients, as well as direct ionic effects on metabolism of plant (Abdel Latef et al., 2017). Heavy metal pollution can be man-made (from fertilizer use, industrial activities, waste burning, sewage and sludge discharges, and so on) or geogenic (from natural air deposition or regional geological methods) (Barbieri et al., 2018). Ye and others (2020) demonstrated that jalapeo pepper seeds (Capsicum annuum

L.) primed with manganese nanoparticles enhanced sprouting and root elongation in high salinity environments, as well as improved salt stress by altering sodium dispersion between shoots and roots via oxidative stress regulation.

According to Abdel-Latef and others (2017), zinc nanoparticles primed lupin seeds with prevented growth metrics deficits (shoot and root elongation) and reduced salinity stress. weights and the total amount of chlorophyll, fresh and dried. Maswada and colleagues (2018) report that sprouting, total chlorophyll, and growth were all enhanced in saline conditions by nano-iron (III) coated sorghum seeds. It could be used to reduce stress while simultaneously increasing seed germination, according to these findings. Additional research indicates that by increasing levels of chlorophyll, phenols, organic compounds, antioxidant enzymes, and pretreatment of lupin seeds with zinc nanoparticles, plant development in saline conditions was enhanced.

In the presence of heavy metal pollution, Rizwan and others (2019) discovered that primed wheat seeds containing iron and zinc nanoparticles reduced cadmium absorption, resulting in low cadmium concentrations in the grains. Zinc nanoparticles reduced cadmium levels in grains, shoots, and roots by 38%, 55%, and 83%, respectively. Iron nanoparticles reduced cadmium levels in grains, shoots, and roots by 54%, 56%, and 84%, respectively. The plants had a lot of zinc and iron after seed priming (Ahuja et al., 2019). Hussain and colleagues (2019) found that silicon nanoparticle seed priming decreased cadmium absorption, increased plant biomass, photosynthetic rate, levels of carotenoids and chlorophyll a and b, and reduced the production of reactive oxygen species when seeds germinated in cadmium-contaminated soil. The production of phytohormones like jasmonic acid and salicylic acid, which are responsible for plant defense responses, may be impacted by soil iron and zinc deficiencies (Choudhary et al., 2019). Kasote and colleagues (2019) used biogenic iron nanoparticles made from onion extract to make watermelon seeds. During the early seedling stage, the plants produced more jasmonic acid and its precursor, cis-(+)-12-oxo-phytodienoic acid, which may also increase stress resistance.

Nguyen and group (2020), assert that priming maize seedlings with cooper nanoparticles might increase the plant's drought resistance. The high water content of the leaves maintained an increase in anthocyanin, chlorophyll, and carotenoids as well as a decrease in oxidative stress. By regulating metabolism to resist stress conditions and improve plant growth, these findings demonstrate that seed-nano priming can reduce plant stress caused by salty conditions, drought, heavy metal presence, or nutritional deficiency. A powerful alternative for reducing field production losses caused by global environmental catastrophes or human and geogenic factors is the use of seed nano-priming to develop plant resistance to abiotic stress.

Seed Priming-Induced Molecular Responses During Germination, Abiotic, and Biotic Stress

Nanoparticles can be absorbed into a variety of cell compartments and interact with plant cells. Throughout the germination process and even over time, seed nano-priming can trigger or modify multiple gene and metabolic pathway expression profiles (Salama et al., 2019). During the imbibition phase of seed germination, water exchange occurs. Nanoparticle seed priming may increase water consumption in seeds by enhancing aquaporin gene expression during the seedling stage (Lahiani et al., 2013). Numerous subclasses of seed-located water channel genes (aquaporins) increased in abundance when multiwalled carbon nanotubes were air-sprayed onto barley, maize, and soybean crop seeds (Lahiani et al., 2013). According to Mahakham et al. (2017), rice seeds (Oriza sativa L. cv. KDML 105) were found to have an increased expression of aquaporin genes. with silver nanoparticles acting as a stimulant. Aquaporins enabled the mobility of gases (CO2, NH3), nutrients, and ROS (particularly H2O2) while also allowing water to diffuse across biological membranes. The researchers found that seed nano-priming caused aquaporin genes to be activated, which led to faster sprouting than in controls.

The transcriptomes of cotton seedlings primed with cerium oxide nanoparticles coated in the antioxidant poly(acrylic acid) were examined by An and others (2020), whether they were subjected to salt stress or not. In the nanoparticle-treated seeds without saline stress, the expression of 7799 distinct genes was higher than in the control. The nano-primed seeds expressed 13 genes related to ROS pathways and ten genes related to salt-induced ion homeostasis, according to the study. Ye and others (2020) discovered that seed priming Capsicum annuum L. with manganese nanoparticles increased seedling growth in saline environments. The results showed that primed seeds raised levels of the MnSOD gene, also known as Mn superoxide dismutase. This gene protects plant cells from ROS damage and reduces phytotoxicity.

It was also demonstrated that the nano-priming treatment for the seeds could alter genes that are resistant to biotic stress. Siddaiah et al. (2018) claim that the biotrophic oomycete Sclerospora graminicola's downy mildew disease was made more resistant to seeds coated with chitosan nanoparticles. The innate immune system was strengthened by seed priming. Phenylalanine ammonia lyase, peroxidase, and polyphenoloxidase genes were all elevated in the plants. The plants also showed an rise in the expression of salicylic acid pathway-related (PR) genes (PR1 and PR5). Plaksenkova and others (2020) found that zinc oxide nanoparticles led to an increase in the synthesis of the microRNAs miR156 and miR159, which play a role in defense responses to both biotic and abiotic stress.

In a nutshell, more research on the molecular effects of seed nano-priming and how nanoparticles alter gene expression is urgently needed. Plant metabolism can be affected in different ways by nanoparticles, treatments, and quantities. In addition, phytohormone-related gene transcription regulation and the response to biotic and abiotic stressors may be regulated by nanoparticles acting as a signal or co-factor (Chandrasekaran et al., 2020).

Effects on Microbiota

The impact of seed nano-priming on the microbiota of plants is a relatively new area that requires more research. The absorption of nutrients, resistance to infections, crop health, and production of crops all depend on interactions between plants and microorganisms (Shcherbakova et al., 2017). The root microbial ecosystem can be altered by applying nanoparticles to the leaves. Raliya and others (2016) demonstrated that foliar application of zinc nanoparticles to mung beans resulted in a 58.9% increase in the development of rhizobial zones. They also found an increase in the

Figure 8. The plant growth-promoting rhizobacteria (PGPR) combines seed nanopriming with a range of microbicidal nanomaterials without affecting the PGPR sparing beneficial soil and plant tissues. This enhances harvest quality and productivity by selectively raising the threshold for biotic stressors.

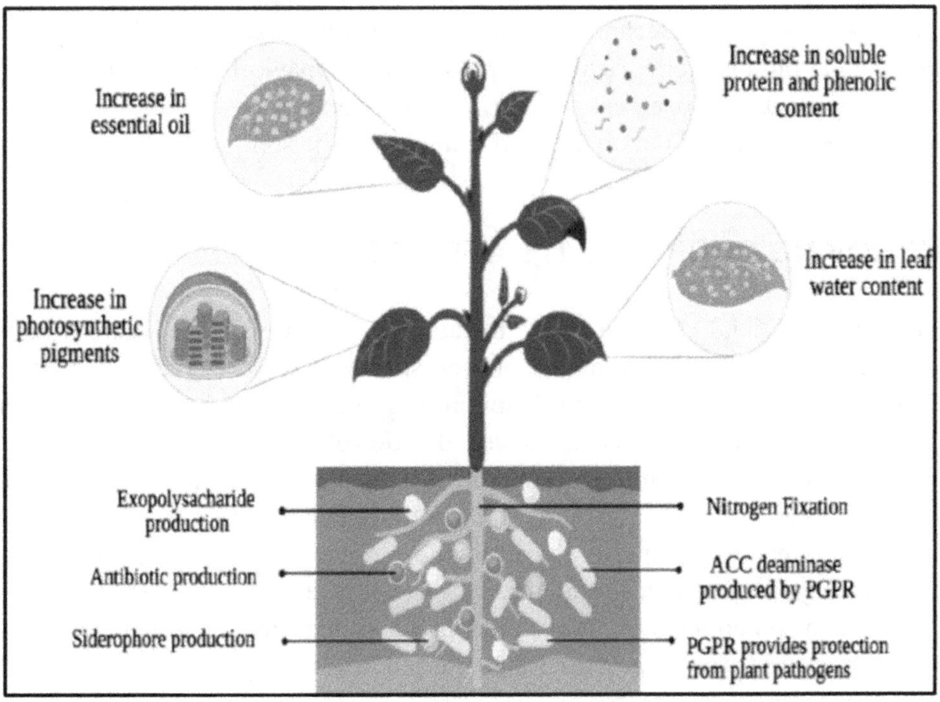

activity of enzymatic enzymes in the rhizosphere, particularly acid phosphatase (98.07 percent), alkaline phosphatase (93.22 percent), and phytase. Crop uptake of phosphorus and nitrogen depends on these parts.

An and others (2020) examined the transcriptomes of cotton seedlings primed with cerium oxide nanoparticles coated in the antioxidant poly(acrylic acid) to determine whether or not they were exposed to salt stress. The expression of 7799 distinct genes was higher in the nanoparticle-treated seeds than in the control, and this was without saline stress. According to the study, the nano-primed seeds expressed ten genes related to salt-induced ion homeostasis and 13 genes related to ROS pathways. In saline environments, Ye and others (2020) discovered that priming Capsicum annuum L. seeds with manganese nanoparticles increased seedling growth. The findings demonstrated that primed seeds raised MnSOD gene (also known as Mn superoxide dismutase) levels. Plant cells are shielded from ROS damage and phytotoxicity is reduced by this gene.

Rahman and others (2020) demonstrated that priming the seed with poly(vinylpyrrolidone-stabilized platinum nanoparticles had negative effects on the root microbiota of pea plants, including a decrease in the establishment of rhizobia and mycorrhizal fungi, whereas the same system with silver or gold nanoparticles had no such effects. In order to comprehend the mechanisms by which seed-nano priming alters the microbiome selected by the roots throughout plant growth and affects seed metabolism, additional research in this area is required. Although numerous nanoparticle technologies may be harmful to soil bacteria, they have shown great promise for agricultural applications. It may offer a safer method for increasing plant establishment while avoiding adverse effects in the soil because the treatment is carried out on the seeds rather than the soil.

IMPROVING CROP QUALITY AND PRODUCTION

Due to its beneficial effects on plant metabolism and growth, seed nano-priming can increase crop yield in a variety of crop species. The plant's ability to obtain nutrients and water is enhanced by rapid root development, which is followed by rapid leaf expansion, which increases the utilization of light energy for plant growth. Because of the effects on the innate immune system, disease resistance may increase, necessitating less pesticide use. As a result, food's remaining levels of agrochemicals are reduced, making them safer to eat. Studies on the effect of seed priming prior to harvesting on agronomically significant species have demonstrated these results.

Yasmeen and colleagues (2017) demonstrated that seed-priming wheat crops with copper and iron nanoparticles increased spike length, plant grain yield, and spike grain count. Rahman et al. say that (2020), seedling priming with poly(vinylpyrrolidone-treated platinum nanoparticles resulted in a 163.5 percent increase in output and a 66.7% decrease in seed bulk. The potential benefits of priming watermelon seeds with biogenic silver nanoparticles derived from onion preparations were emphasized by Acharya and colleagues (2020). The study found that production increased by 31.6% to 35.6% throughout the life of the plant as growth and metabolic reactions increased. Alginate/chitosan nanoparticle priming of tomato seedlings increased fruit production by approximately fourfold, according to Pereira and others (2019).

Joshi and colleagues (2018) demonstrated that treating wheat seeds with multi-walled carbon nanotubes enhanced germination, root length, amount of root hairs, stem length, crop mass, and mean vascular system length (xylem and phloem) length. The number of spikelets increased by 20%, grain yield increased by 32%, and spike length and weight increased by 21% and 27%, respectively, at 90 g/mL. Biofortification is an additional consideration. For instance, the treatment of seeds with nanoparticles of iron, zinc, and manganese produced plants and grains with higher concentrations of these minerals (Sundaria et al., 2018). According to another study, vegetables with higher concentrations of these elements in their roots, shoots, leaves, or grains may have higher levels of essential minerals (Rizwan et al., 2019; Sundaria et al., 2019). The amount of fertilizer applied to crops could be significantly reduced through the use of seed nano-priming. The fact that crops only take in 30-50% of the nitrogen and 45% of the phosphorus provided by the field is a major problem in agriculture (Das et al., 2018). Eutrophication of aquatic systems and pollution of native terrestrial ecosystems are two potential effects of these losses on the environment. Das et al. (2018) conducted field studies to discover that fertilized seedlings grew from nano-pyrite (FeS2) rice seeds primed with the same productivity as untreated seeds. It has been demonstrated that seed nano-priming can boost the synthesis of enzymes necessary for nutrient absorption and promote root growth, allowing for less fertilizer use in the field (Guha et al., 2018).

IMPACT OF COLD PLASMA TECHNOLOGY AND SEED PRIMING TREATMENT ON NANOPARTICLES

Rarely is research done into the potential benefits of combining seed priming with metal-based nanoparticles and cold plasma. Nitrogen species and reactive oxygen, also known as ROS and RNS, are important signaling chemicals that have been linked to gene expression, cellular physiology, seed signaling networks, differentiation, and growth (Spagnoletti et al., 2019). A high-tech eco-agricultural technology that can increase crop yields is seed treatment with cold plasma and nanomaterials (Saharan et al., Nandini et al., 2019). Because the low-level, non-ionizing radiation boosts a seed's viability without causing a DNA mutation, there is no genetic risk. Cold plasma and nanotechnology in agriculture offer a novel strategy for increasing grain output because they are quick, inexpensive, environmentally friendly, and risk-free (Choudhary et al., 2019). According to Shcherbakova et al. (2017), cold plasma may increase peroxidase activity and seed germination rates. Seed priming with cold plasma increased early development and blooming while selenium nanoparticles protected Cichorium intybus, according to research. According to De Oliveira et al. (2019) plasma priming combined with a small amount of nano-Selenium is a successful strategy for promoting plant growth, biochemistry, and protection.

Zinc oxide and selenium nanoparticle responses to non-thermal plasma seed priming were investigated by Babajani and colleagues (2019). Melissa officinalis responds differently to zinc oxide (nZnO) and selenium (nSe) nanoparticles after being cold plasma primed. The seedlings were then grown in a nutrient solution containing nSe and nZnO. Plasma priming improved growth-related characteristics (stem length, root length, and leaf width) and biomass accumulation while simultaneously reducing symptoms of nSe poisoning. The distinct physiology and expression of universal stress protein (USP) and phenylalanine ammonia lyase (PAL) in endangered species has also been demonstrated by some studies.

Phenylalanine ammonia lyase gene expression was highest in plasma and nano silicon (nSi) experiments. While nSi-treated seedlings displayed elevated levels of USP expression, plasma administrations had no effect on the gene's expression. Plasma and nSi (xylem and phloem) were used to increase leaf thickness and vascular tissue growth. Raja et al. (2019) describe plasma and nSi as potential phytotoxicity antagonists that could serve as a theoretical basis for future commercialization. According to A. Iranbakhsh et al., (2018) cold plasma altered capsicum cayenne proliferation, differentiation, and physiology to reduce nano zinc oxide toxicity indicators.

Utilizing a practical technological device and metal-based nanoparticles, this work evaluated the systemically administered cold plasma solution and zinc oxide nanoparticles (nZnO) within a plant (*Capsicum annum*) under in vitro and pot conditions. By dosing both roots with plasma or nZnO, the action of the peroxidase enzyme in either of the growth media was eliminated. Additionally, it was discovered that the plasma and/or nZnO activity of the phenylalanine ammonia-lyase enzyme was meaningly higher. The use of nanomaterials to soak the seeds prior to plasma treatment had the greatest effect on plant growth in the pot experiment (Iranbakhsh et al., 2018).

Figure 9. As an alternate to environmentally harmful chemical treatments, non-thermal and cold plasma-based seed nano priming improves seed vigor and germination by regulating environmental stresses. Better physiological processes are made possible by the strengthened antioxidant system and enhanced defensive response; Seed scarification and plasma-mediated pathogen inactivation alter the surface environment of the seed, enhancing molecular characteristics and the ratio of leaves to roots to shoots.

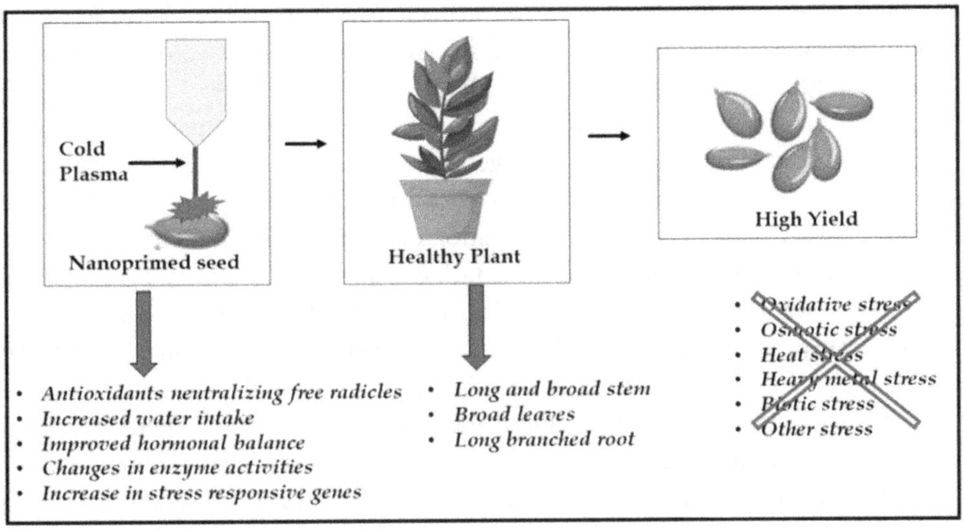

MACHINE LEARNING AND ARTIFICIAL INTELLIGENCE TECHNOLOGY ENABLED NANOPRIMED SEED DIAGNOSTICS

Seeds' increased frequency of seedling emergence and seed germination can ensure uniform breakout in a variety of agricultural settings and contribute significantly to yield potential and homogeneity. At the moment, personal monitoring is used to rate sprouting, but only with limited precision, scale, and frequency. More than one research-based solution, such as germinator, phenoSeeder, and the MultiSense program was developed to alleviate this bottleneck by automating both seedling diagnoses and accompanying phenotypic evaluations (Colmer et al., 2020).

After the seed has been nanoprimed, this software can be used to predict the seed's viability and conduct phenotypic analysis. The Rice Seed Germination Evaluation System (Nguyen et al.,2018) and other cutting-edge computer vision and machine learning methods In recent germination studies, artificial neural networks have been used to evaluate the germination condition of Thai rice species. After the seed has been nanoprimed, this software can be used to predict the seed's viability and conduct phenotypic analysis. In order to automate phenotypic assessments of various organs, such as leaves, roots, and reproductive organs, a number of computer vision and machine learning algorithms that incorporate analytical methods have been developed (Nguyen et al., 2018) Rice Seed Germination Evaluation System, as well as other cutting-edge computer vision and machine learning methods, have been utilized in recent germination studies to evaluate the germination condition of the Thai rice species. Numerous computer vision and machine learning algorithms that incorporate analytical methods have been developed in order to automate phenotypic assessments of various organs, including reproductive organs, leaves, and roots (Zhou et al., 2017; Falk et al., 2020).

Utilizing a seed's colour, texture, form, and development pattern, sprouting nanoprimed seeds can be observed energetically and accurately, allowing for novel scientific findings regarding seed physiology. In addition, standardizing research in seed science can begin with the automation of seed germination scoring. Digital analysis can be used to determine the strength and quality of seeds, and quantitative comparisons of biological testing under a variety of conditions can increase confidence in research findings. Utilizing machine learning, Genze and others (2020) demonstrated efficient germination detection, prediction, and quality evaluation.

Numerous attempts have been made to automate seed testing in order to reduce the number of manual steps required for this strategy to fail. Recent seed evaluations have utilized sophisticated image analytic techniques, which are simple to automate and may provide accurate, impartial data (Dell'Aquila, 2019). In this study, sophisticated convolutional networks were used to identify individual seeds and more precisely distinguish seeds that had germinated from those that had not. The review process might be sped up if AI technology is used to predict germination capability. It performs better and has a lower error rate than manual and conventional procedures. Because of this, it is able to evaluate seed quality and provide germination indices that are more precise. A D Medeiros et al. (2020) used data from FT-NIR spectroscopy and X-ray imaging to develop a method for evaluating the quality of seeds.

X-ray imaging and FT-NIR spectroscopy can be used to foresee seed quality characteristics like sprouting and vitality, which is the focus of this investigation. The suggested method can accurately measure seed germination capability (85% accuracy). Using a machine learning approach that is based on both X-ray images and NIR spectra, it is possible to predict the germination of seeds quickly, effectively, and with confidence. A method for identifying soybean seedlings was developed by A. D. Medeiros and colleagues (2020) on the basis of high-quality data. The shape, size, and physiological potential of soybean seeds and seedlings could be determined using conventional machine learning algorithms, according to this study. Based on images of soybean seedlings and seeds, the models were created using free or inexpensive software. 94% of seeds and seedlings were correctly categorized using the established classification model. High accuracy was demonstrated by the traditional and interactive machine learning models, which were able to quickly and objectively identify soybean seeds based on appearance as well as beans and seedlings based on vigor. A deep learning architecture, a type of Artificial Neural Network (ANN), is used to identify patterns in data or simulate intricate input-output interactions. Classifiers based on multilayer neural networks are used to identify anomalies in wheat grain (seed) grown in a variety of agricultural environmental zones. A lot of data can now be generated by agricultural businesses thanks to advances in science and technology. To analyze this amount of data, automated analytical methods are necessary. It is a highly effective method for assisting professionals at agricultural centers by utilizing cutting-edge computing techniques (Singh et al., 2020). When combined with real-time data processing, it enables farmers to make well-informed decisions regarding fertilizer, planting, and harvesting.

CONCERNS

Researchers have made significant progress, but they still lack a clear knowledge of how these nanomaterials alter the seed's micro- and macroenvironments. Given the current and projected exposure levels, the fact that there is still a shortage of fundamental information concerning the potential wellbeing and security effects of man-made nanomaterials on both human and nonhuman receptors is troubling. Nanomaterials have the potential to be used for priming and coating seeds, but they should be handled with caution. The establishment of appropriate rules based on reliable research is required for the proper application of these technologies, not only in the agricultural sector but also in other industrial fields (Kah et al., 2018). An evaluation of the ecosystem's effects on nanoparticles, taking into account their potential for environmental toxicity and legal restrictions, is required for the commercial production of nanoparticles, their application in agriculture, and the disposal of industrial discharge. Nanoparticles have the potential to have a significant impact on numerous ecosystems that are connected to agricultural operations (Lowry et al., 2019). Therefore, developing nanomaterials that are safe for land and the surrounding environment necessitates an understanding of their mechanism of action. The priming conditions, which include the size and quantity of nanoparticles and the treatment duration, must be evaluated before treating seeds.

Some of the negative effects of improper priming conditions include a decrease in plant development, alterations in the interaction between microbiota and roots, inhibition of germination, detrimental changes in cell metabolism and structure, and germination inhibition (Rahman et al., 2020). Understanding the impact of nanoparticle chemical and physical properties on seeds and associated species is necessary for the development of nanomaterials that are both effective and minimally harmful (Camara et al., 2019). The physical, chemical, and biological properties of each of the aforementioned nanoparticles—polymeric, metallic, and biogenic metallic nanoparticles—are distinct. Biofortification, seed protection, plant tolerance to biotic and abiotic stresses, or a combination of these, are all possible applications for seed nano-priming. Nanoparticle seed priming can have significant advantages. Seed treatments reduce nanoparticle exposure in comparison to soil and foliar sprays. The fact that seed priming only requires a small quantity of nanoparticles is another advantage. To prevent excessive material discharge into the environment, manufacturers may provide it in a controlled manner. Despite the fact that additional research is required to determine how various nanoparticles, such as biogenic metals, polymeric nanoparticles, and metallic nanoparticles, behave and interact with plant growth, the quantity of nanoparticles that are still present in the plant is likely to be negligible, extremely small, or even zero.

CONCLUSION

A promising area for agronomic modification is nanotechnology. One method that could be applied to enhance sustainability in nano-priming in seeds. The increased quantity and quality of food produced by these methods can be attributed to an increase in plant growth and security from a range of strains. As a result, nanotechnology-based seed priming has the potential to transform conventional farming practices, such as the use of agrochemicals, into more environmentally friendly ones. A system that still provides farmers and customers with a product that is superior and more dependable when these characteristics are combined can result in a system that reduces ecological damage caused by traditional methods. Throughout the development of these technical materials for use in businesses and agriculture, numerous considerations must be given to things like seed priming conditions, upgrading, and the negative consequences on plants and other animals that are related to them. However, by reducing pesticide use and the associated dangers, using nanoparticles in crop management can lead to safer practices for farmers, consumers, and the environment. By incorporating nano-bio-fortification into seedlings, less fertilizer can be used. To produce the same amount of nutrient-dense food, less water and resources are needed.

Nano-bio fortification of seeds may also help cut down on the amount of time it takes for plants to mature from seed, allowing the land to produce crops more quickly over time. Seed nano-biofortification's efficacy will be determined by how well it adapts to the environment, particularly the type of land, plant and weather, and region, just like other agricultural technologies. It is necessary to weigh the advantages of nanotechnology against other biofortification processes like plant breeding programs and CRISPR/transgenic technologies, which may offer enduring and cheaper results. It is essential to establish whether nano-based results are viable in comparison to traditional methods. This is in addition to evaluating the overall effects on the environment. Additional cost reductions have resulted from the larger-scale production of nanomaterials. Nanotechnology and seed biofortification processes are very likely to combine in the future if highly focused and programmable nanomaterials emerge. Nano-enabled seed biofortification is an important issue that requires additional research and attention because of its potential benefits and the growing global food insecurity we will face in the coming decades.

REFERENCES

UN. (2017). 2017 HLPF Thematic review of SDG2. *High-level Political Forum on Sustainable Development*. UN.

Abbasi Khalaki, M., Moameri, M., Asgari Lajayer, B., & Astatkie, T. (2020). Influence of nano-priming on seed germination and plant growth of forage and medicinal plants. *Plant Growth Regulation*, *931*(93), 13–28. doi:10.100710725-020-00670-9

Abdel Latef, A. A. H., Abu Alhmad, M. F., & Abdelfattah, K. E. (2017). The Possible Roles of Priming with ZnO Nanoparticles in Mitigation of Salinity Stress in Lupine (Lupinus termis) *Plants. Journal of Plant Growth Regulation*, *36*(1), 60–70. doi:10.100700344-016-9618-x

Abedi, S., Iranbakhsh, A., Oraghi Ardebili, Z., & Ebadi, M. (2020). Seed priming with cold plasma improved early growth, flowering, and protection of Cichorium intybus against selenium nanoparticle. *J. Theor. Appl. Phys.*, *14*(2), 113–119. doi:10.100740094-020-00371-8

Acharya, P., Jayaprakasha, G.K., Crosby, K.M., Jifon, J.L., & Patil, B.S. (2019). Green-Synthesized Nanoparticles Enhanced Seedling Growth, Yield, and Quality of Onion (*Allium cepa* L.). *ACS Sustain. Chem. Eng.* 7, 14580-14590. https://doi.org/SCHEMENG.9B02180/SUPPL_FILE/SC9B02180_SI_001.PDF. doi:10.1021/ACSSU

Acharya, P., Jayaprakasha, G. K., Crosby, K. M., Jifon, J. L., & Patil, B. S. (2020). Nanoparticle-Mediated Seed Priming Improves Germination, Growth, Yield, and Quality of Watermelons (*Citrullus lanatus*) at multi-locations in Texas. *Scientific Reports*, *101*(10), 1–16. doi:10.103841598-020-61696-7 PMID:32193449

Ahuja, R., Sidhu, A., & Bala, A. (2019). Synthesis and Evaluation of Iron (Ii) Sulfide Aqua Nanoparticles (FeS-NPs) against Fusarium Verticillioides Causing Sheath Rot and Seed Discoloration of Rice. *European Journal of Plant Pathology*, *155*(1), 163–171. doi:10.100710658-019-01758-3

Akter, L., Fakir, O. A., Alam, M. K., Islam, M. U., Chakraborti, P., Alam, M. J., Rashid, M. H., Begum, M., & Kader, M. A. (2018). Amelioration of Salinity Stress in Maize Seed Germination and Seedling Growth Attributes through Seed Priming. *Open J. Soil Sci.*, *08*(5), 137–146. doi:10.4236/ojss.2018.85011

An, J., Hu, P., Li, F., Wu, H., Shen, Y., White, J. C., Tian, X., Li, Z., & Giraldo, J. P. (2020). Emerging investigator series: Molecular mechanisms of plant salinity stress tolerance improvement by seed priming with cerium oxide nanoparticles. *Environmental Science. Nano*, *7*(8), 2214–2228. doi:10.1039/D0EN00387E

Avellan, A., Schwab, F., Masion, A., Chaurand, P., Borschneck, D., Vidal, V., Rose, J., Santaella, C., & Levard, C. (2017). Nanoparticle Uptake in Plants: Gold Nanomaterial Localized in Roots of Arabidopsis thaliana by X-ray Computed Nanotomography and Hyperspectral Imaging. *Environmental Science & Technology*, *51*(15), 8682–8691. doi:10.1021/acs.est.7b01133 PMID:28686423

Awasthi, A., Bansal, S., Jangir, L. K., Awasthi, G., Awasthi, K. K., & Awasthi, K. (2017). Effect of ZnO Nanoparticles on Germination of Triticum aestivum Seeds. *Macromolecular Symposia*, *376*(1), 1700043. doi:10.1002/masy.201700043

Azimi, F., Oraei, M., Gohari, G., Panahirad, S., & Farmarzi, A. (2021). Chitosan-selenium nanoparticles (Cs–Se NPs) modulate the photosynthesis parameters, antioxidant enzymes activities and essential oils in Dracocephalum moldavica L. under cadmium toxicity stress. *Plant Physiology and Biochemistry*, *167*, 257–268. doi:10.1016/j.plaphy.2021.08.013 PMID:34391200

Babajani, A., Iranbakhsh, A., Oraghi Ardebili, Z., & Eslami, B. (2019). Seed Priming with Non-thermal Plasma Modified Plant Reactions to Selenium or Zinc Oxide Nanoparticles: Cold Plasma as a Novel Emerging Tool for Plant Science. *Plasma Chemistry and Plasma Processing*, *39*(1), 21–34. doi:10.100711090-018-9934-y

Bailly, C. (2019). The signalling role of ROS in the regulation of seed germination and dormancy. *The Biochemical Journal*, *476*(20), 3019–3032. doi:10.1042/BCJ20190159 PMID:31657442

Barbieri, M., Sappa, G., & Nigro, A. (2018). Soil Pollution: Anthropogenic versus Geogenic Contributions over Large Areas of the Lazio Region. *Journal of Geochemical Exploration*, *195*, 78–86. doi:10.1016/j.gexplo.2017.11.014

Bienert, G. P., & Chaumont, F. (2014). Aquaporin-facilitated transmembrane diffusion of hydrogen peroxide. *Biochimica et Biophysica Acta, 1840*(5), 1596–1604. doi:10.1016/j.bbagen.2013.09.017 PMID:24060746

Bombo, A.B., Pereira, A.E.S., Lusa, M.G., De Medeiros Oliveira, E., De Oliveira, J.L., Campos, E.V.R., De Jesus, M.B., Oliveira, H.C., Fraceto, L.F., & Mayer, J.L.S. (2019). A Mechanistic View of Interactions of a Nanoherbicide with Target Organism. *J. Agric. Food Chem., 67*, 4453-4462. DOI: https://doi.org/. 9B00806.SOCIAL. JPEG_V03 doi:10.1021/ACS.JAFC.9B00806/ASSET/IMAGES/ACS.JAFC

Bravo Cadena, M., Preston, G. M., Van der Hoorn, R. A. L., Flanagan, N. A., Townley, H. E., & Thompson, I. P. (2018). Enhancing Cinnamon Essential Oil Activity by Nanoparticle Encapsulation to Control Seed Pathogens. *Industrial Crops and Products, 124*, 755–764. doi:10.1016/j.indcrop.2018.08.043

Carrillo-Reche, J., Vallejo-Marín, M., & Quilliam, R. S. (2018). Quantifying the potential of 'on-farm' seed priming to increase crop performance in developing countries. A meta-analysis. *Agronomy for Sustainable Development, 38*(6), 1–14. doi:10.100713593-018-0536-0

Chandrasekaran, U., Luo, X., Wang, Q., & Shu, K. (2020). Are There Unidentified Factors Involved in the Germination of Nanoprimed Seeds? *Frontiers in Plant Science, 11*, 832. doi:10.3389/fpls.2020.00832 PMID:32587599

Chau, N. H., Doan, Q. H., Chu, T. H., Nguyen, T. T., Dao Trong, H., & Ngo, Q. B. (2019). Effects of Different Nanoscale Microelement-Containing Formulations for Presowing Seed Treatment on Growth of Soybean Seedlings. *Journal of Chemistry, 2019*, 1–8. Advance online publication. doi:10.1155/2019/8060316

Chen, Y., Wang, D., Zhu, X., Zheng, X., & Feng, L. (2012). Long-term effects of copper nanoparticles on wastewater biological nutrient removal and N2O generation in the activated sludge process. *Environ. Sci. Technol.* 46, 12452-12458. DOI: https:// doi.org/ 302646Q doi:10.1021/ES

Choudhary, R. C., Kumaraswamy, R. V., Kumari, S., Sharma, S. S., Pal, A., Raliya, R., Biswas, P., & Saharan, V. (2019). Zinc encapsulated chitosan nanoparticle to promote maize crop yield. *International Journal of Biological Macromolecules, 127*, 126–135. doi:10.1016/j.ijbiomac.2018.12.274 PMID:30610949

Colmer, J., O'Neill, C. M., Wells, R., Bostrom, A., Reynolds, D., Websdale, D., Shiralagi, G., Lu, W., Lou, Q., Le Cornu, T., Ball, J., Renema, J., Flores Andaluz, G., Benjamins, R., Penfield, S., & Zhou, J. (2020). SeedGerm: A cost-effective phenotyping platform for automated seed imaging and machine-learning based phenotypic analysis of crop seed germination. *The New Phytologist*, *228*(2), 778–793. doi:10.1111/nph.16736 PMID:32533857

Dai, Y., Chen, F., Yue, L., Li, T., Jiang, Z., Xu, Z., Wang, Z., & Xing, B. (2020). Uptake, Transport, and Transformation of CeO2 Nanoparticles by Strawberry and Their Impact on the Rhizosphere Bacterial Community. *ACS Sustainable Chemistry & Engineering*, *8*(12), 4792–4800. doi:10.1021/acssuschemeng.9b07422

Dantas de Medeiros, A., Silva, L., Ribeiro, J. P., Ferreira, K., Fim Rosas, J., Santos, A., & Barboza da Silva, C. (2020). Machine Learning for Seed Quality Classification: An Advanced Approach Using Merger Data from FT-NIR Spectroscopy and X-ray Imaging. *Sensors (Basel)*, *20*(15), 4319. doi:10.339020154319 PMID:32756355

Das, C. K., Jangir, H., Kumar, J., Verma, S., Mahapatra, S. S., Philip, D., Srivastava, G., & Das, M. (2018). Nano-Pyrite Seed Dressing: A Sustainable Design for NPK Equivalent Rice Production. *Nanotechnol. Environ. Eng*, *3*(1), 14. doi:10.100741204-018-0043-1

De La Torre-Roche, R., Cantu, J., Tamez, C., Zuverza-Mena, N., Hamdi, H., Adisa, I. O., Elmer, W., Gardea-Torresdey, J., & White, J. C. (2020). Seed Biofortification by Engineered Nanomaterials: A Pathway to Alleviate Malnutrition? *Journal of Agricultural and Food Chemistry*, *68*(44), 12189–12202. doi:10.1021/acs.jafc.0c04881 PMID:33085897

De Oliveira, J. L., Campos, E. V. R., Germano-Costa, T., Lima, R., Vechia, J. F. D., Soares, S. T., de Andrade, D. J., Gonçalves, K. C., do Nascimento, J., Polanczyk, R. A., & Fraceto, L. F. (2019). Association of Zein Nanoparticles with Botanical Compounds for Effective Pest Control Systems. *Pest Management Science*, *75*(7), 1855–1865. doi:10.1002/ps.5338 PMID:30653814

Dell'Aquila, A. (2009). Digital Imaging Information Technology Applied to Seed Germination Testing: A Review. *Agronomy for Sustainable Development*, *29*(1), 213–221. doi:10.1051/agro:2008039

Dietz, K. J., & Herth, S. (2011). Plant nanotoxicology. *Trends in Plant Science*, *16*(11), 582–589. doi:10.1016/j.tplants.2011.08.003 PMID:21906987

Dileep Kumar, G., Raja, K., Natarajan, N., Govindaraju, K., & Subramanian, K. S. (2020). Invigouration Treatment of Metal and Metal Oxide Nanoparticles for Improving the Seed Quality of Aged Chilli Seeds (*Capsicum annum* L.). *Materials Chemistry and Physics*, *242*, 122492. doi:10.1016/j.matchemphys.2019.122492

Duran, N. M., Savassa, S. M., Lima, R. G., de Almeida, E., Linhares, F. S., van Gestel, C. A. M., & Pereira de Carvalho, H. W. (2017). X-ray Spectroscopy Uncovering the Effects of Cu Based Nanoparticle Concentration and Structure on Phaseolus vulgaris Germination and Seedling Development. *Journal of Agricultural and Food Chemistry*, *65*(36), 7874–7884. doi:10.1021/acs.jafc.7b03014 PMID:28817280

Ernest, V., Shiny, P. J., Mukherjee, A., & Chandrasekaran, N. (2012). Silver nanoparticles: A potential nanocatalyst for the rapid degradation of starch hydrolysis by α-amylase. *Carbohydrate Research*, *352*, 60–64. doi:10.1016/j.carres.2012.02.009 PMID:22405762

Falk, K. G., Jubery, T. Z., Mirnezami, S. V., Parmley, K. A., Sarkar, S., Singh, A., Ganapathysubramanian, B., & Singh, A. K. (2020). Computer vision and machine learning enabled soybean root phenotyping pipeline. *Plant Methods*, *16*(1), 5. doi:10.118613007-019-0550-5 PMID:31993072

Falsini, S., Clemente, I., Papini, A., Tani, C., Schiff, S., Salvatici, M. C., Petruccelli, R., Benelli, C., Giordano, C., Gonnelli, C., & Ristori, S. (2019). When Sustainable Nanochemistry Meets Agriculture: Lignin Nanocapsules for Bioactive Compound Delivery to Plantlets. *ACS Sustainable Chemistry & Engineering*, *7*(24), 19935–19942. doi:10.1021/acssuschemeng.9b05462

Farias, B. V., Pirzada, T., Mathew, R., Sit, T. L., Opperman, C., & Khan, S. A. (2019). Electrospun Polymer Nanofibers as Seed Coatings for Crop Protection. *ACS Sustainable Chemistry & Engineering*, *7*(24), 19848–19856. doi:10.1021/acssuschemeng.9b05200

Fraceto, L. F., Grillo, R., de Medeiros, G. A., Scognamiglio, V., Rea, G., & Bartolucci, C. (2016). Nanotechnology in agriculture: Which innovation potential does it have? *Frontiers in Environmental Science*, *4*, 20. doi:10.3389/fenvs.2016.00020

Friedrichs, S., Takasu, Y., Kearns, P., Dagallier, B., Oshima, R., Schofield, J., & Moreddu, C. (2019). An overview of regulatory approaches to genome editing in agriculture. *Biotechnol. Res. Innov.* 3, 208-220. DOI: https://doi.org/. 2019.07.001 doi:10.1016/J.BIORI

Gaafar, R. M., Diab, R. H., Halawa, M. L., El-Shanshory, A. R., El-Shaer, A., & Hamouda, M. M. (2020). Role of Zinc Oxide Nanoparticles in Ameliorating Salt Tolerance in Soybean. *Egyptian Journal of Botany*, *60*(0), 733–747. doi:10.21608/ejbo.2020.26415.1475

Garg, M., Sharma, N., Sharma, S., Kapoor, P., Kumar, A., Chunduri, V., & Arora, P. (2018). Biofortified Crops Generated by Breeding, Agronomy, and Transgenic Approaches Are Improving Lives of Millions of People around the World. *Frontiers in Nutrition*, *5*, 12. doi:10.3389/fnut.2018.00012 PMID:29492405

Genze, N., Bharti, R., Grieb, M., Schultheiss, S. J., & Grimm, D. G. (2020). Accurate machine learning-based germination detection, prediction and quality assessment of three grain crops. *Plant Methods*, *16*(1), 157. doi:10.118613007-020-00699-x PMID:33353559

Ghafariyan, M. H., Malakouti, M. J., Dadpour, M. R., Stroeve, P., & Mahmoudi, M. (2013). Effects of Magnetite Nanoparticles on Soybean Chlorophyll. *Environmental Science & Technology*, *47*, 10645–10652. doi:10.1021/es402249b PMID:23951999

Ghidan, A. Y., Al Antary, T. M., & Awwad, A. M. (2017b). Aphidicidal potential of green synthesized magnesium hydroxide nanoparticles using *Olea europaea* leaves extract. *Journal of Agricultural and Biological Science*, *12*(10), 293–301.

Ghidan, A. Y., Al-Antary, T. M., Salem, N. M., & Awwad, A. M. (2017a). Facile green synthetic route to the zinc oxide (ZnONPs) nanoparticles: Effect on green peach aphid and antibacterial activity. *Journal of Agricultural Science (Toronto)*, *9*(2), 131–138. doi:10.5539/jas.v9n2p131

Gilbertson, L. M., Pourzahedi, L., Laughton, S., Gao, X., Zimmerman, J. B., Theis, T. L., Westerhoff, P., & Lowry, G. V. (2020). Guiding the design space for nanotechnology to advance sustainable crop production. *Nature Nanotechnology*, *15*(9), 801–810. doi:10.103841565-020-0706-5 PMID:32572231

Gnanasekaran, D. k., Raja, K., Natarajan, N., Govindaraju, K., & Subramanian, K. (2019). Invigouration treatment of metal and metal oxide nanoparticles for improving the seed quality of aged chilli seeds (*Capsicum annum* L.). *Materials Chemistry and Physics*, *242*, 122492.

UN. (n.d.). *Goal 2: End hunger, achieve food security and improved nutrition and promote sustainable agriculture — SDG Indicators*. UN. unstats.un.org.

Gohari, G., Panahirad, S., Sadeghi, M., Akbari, A., Zareei, E., Zahedi, S. M., Bahrami, M. K., & Fotopoulos, V. (2021). Putrescine functionalized carbon quantum dot (put-CQD) nanoparticles effectively prime grapevine (*Vitis vinifera* cv. 'Sultana') against salt stress. *BMC Plant Biology*, *21*(1), 120. doi:10.118612870-021-02901-1 PMID:33639848

Gomes, M. P., & Garcia, Q. S. (2013). Reactive oxygen species and seed germination. *Biologia*, *68*(3), 351–357. doi:10.247811756-013-0161-y

Gross, M. S., Bean, T. G., Hladik, M. L., Rattner, B. A., & Kuivila, K. M. (2020). Uptake, Metabolism, and Elimination of Fungicides from Coated Wheat Seeds in Japanese Quail (*Coturnix japonica*). *Journal of Agricultural and Food Chemistry*, *68*(6), 1514–1524. doi:10.1021/acs.jafc.9b05668 PMID:31977218

Guha, T., Ravikumar, K. V. G., Mukherjee, A., Mukherjee, A., & Kundu, R. (2018). Nanopriming with Zero Valent Iron (NZVI) Enhances Germination and Growth in Aromatic Rice Cultivar (*Oryza sativa* Cv. Gobindabhog L.). *Plant Physiology and Biochemistry*, *127*, 403–413. doi:10.1016/j.plaphy.2018.04.014 PMID:29679934

Hojjat, S. S., & Kamyab, M. (2017). The effect of silver nanoparticle on Fenugreek seed germination under salinity levels. *Russian Agricultural Sciences*, *43*(1), 61–65. doi:10.3103/S1068367417010189

Hu, P., An, J., Faulkner, M.M., Wu, H., Li, Z., Tian, X., Giraldo, J.P. (2020). Nanoparticle Charge and Size Control Foliar Delivery Efficiency to Plant Cells and Organelles. *ACS Nano, 14*, 7970-7986. https://doi.org/ PL_FILE/NN9B09178_SI_017.AVI doi:10.1021/ACSNANO.9B09178/SUP

Huang, S., Wang, L., Liu, L., Hou, Y., & Li, L. (2015). Nanotechnology in agriculture, livestock, and aquaculture in China. A review. *Agronomy for Sustainable Development*, *35*(2), 369–400. doi:10.100713593-014-0274-x

Hussain, A., Rizwan, M., Ali, Q., & Ali, S. (2019). Seed priming with silicon nanoparticles improved the biomass and yield while reduced the oxidative stress and cadmium concentration in wheat grains. *Environmental Science and Pollution Research International*, *268*(26), 7579–7588. doi:10.100711356-019-04210-5 PMID:30661166

Iranbakhsh, A., Oraghi Ardebili, Z., Oraghi Ardebili, N., Ghoranneviss, M., & Safari, N. (2018). Cold plasma relieved toxicity signs of nano zinc oxide in *Capsicum annuum* cayenne via modifying growth, differentiation, and physiology. *Acta Physiologiae Plantarum*, *40*(8), 154. doi:10.100711738-018-2730-8

Itroutwar, P. D., Govindaraju, K., Tamilselvan, S., Kannan, M., Raja, K., & Subramanian, K. S. (2019). Seaweed-Based Biogenic ZnO Nanoparticles for Improving Agro-morphological Characteristics of Rice (Oryza sativa L.). *Journal of Plant Growth Regulation*, *392*(39), 717–728. doi:10.100700344-019-10012-3

Itroutwar, P. D., Kasivelu, G., Raguraman, V., Malaichamy, K., & Sevathapandian, S. K. (2020). Effects of biogenic zinc oxide nanoparticles on seed germination and seedling vigor of maize (*Zea mays*). *Biocatalysis and Agricultural Biotechnology*, *29*, 101778. Advance online publication. doi:10.1016/j.bcab.2020.101778

Joshi, A., Kaur, S., Dharamvir, K., Nayyar, H., & Verma, G. (2018). Multi-walled carbon nanotubes applied through seed-priming influence early germination, root hair, growth and yield of bread wheat (*Triticum aestivum* L.). *Journal of the Science of Food and Agriculture*, *98*, 3148–3160. doi:10.1002/jsfa.8818 PMID:29220088

Kah, M., & Hofmann, T. (2014). Nanopesticide: Current trends and future priorities. *Environment International*, *63*, 224–235. doi:10.1016/j.envint.2013.11.015 PMID:24333990

Kah, M., Tufenkji, N., & White, J. C. (2019). Nano-enabled strategies to enhance crop nutrition and protection. *Nature Nanotechnology*, *146*(14), 532–540. doi:10.103841565-019-0439-5 PMID:31168071

Kasote, D. M., Lee, J. H. J., Jayaprakasha, G. K., & Patil, B. S. (2019). Seed Priming with Iron Oxide Nanoparticles Modulate Antioxidant Potential and Defense-Linked Hormones in Watermelon Seedlings. [CrossRef]. *ACS Sustainable Chemistry & Engineering*, *7*(5), 5142–5151. doi:10.1021/acssuschemeng.8b06013

Khodakovskaya, M., Dervishi, E., Mahmood, M., Xu, Y., Li, Z., Watanabe, F., & Biris, A. S. (2009). Carbon nanotubes are able to penetrate plant seed coat and dramatically affect seed germination and plant growth. *ACS Nano*, *3*(10), 3221–3227. doi:10.1021/nn900887m PMID:19772305

Komyshev, E., Genaev, M., & Afonnikov, D. (2017). Evaluation of the SeedCounter, A Mobile Application for Grain Phenotyping. *Frontiers in Plant Science*, *7*, 1990. doi:10.3389/fpls.2016.01990 PMID:28101093

Kumar, S., Nehra, M., Dilbaghi, N., Marrazza, G., Tuteja, S. K., & Kim, K.-H. (2020). Nanovehicles for Plant Modifications towards Pest- and Disease-Resistance Traits. *Trends in Plant Science*, *25*(2), 198–212. doi:10.1016/j.tplants.2019.10.007 PMID:31780333

Lahiani, M. H., Dervishi, E., Chen, J., Nima, Z., Gaume, A., Biris, A. S., & Khodakovskaya, M. V. (2013). Impact of Carbon Nanotube Exposure to Seeds of Valuable Crops. *ACS Applied Materials & Interfaces*, *5*(16), 7965–7973. doi:10.1021/am402052x PMID:23834323

Lemmens, E., Deleu, L.J., De Brier, N., De Man, W.L., De Proft, M., Prinsen, E., & Delcour, J.A. (2019). The Impact of Hydro-Priming and Osmo-Priming on Seedling Characteristics, Plant Hormone Concentrations, Activity of Selected Hydrolytic Enzymes, and Cell Wall and Phytate Hydrolysis in Sprouted Wheat (*Triticum aestivum* L.). *ACS Omega, 4,* 22089-22100. https://doi.org/ _SI_001.PDF doi:10.1021/ACSOMEGA.9B03210/SUPPL_FILE/AO9B03210

Leng, X., Jia, H., Sun, X., Shangguan, L., Mu, Q., Wang, B., & Fang, J. (2015). Comparative transcriptome analysis of grapevine in response to copper stress. *Scientific Reports*, *5*(1), 17749. doi:10.1038rep17749 PMID:26673527

Li, M., Liu, H., Dang, F., Hintelmann, H., Yin, B., & Zhou, D. (2020). Alteration of Crop Yield and Quality of Three Vegetables upon Exposure to Silver Nanoparticles in Sludge-Amended Soil. *ACS Sustainable Chemistry & Engineering*, *8*(6), 2472–2480. doi:10.1021/acssuschemeng.9b06721

Li, R., He, J., Xie, H., Wang, W., Bose, S. K., Sun, Y., Hu, J., & Yin, H. (2019). Effects of chitosan nanoparticles on seed germination and seedling growth of wheat (*Triticum aestivum* L.*). International Journal of Biological Macromolecules*, *126*, 91–100. doi:10.1016/j.ijbiomac.2018.12.118 PMID:30557637

Lowry, G. V., Avellan, A., & Gilbertson, L. M. (2019). Opportunities and challenges for nanotechnology in the agri-tech revolution. *Nature Nanotechnology*, *14*(6), 517–522. doi:10.103841565-019-0461-7 PMID:31168073

Mahakham, W., Sarmah, A. K., Maensiri, S., & Theerakulpisut, P. (2017). Nanopriming technology for enhancing germination and starch metabolism of aged rice seeds using phytosynthesized silver nanoparticles. *Scientific Reports*, *7*(1), 8263. doi:10.103841598-017-08669-5 PMID:28811584

Mahakham, W., Theerakulpisut, P., Maensiri, S., Phumying, S., & Sarmah, A. K. (2016). Environmentally benign synthesis of phytochemicals-capped gold nanoparticles as nanopriming agent for promoting maize seed germination. *The Science of the Total Environment*, *573*, 1089–1102. doi:10.1016/j.scitotenv.2016.08.120 PMID:27639594

Maity, A., Natarajan, N., Pastor, M., Vijay, D., Gupta, C. K., & Wasnik, V. K. (2018). Nanoparticles influence seed germination traits and seed pathogen infection rate in forage sorghum (Sorghum bicolour) and cowpea (*Vigna unguiculata*). *Indian Journal of Experimental Biology*, *56*, 363–372.

Malik, A., Mor, V. S., Tokas, J., Punia, H., Malik, S., Malik, K., Sangwan, S., Tomar, S., Singh, P., Singh, N., Himangini, Vikram, Nidhi, Singh, G., Vikram, Kumar, V., Sandhya, & Karwasra, A. (2020). Biostimulant-Treated Seedlings under Sustainable Agriculture: A Global Perspective Facing Climate Change. *Agronomy (Basel)*, *11*(1), 14. doi:10.3390/agronomy11010014

Manjaiah, K.M., Mukhopadhyay, R., Paul, R., Datta, S.C., Kumararaja, P., Sarkar, B. (2018). Clay minerals and zeolites for environmentally sustainable agriculture. Modif. Clay Zeolite Nanocomposite Mater. *Environ. Pharm. Appl.*, 309-329. doi:10.1016/B978-0-12-814617-0.00008-6

Marmiroli, N., & White, J. C. (2016). Editorial: Nanotoxicology and Environmental Risk Assessment of Engineered Nanomaterials (ENMs) in Plants. *Frontiers in Plant Science*, *7*, 1370. doi:10.3389/fpls.2016.01370 PMID:27672390

Marthandan, V., Geetha, R., Kumutha, K., Renganathan, V. G., Karthikeyan, A., & Ramalingam, J. (2020). Seed Priming: A Feasible Strategy to Enhance Drought Tolerance in Crop Plants. *International Journal of Molecular Sciences*, *21*(21), 8258. doi:10.3390/ijms21218258 PMID:33158156

Maswada, H. F., Djanaguiraman, M., & Prasad, P. V. V. (2018). Seed treatment with nano-iron (III) oxide enhances germination, seeding growth and salinity tolerance of sorghum. *Journal Agronomy & Crop Science*, *204*(6), 577–587. doi:10.1111/jac.12280

Matyjaszczyk, E. (2017). Comparison between Seed and Foliar Treatment as a Tool in Integrated Pest Management. *Journal of Agricultural and Food Chemistry*, *65*(30), 6081–6086. doi:10.1021/acs.jafc.7b01095 PMID:28686448

Misson, M., Zhang, H., & Jin, B. (2015). Nanobiocatalyst advancements and bioprocessing applications. *Journal of the Royal Society, Interface*, *12*(102), 20140891. doi:10.1098/rsif.2014.0891 PMID:25392397

Montanha, G. S., Rodrigues, E. S., Marques, J. P. R., de Almeida, E., Colzato, M., & Pereira de Carvalho, H. W. (2020). Zinc nanocoated seeds: An alternative to boost soybean seed germination and seedling development. *SN Applied Sciences*, *2*(5), 1–11. doi:10.100742452-020-2630-6

Müller, K., Linkies, A., Vreeburg, R. A. M., Fry, S. C., Krieger-Liszkay, A., & Leubner-Metzger, G. (2009). In Vivo Cell Wall Loosening by Hydroxyl Radicals during Cress Seed Germination and Elongation Growth. *Plant Physiology*, *150*(4), 1855–1865. doi:10.1104/pp.109.139204 PMID:19493972

Naguib, D. M., & Abdalla, H. (2019). Metabolic Status during Germination of Nano Silica Primed Zea mays Seeds under Salinity Stress. *Journal of Crop Science and Biotechnology*, *22*(5), 415–423. doi:10.100712892-019-0168-0

Najafi Disfani, M., Mikhak, A., Kassaee, M. Z., & Maghari, A. (2017). Effects of nano Fe/SiO2 fertilizers on germination and growth of barley and maize. *Archives of Agronomy and Soil Science*, *63*(6), 817–826. doi:10.1080/03650340.2016.1239016

Nandhini, M., Rajini, S. B., Udayashankar, A. C., Niranjana, S. R., Lund, O. S., Shetty, H. S., & Prakash, H. S. (2019). Biofabricated Zinc Oxide Nanoparticles as an Eco-Friendly Alternative for Growth Promotion and Management of Downy Mildew of Pearl Millet. *Crop Protection (Guildford, Surrey)*, *121*, 103–112. doi:10.1016/j.cropro.2019.03.015

Nguyen, D. V., Nguyen, H. M., Le, N. T., Nguyen, K. H., Le, H. M., Nguyen, A. T., Dinh, N. T. T., Hoang, S. A., & Ha, C. V. (2020). Copper Nanoparticle Application Enhances Plant Growth and Grain Yield in Maize under Drought Stress Conditions. *Plant Biology*. doi:10.1101/2020.02.24.963132

Nguyen, T., Hoang, V.-N., Le, T., Tran, T.-H., & Hai, V. (2018). A Vision Based Method for Automatic Evaluation of Germination Rate of Rice Seeds. In *1st International Conference on Multimedia Analysis and Pattern Recognition (MAPR)*, Ho Chi Minh, Vietnam. 10.1109/MAPR.2018.8337511

Nonogaki, H. (2014). Seed dormancy and germination-emerging mechanisms and new hypotheses. *Frontiers in Plant Science*, *5*, 233. doi:10.3389/fpls.2014.00233 PMID:24904627

Nonogaki, H., Bassel, G. W., & Bewley, J. D. (2010). Germination—Still a mystery. *Plant Science*, *179*(6), 574–581. doi:10.1016/j.plantsci.2010.02.010

Noorhosseini, S. A., Jokar, N. K., & Damalas, C. A. (2017). Improving Seed Germination and Early Growth of Garden Cress (Lepidium sativum) and Basil (Ocimum basilicum) with Hydro-priming. *Journal of Plant Growth Regulation*, *371*(37), 323–334. doi:10.100700344-017-9728-0

Panpatte, D.G., Jhala, Y.G., Shelat, H.N., & Vyas, R. V. (2016). Nanoparticles: The Next Generation Technology for Sustainable Agriculture. *Microb. Inoculants Sustain. Agric. Product,* 289-300. doi:10.1007/978-81-322-2644-4_18

Panyuta, O., Belava, V., Fomaidi, S., Kalinichenko, O., Volkogon, M., & Taran, N. (2016). The Effect of Pre-Sowing Seed Treatment with Metal Nanoparticles on the Formation of the Defensive Reaction of Wheat Seedlings Infected with the Eyespot Causal Agent. *Nanoscale Research Letters*, *11*(1), 92. doi:10.118611671-016-1305-0 PMID:26880729

Paparella, S., Araújo, S. S., Rossi, G., Wijayasinghe, M., Carbonera, D., & Balestrazzi, A. (2015). Seed priming: State of the art and new perspectives. *Plant Cell Reports*, *34*(8), 1281–1293. doi:10.100700299-015-1784-y PMID:25812837

Parveen, A., & Rao, S. (2015). Effect of Nanosilver on Seed Germination and Seedling Growth in Pennisetum glaucum. *Journal of Cluster Science*, *26*(3), 693–701. doi:10.100710876-014-0728-y

Pereira, A. E. S., Oliveira, H. C., & Fraceto, L. F. (2019). Polymeric nanoparticles as an alternative for application of gibberellic acid in sustainable agriculture: A field study. *Scientific Reports*, *9*(1), 7135. doi:10.103841598-019-43494-y PMID:31073210

Pérez-de-Luque, A. (2017). Interaction of nanomaterials with plants: What do we need for real applications in agriculture? *Frontiers in Environmental Science*, *5*(12). doi:10.3389/fenvs.2017.00012

Pilarska, A., Wysokowski, M., Markiewicz, E., & Jesionowski, T. (2013). Synthesis of magnesium hydroxide and its calcinates by a precipitation method with the use of magnesium sulfate and poly (ethylene glycols). *Powder Technology*, *235*, 148–157. doi:10.1016/j.powtec.2012.10.008

Pirzada, T., de Farias, B. V., Mathew, R., Guenther, R. H., Byrd, M. V., Sit, T. L., Pal, L., Opperman, C. H., & Khan, S. A. (2020). Recent advances in biodegradable matrices for active ingredient release in crop protection: Towards attaining sustainability in agriculture. *Current Opinion in Colloid & Interface Science*, *48*, 121–136. doi:10.1016/j.cocis.2020.05.002 PMID:33013179

Plaksenkova, I., Kokina, I., Petrova, A., Jermalonoka, M., Gerbreders, V., & Krasovska, M. (2016). The Impact of Zinc Oxide Nanoparticles on Cytotoxicity, Genotoxicity, and MiRNA Expression in Barley (*Hordeum vulgare* L.) Seedlings. *Hindawi*. https://www. hindawi.com/journals/tswj/2020/6649746/

Prasad, R., Bhattacharyya, A., & Nguyen, Q. D. (2017). Nanotechnology in Sustainable Agriculture: Recent Developments, Challenges, and Perspectives. *Frontiers in Microbiology*, *8*, 1014. doi:10.3389/fmicb.2017.01014 PMID:28676790

Rahman, M. S., Chakraborty, A., Mazumdar, S., Nandi, N. C., Bhuiyan, M. N. I., Alauddin, S. M., Khan, I. A., & Hossain, M. J. (2020). Effects of Poly (Vinylpyrrolidone) Protected Platinum Nanoparticles on Seed Germination and Growth Performance of Pisum Sativum. *Nano Struct. Nano Objects*, *21*, 100408. doi:10.1016/j.nanoso.2019.100408

Raja, K., Sowmya, R., Sudhagar, R., Moorthy, P. S., Govindaraju, K., & Subramanian, K. S. (2019). Biogenic ZnO and Cu Nanoparticles to Improve Seed Germination Quality in Blackgram (*Vigna mungo*). *Materials Letters*, *235*, 164–167. doi:10.1016/j.matlet.2018.10.038

Rajput, V., Minkina, T., Fedorenko, A., Sushkova, S., Mandzhieva, S., Lysenko, V., Duplii, N., Fedorenko, G., Dvadnenko, K., & Ghazaryan, K. (2018). Toxicity of copper oxide nanoparticles on spring barley (Hordeum sativum distichum). *The Science of the Total Environment*, *645*, 1103–1113. doi:10.1016/j.scitotenv.2018.07.211 PMID:30248835

Rajput, V., Minkina, T., Mazarji, M., Shende, S., Sushkova, S., Mandzhieva, S., Burachevskaya, M., Chaplygin, V., Singh, A., & Jatav, H. (2020a). Accumulation of nanoparticles in the soil-plant systems and their effects on human health. *Annals of Agricultural Science*, *65*(2), 137–143. doi:10.1016/j.aoas.2020.08.001

Rajput, V. D., Chen, Y., & Ayup, M. (2015). Effects of high salinity on physiological and anatomical indices in the early stages of Populus euphratica growth. *Russian Journal of Plant Physiology: a Comprehensive Russian Journal on Modern Phytophysiology*, *622*(62), 229–236. doi:10.1134/S1021443715020168

Rajput, V.D., Singh, A., Minkina, T.M., Shende, S.S., Kumar, P., Verma, K.K., Bauer, T., Gorobtsova, O., Deneva, S., Sindireva, A. (2021c). Potential Applications of Nanobiotechnology in Plant Nutrition and Protection for Sustainable Agriculture. *Nanotechnol. Plant Growth Promot. Prot.* 79-92. doi:10.1002/9781119745884.ch5

Raliya, R., Tarafdar, J. C., & Biswas, P. (2016). Enhancing the Mobilization of Native Phosphorus in the Mung Bean Rhizosphere Using ZnO Nanoparticles Synthesized by Soil Fungi. *Journal of Agricultural and Food Chemistry*, *64*(16), 3111–3118. doi:10.1021/acs.jafc.5b05224 PMID:27054413

Rastogi, A., Tripathi, D.K., Yadav, S., Chauhan, D.K., Živčák, M., Ghorbanpour, M., El-Sheery, N.I., & Brestic, M. (2019). Application of silicon nanoparticles in agriculture. *3 Biotech., 9.* doi:10.1007/s13205-019-1626-7

Rizwan, M., Ali, S., Ali, B., Adrees, M., Arshad, M., Hussain, A., Zia ur Rehman, M., & Waris, A. A. (2019). Zinc and Iron Oxide Nanoparticles Improved the Plant Growth and Reduced the Oxidative Stress and Cadmium Concentration in Wheat. *Chemosphere*, *214*, 269–277. doi:10.1016/j.chemosphere.2018.09.120 PMID:30265934

Saharan, V., Kumaraswamy, R. V., Choudhary, R. C., Kumari, S., Pal, A., Raliya, R., & Biswas, P. (2016). Cu-Chitosan Nanoparticle Mediated Sustainable Approach To Enhance Seedling Growth in Maize by Mobilizing Reserved Food. *Journal of Agricultural and Food Chemistry*, *64*(31), 6148–6155. doi:10.1021/acs.jafc.6b02239 PMID:27460439

Salama, D. M., Osman, S. A., Abd El-Aziz, M. E., Abd Elwahed, M. S. A., & Shaaban, E. A. (2019). Effect of Zinc Oxide Nanoparticles on the Growth, Genomic DNA, Production and the Quality of Common Dry Bean (*Phaseolus vulgaris*). *Biocatalysis and Agricultural Biotechnology*, *11*, 101083. doi:10.1016/j.bcab.2019.101083

Santaella, C., & Plancot, B. (2020). Interactions of Nanoenabled Agrochemicals with Soil Microbiome. In V. L. de Castro, R. Grillo, D. Ávila, H. Caixeta Oliveira, & R. Lima (Eds.), *Nanopesticides: From Research and Development to Mechanisms of Action and Sustainable Use in Agriculture; Fraceto, L.F., S.S* (pp. 137–163). Springer International Publishing. doi:10.1007/978-3-030-44873-8_6

Sanzari, I., Leone, A., & Ambrosone, A. (2019). Nanotechnology in Plant Science: To Make a Long Story Short. *Frontiers in Bioengineering and Biotechnology*, *7*, 120. doi:10.3389/fbioe.2019.00120 PMID:31192203

Sathiyabama, M., & Muthukumar, S. (2020). Chitosan guar nanoparticle preparation and its in vitro antimicrobial activity towards phytopathogens of rice. *International Journal of Biological Macromolecules*, *153*, 297–304. doi:10.1016/j.ijbiomac.2020.03.001 PMID:32135260

Savassa, S. M., Duran, N. M., Rodrigues, E. S., de Almeida, E., van Gestel, C. A. M., Bompadre, T. F. V., & de Carvalho, H. W. P. (2018). Effects of ZnO Nanoparticles on *Phaseolus vulgaris* Germination and Seedling Development Determined by X-ray Spectroscopy. *ACS Applied Nano Materials*, *1*, 6414–6426. doi:10.1021/acsanm.8b01619

Schmitt, C. C., Moreira, R., Neves, R. C., Richter, D., Funke, A., Raffelt, K., Grunwaldt, J.-D., & Dahmen, N. (2020). From Agriculture Residue to Upgraded Product: The Thermochemical Conversion of Sugarcane Bagasse for Fuel and Chemical Products. *Fuel Processing Technology*, *197*, 106199. doi:10.1016/j.fuproc.2019.106199

Scott, N. R., Chen, H., & Cui, H. (2018). Nanotechnology Applications and Implications of Agrochemicals toward Sustainable Agriculture and Food Systems. *Journal of Agricultural and Food Chemistry*, *66*(26), 6451–6456. doi:10.1021/acs.jafc.8b00964 PMID:29879842

Sharma, D., Afzal, S., & Singh, N. K. (2021). Nanopriming with phytosynthesized zinc oxide nanoparticles for promoting germination and starch metabolism in rice seeds. *Journal of Biotechnology*, *336*, 64–75. doi:10.1016/j.jbiotec.2021.06.014 PMID:34116127

Shcherbakova, E. N., Shcherbakov, A. V., Andronov, E. E., Gonchar, L. N., Kalenskaya, S. M., & Chebotar, V. K. (2017). Combined Pre-Seed Treatment with Microbial Inoculants and Mo Nanoparticles Changes Composition of Root Exudates and Rhizosphere Microbiome Structure of Chickpea (*Cicer arietinum* L.) Plants. *Symbiosis*, *73*(1), 57–69. doi:10.100713199-016-0472-1

Siddaiah, C. N., Prasanth, K. V. H., Satyanarayana, N. R., Mudili, V., Gupta, V. K., Kalagatur, N. K., Satyavati, T., Dai, X. F., Chen, J. Y., Mocan, A., Singh, B. P., & Srivastava, R. K. (2018). Chitosan nanoparticles having higher degree of acetylation induce resistance against pearl millet downy mildew through nitric oxide generation. *Scientific Reports*, *81*(8), 1–14. doi:10.103841598-017-19016-z PMID:29410438

Simonin, M., Colman, B. P., Tang, W., Judy, J. D., Anderson, S. M., Bergemann, C. M., Rocca, J. D., Unrine, J. M., Cassar, N., & Bernhardt, E. S. (2018). Plant and Microbial Responses to Repeated Cu (OH)2 Nanopesticide Exposures under Different Fertilization Levels in an Agro-Ecosystem. *Frontiers in Microbiology*, *9*, 9. doi:10.3389/fmicb.2018.01769 PMID:30108580

Singh, A., Rajput, V., Rawat, S., Singh, A. K., Bind, A., & Singh, A. (2020). Monitoring Soil Salinity and Recent Advances in Mechanism of Salinity Tolerance in Plants. *Biogeosystem Technique*, *7*(2), 66–87. doi:10.13187/bgt.2020.2.66

Singh, A., Rakesh Singh Sengar, R. S., Sharma, R., Rajput, P., & Singh, A. K. (2021a). Nano-Priming Technology for Sustainable Agriculture. *Biogeosystem Technique.*, *8*(2). doi:10.13187/bgt.2021.2.79

Singh, A., Sengar, R. S., Rajput, V. D., Minkina, T., & Singh, R. K. (2022). Zinc Oxide Nanoparticles Improve Salt Tolerance in Rice Seedlings by Improving Physiological and Biochemical Indices. *Agriculture*, *12*(7), 1014. doi:10.3390/agriculture12071014

Singh, A., Sengar, R. S., Shahi, U. P., Rajput, V. D., Minkina, T., & Ghazaryan, K. A. (2023). Prominent Effects of Zinc Oxide Nanoparticles on Roots of Rice (*Oryza sativa* L.) Grown under Salinity Stress. *Stresses*, *3*(1), 33–46. doi:10.3390tresses3010004

Singh, A. V., Chandrasekar, V., Janapareddy, P., Mathews, D. E., Laux, P., Luch, A., Yang, Y., Garcia-Canibano, B., Balakrishnan, S., Abinahed, J., Al Ansari, A., & Dakua, S. P. (2021). Emerging Application of Nanorobotics and Artificial Intelligence To Cross the BBB: Advances in Design, Controlled Maneuvering, and Targeting of the Barriers. *ACS Chemical Neuroscience*, *12*(11), 1835–1853. doi:10.1021/acschemneuro.1c00087 PMID:34008957

Spagnoletti, F. N., Spedalieri, C., Kronberg, F., & Giacometti, R. (2019). Extracellular Biosynthesis of Bactericidal Ag/AgCl Nanoparticles for Crop Protection Using the Fungus *Macrophomina Phaseolina. Journal of Environmental Management*, *231*, 457–466. doi:10.1016/j.jenvman.2018.10.081 PMID:30388644

Spielman-Sun, E., Avellan, A., Bland, G. D., Tappero, R. V., Acerbo, A. S., Unrine, J. M., Giraldo, J. P., & Lowry, G. V. (2019). Nanoparticle surface charge influences translocation and leaf distribution in vascular plants with contrasting anatomy. *Environmental Science. Nano*, *6*(8), 2508–2519. doi:10.1039/C9EN00626E

Sundaria, N., Singh, M., Upreti, P., Chauhan, R., Jaiswal, J. P., & Kumar, A. (2019). Seed Priming with Iron Oxide Nanoparticles Triggers Iron Acquisition and Biofortification in Wheat (Triticum aestivum L.) Grains. *Journal of Plant Growth Regulation*, *38*(1), 122–131. doi:10.100700344-018-9818-7

Tanabata, T., Shibaya, T., Hori, K., Ebana, K., & Yano, M. (2012). SmartGrain: High-throughput phenotyping software for measuring seed shape through image analysis. *Plant Physiology*, *160*(4), 1871–1880. doi:10.1104/pp.112.205120 PMID:23054566

Taran, N. Y., Gonchar, O. M., Lopatko, K. G., Batsmanova, L. M., Patyka, M. V., & Volkogon, M. V. (2014). The effect of colloidal solution of molybdenum nanoparticles on the microbial composition in rhizosphere of *Cicer arietinum* L. *Nanoscale Research Letters*, *9*(1), 289. doi:10.1186/1556-276X-9-289 PMID:25024677

United Nations. (2015) *Resolution adopted by the General Assembly on 25 September 2015, Transforming our world: the 2030 Agenda for Sustainable Development (A/RES/70/1)*. UN.

Vejan, P., Abdullah, R., Khadiran, T., Ismail, S., & Nasrulhaq Boyce, A. (2016). Role of Plant Growth Promoting Rhizobacteria in Agricultural Sustainability-A Review. *Molecules (Basel, Switzerland)*, *21*(5), 573. doi:10.3390/molecules21050573 PMID:27136521

Villagarcia, H., Dervishi, E., De Silva, K., Biris, A. S., & Khodakovskaya, M. V. (2012). Surface Chemistry of Carbon Nanotubes Impacts the Growth and Expression of Water Channel Protein in Tomato Plants. *Small*, *8*(15), 2328–2334. doi:10.1002mll.201102661 PMID:22514121

Wang, H., Kou, X., Pei, Z., Xiao, J. Q., Shan, X., & Xing, B. (2011). Physiological effects of magnetite (Fe3O4) nanoparticles on perennial ryegrass (*Lolium perenne* L.) and pumpkin (*Cucurbita mixta*) plants. *Nanotoxicology*, *5*(1), 30–42. doi:10.3109/17435390.2010.489206 PMID:21417686

World Population Clock. (n.d.). *World Population.* Worldometers.

Yasmeen, F., Raja, N. I., Razzaq, A., & Komatsu, S. (2017). Proteomic and Physiological Analyses of Wheat Seeds Exposed to Copper and Iron Nanoparticles. *Biochimica et Biophysica Acta. Proteins and Proteomics*, *1865*(1), 28–42. doi:10.1016/j.bbapap.2016.10.001 PMID:27717896

Ye, Y., Cota-Ruiz, K., Hernández-Viezcas, J. A., Valdés, C., Medina-Velo, I. A., Turley, R. S., Peralta-Videa, J. R., & Gardea-Torresdey, J. L. (2020). Manganese Nanoparticles Control Salinity-Modulated Molecular Responses in *Capsicum annuum* L. Through Priming: A Sustainable Approach for Agriculture. *ACS Sustainable Chemistry & Engineering*, *8*(3), 1427–1436. doi:10.1021/acssuschemeng.9b05615

Zahedifar, M., & Zohrabi, S. (2016). Germination and seedling characteristics of drought-stressed corn seed as influenced by seed priming with potassium nano-chelate and sulfate fertilizers. *Acta Agriculturae Slovenica*, *107*(1), 113–128. doi:10.14720/aas.2016.107.1.12

Zhai, Y., Hunting, E. R., Liu, G., Baas, E., Peijnenburg, W. J. G. M., & Vijver, M. G. (2019). Compositional Alterations in Soil Bacterial Communities Exposed to TiO2 Nanoparticles Are Not Reflected in Functional Impacts. *Environmental Research*, *178*, 108713. doi:10.1016/j.envres.2019.108713 PMID:31518961

Zhang, H., Huang, M., Zhang, W., Gardea-Torresdey, J. L., White, J. C., Ji, R., & Zhao, L. (2020). Silver Nanoparticles Alter Soil Microbial Community Compositions and Metabolite Profiles in Unplanted and Cucumber-Planted Soils. *Environmental Science & Technology*, *54*(6), 3334–3342. doi:10.1021/acs.est.9b07562 PMID:32088952

Zhou, J., Applegate, C., Alonso, A., Reynolds, D., Orford, S., Mackiewicz, M., Griffiths, S., Penfield, S., & Pullen, N. (2017). Leaf-GP: An open and automated software application for measuring growth phenotypes for arabidopsis and wheat. *Plant Methods*, *13*(1), 117. doi:10.118613007-017-0266-3 PMID:29299051

APPENDIX

List of Abbreviations

Ag: Silver
CNT: Carbon Nano Tubes
Cu: Copper
Fe: Iron
H_2O_2: Hydrogen Peroxide
nm: nanometer
NPs: Nanoparticles
NT: Nanotechnology
QD: Quantum Dots
ROS: Reactive Oxygen Species
SDGs: Sustainable Development Goals
Si: Silicon
SWCNT: Single Walled Carbon Nano tubes
UN: United Nations
WHO: World Health Organization
Zn: Zinc
ZnO: Zinc Oxide
UV: Ultra Violet radiations

Section 3

Role of Nanopriming for Crop Production and Protection

Chapter 4
Carbon Nano Tubes (CNTs) as a Tool of Seed Quality Enhancement Using Nanopriming Approach:
A Review

Manish Kumar Vijay
https://orcid.org/0000-0002-1002-8563
ICFRE-Tropical Forest Research Institute, Jabalpur, India

Radheshyam Sharma
Jawaharlal Nehru KrishiVishwaVidyalaya, India

ABSTRACT

Nano seed technology is a new platform with endless possibilities and impacts on applied agriculture and forestry research. The implementation of nanotechnology in seed research is in its infancy and needs more pace to meet the needs of current global demand. Carbon nanotubes (CNTs) are allotropes of carbon and engineered nanomaterials. Based on the number of concentric layers of rolled graphene sheets, they are classified into single-walled CNTs (SWCNTs) and multi-walled CNTs (MWCNTs). CNTs exhibit both beneficial and toxic effects on plant species by altering the morphological and physiological properties of plant cells. In most of the studies, low concentrations of CNTs are considered safe. Therefore, there is plenty of room to use these cutting-edge technologies to alleviate seed dormancy, boost vigor, and solves the problem of slow growth in seed propagated species. This chapter summarizes the results of various studies investigating the effects of carbon nanotubes on seed viability, as well as seedling growth and biomass.

DOI: 10.4018/978-1-6684-7232-3.ch004

INTRODUCTION

Seed is the most important and effective input for plant propagation. The quality of seed, which plays a critical role in the successful establishment of a plant population, determines the success of future plant development. Many factors influence the quality of a seed, including its physical, physiological, genetic, and health status. Good processing procedures can increase the physical quality of any seed. Similarly, good breeding procedures and proper rouging of undesired plants, as well as scientific seed generation chain maintenance, can improve the genetic and health quality of a seed. The planting value (germination and vigour) of a seed determines its physiological quality, which can be improved using various seed quality enhancement technologies (Villa et al., 2019). Seed germination is the most important step in initiating a physiological process that requires plant growth and water intake (Adhikari et al., 2013). This is determined, among other things, by the amount of water accessible during the seed coat penetration and germination cycle seed maturity, and surrounding growth conditions (Khodakovskaya et al., 2012). Seed vigour is a key metric for evaluating seedling quality, storability, emergence, and establishment under a variety of environmental and natural circumstances.

Seed dormancy is a situation in which seeds cannot germinate under otherwise ideal germination conditions.. There are three types of dormancy in seeds: a) strong seed coatings that are not only impervious to water and gases, but also restrict embryo expansion (physical dormancy). b) The second is embryo or internal dormancy, which occurs when seeds have small and underdeveloped embryos that do not germinate after ripening. c) Germination inhibitors, which prevent seed germination and plant growth, induce the third seed dormancy. These inhibitors must be laeched out (Kobayashi et al., 2010). Finding efficient techniques to relieve seed dormancy in numerous forest species and enhance their use and success in forest restoration is essential. Germination rates of rosemary seeds with hard seed coats, embryos and physiological seed dormancy are increased by low temperature stratification after maturation, incubation with gibberellic acid (hormone priming), seed coat scarring and inhibitor leaching. Acid scarification, which is used to break seed coat-induced dormancy, can harm seedling growth and is not environmentally friendly. Yucel (2000a) studied the effects of different NaCl, $KN0_3$, and H_2SO_4 concentrations on the germination rate of six Salvia species. The proportion and rate of H_2SO_4 on germination that is hazardous is reported. As a result, acids must be replaced with safer chemicals.

Pre-sowing, pre-storage, and mid-storage treatments are examples of seed enhancement technologies (SETs) that are widely employed in today's seed science and technology (Sharma et al., 2015). Scarification, stratification, seed pelleting, seed priming, seed coating, and seed protection are all used as pre-planting treatments to break dormancy, increase germination, and for precision sowing of seeds (Ahmed and Kumar 2020). Pre-storage and mid-storage treatments are used to improve or maintain the viability and vigour of seeds while they are being stored. Seed priming is a pre-sowing seed preparation that helps seeds to be hydrated and proceed through the first stage of germination while preventing radical protrusion through the seed coat (McDonald, 2000). It comes in a variety of forms, including hydropriming, osmopriming, halopriming, hormonal priming, and biopriming (See Figure 2). Despite the fact that these seed treatment methods have been used for a long time to improve seed growth, they have a number of drawbacks. Traditional seed priming systems are plagued by inconsistency, the need for repeated treatments, and a huge waste of priming material (Tsuji, 2001). To overcome the challenges given by existing approaches, a prospective alternative seed growth enhancement methodology is required.

Several recent studies have demonstrated effectiveness in overcoming embryo and seed coat dormancy and boosting germination in agriculture crop species utilising nano-priming employing nanoparticles as an unique and efficient seed priming and growth enhancer agent (Mahakham et al.,2017). Nanoparticles (also known as nano-scale particles or NPs) are molecules or atoms with a diameter of 1 to 100 nanometers. USEPA, 2005 divides NPs into four categories: (1) carbon-based materials (2) metal-based materials (3) dendrimers (nano-sized polymers made up of branching units) and (4) composites (nanoparticles combined with

Figure 1. Types of seed priming

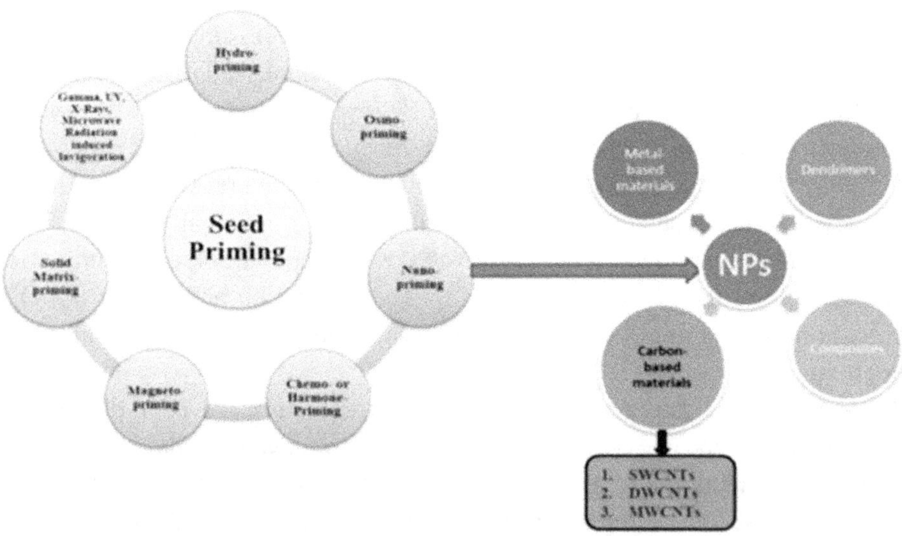

other nanoparticles or larger, bulk-type materials). The first two types are the most common and are frequently researched.

CARBON NANO TUBES (CNTS)

Carbon is a a crucial component of plant molecules including carbohydrates, proteins, and lipids. Plants use carbon in the form of carbon dioxide (CO_2) to create food and oxygen (O_2) through the photosynthesis process, which converts sunlight. Carbon's four valence electrons can participate in the production of compounds with other elements or with itself. The three allotropes of carbon, diamond, graphite, and fullerenes, have diverse physical properties due to differences in their molecular environments. CNTs were discovered and synthesised in 1991 (Iijima,1991), and the graphene family in 2004, both of which have a wide range of uses (Bergmann & Machado, 2015; Hong et al., 2015). Diamond is the hardest substance known, while graphite is a lubricant. Each layer of graphite is known as "graphene," and when rolled, it transforms into carbon nanotubes (CNT). Carbon nanotubes (CNT), carbon nanofibers, and fullerenes are three primary carbon nanostructures. The first two are linear structures, but the third has spherical dimensions. Chemical vapour deposition, laser ablation, and arc discharge techniques are all common synthetic ways for making CNTs. The nanostructure, the diameter and length of the tubes, and the arrangement of the graphene sheets (how the sheets are "rolled") all influence the properties of CNTs. Single-walled carbon nanotubes (SWCNT) are made up of only one graphene layer, whereas multi-walled carbon nanotubes (MWCNT) are made up of many graphene layers. Both the SWCNT and MWCNT have chemically inert and hydrophobic sidewalls because CNTs are made up of graphite basal planes. The presence of carboxylic groups on the walls of MWCNTs or SWCNTs boosts their biocompatibility capabilities, particularly in terms of adsorption capacity, electrostatic contacts, and covalent bonding, allowing them to be functionalized through the reflux process. CNTS can be functionalized with different hydrophilic groups, by using a surfactant (PEG), or by using different protocols for ultrasonication. Furthermore, variable durations or intensities of ultrasonication might affect the degree of dispersion of the CNT, bringing its component to the variability. Ultrasonication at too high a frequency is thought to harm CNTs (Vaisman et al., 2006).

RESPONSE OF PLANTS TO CNTS TREATMENTS

Plant response to nanoparticles appears to be a complex process that is influenced by three primary factors: plant species, growth stages, and nanomaterial nature

(Nair et al., 2010), all of which are interconnected. As a result, every variation in one of these characteristics is a source of discrepancies and variability among research (Zaytseva & Neumann, 2016). Several recent research on CNTs and cold plasma have found increased germination, seedling growth, and physiological activities, including photosynthetic activity, as well as an increase in the level of essential enzymes in metabolic pathways. These also cause beneficial changes in gene expression that are required for cell division and plant development, implying that they could be used to boost plant growth. Fullerol, grapheme, functionalized and non-functionalized MWCNTs, and SWCNTs were used to study the effects of CNTs on plants. Different types of culture media such as wet filter paper, agar with MS nutrients, and hydroponics with or without Hoagland's nutrients have been employed in studies (Tripathi et al., 2011). Plant response to CNTs is principally influenced by their size, concentration, and solubility.

Size: The diameter and size of a plant cell are shown to be the limiting criteria for CNT penetration (Serag et al., 2015). The pace of plant growth and fruit development is inversely proportional to the penetration of carbon nanotubes into the plant system. Plant cell walls operate as a barrier to prevent exogenous substances, such as nanoparticles, from entering the cells. The pore diameter, which ranged from 5 to 20 nm, determined the sieving capabilities of the cell wall. As a result, only nanoparticles or nanoparticle aggregates with a diameter smaller than the cell wall can pass through the cell wall and reach the plasma membrane. Plant detection and internal transfers have also been documented in MWCNTs, despite the fact that their width and length are generally bigger than fullerenes and SWCNTs. They are carried to areas where the pathway is wider than their size through capillary action. Taller MWCNT (more than 200 nm) is found in fewer cellular organelles, while shorter MWCNTs are found in vacuoles, nuclei, and plastids, according to HRTEM investigations (Serag et al., 2012). Small MWCNTs (13 nm) detected in the germination zone can penetrate cell walls and subsequently be found in wheat roots (*T. aestivum*) (Wang et al., 2012) and red spinach seedlings (*A. tricolour*). (Begum & Fugetsu, 2011). MWCNTs with a wavelength of 110–170 nm have been found to penetrate the epidermal cell wall and consequently up to 4 µm of the cytoplasm of wheat (T. aestivum) root hairs by Wild and Jones (Wild & Jones, 2009). Plant roots are used to make MWCNTs, which are then discovered in xylem and phloem cells (Zhai et al., 2015a). As revealed by wheat (*T. aestivum*) and rape seeds (*B. napus*), the origin of MWCNT shoot transfer is likely to be driven by respiration (Chen et al., 2015). MWCNTs are discovered in plant material and even flowers in soil-grown tomato plants (Khodakovskaya et al., 2013).

Concentration: Carbon-based nanoparticles, like other nanoparticles, have benefits and drawbacks in terms of seed germination and seedling vigour. Higher MWCNT concentrations were shown to be less effective, presumably due to increased

CNT clumping, which reduced effectiveness and caused toxicity (Tiwari et al., 2013). As a result, the concentration of carbon nanotubes in diverse types of plants must be optimised. Low doses of SWCNT, MWCNT, carbon nanotubes, fullerenes, and graphene have a favourable effect when used at low quantities. Low doses of SWCNT, MWCNT, carbon nanotubes, fullerenes, and graphene have a favourable effect when used at low quantities (Khodakovskaya et al., 2009; Mondal et al., 2011; Tripathi et al., 2011; Khodakovskaya et al., 2011; Larue et al., 2012;Villagarcia et al., 2012; Khodakovskaya et al., 2012; Haghighi & Silva, 2014; Lahiani et al., 2015; Mukherjee et al., 2016; Zaytseva & Neumann, 2016; Vithanage et al., 2017; Yousefi et al., 2017).

Charge: Nanomaterials' charge has an impact on their toxicity, bioavailability, environmental behaviour, and destiny. Many studies have demonstrated that MWCNTs with higher negative charges are more effective in increasing seed germination and fresh weight than MWCNTs with lower negative charges. In six crop species (cabbage, carrot, cucumber, lettuce, onion, and tomato), SWCNTs caused root elongation and showed charge and plant species dependence (Haghighi et al., 2014). SWCNTs (functionalized with poly-3 amino benzene sulfonic acid) with a negative charge hindered root elongation in lettuce, but had no effect on cabbage or carrots. (Villagarcia et al., 2012)

IMPACT OF CNTS ON SEED GERMINATION

The effects of CNT on seed coats could lead to the development of treatments for physical and mechanical seed dormancy. The nanomaterial can enter the plant through binding to carrier proteins, aquaporin, ion channels, endocytosis, sprouting new pores, or binding to an organic molecule in the surrounding environment (Rico et al., 2011). Because of the thick seed coat, nanoparticle penetration into seeds is expected to be difficult compared to plant cell walls and membranes. Despite this, carbon nanotubes penetrate seed coverings effectively and have an impact on seed germination and plant growth. Through perforation and the emergence of new pores, multiwall nanotubes can penetrate not just cells of growing seedlings, but also rigid seed coverings. After passing through the seed coat, nanotubes can make contact with the radicle and other seedling organs. Because aquaporins play such a crucial role in germination, it's been hypothesised that the beneficial effects of MWCNTs on seed water uptake and germination could be mediated by the identified aquaporin (Liu et al., 2009; Pourkhaloee et al., 2011; Tripathi et al., 2011; Serag et al., 2012; Villagarcia et al., 2012; Haghighi & Silva, 2014). This notion, however, is dubious because aquaporins are involved in a variety of physiological processes, including stress responses, which are also induced by CNTs, necessitating a more detailed investigation of the involved aquaporin genes. Penetrating MWCNTs in seed as well as the root system has a significant impact on their biological interest by increasing the amount of water present in the seed during germination. Seed priming with CNP could help break seed dormancy in forest species that are being considered for reclamation. It is a process in which seeds are soaked in CNP solutions to improve seed germination (percentage or pace), establish uniform seedlings, and overcome seed dormancy issues (Farahani et al., 2012; Mahakham et al., 2017). Canas et al.,(2008) found that functionalized CNTs were more effective in increasing seed germination rate in *Onobrychis arenaria* than non-functionalized CNTs. The germination of rice seeds treated with SWCNTs and MWCNTs was also improved (Stamphoulis et al.,2009). Because mature seeds are exceptionally dry and require a significant amount of water to commence cellular metabolism and growth, water intake is critical during seed germination. Carbon nanotubes can help increase water intake by increasing the number of new pores in the seed coat. CNTs may potentially facilitate water ingress into cells by influencing water routes in the seed coat and controlling their performance. CNTs should effectively penetrate seed coats, impacting seed germination and plant growth, according to Khodakovskaya et al.,(2009). Tomato seed germination and growth rate were improved when seeds were exposed to CNTs (40 μ g/ml in MS media). The Raman spectroscopy and transmission electron microscope both confirmed the existence of CNTs in the seeds. The CNTs form pores within the cellular wall, allowing for better water absorption and germination. Carbon nanoparticles boost the germination rate and seedling growth of soybeans(*Glycine max*), according to Liu et al., (2009). Single-walled carbon nanotubes can cross both the plant cellular wall and the cellular membrane, according to confocal fluorescence imaging studies. Nair et al.,(2010) investigated the effects of single-walled carbon nanotubes (SWCNTs) and multi-walled carbon nanotubes (MWCNTs) on rice seed germination and discovered that carbon nanotubes improved seed germination. The amount of water consumed increased in tomato seeds treated with CNTs, according to Srinivasan and Saraswathi (2010), which contributed to the accelerated seedling vigour. When MWCNTs were used at 2.3 mg/L, Mondal et al.,(2011) found that CNTs improved mustard (*Brassica juncea* L.) germination and growth. MWCNTs increased the moisture content of the material and improved the water absorption mechanism of root tissues, which resulted in increased germination. Seed germination stimulation has been reported for salvia (*Salvia macrosiphon*), pepper (*Capsicum annuum*), and tall fescue (*Festuca arundinacea*) in response to SWCNTs treatments (10–40 mg/L), most likely due to seed coat perforation. The greatest germination rates were obtained with mild SWCNT concentrations of 10 mg/L for pepper and 30 mg/L for salvia and tall fescue, respectively, among the tested remedies (Pourkhaloee et al., 2011). Yan et al.,(2013) found that a similar concentration of SWCNTs (20 mg/L) had no influence on the germination of maize (Z. mays) seeds. In wheat, maize, peanut, and garlic, it was reported that multi-wall carbon nanotubes (MWCNTs) were able to penetrate the seed coat by generating new holes, boosting water intake and significantly increasing seed germination, plant development, and biomass when compared to controls (Anita and Rao,2014). MWCNTs have been shown to enhance maize growth by promoting biomass accumulation and seed germination. Srivastava and Rao (2014) investigated the effect of 0–50 g/mL MWCNT on wheat, maize, peanut, and garlic seeds at various concentrations. Low doses of MWCNTs have been shown to improve water absorption and hasten the germination process, resulting in a shorter germination period. By increasing ion-CNT transient-dipole interactions, MWCNTs improve water imbibition and mineral nutrient delivery in maize, enhancing seed germination and seedling growth (Dasgupta et al.,2017; Tiwari et al.,2013; Zaytseva 2016; Zhai et al.,2015). Zhang et al.,(2015) found that the use of graphene oxide (GO) (40mg/L) increased tomato seed germination because the GO pierced the seed husks, allowing water to be captured more easily. Furthermore, the activity of the enzymes glutathione reductase, glutathione peroxidase, and glutathione sulfotransferase was raised, indicating that the content of reduced glutathione was increased. Tobacco, maize, soybean, switchgrass, tomato, rice, and barley can all benefit from SWCNT, according to Milewska-Hendel et al.,(2016). Seed priming using MWCNTs improved seed germination %, mean germination time, root and stem lengths, as well as the fresh and dry weights of the roots and stems of Hopbush (*Dodonaea viscosa* L.) under drought stress, according to Yousefi et al.,(2017). The effects of -OH functionalized MWCNTs on brinjal seeds have been reported (Mukherjee et al.,2018). MWCNTs improved seed germination and growth characteristics, although toxicity was discovered at concentrations higher than 50 mg/L. Low concentrations of graphene (25-50 mg/L) were shown to increase the dry matter of the root, stem, and leaves, as well as root length, surface area, volume, and average diameter in seedlings of forest species such as Larix oogenesis; however, concentrations greater than a hundred mg/L had the opposite effect (Song et al.,2020). Ali et al., (2020) discovered that MWCNTs functionalized with carboxylic acid (MWCNT-COOH) were superior in breaking both embryo and seed coat dormancy, resulting in increased seed germination in both bog birch and Labrador tea species. Apart from the positive effects of MWCNTs on germination, numerous reviews have claimed that MWCNTs have no effect on a wide range of plant species, including Wheat (T. aestivum) (Wang et al.,2012), radish (R. sativus), rape (B. napus), ryegrass (Loliumperenne), lettuce (L. sativa), maize (Z. mays), cucumber (C. sativus) (Lin & Xing, 2007), mustard (B. juncea), black lentil (P. mungo), and zucchini (C. pepo) (Pourkhaloee et al.,2011). At lower concentrations, multiwall carbon nanotubes (MWCNTs) improved seed germination and seedling growth of crops such as maize, peanut, garlic, and soybean, but at greater concentrations, they lowered early growth (Lahiani et al.,2013; Rao and Srivastava 2014; Tiwari et al.,2013; Zaytseva 2016). The ability of carbon nanotubes to penetrate the seed coat has been hypothesised as a possible mechanism by which nanoparticles help stimulate the dormant embryo, resulting in better seed germination. This discrepancy can be attributable to genotypic variations or variability in seed lot quality of the investigated seed material, but it can also be linked to the test settings.

IMPACT OF CNTS ON EARLY SEEDLING GROWTH

Canas et al.,(2008) investigated the effects of functionalized SWCNTs (fCNTs; functionalized with poly-3-aminobenzenesulfonic for high dispersibility) and non-functionalized SWCNTs (CNTs) on root elongation in six different crop species (cabbage (*Brassica oleracea*), carrot (*Daucus carota*), cucumber (*Cucumis sativus*), lettuce (*Lactuca sativa),* onion (*Allium cepa*) and tomato (*Solanum Lycopersicum*). CNTs increased root elongation in onion and cucumber, and both fCNTs and CNTs generated nanotube sheets on the cucumber root surface as a result of their contact with the root surface; nevertheless, none of them entered the roots. Both types of nanotubes had little effect on cabbage and carrots. Liu et al.,(2009) showed that single-walled carbon nanotubes may pass through plant cell walls and membranes, as well as intracellular organelles. CNTs were found to affect growth-related features in Triticumaestivum (Joshi, 2018), Zea mays (yan, 2013), Oryzasativa (yan, 2016), and broccoli plants (Brassica oleracea L. var. Italica) (Martinez-Ballesta et al.,2016). MWCNTs are delivered to the stem after being absorbed by the roots, improving growth by enhancing shoot and root elongation, biomass, and vigour indices of diverse crops' seeds (Joshi et al.,2018; Seddighinia et al.,2020).

IMPACT OF CNTS ON GENE EXPRESSION

On transcriptional responses to CNTs, there is scanty molecular evidence. Furthermore, CNTs have been linked to overexpression of genes involved in secondary metabolism in various studies. In this regard, previous research has linked CNT treatment to alterations in growth and development-related genes, such as the SLR1, RTCS, RTH1, and RTH3 genes in maize (Yan et al.,2013) and CycB, NtLRX1, and NtPIP1 genes in tomato (khodakovskaya et al.,2012). CNT was utilised on tobacco cells by Khodakovskaya et al.,(2012) who found that the expression of some genes, such as NtPIP1, which regulates the water channels in tobacco cells, increased. MWCNTs have been shown to have a concentration-dependent effect on the expression of aquaporin genes in sprouting soybean (*Glysine max*), barley (*H.*

vulgare), and maize (*Z. mays*) seeds (Lahiani et al.,2013). Furthermore, Hossain et al.,(2015) discovered that MWCNT can activate a stress-related gene in tomatoes, resulting in improved seed germination. MWCNTs activate the expression of antioxidant genes and increase the activities of antioxidants such as catalase (CAT) and peroxidase (POD) in plants (Fan et al.,2018; Joshi et al.,2018; Seddighinia et al.,2020). CNT treatments affected secondary metabolism by upregulating many genes, including DAT in *Catharanthus roseus* (Ghasempour, 2019), PAL, TAT, and RAS in *Salvia verticillat* and HPPR, RAS, PAL, and TAT genes in *Satureja khuzistanica* (Rahmani, 2020). CNT supplementation and seed priming with plasma synergistically increased the plant's early growth and performance, according to Seighinia et al.,(2021). The applied therapies induced differences in transcriptions of the phytoene-synthase gene (McPSY) and phytoenedesaturase gene, according to quantitative real-time PCR analysis (McPDS). The McPDSand McPSYgenes showed a similar expression pattern, with the CNT50+Plasma 60 group having the greatest expression levels. The CNT50+Plasma60 treatment was the most efficient strategy to boost -carotene concentrations, according to HPLC analysis. The zeaxanthin concentration was enhanced by the treatments, which varied in dose and treatment manner. CNT50+Plasma 60 and CNT100+Plasma 60 had considerably greater amounts of -carotene than the other treatment groups. These findings showed that, like other nanomaterials, carbon nanotubes may induce signalling, altering gene transcription and metabolism. More molecular research is needed, however, to complete information gaps and understand the pathways that may be implicated in plant responses to CNTs.

TOXICITY

The toxicity of carbon nanotubes is still a barrier to their widespread use. The type of plant investigated, the type of growth medium and ambient growing conditions, the type of CNT (pristine or functionalized), its concentration, and the ultrasonication and associated processes employed to create the aqueous dispersion of the MWCNT will all influence the effect of the CNT. In plant physiology, the first three factors are recognised to influence seedling growth. Some cells that come into direct contact with nanotubes operate as a barrier, protecting the other cells. The binding of MWCNTs to cell wall proteins and polysaccharides triggers a signalling cascade that boosts the production of specific chemicals required for cell wall thickening. Such hypersensitivity reactions are prevalent in plants when they are subjected to other biotic and abiotic stimuli that have a negative impact on their growth. The detachment of the plasma membrane from the cell wall of those cells that interacted with MWCNTs resulted in cell death, according to research.

MWCNTs caused the accumulation of reactive oxygen species (ROS), which resulted in increased oxidative stress and decreased cell growth, ultimately leading to cell death. MWCNTs were found to be hazardous to Arabidopsis T87 suspension cells, just as they were to rice suspension cells (Ten et al.,2009). The cells cultured in media containing loose agglomerates of MWCNTs (as received product without any further treatment) and fine-agglomerates of MWCNTs (obtained by ultra sonicating the loose agglomerates of MWCNTs) had lower dry weight, viability, chlorophyll content, and superoxide dismutase activity, indicating that MWCNTs have negative effects on Arabidopsis suspension cells. Cell development was suppressed more by fine agglomerates of MWCNTs than by loose MWCNTs. The fact that plant cell cultures have a higher tendency to produce clumps with a varied number of cells explains this. Fine agglomerates of nanotubes are easier to distribute in clumps than loose agglomerates of nanotubes of larger diameters because of their tiny size and irreversible interaction with cell wall proteins and polysaccharides. SWCNTs appear to be more poisonous than MWCNTs due to their smaller size, and toxicity is increased further by nanotube functionalization. Pristine SWCNTs increased root elongation in onions and cucumbers while inhibiting it in tomatoes, but had no effect on root elongation in cabbage and carrots. (Canas et al.,2008). In the tested range of MWCNTs, no detrimental effects on seed germination or root extension were seen in zucchini plants, however a decrease in plant biomass was observed during subsequent growth in the presence of SWCNTs (Stampoulis et al.,2009). Because CNTs have a high tensile strength, piercing impacts may cause mechanical injury to tissues (Samaj et al.,2004; Khodakovskaya et al.,2009; Liu et al.,2009). With 2,000 mg/l of MWCNT, maize and ryegrass germination rates declined, but their Root length increased (Lin and Xing 2007). Higher doses of MWCNTs (1000-2000 mg/L) were hazardous, resulting in a 38 percent loss in zucchini biomass and in lettuce root length (Begum and Fugetsu, 2012; Lin and Xing, 2007). MWCNTs inhibited the growth of rice cells (Oryza sativa L.) by increasing reactive oxygen species (ROS) levels, resulting in a decrease in cell viability and density. (Tan et al.,2009) MWCNT interactions with rice suspension cells resulted in cell aggregation, resulting in a self-defense response among the cells to protect themselves from the harmful effects of MWCNTs. It has been discovered that pristine CNT causes toxicity in plants, and that SWCNT causes toxicity in Arabidopsis and rice (Shen et al.,2010). This is because they encourage the creation of ROS by interacting with organelles such as mitochondria; also, ROS produce an imbalance in the concentration of antioxidant chemicals and the level of oxidative stress in plants, resulting in damage. Carbon nanotubes have been shown to be hazardous to a variety of creatures. MWCNT decreased the root fresh weights of rice and cucumber seedlings, while root lengths fluctuated in a non-orderly manner (rice) or declined (cucumber) (Begum et al.,2012).

CONCLUSION AND FUTURE PERSPECTIVES

Nanoparticles can effectively alleviate seed dormancy, speed up germination, vitality, physiology, and the early growing process of plants. Seed coat perforation by nanotubes and enhanced seed water intake appear to be linked to beneficial seed germination effects. Although this technology's promise for enhancing seed quality, more research is needed to widen the application and establish the toxicity level for the positive effect of nanoparticles, which will provide a clearer picture of the technology's potential. Many research have been published in the scientific literature that show how nanotechnology might help agricultural species with seed dormancy and low germination. To our knowledge, however, relatively few research have been conducted to evaluate the benefits of Nano priming, particularly when combined with stratification and hormone priming, in resolving seed dormancy in forest species. Furthermore, little is known about the type of dormancy present in some of these native species, as well as what approaches or seed treatments would be effective in breaking this dormancy. Detailed investigations on the impact of nanotubes on seed germination and seedling development are required to meet the goals of nano-seed technology. Despite these encouraging results, the precise mechanism of action of these nanomaterials is unknown, necessitating further investigation. Further research should look into the effects of nanoparticle type, concentration, size, and specific surface on different plant species, as well as the conditions of their application, such as seed and plant uptake.

REFERENCES

Adhikari, T., Kundu, S., & Rao, A. S. (2013). Impact of SiO2 and Mo nano particles on seed germination of rice (Oryza sativa L.). *Int J Agric Food Sci Technol*, *4*(8), 809–816.

Ahmed, S., & Kumar, S. (2020). Seed coating with fungicides and various treatments for protection of crops: A review. *Int J Agric Environ Sustain*, *2*(1), 6–13.

Ali, M. H., Sobze, J. M., Pham, T. H., Nadeem, M., Liu, C., Galagedara, L., Cheema, M., & Thomas, R. (2020). Carbon Nanotubes Improved the Germination and Vigor of Plant Species from Peatland Ecosystem Via Remodeling the Membrane Lipidome. *Nanomaterials (Basel, Switzerland)*, *10*(9), 1852. doi:10.3390/nano10091852 PMID:32947854

Anita, S., & Rao, D. P. (2014). Enhancement of seed germination and plant growth of wheat, maize, peanut and garlic using multiwalled carbon nanotubes. *European Chemical Bulletin*, *3*(5), 502–504.

Begum, P., & Fugetsu, B. (2012). Phytotoxicity of multi-walled carbon nanotubes on red spinach (Amaranthus tricolor L.) and the role of ascorbic acid as an antioxidant. *Journal of Hazardous Materials*, *243*, 212–222. doi:10.1016/j.jhazmat.2012.10.025 PMID:23146354

Begum, P., & Fugetsu, B. (2013). Induction of cell death by graphene in Arabidopsis thaliana (Columbia ecotype) T87 cell suspensions. *Journal of Hazardous Materials*, *260*, 1032–1041. doi:10.1016/j.jhazmat.2013.06.063 PMID:23892171

Begum, P., Ikhtiari, R., & Fugetsu, B. (2011). Graphene phytotoxicity in the seedling stage of cabbage, tomato, red spinach, and lettuce. *Carbon*, *49*(12), 3907–3919. doi:10.1016/j.carbon.2011.05.029

Bentsink, L., & Koornneef, M. (2008). Seed dormancy and germination. *The Arabidopsis Book / American Society of Plant Biologists*, *6*, e0119. doi:10.1199/tab.0119 PMID:22303244

Bergmann, C. P., & Machado, F. (2015). *Carbon Nanomaterials as Adsorbents for Environmental and Biological Applications*. Springer. doi:10.1007/978-3-319-18875-1

Canas, J. E., Long, M., Nations, S., Vadan, R., Dai, L., Luo, M., Ambikapathi, R., Lee, E. H., & Olszyk, D. (2008). Effects of functionalized and nonfunctionalized single-walled carbon nanotubes on root elongation of select crop species. *Environmental Toxicology and Chemistry*, *27*(9), 1922–1931. doi:10.1897/08-117.1 PMID:19086209

Chauhan, P. S., Singh, J., & Kavita, A. (2012). Chironji: A promising tree fruits subtropics. *HortFlora Research Spectrum*, *1*(3), 375–379.

Chen, G., Qiu, J., Liu, Y., Jiang, R., Cai, S., Liu, Y., Zhu, F., Zeng, F., Luan, T., & Ouyang, G. (2015). Carbon Nanotubes Act as Contaminant Carriers and Translocate within Plants. *Scientific Reports*, *5*(1), 1–9. doi:10.1038rep15682 PMID:26498499

Fan, Y., Clavel, G., & Pinna, N. (2018). Effect of passivating Al2O3 thin films on MnO2/carbon nanotube composite lithium-ion battery anodes. *Journal of Nanoparticle Research*, *20*(8), 216. doi:10.100711051-018-4315-2

Farahani, H. A., Sani, B., & Maroufi, K. (2012). The germination variations in fleawort (Plantago psyllium L.) by nano-particle. Research Gate.

Filho, J. M. (2015). Seed vigor testing: An overview of the past, present and future perspectives. *Scientia Agrícola*, *72*(4), 363–374. doi:10.1590/0103-9016-2015-0007

Ghasempour M, Iranbakhsh A, Ebadi M, & Oraghi Ardebili Z. (2019). Multi-walled carbon nanotubes improved growth, anatomy, physiology, secondary metabolism, and callus performance in *Catharanthus roseus*: an in vitro study. *Biotech.*. doi:10.1007/s13205-019-1934-y

Gs, G., Seo, Y., Park, D., & Lee, D. (2010). Phytotoxicity of Carbon Nanotubes Assessed by Brassica Juncea and Phaseolus Mungo. *Journal of Nanoelectronics and Optoelectronics*, *5*(2), 157–160. doi:10.1166/jno.2010.1084

Haghighi, M., & da Silva, T. (2014). The effect of N-TiO2 on tomato, onion, and radish seed germination. *Journal of Crop Science and Biotechnology*, *17*(4), 221–227. doi:10.100712892-014-0056-7

Haghighi, M., & Silva, J. A. T. (2014). The effect of carbon nanotubes on the seed germination and seedling growth of four vegetable species. *Journal of Crop Science and Biotechnology*, *17*(4), 201–208. doi:10.100712892-014-0057-6

Hong, G., Diao, S., Antaris, A. L., & Dai, H. (2015). Carbon nanomaterials for biological imaging and nanomedicinal therapy. *Chemical Reviews*, *115*(19), 10816–10906. doi:10.1021/acs.chemrev.5b00008 PMID:25997028

Hossain, Z., Mustafa, G., & Komatsu, S. (2015). Plant Responses to Nanoparticle Stress. *International Journal of Molecular Sciences*, *16*(11), 26644–26653. doi:10.3390/ijms161125980 PMID:26561803

Iijima, S. (1991). Helical microtubules of graphitic carbon. *Nature*, *354*(6348), 56–58. doi:10.1038/354056a0

Joshi, A., Kaur, S., Dharamvir, K., Nayyar, H., & Verma, G. (2018). Multi-walled carbon nanotubes applied through seed-priming influence early germination, root hair, growth and yield of bread wheat (*Triticumaestivum* L.) [PubMed] [Google Scholar]. *Journal of the Science of Food and Agriculture*, *98*(8), 3148–3160. PMID:29220088

Khodakovskaya, M., Dervishi, E., Mahmood, M., Xu, Y., Li, Z., Watanabe, F., & Biris, A. S. (2009). Carbon nanotubes are able to penetrate plant seed coat and dramatically affect seed germination and plant growth. *ACS Nano*, *3*(10), 3221–3227. doi:10.1021/nn900887m PMID:19772305

Khodakovskaya, M. V., De Silva, K., Biris, A. S., Dervishi, E., & Villagarcia, H. (2012). Carbon nanotubes induce growth enhancement of tobacco cells. *ACS Nano*, *6*(3), 2128–2135. doi:10.1021/nn204643g PMID:22360840

Khodakovskaya, M. V., De Silva, K., Nedosekin, D. A., Dervishi, E., Biris, A. S., Shashkov, E. V., Galanzha, E. I., & Zharov, V. P. (2011). Complex genetic, photothermal, and photoacoustic analysis of nanoparticle-plant interactions. *Proceedings of the National Academy of Sciences of the United States of America, 108*(3), 1028–1033. doi:10.1073/pnas.1008856108 PMID:21189303

Khodakovskaya, M. V., Kim, B.-S., Kim, J. N., Alimohammadi, M., Dervishi, E., Mustafa, T., & Cernigla, C. E. (2013). Carbon Nanotubes as Plant Growth Regulators: Effects on Tomato Growth, Reproductive System, and Soil Microbial Community. *Small, 9*(1), 115–123. doi:10.1002mll.201201225 PMID:23019062

Kobayashi, H., Yanaka, M., & Ikeda, T. M. (2015). Interaction of carbon nanohorns with plants: Uptake and biological effects. *Carbon, 81*, 607–619. doi:10.1016/j.carbon.2014.09.095

Lahiani, M. H., Dervishi, E., Chen, J., Nima, Z., Gaume, A., Biris, A. S., & Khodakovskaya, M. V. (2013, August 28). Impact of carbon nanotube exposure to seeds of valuable crops. *ACS Applied Materials & Interfaces, 5*(16), 7965–7973. doi:10.1021/am402052x PMID:23834323

Larue, C., Pinault, M., Czarny, B., Georgin, D., Jaillard, D., Bendiab, N., Mayne-L'Hermite, M., Taran, F., Dive, V., & Carrière, M. (2012). 'Quantitative evaluation of multi-walled carbon nanotube uptake in wheat and rapeseed', Journal of Hazardous Materials. *Journal of Hazardous Materials, 227–228*, 155–163. doi:10.1016/j.jhazmat.2012.05.033 PMID:22652322

Li, X., Baskin, J. M., & Baskin, C. C. (1999). Comparative morphology and physiology of fruit and seed development in the two shrubs Rhus aromatic and *R. glabra* (Anacardiaceae). *American Journal of Botany, 86*(9), 1217–1225. doi:10.2307/2656769 PMID:10487809

Lin, D., & Xing, B. (2007). Phytotoxicity of nanoparticles: Inhibition of seed germination and root growth. *Environmental Pollution, 150*(2), 243–250. doi:10.1016/j.envpol.2007.01.016 PMID:17374428

Liu, Q., Chen, B., Wang, Q., Shi, X., Xiao, Z., Lin, J., & Fang, X. (2009). Carbon nanotubes as molecular transporters for walled plant cells. *Nano Letters, 9*(3), 1007–1010. doi:10.1021/nl803083u PMID:19191500

Mahakham, W., Sarmah, A. K., Maensiri, S., & Theerakulpisut, P. (2017a). Nanopriming technology for enhancing germination and starch metabolism of aged rice seeds using phytosynthesized silver nanoparticles. *Scientific Reports, 7*(1), 1–21. doi:10.103841598-017-08669-5 PMID:28811584

Mahakham, W., Sarmah, A. K., Maensiri, S., & Theerakulpisut, P. (2017b). Nanopriming technology for enhancing germination and starch metabolism of aged rice seeds using phytosynthesized silver nanoparticles. *Scientific Reports*, *7*(1), 8263. doi:10.103841598-017-08669-5 PMID:28811584

Mahakham, W., Theerakulpisut, P., Maensiri, S., Phumying, S., & Sarmah, A. K. (2016). Environmentally benign synthesis of phytochemicals-capped gold nanoparticles as nanopriming agent for promoting maize seed germination. *The Science of the Total Environment*, *573*, 1089–1102. doi:10.1016/j.scitotenv.2016.08.120 PMID:27639594

Malik, S. K., Chaudhary, R., Panwar, N. S., & Dhariwal, O. P. (2012). Genetic resources of Chironji (*Buchanania lanzan* Spreng.): A socio-economically important tree species of central India tribal population. *Genetic Resources and Crop Evolution*, *59*(4), 615–623. doi:10.100710722-012-9801-2

Martínez-Ballesta, M. C., Zapata, L., Chalbi, N., & Carvajal, M. (2016). Multiwalled carbon nanotubes enter broccoli cells enhancing growth and water uptake of plants exposed to salinity. *Journal of Nanobiotechnology*, *14*(1), 42. doi:10.118612951-016-0199-4 PMID:27278384

McDonald, M. B. (2000). Seed Priming. In M. Black & J. D. Bewley (Eds.), *Seed Technology and its Biological Basis* (pp. 287–325). Shefield Academic Press.

Mishra, S. N., Kumar, D., Kumar, B., & Tiwari, S. (2021). Assessing impact of varying climatic conditions on distribution of *Buchanania Cochinchinensis* in Jharkhand using species distribution modeling approach. *Current Research in Environmental Sustainability*, *V*, 3. doi:10.1016/j.crsust.2021.100025

Mondal, A., Basu, R., Das, S., & Nandy, P. (2011). Beneficial role of carbon nanotubes on mustard plant growth: An agricultural prospect. *Journal ofNanoparticle Research*, *13*(10), 4519–4528. doi:10.100711051-011-0406-z

Mukherjee, A., Majumdar, S., Servin, A. D., Pagano, L., Dhankher, O. P., & White, J. C. (2016). Carbon nanomaterials in agriculture: A critical review. *Frontiers in Plant Science*, *7*, 172. doi:10.3389/fpls.2016.00172 PMID:26941751

Nair, R., Mohamed, M. S., Gao, W., Maekawa, T., Yoshida, Y., Ajayan, P. M., & Kumar, D. S. (2012). Effect of carbon nanomaterials on the germination and growth of rice plants. *Journal of Nanoscience and Nanotechnology*, *12*(3), 2212–2220. doi:10.1166/jnn.2012.5775 PMID:22755040

Nair, R., Varghese, S. H., Nair, B. G., Maekawa, T., Yoshida, Y., & Kumar, D. S. (2010). Nanoparticulate material delivery to plants. *Plant Science*, *179*(3), 154–163. doi:10.1016/j.plantsci.2010.04.012

Niratker, C., & Sailaja, D. (2018). Evaluation of genetic diversity of Chironji (*Buchnania Lanzan*) in Chhattisgarh. *International Journal of Research and Analytical Reviews*, *5*, 4.

Pourkhaloee, A., Haghighi, M., Saharkhiz, M., Jouzi, H., & Mahdi, M. (2011). Carbon Nanotubes Can Promote Seed Germination via Seed Coat Penetration. *Seed Technology*, *33*, 155–169.

Rahmani, N., Radjabian, T., & Soltani, B. M. (2020). Impacts of foliar exposure to multi-walled carbon nanotubes on physiological and molecular traits of Salvia verticillata L., as a medicinal plant. *Plant Physiology and Biochemistry*, *150*, 27–38. doi:10.1016/j.plaphy.2020.02.022 PMID:32109787

Rico, C. M., Majumdar, S., Duarte-Gardea, M., Peralta-Videa, J. R., & Gardea-Torresdey, J. L. (2011). Interaction of nanoparticles with edible plants and their possible implications in the food chain. *Journal of Agricultural and Food Chemistry*, *59*(8), 3485–3498. doi:10.1021/jf104517j PMID:21405020

Samaj, J., Baluska, F., Voigt, B., Schlicht, M., Volkmann, D., & Menzel, D. (2004). Endocytosis, actin cytoskeleton, and signaling. *Plant Physiology*, *135*(3), 1150–1161. doi:10.1104/pp.104.040683 PMID:15266049

Seddighinia, F., Iranbakhsh, A., & Ardebili, O. (2020). Seed Priming with Cold Plasma and Multi-walled Carbon Nanotubes Modified Growth, Tissue Differentiation, Anatomy, and Yield in Bitter Melon (Momordica charantia). *Journal of Plant Growth Regulation*, *39*(1), 87–98. Advance online publication. doi:10.100700344-019-09965-2

Seddighinia, F. S., Iranbakhsh, A., Oraghi Ardebili, Z., & Soleimanpour, S. (2021). Seed-priming with cold plasma and supplementation of nutrient solution with carbon nanotube enhanced carotenoid contents and the expression of psy and pds in Bitter melon (Momordica charantia). *Journal of Applied Botany and Food Quality*, *94*, 7–14. doi:10.5073/JABFQ.2021.094.002

Serag, M. F., Kaji, N., Tokeshi, M., & Baba, Y. (2012). Introducing carbon nanotubes into living walled plant cells through cellulase-induced nanoholes. *RSC Advances*, *2*(2), 398–400. doi:10.1039/C1RA00760B

Serag, M. F., Kaji, N., Tokeshi, M., & Baba, Y. (2015). In H. Manzer, H. Mohamed, & M. Firoz (Eds.), *Carbon nanotubes and modern nanoagriculture," in Nanotechnology and Plant Sciences* (pp. 183–201). Springer International Publishing.

Sharma, K., Singh, U. S., Sharma, P., Kumar, A., & Sharma, L. (2015). Seed treatments for sustainable agriculture-A review. *Journal of Applied and Natural Science*, *7*(1), 521–539. doi:10.31018/jans.v7i1.641

Sharma, K., Singh, U. S., Sharma, P., Kumar, A., & Sharma, L. (2015). Seed treatments for sustainable agriculture-A review. *Journal of Applied and Natural Science*, *7*(1), 521–539. doi:10.31018/jans.v7i1.641

Shen, C. X., Zhang, Q. F., Li, J., Bi, F. C., & Yao, N. (2010). Induction of programmed cell death in Arabidopsis and rice by single-wall carbon nanotubes. *American Journal of Botany*, *97*(10), 1602–1609. doi:10.3732/ajb.1000073 PMID:21616795

Shukla, S. K., & Solanki, K. R. (2000). Studies on seed germination, plant survival and growth of Chironjee (Buchanania lanzan Spreng.). *Journal of Tropical Forestry*, *16*, 44–49.

Shukla, S. K., Solanki, K. R., Kumar, R. V., & Dwivedi, R. P. (1999). Stud ies on air layer ing in Buchanania lanzan (Chironji) and Basia latifolia (Mahua). Annual Report, NRCAF, Jhansi, p. 47.

Srinivasan, C., & Saraswathi, R. (2010). Nano-agriculture—Carbon nanotubes enhance tomato seed germination and plant growth. *Current Science*, *99*, 274–275.

Srivastava, A., & Rao, D. P. (2014). Enhancement of seed germination and plant growth of wheat, maize, peanut and garlic using multiwalled carbon nanotubes. *Chem. Bull*, *3*, 502–504.

Stampoulis, D., Sinha, S. K., & White, J. C. (2009). Assay-dependent phytotoxicity of nanoparticles to plants. *Environmental Science & Technology*, *43*(24), 9473–9479. doi:10.1021/es901695c PMID:19924897

Tan, X., Lin, C., & Fugetsu, B. (2009). Bunshi Fugetsu, Studies on toxicity of multi-walled carbon nanotubes on suspension rice cells. *Carbon*, *47*(15), 3479–3487. doi:10.1016/j.carbon.2009.08.018

Tan, X. M., Lin, C., & Fugetsu, B. (2009). Studies on toxicity of multi-walled carbon nanotubes on suspension rice cells. *Carbon*, *47*(15), 3479–3487. doi:10.1016/j.carbon.2009.08.018

Tiwari, D. K., Dasgupta-Schubert, N., Villaseñor Cendejas, L. M., Villegas, J., Carreto Montoya, L., & Borjas García, S. E. (2013). Interfacing carbon nanotubes (CNT) with plants: Enhancement of growth, water and ionic nutrient uptake in maize (Zea mays) and implications for nanoagriculture. *Applied Nanoscience*, *4*(5), 577–591. doi:10.100713204-013-0236-7

Tripathi, S., Sonkar, S. K., & Sarkar, S. (2011). Growth stimulation of gram (Cicer arietinum) plant by water soluble carbon nanotubes. *Nanoscale, 3*(3), 1176–1181. doi:10.1039/c0nr00722f PMID:21253651

Tsuji, K. (2001). Microencapsulation of pesticides and their improved handling safety. *Journal of Microencapsulation, 18*(2), 137–147. doi:10.1080/02652040175006385б PMID:11253931

Vaisman, L. & Wagner, H.D. (2006). The role of surfactants in dispersion of carbon nanotubes. *Advances in Colloid and Interface Science, 128,* 37-46.

Villa, F., Silva, D. F. D., Rotili, M. C. C., Herzog, N. F. M., & Malavasi, M. D. M. (2019). Seed physiological quality and harvest point of dovyalis fruits. *Pesquisa Agropecuária Tropical, 49*, 49. doi:10.1590/1983-40632019v4954520

Villagarcia, H., Dervishi, E., de Silva, K., Biris, A. S., & Khodakovskaya, M. V. (2012). Surface chemistry of carbon nanotubes impacts the growth and expression of water channel protein in tomato plants. *Small, 8*(15), 2328–2334. doi:10.1002mll.201102661 PMID:22514121

Vithanage, M., Herath, I., Almaroai, Y. A., Rajapaksha, A. U., Huang, L., Sung, J. K., Lee, S. S., & Ok, Y. S. (2017). Effects of carbon nanotube and biochar on bioavailability of Pb, Cu and Sb in multi-metal contaminated soil. *Environmental Geochemistry and Health, 39*(6), 1409–1420. doi:10.100710653-017-9941-6 PMID:28332174

Wang, X., Han, H., Liu, X., Gu, X., Chen, K., & Lu, D. (2012). Multi-walled carbon nanotubes can enhance root elongation of wheat (Triticum aestivum) plants. *Journal of Nanoparticle Research, 14*(6), 841. doi:10.100711051-012-0841-5

Wild, E., & Jones, K. C. (2009). Novel method for the direct visualization of in vivo nanomaterials and chemical interactions in plants. *Environmental Science & Technology, 43*(14), 5290–5294. doi:10.1021/es900065h PMID:19708355

Yan, S., Zhao, L., Li, H., Zhang, Q., Tan, J., Huang, M., He, S., & Li, L. (2013, February 15). Single-walled carbon nanotubes selectively influence maize root tissue development accompanied by the change in the related gene expression. *Journal of Hazardous Materials, 246-247,* 110–118. doi:10.1016/j.jhazmat.2012.12.013 PMID:23291336

Yousefi, E., Sheidaei, A., Mahdavi, M., Baniassadi, M., Baghani, M., & Faraji, G. (2017). Azadeh Sheidaei, Mostafa Mahdavi, Majid Baniassadi, Mostafa Baghani, Ghader Faraji, Effect of nanofiller geometry on the energy absorption capability of coiled carbon nanotube composite material. *Composites Science and Technology*, *153*, 222–231. doi:10.1016/j.compscitech.2017.10.025

Yousefi, S., Kartoolinejad, D., & Naghdi, R. (2017). Effects of priming with multi-walled carbon nanotubes on seed physiological characteristics of hopbush (Dodonaeaviscosa L.) under drought stress. *The International Journal of Environmental Studies*, *74*(4), 528–539. doi:10.1080/00207233.2017.1325627

Yücel, E. (2000a). Effects of different salt (NaCl), nitrate (KNO3) and acid (H2S04) concentrations on the germination of some Salvia species seeds. *Seed Science and Technology*, *28*, 853–860.

Zaytseva, O., & Neumann, G. (2016). Carbon nanomaterials: Production, impact on plant development, agricultural and environmental applications. *Chemical and Biological Technologies in Agriculture*, *3*(1), 17. doi:10.118640538-016-0070-8

Zhai, G., Gutowski, S. M., Walters, K. S., Yan, B., & Schnoor, J. L. (2015a). Charge, Size, and Cellular Selectivity for Multiwall Carbon Nanotubes by Maize and Soybean. *Environmental Science & Technology*, *49*(12), 7380–7390. doi:10.1021/acs.est.5b01145 PMID:26010305

Chapter 5
Nanopriming for Crop Management for Sustainable Agriculture:
An Overview

Muhammad Majeed
University of Gujrat, Pakistan

Murad Muhammad
Chinese Academy of Sciences, China & University of Peshawar, Pakistan

Sehar Nawaz
University of Agriculture, Faisalabad, Pakistan

Tayyaba Naz
University of Agriculture, Faisalabad, Pakistan

Muhammad Mazhar Iqbal
Ayub Agricultural Research Institute, Department of Agriculture, Government of Punjab, Pakistan

Nafeesa Zahid
University of Kotli, Pakistan

Mumtaz Hussain
Arid Zone Research Institute, Pakistan

Allah Nawaz Khan
Chinese Academy of Sciences, China

Ghulam Abbas
University of Agriculture, Pakistan

Allah Bakhsh Gulshan
Ghazi University, Pakistan

Maria Mehboob
University of Poonch, Pakistan

ABSTRACT

The purpose of this study is to inform scholars in the aforementioned fields on the state of the art in seed nano-priming methods and the ongoing research that is shaping sustainable agricultural practices for the modern day. As things are, agriculture is hampered by a number of abiotic and variables that reduce yields. These stressors

DOI: 10.4018/978-1-6684-7232-3.ch005

on crops are made worse by a lack of freshwater supplies, changes in the climate, and the inefficiency of currently available agrochemicals. The agricultural sector has recently shown an interest in nanomaterials such carbon tubes, nanofiber, and nanoparticles. Coating the nanoparticle with the target substance before releasing it into the surrounding soil or plant increases absorption of the target substance relative to the non-engineered nanomaterial. In order to strike an evolving equilibrium between agricultural output and environmental sustainability, nanotechnology undoubtedly offers a fresh platform. Nano-priming may also be utilized for targeted biofortification of seeds, which can improve crop yields and food quality.

INTRODUCTION

One of the greatest difficulties of our time will be to provide food for the whole world's population in a way that is both safe and sustainable (Willett et al., 2019). With the global population expected to hit 8.5 billion in 2030, the UN estimates that 50 percent more food would be needed to fulfil the rising demand (Organization, 2020). It's an endless process of rising fertilizer use and pollution because of inefficient application (Teh et al., 2016). Grain quality and production are diminished by diseases spread by bacteria, fungus, pests, and viruses (Haggag et al., 2015). Plant development is stunted by a variety of abiotic stress situations caused by both natural and human-made factors (Hussain et al., 2019). To solve this agricultural problem, we need cutting-edge technology that is both effective and environmentally friendly(Majeed, Bhatti, & Amjad, 2021).

Population projections indicate that by 2050, the world's population will have increased to over 9.6 billion. So, in order to sustain the expanding population, we must see a 70-100% increase in agricultural output (Oweis, 2018). Currently, agriculture uses around 70% of the world's freshwater, but that number is expected to rise to 83% by 2050 as food production tries to keep up with rising demand (Rosegrant et al., 2009). Crop losses due to salinity and drought amount to billions of dollars per year. The usage of agrochemicals has become more important in today's farming practices.

The yearly application of 2.5 million tons of pesticides is ineffective due to loss to the air and runoff (Poudel et al., 2020). Heavy-metals pollution is a global issue that threatens ecosystems, agricultural output, and human health. About 70% most HMs and related amalgams that reach the human body are ingested, according to reports (Filipoiu et al., 2022). Additionally, the pH, salinity, and fertility of agricultural soil are all negatively affected by the persistent discharge of metallic waste material. Crop productivity and food security have been further hampered by the deterioration brought on by the world's changing climate (Adugna, 2016;

Ukwattage et al., 2013). These environmental limitations have a greater impact on plants during their early developmental phases, like seed germination with seedling development, and so result in slower development and lower yields overall (Veneklaas et al., 2012). Sustainable crop production and increased harvest yields need cutting-edge technology in today's agriculture(Majeed et al., 2022). The agricultural sector is vital to emerging economies and the provision of food for the world's rapidly expanding population of over 7.5 billion (Wright, 2014). Ninety percent of all food crops begin as seeds, making seed a crucial component of agricultural sustainability (Itelima et al., 2018). A strong agro seed will result in a crop that is more likely to germinate and grow into a strong plant. Seed germination, seedling growth, and final crop output may all be negatively impacted by a number of current environmental circumstances such as drought, salinity, heavy metal deposition in soil, climate change (Singh et al., 2022). Seed-borne illnesses or the destruction of seeds by bugs and other pests may also lower their quality (Ain et al., 2023; Pandey et al., 2018). This review summarizes that, the use of nanotechnology to prime crops for optimal growth and harvest is a relatively new idea in sustainable agriculture. Coating seeds with nanomaterials like nanoparticles and nanotubes is a step in the process. To guarantee the safe and responsible application of this technology in agriculture, extensive testing and regulation are necessary.

SUSTAINABLE AGRICULTURE

One of the most important concepts in the rapidly developing field of precision agriculture, which helps farmers maximize the value of their inputs like fertilizer and water. The creation of novel nanotools to increase crop tolerance to stress and enhance plant nutrient absorption is one example of how nanotechnology may one day benefit the agricultural and food sectors. Finally, the limitations of using synthesized nanomaterials in agriculture and the need of considering their biosafety have been examined in detail in order to promote long-term agricultural sustainability. Long-term environmental and food security depend on the adoption of sustainable agricultural techniques. Nano-priming shows promise as a method for accomplishing these objectives by boosting crop output with little chemical inputs. Therefore, it is crucial to keep looking for and creating secure and efficient uses of nano-priming in agriculture.

Agriculture supports the economy and people by providing food for a higher standard of living. Today's farmers (Majeed, Tariq, et al., 2022) must contend with a number of issues, including climate change's unpredictability(Jamil et al., 2022), soil contamination from environmental pollutants like fertilizers and pesticides, and soaring food demands as the world's population rises(Singh et al., 2021). The United

Nations has released a research predicting that by 2030 the world's population would reach 8.5 billion, and by 2050 it will reach about 9 billion (Samir & Lutz, 2017). Therefore, there is an immediate need to boost food production by more than 50% to fulfil the needs of the consistently increasing population (Willett et al., 2019).

Negative effects on crop production and output might result from aberrant dormancy of seeds, non-viability, and poor water absorption (Smith & Berjak, 2017). For successful germination, establishment of seedlings, and crop development, preserving seed quality is essential. Seeds may be protected from both abiotic and biotic pressures with the use of agrochemicals, avoiding problems and improving seed quality. Multiple agrochemicals, both alone and in combination, are used for seed treatments to combat and prevent a wide range of illnesses, pests, and nutritional deficiencies (Shelar et al., 2023). Fungicides, insecticides, fertilizers, and fertilizer boosters are all examples. However, these chemicals are expensive, harmful to human health, toxic releasing happens in soil, seeds pathogens have developed resistance, and chemicals can end up in water sources including rivers or the sea, where they cause eutrophication, decrease soil fertility, lessen helpful microbial activity, and change the soil's pH (Oliver & Gregory, 2015).

Pollution of the natural world and soil is a direct result of the overuse of traditional agrochemicals and the discharge of their wastes, which causes nutrient and nutritional imbalances in ecosystems (Islam & Yasmin, 2017). To keep seeds safe from insects and other pests while also preserving the agroecosystem, sustainable agriculture practices must be used because of problems with leaching, degradation, and hydrolysis, conventional agrochemicals are not recommended for use in sustainable agricultural seed treatment practices (Altieri et al., 2017). To alleviate environmental stress on soil, new solutions that utilize green chemistry principles that are both safe and cost-effective are required immediately.

Agro seeds have long been a source of contention, but new methods and approaches are continually being developed to combat these problems. Products based on nanomaterials are being offered to modernize agricultural methods (Dhewa, 2015). These materials are able to keep up with rising demand because of their high surface area to volume ratio and new physicochemical features.

NANOTECHNOLOGY

Nanotechnology in agriculture can only be used if trained workers are readily available in food-producing regions. In order to improve degraded soil health and guarantee balanced nutrition to plants and its microenvironment, this review discusses the latest developments on uses for nanotechnology in agriculture, including the roles of multiple nanomaterials like nano biochar, nano ceramics, nano-nutrients.

Green synthesis of nanomaterials, which is gentler on the environment, has also been covered. There is a growing need for rapid, non-destructive, and objective techniques of determining the quality of nano primed seeds. Similarly, this chapter discusses the use of nanoparticles in sustainable farming, particularly in the form of nano fertilizers, nano pesticides, nano herbicides, biostimulators, nanocarriers for nutrition delivery, nano biosensors, and nano-enabled remediation techniques. The discipline of nanoscience is expanding, particularly in the area of nanotechnology. Nanotechnology is a new revolution in science that has spread to every corner of our lives. The environmental impact of nanotechnology and nanomaterials has been greatly reduced. Due to increasing agricultural output and taking agronomic measures are crucial to world hunger. Scientists and farmers have been attempting to find conventional ways to boost crop yields for decades. However, progress in sustainable agriculture faces obstacles from a number of directions. Using knowledge from physics, chemistry, and biology, nanotechnology is able to modify and monitor substances at the nanoscale (Klaine et al., 2012). We should anticipate the globe to transform into a "NPs world" as a result of the rapid advancements in nanotechnology over the previous several decades. Due to their tiny size (100 nm) and great specific surface area, NPs exhibit distinctive optical, electrical, magnetic, chemical, and mechanical capabilities (Khan et al., 2019). Agricultural by products like soy hulls and wheat straw may be recycled into high-tech bio-nano-composites with enhanced mechanical and physical qualities that can be put to use in a variety of manufacturing processes. Gaining the real-world benefits of nanotechnology will need more development in areas like process design, risk evaluation of small-molecule pesticide and nano-fertilizers, and commercialization laws (Nasrollahzadeh et al., 2019). New crop types that can withstand heat, drought, and other stressors in order to communicate the whole gamut of current agricultural practices around the globe is another difficult problem that has yet to be addressed(T. Mangala & S. Bhirud, n.d.). To achieve the intended result of bolstering agriculture throughout the world, we think a holistic view of nanotechnology is necessary.

Nanotechnology in agriculture has the potential to produce both beneficial and negative changes, as is the case with every technological advancement (Kumar et al., 2019). Both the benefits of employing nano fertilizers and some of the problems that have been linked to them have been emphasized. The active chemicals used in the pest control industry are lipid-soluble. Because of this, they are not as bioavailable (Ragaei & Sabry, 2014). As a result of their enhanced bioavailability and smaller particle size, nano pesticides provide a viable solution to this issue. Pesticides may come in either a solid or a liquid form (Nuruzzaman et al., 2016). Both emulsifiable concentrate (EC) and wetter powder (WP) are examples of modern pesticide formulations. Nano fertilizers, nano pesticides, and nano biofertilizers are just a few examples of how nanotechnology is being put to use in the agricultural sector, and they are all included in this overview. Both the benefits and drawbacks of using nanoparticles in farming were discussed. Reforms at the policy making level are needed in all agricultural countries throughout the world to educate farmers on the proper use of new technologies (Quisumbing & Pandolfelli, 2010).

Nanotechnology focuses on substances between 1 and 100 nm in size (Sanna et al., 2014). In addition, the nanomaterial's reactivity might be affected by its surroundings since the area of its surface per mass has increased. When compared to materials on a broader scale, nanoparticles have a substantially higher surface area for their weight. The study of how nanotechnology may be used to treat seeds is an emerging field of study (Qureshi et al., 2018). Nanoparticle induced processes and seeds have not advanced very far at the molecular level, which is a necessary step in assessing putative mechanisms. It is vital for knowing how seeds react to nanoparticles in order to comprehend their underlying processes, responses, and modifications in gene expression using molecular techniques. The past 10 years have seen an explosion of research into the chemical interactions between nanomaterials and agro seed systems for a variety of applications, including disease control, yield increases, and reduced environmental impact (Manjunatha et al., 2016).

The fundamental unit of nanotechnology is the man-made nanoparticle, with at least one dimension less than 100 nm (Mu et al., 2014). Despite its little size, it has proven useful to the agro-technological uprising by monitoring an important step in crop management (Sinha & Dhanalakshmi, 2022). The several advantages it may have, such as improved resistance to stress and higher crop yields, enhanced food safety and following quality, better uptake of soil nutrients, reduced reliance on artificial fertilizers and pesticides, and so on, have interested scientists (Bhardwaj et al., 2014). Reduced chemical usage, reduced nutrient loss during fertilization, and increased crop tolerance to biotic and abiotic stress are all primary goals of nanotechnology in agriculture (Sajid et al., 2021).

Nanotechnology has the potential to efficiently increase agricultural output and product quality while decreasing the adverse ecological effects of fertilizers and pesticides in light of the many threats to agriculture posed by severe environmental stress, climate change, and food security (Shang et al., 2019). Moreover, nanotechnology may reduce abiotic pressures by enhancing the quality of both soil and agricultural products and by boosting biological processes during plant development(Majeed et al., 2023; Majeed, Lu, et al., 2022). Due to their unique characteristics, NPs are an essential tool for addressing issues in the agriculture sector (Manjunatha et al., 2019).

The use of nanotechnology in agriculture might help bring about the next technological revolution in food production (Elizabath et al., 2019). Agricultural uses of nano-materials include innovative approaches to water and soil rehabilitation, and also nano-fertilizers or nano-pesticides that aim to decrease fertilizer and pesticide use without compromising crop yield or quality (Pramanik et al., 2020). In addition, the usage of Nano enabled goods for crop protection is an ecofriendly solution since it may greatly lessen the damage generated by agriculture on the environment (Mukhopadhyay, 2014).

Nanotechnology has allowed for the continuing development of nano formulations for environmentally friendly farming that are more effective and contain fewer contaminants (H. Singh et al., 2021). It is easy to take advantage of the fact that these modern compounds may immediately affect plant physiology upon entering the intricate plant-soil system. Furthermore, crucial understanding of the positive or negative interactions of these NMs on plants is necessary for regulated delivery of active chemicals (Adisa et al., 2019). As a result, they may provide fresh opportunities for improving goods made from nanomaterials. Natural NM concentrations are reportedly significantly lower than those deemed dangerous, as well. However, there are still certain blanks that need to be filled up with thorough safety evaluations (Organization, 2019).

Today, nano-agrochemicals for treatment of seeds have the potential to become widely used since they are more effective than traditional agrochemicals (Gomez et al., 2021). The efficiency of farming inputs may be increased, and the creation of nanoscale agricultural chemicals for seed treatment can be made more sustainable, thanks to advances in nanotechnology (Shelar et al., 2023). There is evidence that seedlings treated with nanoparticles germinate better and produce more biomass. The seed's resilience to biotic and abiotic stressors has also been improved by the nanoparticles. Seeds rely on molecular processes to carry out their biological roles (Mildaziene et al., 2022).

NANO-PRIMING

Nanoparticles in the forms of microorganisms nano-pesticides, and nano-sensors are used to monitor crop output and nutrient levels, respectively, in order to boost productivity and protect plants from biotic and abiotic stressors (Jasrotia et al., 2018).

The seeds, nano-priming can protect them while they're stored, boost germination and synchronization rates, boost plant growth, and increase crop resistance to abiotic and biotic stress, all of which can lead to lower pesticide and fertilizer usage (Nile et al., 2022). Recent research has shown that seed nano-priming may activate several genes during germination, including those involved in the plant's response to stress (Nile et al., 2022). Although studies on the use of nanotechnology in seed priming are still in their early stages, they have showed encouraging outcomes(Liu et al., 2013). Many nanoparticles possess antimicrobial capabilities and may be loaded with antimicrobial compounds, hence seed nano-priming can also be employed for seed protection (do Espirito Santo Pereira et al., 2021).

Nano priming are an ongoing initiative to develop nano-agrochemicals for the timed release of targeted nutrients, which helps to preserve soil fertility (H. Singh et al., 2021). Since the quality of seeds is crucial to agricultural productivity, using high-quality seed identification technology increases the likelihood of identifying a wide range of improved crops. Machine learning technology for detection and categorization of nano primed seeds to boost crop yield is discussed, as is the possible use of tailored nanomaterials in seeds nano priming for sustainable agriculture (Xi et al., 2022).

As our knowledge of nanotechnology grows, we will find more and more uses for nanomaterials in the agricultural sector. Abiotic stress conditions, such as salt and drought, have been seen to restrict plant dispersal throughout most agricultural fields (Zhu, 2016). Revolutionary for ecosystems, agricultural development, the well-being of animals, and pesticide usage, nanotechnology allows for the release of priming chemicals at precise areas and in regulated ways (Kim et al., 2018).

Seed germination and fixation of nitrogen have both been improved by SiO_2 and TiO_2, and Glycine max growth has been stimulated. CNTs are utilized as a fertilizer in many different agricultural contexts, including the cultivation of vegetables (such as cucumbers, tomato, etc.) and cereal grains (such as maize and soybeans)(Singh et al., 2015). In agriculture, nanoparticles are often employed as a pesticide to combat biotic stress. Since they are used less often and in smaller doses than traditional pesticides, they help save resources like water and electricity.

Fusarium culmorum, *Botrytis cinerea*, *Biploaris sorokinniana*, and *Megnaporthe grisea* have all been shown to be susceptible to control by silver (Ag) NPs has reported to be effective against pest illnesses like powdery mildew, other nano pesticides are copper (Cu) NPs, silica NMs, and Si-Ag NPs (Bamal et al., 2021). As an added bonus, nano biosensors made from gold (Au) NPs, nanodots, magnetic NPs, and carbon nanostructures have been described. When it comes to analyzing biomolecules like DNA and proteins, nano biosensors are classified according to the specific biological receptors they use (Huang et al., 2021). To detect single or multiplex solutes, nano biosensors use nanoscale components that operate as bio-receptors and transmit signals to recognition elements.

NANO AGRICULTURE

Therefore, before being used in agriculture, nanomaterials should be tested for biocompatibility, environmental friendliness, and therapeutic suitability in the human body. Nanomaterial engineering's use of biocompatible materials and environmentally benign green-synthesis techniques provides the breadth of space needed for agriculture's realistic growth (Rai et al., 2022). Many uncommon problems, such as rare illnesses, abnormally high temperatures, abnormally high salinity, heavy metal pollutions, nutritional imbalances, and environmental pollution, threaten the viability of conventional agriculture today (Arora et al., 2018). The use of nanotechnology has led to the development of smart, environmentally friendly farming practices. Farm production may be raised with the aid of recent advances in wireless communication and sensor-based monitoring systems for diseases, pests, and irrigation systems (Rehman et al., 2022). Despite a wealth of information on several agricultural uses for nanomaterials, a comprehensive literature review on the subject that takes into account all of these fields are still lacking.

Formulations and delivery systems for nano-fertilizers that maximize their absorption and use in plants are crucial in contemporary agriculture. By investigating NPs based on different metals or metal oxides for use in agriculture, we can improve nutrient usage efficiency and the quality of the environment while decreasing nutrient losses owing to leaching (Rahman et al., 2021). Nano-fertilizers are able to boost nutrient uptake by roots because their decreased size, achieved via physical/chemical processes, increases their surface-mass ratio (Verma et al., 2022).

Nano fertilizers boost agricultural productivity by increasing crop quality and yield via increased nutrient absorption and decreasing production expenses. Nanomaterials (NMs) supplied to plants either endogenously (via the seed) or exogenously (through the stomata of the leaves) are taken up by the plant and the effectiveness of nano fertilizers exceeded that of traditional fertilizers by a factor of 18-29% (Lastochkina et al., 2022). The usage of phosphorus and calcium hydroxyapatite NPs had a similar effect, increasing Glycine max output by 20-33% compared to conventional phosphorus application methods (Dhir, 2021). Applying phosphorus, iron, and nitrogen to the southern pea plant increased leaf chlorophyll levels by 10% (Koohkan & Maftoun, 2016). Other NPs that have been shown to increase plant development and stress tolerance include zinc oxide (ZnO), cerium dioxide (CeO_2), silicone oxide (SiO_2), carbon nanotubes (CNTs), and titanium oxide (TiO_2) (Das & Das, 2019).

Recent years have seen a proliferation of studies that compile the numerous uses of GS-MNPs in biomedicine, antimicrobials, and the cleanup of heavy metal polluted soils. However, there aren't many comprehensive reviews on GS-MNP use in farming. The literature from the last decade is screened and analyzed for the very first time in this review. Then, the many variables, such as temperature, pH, reaction duration, and extraction solvent concentration that influence the final NPs' size and form are discussed. Furthermore, the benefits of GS-MNPs over conventional NPs are highlighted, including their lower toxicity as well as their antimicrobial and antioxidant qualities.

Next, the current standing of GS-MNPs in agricultural applications is presented. Finally, the benefits of GS-MNPs for plants are discussed in detail, including their ability to stimulate growth, inhibit the development of bacteria, and reduce the impact of heavy metal stress (Jiang et al., 2022). The impact of GS-MNPs on soils and plant health, as well as heavy metal nutrient transport, may be summed up in this review. However, the mechanism of interaction between GS-MNPs and plants is not fully understood, including the absorption and transport of GS-MNPs by plants. That's why there's so much room for growth and opportunity in using GS-MNPs in farming.

The agricultural industry is reaping the rewards of scientific and technological progress, which is providing us with new approaches and tools to address longstanding challenges. With the aid of these cutting-edge materials, modern agriculture is evolving into precision agriculture, which maximizes production while minimizing inputs (Singh et al., 2022). Nano-clays, nanotubes and nanowires are only a few examples of various NMs that have their own distinct surface chemical, electrical, and optical characteristics that allow them to be more sensitive, have lower detection limits, and react quickly.

Fertilizers raise crop yields but have a double-edged sword: they reduce soil fertility by upsetting the mineral balance of the soil. Spraying pesticides, fertilizers, and antibiotics is common practice; these substances are easily washed away (Mie et al., 2017).

The high manufacturing costs of insecticides and fertilizers also need regulation. In order to maximize output while minimizing costs, farmers are increasingly turning to NMs to cut down on wasteful practices such fertilizer loss and pesticide use (Fernández-Luqueño et al., 2018). NMs provide a more effective means to enhance agricultural and dietary practices. We stress, however, that before releasing goods based on NM to the public, it is critical to determine their level of toxicity. Smart agronomic delivery systems that use NMs for a carrier of active substances are only one example of the kinds of nanotechnology-based solutions that are constantly being created.

To feed the world's projected 9 billion inhabitants in 2050, global food production would need to grow by 50 percent. Future generations need technology solutions which can provide food security in the face of increasing demand for food and rapidly changing climatic circumstances across the planet (Shiferaw et al., 2011). Nanomaterials, and nanoparticles in particular, offer several potential uses in crop protection. Businesses in the agricultural industry have taken an interest in this significant area of study, leading to the use of nanoparticles into agricultural formulations (Kim et al., 2018).

NANO PESTICIDES

Over fifty per cent of all people on Earth use plants and plant products for subsistence, making it imperative that plant production and health be prioritized. Pathogens, insufficient nutrients, and soil and air pollution are just a few examples of biotic and abiotic variables that may negatively impact plant health and ultimately result in lower crop and fruit yields (Roberts & Mattoo, 2018). Many different kinds of pests and pathogens may cause illness in plants. As a result, stopping them from preventing more plant withering is an urgent need (Gamalero & Glick, 2020). Insects are a major problem in farming because they destroy crops and consume all of the stored food, lowering food quality and spreading plant illnesses. If you go by the official American definition, that is. According to the EPA, pesticides are any compounds used for the purpose of preventing, repelling, or destroying pests (Saravi & Shokrzadeh, 2011).

Nanotechnology has far-reaching implications beyond pest control in agriculture, including the reduction of food waste, the tracking of plant development, the assurance of higher-quality food, and the safe increase of worldwide harvests (Lockie et al., 2020). Nano herbicides, nano insecticides, nano nematicides, and nano fungicides are the nano pesticides tested most often in clinical studies (Mittal et al., 2020).

Industrialization rapidly reduces natural resources like seas, forests, and biological variety that people rely on for survival (Shimada, 2015). Therefore, it argues for a rise in agricultural output and safer food supplies. Strengthening agricultural resistance against environmental stress that might lead to crop failure, ecological biodiversity plays a crucial role in preserving the delicate balance between the natural world and food production (Altieri et al., 2015). Animals, plants, trees, and fish are all part of the agricultural and food biodiversity (Majeed, Bhatti, Pieroni, et al., 2021) that keeps people alive. A variety of plants and animals found in the oceans and woods provide substantial support for human needs such as food, feed, fuel, and fiber.

Nano-pesticides and nano-fertilizers have the potential to improve insect management, plant nutrition, and green manufacturing practices. Pests causing losses in crop yields, climate change's dire consequences, and dwindling natural resources are just a few of the problems that farmers have to contend with today (Altieri et al., 2015). Another issue is that conventional farming methods have long relied on the constant use of pesticides and fertilizers, which has led to widespread pollution.

NANO HERBICIDES

Weeds are often regarded as the most significant crop-damaging hazard because they divert valuable resources away from desirable plants. Using traditional techniques, such as pulling weeds by hand, to get rid of weeds is a laborious and time-consuming process (Woyessa, 2022). There are a wide variety of herbicides on the market today, each with the ability to eradicate weeds but also to damage crops and reduce soil fertility. When it comes to weed management that doesn't leave behind hazardous residues in the soil, nano herbicides may be a more sustainable option (Parra-Arroyo et al., 2022). If the same herbicide is used repeatedly over a longer length of time, the weeds may develop resistance. The conventional herbicide atrazine has been loaded into poly (epsilon-caprolactone) (PCL) nano capsules. When tested against a commercial atrazine formulation, the atrazine-containing nano capsules demonstrated superior post-emergence herbicidal action against mustard plants (Preisler et al., 2022). Through disulfide bond cross-linking, chitosan NPs were synthesized for its regulated administration dependent on glutathione concentration.

NANO FUNGICIDES

More than 70 percent of the damage to important crops including rice, wheat, barley, peanuts, and cotton is caused by fungal infections (Afsah-Hejri et al., 2013). Crop yields and economic output are negatively impacted as a result of these illnesses. Traditional fungicides may help reduce these losses, but they have the unintended side effect of killing off non-target species, which can throw the ecosystem out of whack. Thus, we need to investigate other methods for more accurate control of fungal diseases. The creation of NPs is a promising option for combating fungal infections (Yang et al., 2021). The antibacterial characteristics of AgNPs have led to their widespread use as a disinfectant.

NANO FERTILIZERS

There are several advantages to using nanoparticles rather than large particles or ionized salts because of their smaller size and the potential for a distinct absorption kinetics. Since nano-fertilizers have shown an improvement in productivity by guaranteeing targeted transport/gradual release of nutrients and decreasing fertilizer application with an increase in NUE, their use may enhance nutrient delivery efficiency in plant. Fertilizers, insecticides, enzymes, and hormones all fall under the category of "desired material" here. Increased plant growth and nutrient absorption are two reasons why is intriguing. This review takes into account the benefits of nanotechnology for agriculture and food security by focusing on the several fields of study that make use of nanotechnology. The massive manufacturing of traditional fertilizers to improve agricultural output and crop protection has been motivated by the unchecked population growth, yet this has a negative impact on soil fertility and food quality (Sekhon, 2014). Unused chemical fertilizers are hazardous to human health and contribute to problems in the delicate ecology.

The United Nations' Food and Agriculture Organization predicts that fertilizer consumption will increase worldwide in the years ahead (Niggli et al., 2009). Therefore, it is crucial to use smart solutions, such as nanotechnology, for sustainable agriculture. Nano fertilizers are smart fertilizers that may enhance fertilizer application rates while decreasing nutrient loss, especially of phosphorus and nitrogen (Dimkpa et al., 2020). It provides a method for the slow and controlled dispersion of nutrients at the desired location, which lessens the likelihood of environmental and water pollution.

Nano fertilizers are ones that can be given in lower doses yet still supply plants with the nutrients they need. Encapsulating fertilizers in nanoparticles increases nutrient uptake, decreases nutrient waste, boosts crop quality and production, and lessens environmental damage. It has also been shown that these nano fertilizers may reduce plant stress when applied to the foliage (Vejan et al., 2021). Depending on what plants need to grow, you may choose from three types of nano fertilizers: (1) those high in macronutrients, (2) those high in micronutrients, and (3) those high in nanoparticles. Potassium (K), magnesium (Mg), nitrogen (N), calcium (Ca), and phosphorus (P) are only few of the components that make up macronutrient nano fertilizers (Sajyan et al., 2020). By 2050, it is expected that 263 Mt would have been used worldwide, demonstrating the urgent demand for macronutrient fertilizers in the agriculture industry (Nagargade et al., 2022).

Micronutrients are essential for plant and animal development and growth, and they include elements like zinc (Zn), chloride (Cl), manganese (Mn), copper (Cu), selenium (Se), silicate (Si), and iron (Fe) (Kirkby, 2023). It serves a tangential purpose in human sustenance. Many forms of malnutrition may result from a lack of necessary micronutrients. China, Australia, Africa, the United States, Iran, Brazil, Pakistan, and Iran were all named as places where people were lacking in certain micronutrients. In order to increase agricultural yields, chemical fertilizers are often applied to soil (depleted of minerals due to pollution). Increased food crop output made possible by the use of synthetic fertilizers allows the nation to diversify its exports. However, the use of chemical fertilizers has negative effects on the soil and leads to water seepage, which has knock-on effects for the aquatic ecology and the people who eat the food directly (Dubey & Thakur, 2020; Smith et al., 2013).

The soil's long-term fertility and water-holding ability are negatively impacted. Biofertilizers may be the ideal alternative to chemical fertilizers since they are less expensive, safer for the environment, and don't disrupt the soil's microbial community (Iqbal et al., 2021). The physiological and metabolic processes of plants, as well as the stimulation of photosynthesis, rely heavily on zinc. Deficiency occurs when intake falls below the recommended range of 5–100 mg/kg. Zinc occurs in nature as a calcareous, crumbly substance at high pH in the soil; selenium (Se) is a necessary micronutrient for human health; its deficiency causes oxidative stress; Se is found in foods like cabbage, cauliflower, broccoli, among others, beans, peas, pork, beef, and fish (Shukla et al., 2018).

Inorganic Se may also Be found in trace amounts in soil. Silicate, like phosphate and potassium, is an insoluble compound that is essential to plant growth and the absorption of carbon dioxide (Soetan et al., 2010). Luckily, there are microbes and other biological aids that can solubilize the insoluble Zn and Si, restoring the plant's health. It's a green movement for the food staple that benefits the environment. Zinc and silicates like phosphates and potash are mobilized by a community of microorganisms (Kumawat et al., 2017). Through the synthesis of organic acids like citrus acid, oxalic acid, fumaric acid, and many others, microorganisms are able to solubilize insoluble material.

Cerium oxide is a type of semiconductor that has numerous potential uses in the scientific community (Chahal et al., 2020). Because of their surface charge, cerium oxide nanocomposite transports in a very particular way. In a research aimed at determining the relative importance of charges, it was shown that positively charged nano-sized cerium oxide was trapped in the epidermal the roots, but cerium oxide with neutral or negative charges made it all the way to the shoot system (Dahle & Arai, 2015). Soil heavy metals including cadmium were extracted using nanosized Cerium oxide. Cerium oxide particles have a major impact on the environment, helping to improve conditions and introducing a more environmentally friendly scientific perspective (Zhang et al., 2017). Cerium oxide particles, however, have been shown in experiments to impede photosynthesis and stunt plant development (Abbas et al., 2020).

Agriculture not only supplies food for improved worldwide living standards, but is also the economic foundation of emerging nations (Hamilton et al., 2014). Crop failure may result from biotic and abiotic pressures introduced by climate change to ecosystems, which disrupt the delicate environmental balance necessary for food production. Our worldwide need for food is expected to increase by that much. 70% by the year 2050 as a consequence of population growth (van Dijk et al., 2021). Inefficient farming operations based on traditional fertilization procedures are a potential threat to the environment, which is why it is crucial to approve cutting edge Agri-technologies if we are striving for global food security with increased plant yield (Weeks Jr & Hettiarachchi, 2019). It seems that the NUE is a useful indicator of both agricultural output and plant nutrient bioavailability. Fertilizers provide nutrients to boost crop yields, but their widespread usage by broadcasting on the soil's surface or mixing with irrigation poses risks to the environment and water supplies (Lu et al., 2015). Since the use of fertilizers can lead to eutrophication and water from the ground pollution, which may have detrimental effects on public health, hydroponic methods are used to grow some crops in greenhouse conditions, despite the cost of about 10 times higher water and energy consumption (Weidner et al., 2021).

To guarantee that the world's population has enough to eat while also protecting biodiversity, ecosystems, and the climate, innovative agricultural solutions are needed to reduce fertilizer consumption by 20% by 2030 (Alexoaei et al., 2022) 85 percent of the world's food supply during the next decade. Increasing acceptance of nanotechnologies in agriculture suggests that they will be used in the future to increase crop yields while maintaining a sustainable agroecosystem, even in the face of environmental challenges (Kah et al., 2019).

The FAO estimates that by 2050 the global population would have reached 10 billion, with an accompanying rise in demand for food of at least 50 percent, most notably in low-income countries (Alexandratos & Bruinsma, 2012). About 815 million people are malnourished right now, and another two billion people are expected to join them in the future (Van der Elst & Williams, 2019). Growing populations, extreme weather, pollution, and increased demand for clean water and power all pose serious challenges to the world's ability to produce and distribute food. The global food production system has to be rethought in light of the current crisis. There is an excessive amount of waste in the current agriculture system (Pulighe & Lupia, 2020).

More than two quadrillion British thermal units (BTU) of energy, 2.7 trillion cubic meters of liquid matter (70% of the globe's total freshwater consumption), and 187 million tons of fertilizers are needed to produce the annual worldwide agricultural production of more than 3 billion tons (Crane et al., 2010). Therefore, current agricultural farming methods are insufficient to reach the SDGs. Since ending global hunger is one of the United Nations' 17 sustainable development objectives, sustainable agriculture represents a crucial industry across the globe. Crop production, quality, and yield may all be improved with the use of a wide variety of agrochemicals, including but not limited to fungicides, insecticides, herbicides as well rodenticides, chemical fertilizers, growth regulators for plants, etc (Lal et al., 2021). Agrochemicals also reduce the amount of damage done to crops and help them fight off diseases.

NANO BIOFERTILIZERS

Nanoparticles (NPs) have had a profound impact on medicine and have many potential uses in other industries, such as agriculture. The agricultural industry of India is expected to increase its food grain output to 291.95 million tons in 2019, as stated by the economic times (S. Singh et al., 2021). The end of world hunger and extreme poverty is one of the UN's sustainable development objectives. NPs refer to particles with sizes between 1 and 100 nm. Nanomaterials, or engineered nanomaterials (ENMs), are used in both crop enhancement products and pest control (Verma et al., 2018). Nanoparticles made of inorganic materials, including copper (CuNP), aluminum (AlNP), silver (AgNP), zinc oxide (ZnO), silica (SiNP), cadmium (Ce2O3), and titanium (TiO2) (Kaushal et al., 2020). There are benefits to using nanoparticles as a fertilizer source for plants. For instance, silica's porous nature enhances the soil's aeration and moisture-holding capability, and the iron particles it encapsulates increases its ability to separate and reuse macro and micronutrients via its magnetic characteristics (Abideen et al., 2022).

Polymers, lipids, and carbon nanotubes are the organic nanomaterials put to use in farming. Biodegradable polymers may help advance sustainability in a number of ways. Plants need both macro and micronutrients for optimal growth and development (Jakhar et al., 2022). Chemical fertilizers provide both macro- and micronutrients, both of which are crucial to plant development. As a result of land dispersion and the encouragement of home agriculture, they are being employed more often in agricultural settings like China. Nitrogen (N), phosphorus (P), and potassium (K) are the major nutrients supplied to plants by chemical fertilizers (Van Wesenbeeck et al., 2021). Zinc, iron, copper, and molybdenum are the primary micronutrients supplied by fertilizers. When it comes to conventional fertilizers and insecticides, nanobiotechnology provides answers for both problems.

Biofertilizers were developed as a solution to the problems caused by chemical fertilizers. Chemical fertilizers have been shown to have a number of drawbacks, including releasing out into bodies of water and ground water, harm to soil fertility if applied in large amounts, high transportation costs, and a vicious cycle of repeated applications because plants cannot use them all at once (Savci, 2012). Because of this nutritional mismatch, the NUE of chemical fertilizers is relatively low. Precision fertilization, divided or localized use, fertigation, and the application of nano fertilizers and nano biofertilizers are only few of the strategies suggested to boost NUE. Forty to seventy percent of applied nitrogen, eighty to ninety percent of applied phosphorus, and fifty to ninety percent of applied potassium are lost or fixed in the soil as a result of using chemical fertilizers (Favaretto et al., 2006). The ecological harm produced by eutrophication must also be considered. According to studies, 50% of all nitrogen fertilizers end up in the water and air after being sprayed to crops (Sahle & Potting, 2013). Greenhouse gases like N-oxide that are emitted into the atmosphere contribute to global warming.

Overuse of chemical fertilizers has a negative impact on soil fertility and other physicochemical qualities (Pahalvi et al., 2021). One of the most sustainable sources of fertilizers, biofertilizers have been saving the day for farmers for a number of years now. Biofertilizers, such as bacteria and fungus, are categorized according to the nutrients they provide; yet, they are not without their drawbacks. To maintain a healthy soil environment, biofertilizers enhance the soil's microbiological condition and aeration, and boost the soil's capacity to store moisture and nutrients (nitrogen and phosphorus) for plant uptake (Bisht & Chauhan, 2020). Nitrogen-fixing biofertilizers, phosphorus-solubilizing biofertilizers, phosphorus-mobilizing biofertilizers, zinc-solubilizing biofertilizers, potassium-solubilizing biofertilizers, silicon-solubilizing biofertilizers, and composite biofertilizers are just a few examples (Bisht & Chauhan, 2020).

Examples of both types of nano fertilizers those containing macronutrients and those containing micronutrients—are included in the review. In addition, the term "nano biofertilizer" is used to describe an application that combines a biofertilizer with a nano fertilizer. The term "nano biofertilizer" is sometimes used interchangeably with "microorganisms encased in NP," despite the fact that NPs are far smaller in size than microorganisms. Therefore, biofertilizers from ferment/broth may be collected in multi-walled carbon nanotubes and then applied to soils. Nano fertilizers are useful in agriculture because of their unique features. Because of their small size and increased surface area, they release nutrients slowly and steadily, wasting fewer of them in the process (Al-Juthery et al., 2021). Furthermore, their high surface tension assures greater adsorption and long-term absorption, and formulations like nanoencapsulation protect them from heat, UV, and oxidative damage. Nano fertilizers may be used in two ways: on the leaves or in the soil before planting (Pavoni et al., 2020).

Nano biofertilizers are useful in the age of climate change because they increase the nutrition of plants and stress tolerance. Small-scale fertilizer applications boost agricultural yields while cutting fertilizer prices and minimizing logistics expenses (Phelps et al., 2013).

Nanotechnology enabled agrochemicals, on the other hand, may be precisely targeted to specific areas of a plant, increasing their efficacy and use in the management of pests and diseases (Prasad et al., 2014). As a result, they provide a novel strategy for mitigating the negative impacts of agrochemicals on public health by lessening their residual effect and on the environment by lessening their vaporization, leaching, and drainage (Tudi et al., 2021). A substitute for pesticides and fungicides that is both effective and environmentally benign might be developed using nanotechnology-enabled agrochemicals. Depending on the particle's chemical characteristics, size, shape, charge at the surface, and dosage, metal nanoparticles may have either beneficial or detrimental effects on plants (Jamkhande et al., 2019).

CONCLUSION

Nutrient deficiency, stagnant crop yields, reducing organic matter in the soil, low water accessibility, lack of nutrients in the soil, reduced land area due to increasing urbanization and degradation of land, and labor shortages are just some of the problems that farmers around the world are currently facing. The use of nanotechnology and nanoscience is expanding rapidly, and new approaches are being presented all the time to create useful and unique materials for agriculture. Recent developments in the use of engineered nanomaterials (ENMs) for agricultural purposes were analyzed; specifically, the increased efficacy of "nano-fertilizers" and "nano-pesticides" made possible by advances in nanotechnology were highlighted. Plants can quickly absorb nano fertilizers and quickly deplete them, so they don't harm the environment. Increases in seedling development, germination, nitrogen metabolism, carbohydrate and protein synthesis, and overall crop output have resulted from the use of nano fertilizers. There are benefits to using various nano agricultural approaches. However, only 0.1% of applied agrochemicals really make it to the intended plant portions, where they may effectively combat the pathogen in question; the remainder of the chemical ultimately ends up in the non-specific desired regions of plants and surroundings. It's also possible that these nanoparticles interacted with vegetation, soil, and air. Because of their static nature, they might be released into the environment, posing a threat to public health that cannot be disregarded. The deterioration, volatilization, photolysis, and leaching caused by agrochemicals pose a significant risk to all forms of life.

REFERENCES

Abbas, Q., Liu, G., Yousaf, B., Ali, M. U., Ullah, H., Munir, M. A. M., Ahmed, R., & Rehman, A. (2020). Biochar-assisted transformation of engineered-cerium oxide nanoparticles: Effect on wheat growth, photosynthetic traits and cerium accumulation. *Ecotoxicology and Environmental Safety*, *187*, 109845. doi:10.1016/j.ecoenv.2019.109845 PMID:31654865

Abideen, Z., Waqif, H., Munir, N., El-Keblawy, A., Hasnain, M., Radicetti, E., Mancinelli, R., Nielsen, B. L., & Haider, G. (2022). Algal-mediated nanoparticles, phycochar, and biofertilizers for mitigating abiotic stresses in plants: A review. *Agronomy (Basel)*, *12*(8), 1788. doi:10.3390/agronomy12081788

Adisa, I. O., Pullagurala, V. L. R., Peralta-Videa, J. R., Dimkpa, C. O., Elmer, W. H., Gardea-Torresdey, J. L., & White, J. C. (2019). Recent advances in nano-enabled fertilizers and pesticides: A critical review of mechanisms of action. *Environmental Science. Nano*, *6*(7), 2002–2030. doi:10.1039/C9EN00265K

Adugna, G. (2016). A review on impact of compost on soil properties, water use and crop productivity. *Academic Research Journal of Agricultural Science and Research*, *4*(3), 93–104.

Afsah-Hejri, L., Jinap, S., Hajeb, P., Radu, S., & Shakibazadeh, S. (2013). A review on mycotoxins in food and feed: Malaysia case study. *Comprehensive Reviews in Food Science and Food Safety*, *12*(6), 629–651. doi:10.1111/1541-4337.12029 PMID:33412719

Ain, Q. T., Siddique, K., Bawazeer, S., Ali, I., Mazhar, M., Rasool, R., Mubeen, B., Ullah, F., Unar, A., & Jafar, T. H. (2023). Adaptive mechanisms in quinoa for coping in stressful environments: An update. *PeerJ*, *11*, e14832. doi:10.7717/peerj.14832 PMID:36883058

Al-Juthery, H. W., Lahmod, N. R., & Al-Taee, R. A. (2021). Intelligent, nano-fertilizers: a new technology for improvement nutrient use efficiency (article review). *IOP Conference Series: Earth and Environmental Science*. IOP Science.

Alexandratos, N., & Bruinsma, J. (2012). *World agriculture towards 2030/2050: the 2012 revision*. FAO.

Alexoaei, A. P., Robu, R. G., Cojanu, V., Miron, D., & Holobiuc, A.-M. (2022). Good Practices in Reforming the Common Agricultural Policy to Support the European Green Deal–A Perspective on the Consumption of Pesticides and Fertilizers. *Amfiteatru Economic*, *24*(60), 525–545. doi:10.24818/EA/2022/60/525

Altieri, M. A., Nicholls, C. I., Henao, A., & Lana, M. A. (2015). Agroecology and the design of climate change-resilient farming systems. *Agronomy for Sustainable Development*, *35*(3), 869–890. doi:10.100713593-015-0285-2

Altieri, M. A., Nicholls, C. I., & Montalba, R. (2017). Technological approaches to sustainable agriculture at a crossroads: An agroecological perspective. *Sustainability (Basel)*, *9*(3), 349. doi:10.3390u9030349

Arora, N. K., Fatima, T., Mishra, I., Verma, M., Mishra, J., & Mishra, V. (2018). Environmental sustainability: Challenges and viable solutions. *Environmental Sustainability*, *1*(4), 309–340. doi:10.100742398-018-00038-w

Bamal, D., Singh, A., Chaudhary, G., Kumar, M., Singh, M., Rani, N., Mundlia, P., & Sehrawat, A. R. (2021). Silver nanoparticles biosynthesis, characterization, antimicrobial activities, applications, cytotoxicity and safety issues: An updated review. *Nanomaterials (Basel, Switzerland)*, *11*(8), 2086. doi:10.3390/nano11082086 PMID:34443916

Bhardwaj, D., Ansari, M. W., Sahoo, R. K., & Tuteja, N. (2014). Biofertilizers function as key player in sustainable agriculture by improving soil fertility, plant tolerance and crop productivity. *Microbial Cell Factories*, *13*(1), 1–10. doi:10.1186/1475-2859-13-66 PMID:24885352

Bisht, N., & Chauhan, P. S. (2020). Excessive and disproportionate use of chemicals cause soil contamination and nutritional stress. *Soil Contamination-Threats and Sustainable Solutions*, 1-10.

Chahal, S., Singh, S., Kumar, A., & Kumar, P. (2020). Oxygen-deficient lanthanum doped cerium oxide nanoparticles for potential applications in spintronics and photocatalysis. *Vacuum*, *177*, 109395. doi:10.1016/j.vacuum.2020.109395

Crane, H., Kinderman, E., & Malhotra, R. (2010). *A cubic mile of oil: realities and options for averting the looming global energy crisis*. Oxford University Press. doi:10.1093/oso/9780195325546.001.0001

Dahle, J. T., & Arai, Y. (2015). Environmental geochemistry of cerium: Applications and toxicology of cerium oxide nanoparticles. *International Journal of Environmental Research and Public Health*, *12*(2), 1253–1278. doi:10.3390/ijerph120201253 PMID:25625406

Das, A., & Das, B. (2019). Nanotechnology a potential tool to mitigate abiotic stress in crop plants. *Abiotic and biotic stress in plants*. InTech Open.

Dhewa, T. (2015). Nanotechnology applications in agriculture: an update. *Octa Journal of Environmental Research, 3*(2).

Dhir, B. (2021). Nanofertilizers and their Applications. *New frontiers of nanomaterials in environmental science*, 229-241.

Dimkpa, C. O., Fugice, J., Singh, U., & Lewis, T. D. (2020). Development of fertilizers for enhanced nitrogen use efficiency–Trends and perspectives. *The Science of the Total Environment*, *731*, 139113. doi:10.1016/j.scitotenv.2020.139113 PMID:32438083

do Espirito Santo Pereira, A., Caixeta Oliveira, H., Fernandes Fraceto, L., & Santaella, C. (2021). Nanotechnology potential in seed priming for sustainable agriculture. *Nanomaterials (Basel, Switzerland)*, *11*(2), 267. doi:10.3390/nano11020267 PMID:33498531

Dubey, J., & Thakur, M. (2020). Pollution: Pesticides in Natural Ecosystems. In Managing Global Resources and Universal Processes (pp. 111-118). CRC Press.

Elizabath, A., Babychan, M., Mathew, A. M., & Syriac, G. M. (2019). Application of nanotechnology in agriculture. *Int. J. Pure Appl. Biosci*, *7*(2), 131–139. doi:10.18782/2320-7051.6493

Favaretto, N., Norton, L., Joern, B., & Brouder, S. (2006). Gypsum amendment and exchangeable calcium and magnesium affecting phosphorus and nitrogen in runoff. *Soil Science Society of America Journal*, *70*(5), 1788–1796. doi:10.2136ssaj2005.0228

Fernández-Luqueño, F., Medina-Pérez, G., López-Valdez, F., Gutiérrez-Ramírez, R., Campos-Montiel, R. G., Vázquez-Núñez, E., Loera-Serna, S., Almaraz-Buendía, I., Del Razo-Rodríguez, O. E., & Madariaga-Navarrete, A. (2018). Use of agronanobiotechnology in the agro-food industry to preserve environmental health and improve the welfare of farmers. *Agricultural Nanobiotechnology: Modern Agriculture for a Sustainable Future*, 3-16.

Filipoiu, D. C., Bungau, S. G., Endres, L., Negru, P. A., Bungau, A. F., Pasca, B., Radu, A.-F., Tarce, A. G., Bogdan, M. A., Behl, T., Nechifor, A. C., Hassan, S. S., & Tit, D. M. (2022). Characterization of the Toxicological Impact of Heavy Metals on Human Health in Conjunction with Modern Analytical Methods. *Toxics*, *10*(12), 716. doi:10.3390/toxics10120716 PMID:36548549

Gamalero, E., & Glick, B. R. (2020). The use of plant growth-promoting bacteria to prevent nematode damage to plants. *Biology (Basel)*, *9*(11), 381. doi:10.3390/biology9110381 PMID:33171782

Gomez, A., Narayan, M., Zhao, L., Jia, X., Bernal, R. A., Lopez-Moreno, M. L., & Peralta-Videa, J. R. (2021). Effects of nano-enabled agricultural strategies on food quality: Current knowledge and future research needs. *Journal of Hazardous Materials*, *401*, 123385. doi:10.1016/j.jhazmat.2020.123385 PMID:32763688

Haggag, W. M., Abouziena, H., Abd-El-Kreem, F., & El Habbasha, S. (2015). Agriculture biotechnology for management of multiple biotic and abiotic environmental stress in crops. *Journal of Chemical and Pharmaceutical Research*, *7*(10), 882–889.

Hamilton, A. J., Burry, K., Mok, H.-F., Barker, S. F., Grove, J. R., & Williamson, V. G. (2014). Give peas a chance? Urban agriculture in developing countries. A review. *Agronomy for Sustainable Development*, *34*(1), 45–73. doi:10.100713593-013-0155-8

Huang, X., Zhu, Y., & Kianfar, E. (2021). Nano biosensors: Properties, applications and electrochemical techniques. *Journal of Materials Research and Technology*, *12*, 1649–1672. doi:10.1016/j.jmrt.2021.03.048

Hussain, S., Shaukat, M., Ashraf, M., Zhu, C., Jin, Q., & Zhang, J. (2019). Salinity stress in arid and semi-arid climates: Effects and management in field crops. *Climate change and agriculture, 13*.

Iqbal, S., Riaz, U., Murtaza, G., Jamil, M., Ahmed, M., Hussain, A., & Abbas, Z. (2021). Chemical fertilizers, formulation, and their influence on soil health. *Microbiota and Biofertilizers: A Sustainable Continuum for Plant and Soil Health*, 1-15.

Islam, M., & Yasmin, R. (2017). Impact of aquaculture and contemporary environmental issues in Bangladesh. *International Journal of Fisheries and Aquatic Studies*, *5*, 100–107.

Itelima, J., Bang, W., Onyimba, I., Sila, M., & Egbere, O. (2018). Bio-fertilizers as key player in enhancing soil fertility and crop productivity. *RE:view*.

Jakhar, A. M., Aziz, I., Kaleri, A. R., Hasnain, M., Haider, G., Ma, J., & Abideen, Z. (2022). Nano-fertilizers: A sustainable technology for improving crop nutrition and food security. *NanoImpact*, *27*, 100411. doi:10.1016/j.impact.2022.100411 PMID:35803478

Jamil, M. D., Waheed, M., Akhtar, S., Bangash, N., Chaudhari, S. K., Majeed, M., Hussain, M., Ali, K., & Jones, D. A. (2022). Invasive Plants Diversity, Ecological Status, and Distribution Pattern in Relation to Edaphic Factors in Different Habitat Types of District Mandi Bahauddin, Punjab, Pakistan. *Sustainability (Basel)*, *14*(20), 13312. doi:10.3390u142013312

Jamkhande, P. G., Ghule, N. W., Bamer, A. H., & Kalaskar, M. G. (2019). Metal nanoparticles synthesis: An overview on methods of preparation, advantages and disadvantages, and applications. *Journal of Drug Delivery Science and Technology*, *53*, 101174. doi:10.1016/j.jddst.2019.101174

Jasrotia, P., Kashyap, P. L., Bhardwaj, A. K., Kumar, S., & Singh, G. (2018). Scope and applications of nanotechnology for wheat production: A review of recent advances. *Wheat Barley Res*, *10*(1), 1–14.

Jiang, Y., Zhou, P., Zhang, P., Adeel, M., Shakoor, N., Li, Y., Li, M., Guo, M., Zhao, W., Lou, B., Wang, L., Lynch, I., & Rui, Y. (2022). Green synthesis of metal-based nanoparticles for sustainable agriculture. *Environmental Pollution*, *309*, 119755. doi:10.1016/j.envpol.2022.119755 PMID:35839973

Kah, M., Tufenkji, N., & White, J. C. (2019). Nano-enabled strategies to enhance crop nutrition and protection. *Nature Nanotechnology*, *14*(6), 532–540. doi:10.103841565-019-0439-5 PMID:31168071

Kaushal, S., Nanda, S. S., Samal, S., & Yi, D. K. (2020). Strategies for the development of metallic-nanoparticle-based label-free biosensors and their biomedical applications. *ChemBioChem*, *21*(5), 576–600. doi:10.1002/cbic.201900566 PMID:31634410

Khan, I., Saeed, K., & Khan, I. (2019). Nanoparticles: Properties, applications and toxicities. *Arabian Journal of Chemistry*, *12*(7), 908–931. doi:10.1016/j.arabjc.2017.05.011

Kim, D. Y., Kadam, A., Shinde, S., Saratale, R. G., Patra, J., & Ghodake, G. (2018). Recent developments in nanotechnology transforming the agricultural sector: A transition replete with opportunities. *Journal of the Science of Food and Agriculture*, *98*(3), 849–864. doi:10.1002/jsfa.8749 PMID:29065236

Kirkby, E. A. (2023). Introduction, definition, and classification of nutrients. In *Marschner's Mineral Nutrition of Plants* (pp. 3–9). Elsevier. doi:10.1016/B978-0-12-819773-8.00016-2

Klaine, S. J., Koelmans, A. A., Horne, N., Carley, S., Handy, R. D., Kapustka, L., Nowack, B., & von der Kammer, F. (2012). *Paradigms to assess the environmental impact of manufactured nanomaterials* (Vol. 31). Citeseer.

Koohkan, H., & Maftoun, M. (2016). Effect of nitrogen–boron interaction on plant growth and tissue nutrient concentration of canola (Brassica napus L.). *Journal of Plant Nutrition*, *39*(7), 922–931. doi:10.1080/01904167.2016.1143492

Kumar, A., Gupta, K., Dixit, S., Mishra, K., & Srivastava, S. (2019). A review on positive and negative impacts of nanotechnology in agriculture. *International Journal of Environmental Science and Technology*, *16*(4), 2175–2184. doi:10.100713762-018-2119-7

Kumawat, N., Kumar, R., Kumar, S., & Meena, V. S. (2017). *Nutrient solubilizing microbes (NSMs): its role in sustainable crop production. Agriculturally Important Microbes for Sustainable Agriculture* (Vol. 2). Applications in Crop Production and Protection.

Lal, R., Bouma, J., Brevik, E., Dawson, L., Field, D. J., Glaser, B., Hatano, R., Hartemink, A. E., Kosaki, T., Lascelles, B., Monger, C., Muggler, C., Ndzana, G. M., Norra, S., Pan, X., Paradelo, R., Reyes-Sánchez, L. B., Sandén, T., Singh, B. R., & Zhang, J. (2021). Soils and sustainable development goals of the United Nations: An International Union of Soil Sciences perspective. *Geoderma Regional*, *25*, e00398. doi:10.1016/j.geodrs.2021.e00398

Lastochkina, O., Aliniaeifard, S., SeifiKalhor, M., Bosacchi, M., Maslennikova, D., & Lubyanova, A. (2022). Novel Approaches for Sustainable Horticultural Crop Production: Advances and Prospects. *Horticulturae*, *8*(10), 910. doi:10.3390/horticulturae8100910

Liu, C., Jian, Z., & Meng, X. (2013). Combining Float Car Data and Multispectral Satellite Images to Extract Road Features and Networks. In *Lecture Notes in Geoinformation and Cartography* (pp. 29–43). Springer Berlin Heidelberg. doi:10.1007/978-3-642-34203-5_

Lockie, S., Fairley-Grenot, K., Ankeny, R., Botterill, L., Howlett, B., Mcbratney, A., Probyn, E., Sorrell, T., Sukkarieh, S., & Woodhead, I. (2020). *The future of agricultural technologies.* Australian Council of Learned Academies (ACOLA).

Lu, Y., Song, S., Wang, R., Liu, Z., Meng, J., Sweetman, A. J., Jenkins, A., Ferrier, R. C., Li, H., Luo, W., & Wang, T. (2015). Impacts of soil and water pollution on food safety and health risks in China. *Environment International*, *77*, 5–15. doi:10.1016/j.envint.2014.12.010 PMID:25603422

Majeed, M., Bhatti, K. H., & Amjad, M. S. (2021). Impact of climatic variations on the flowering phenology of plant species in jhelum district, punjab, pakistan. *Applied Ecology and Environmental Research*, *19*(5), 3343–3376. doi:10.15666/aeer/1905_33433376

Majeed, M., Bhatti, K. H., Pieroni, A., Sõukand, R., Bussmann, R. W., Khan, A. M., Chaudhari, S. K., Aziz, M. A., & Amjad, M. S. (2021). Gathered wild food plants among diverse religious groups in Jhelum District, Punjab, Pakistan. *Foods*, *10*(3), 1–28. doi:10.3390/foods10030594 PMID:33799901

Majeed, M., Lu, L., Anwar, M. M., Tariq, A., Qin, S., El-Hefnawy, M. E., El-Sharnouby, M., Li, Q., & Alasmari, A. (2023). Prediction of flash flood susceptibility using integrating analytic hierarchy process (AHP) and frequency ratio (FR) algorithms. *Frontiers in Environmental Science*, *10*, 1037547. doi:10.3389/fenvs.2022.1037547

Majeed, M., Lu, L., Haq, S. M., Waheed, M., Sahito, H. A., Fatima, S., Aziz, R., Bussmann, R. W., Tariq, A., & Ullah, I. (2022). *Spatiotemporal Distribution Patterns of Climbers along an Abiotic Gradient in Jhelum District.* doi:10.3390/f13081244

Majeed, M., Mahmood, A., Habib, T., Mushahid, M., Ali, H., Khan, N., & Ali, K. (2022). Vegetation analysis and environmental indicators of an arid tropical forest ecosystem of Pakistan. *Ecological Indicators*, *142*(May), 109291. doi:10.1016/j.ecolind.2022.109291

Majeed, M., Tariq, A., Haq, S. M., Waheed, M., Anwar, M. M., Li, Q., Aslam, M., Abbasi, S., Mousa, B. G., & Jamil, A. (2022). A Detailed Ecological Exploration of the Distribution Patterns of Wild Poaceae from the Jhelum District (Punjab), Pakistan. *Sustainability (Basel), 14*(7), 3786. doi:10.3390u14073786

Mangala, T. & Bhirud, S. (n.d.). *Road Network Extraction from High Resolution Satellite Images based on LSE LBF Model*. Academia.

Manjunatha, R., Naik, D., & Usharani, K. (2019). Nanotechnology application in agriculture: A review. *Journal of Pharmacognosy and Phytochemistry, 8*(3), 1073–1083.

Manjunatha, S., Biradar, D., & Aladakatti, Y. R. (2016). Nanotechnology and its applications in agriculture: A review. *J farm Sci, 29*(1), 1-13.

Mie, A., Andersen, H. R., Gunnarsson, S., Kahl, J., Kesse-Guyot, E., Rembiałkowska, E., Quaglio, G., & Grandjean, P. (2017). Human health implications of organic food and organic agriculture: A comprehensive review. *Environmental Health, 16*(1), 1–22. doi:10.118612940-017-0315-4 PMID:29073935

Mildaziene, V., Ivankov, A., Sera, B., & Baniulis, D. (2022). Biochemical and physiological plant processes affected by seed treatment with non-thermal plasma. *Plants, 11*(7), 856. doi:10.3390/plants11070856 PMID:35406836

Mittal, D., Kaur, G., Singh, P., Yadav, K., & Ali, S. A. (2020). Nanoparticle-based sustainable agriculture and food science: Recent advances and future outlook. *Frontiers in Nanotechnology, 2*, 579954. doi:10.3389/fnano.2020.579954

Mu, Q., Jiang, G., Chen, L., Zhou, H., Fourches, D., Tropsha, A., & Yan, B. (2014). Chemical basis of interactions between engineered nanoparticles and biological systems. *Chemical Reviews, 114*(15), 7740–7781. doi:10.1021/cr400295a PMID:24927254

Mukhopadhyay, S. S. (2014). Nanotechnology in agriculture: Prospects and constraints. *Nanotechnology, Science and Applications*, 63–71. doi:10.2147/NSA.S39409 PMID:25187699

Nagargade, M., Tyagi, V., Kumar, D., Shukla, S., & Pathak, A. (2022). Nanofertilizers: Importance in Nutrient Management. In Nanotechnology in Agriculture and Environmental Science (pp. 69-80). CRC Press.

Nasrollahzadeh, M., Sajadi, S. M., Sajjadi, M., & Issaabadi, Z. (2019). An introduction to nanotechnology. In *Interface science and technology* (Vol. 28, pp. 1–27). Elsevier.

Niggli, U., Fließbach, A., Hepperly, P., & Scialabba, N. (2009). *Low greenhouse gas agriculture: mitigation and adaptation potential of sustainable farming systems* (Vol. 141). Food and Agriculture Organization of the United Nations.

Nile, S. H., Thiruvengadam, M., Wang, Y., Samynathan, R., Shariati, M. A., Rebezov, M., Nile, A., Sun, M., Venkidasamy, B., Xiao, J., & Kai, G. (2022). Nano-priming as emerging seed priming technology for sustainable agriculture—Recent developments and future perspectives. *Journal of Nanobiotechnology*, *20*(1), 1–31. doi:10.118612951-022-01423-8 PMID:35659295

Nuruzzaman, M., Rahman, M. M., Liu, Y., & Naidu, R. (2016). Nanoencapsulation, nano-guard for pesticides: A new window for safe application. *Journal of Agricultural and Food Chemistry*, *64*(7), 1447–1483. doi:10.1021/acs.jafc.5b05214 PMID:26730488

Oliver, M. A., & Gregory, P. (2015). Soil, food security and human health: A review. *European Journal of Soil Science*, *66*(2), 257–276. doi:10.1111/ejss.12216

Oweis, T. (2018). Use of brackish and marginal water for irrigation in water-scarce areas Z. Gao, China Institute of Water Resources and Hydropower Research, China. In Water management for sustainable agriculture (pp. 253-276). Burleigh Dodds Science Publishing.

Pahalvi, H. N., Rafiya, L., Rashid, S., Nisar, B., & Kamili, A. N. (2021). *Chemical fertilizers and their impact on soil health. Microbiota and Biofertilizers* (Vol. 2). Ecofriendly Tools for Reclamation of Degraded Soil Environs.

Pandey, S., Giri, K., Kumar, R., Mishra, G., & Raja Rishi, R. (2018). Nanopesticides: Opportunities in crop protection and associated environmental risks. *Proceedings of the National Academy of Sciences. India. Section B, Biological Sciences*, *88*(4), 1287–1308. doi:10.100740011-016-0791-2

Parra-Arroyo, L., González-González, R. B., Castillo-Zacarías, C., Martínez, E. M. M., Sosa-Hernández, J. E., Bilal, M., Iqbal, H. M., Barceló, D., & Parra-Saldívar, R. (2022). Highly hazardous pesticides and related pollutants: Toxicological, regulatory, and analytical aspects. *The Science of the Total Environment*, *807*, 151879. doi:10.1016/j.scitotenv.2021.151879 PMID:34826476

Pavoni, L., Perinelli, D. R., Bonacucina, G., Cespi, M., & Palmieri, G. F. (2020). An overview of micro-and nanoemulsions as vehicles for essential oils: Formulation, preparation and stability. *Nanomaterials (Basel, Switzerland)*, *10*(1), 135. doi:10.3390/nano10010135 PMID:31940900

Phelps, J., Carrasco, L. R., Webb, E. L., Koh, L. P., & Pascual, U. (2013). Agricultural intensification escalates future conservation costs. *Proceedings of the National Academy of Sciences of the United States of America*, *110*(19), 7601–7606. doi:10.1073/pnas.1220070110 PMID:23589860

Poudel, S., Poudel, B., Acharya, B., & Poudel, P. (2020). Pesticide use and its impacts on human health and environment. *Environ Ecosyst Sci*, *4*(1), 47–51. doi:10.26480/ees.01.2020.47.51

Pramanik, P., Krishnan, P., Maity, A., Mridha, N., Mukherjee, A., & Rai, V. (2020). Application of nanotechnology in agriculture. *Environmental Nanotechnology*, *4*, 317–348.

Prasad, R., Kumar, V., & Prasad, K. S. (2014). Nanotechnology in sustainable agriculture: Present concerns and future aspects. *African Journal of Biotechnology*, *13*(6), 705–713. doi:10.5897/AJBX2013.13554

Preisler, A. C., Carvalho, L. B., Saraiva-Santos, T., Verri, W. A. Jr, Mayer, J. L. S., Fraceto, L. F., Dalazen, G., & Oliveira, H. C. (2022). Interaction of Nanoatrazine and Target Organism: Evaluation of Fate and Photosystem II Inhibition in Hydroponically Grown Mustard (Brassica juncea) Plants. *Journal of Agricultural and Food Chemistry*, *70*(25), 7644–7652. doi:10.1021/acs.jafc.2c01601 PMID:35675570

Pulighe, G., & Lupia, F. (2020). Food first: COVID-19 outbreak and cities lockdown a booster for a wider vision on urban agriculture. *Sustainability (Basel)*, *12*(12), 5012. doi:10.3390u12125012

Quisumbing, A. R., & Pandolfelli, L. (2010). Promising approaches to address the needs of poor female farmers: Resources, constraints, and interventions. *World Development*, *38*(4), 581–592. doi:10.1016/j.worlddev.2009.10.006

Qureshi, A., Singh, D., & Dwivedi, S. (2018). Nano-fertilizers: A novel way for enhancing nutrient use efficiency and crop productivity. *International Journal of Current Microbiology and Applied Sciences*, *7*(2), 3325–3335. doi:10.20546/ijcmas.2018.702.398

Ragaei, M., & Sabry, A.-H. (2014). Nanotechnology for insect pest control. *International Journal of Science, Environment and Technology*, *3*(2), 528–545.

Rahman, M. H., Haque, K. S., & Khan, M. Z. H. (2021). A review on application of controlled released fertilizers influencing the sustainable agricultural production: A Cleaner production process. *Environmental Technology & Innovation*, *23*, 101697. doi:10.1016/j.eti.2021.101697

Rai, R. S., Bajpai, V., Khan, M. I., Elboughdiri, N., Shanableh, A., & Luque, R. (2022). An eco-friendly approach on green synthesis, bio-engineering applications, and future outlook of ZnO nanomaterial: A critical review. *Environmental Research*, 114807. PMID:36455633

Rehman, A., Saba, T., Kashif, M., Fati, S. M., Bahaj, S. A., & Chaudhry, H. (2022). A revisit of internet of things technologies for monitoring and control strategies in smart agriculture. *Agronomy (Basel)*, *12*(1), 127. doi:10.3390/agronomy12010127

Roberts, D. P., & Mattoo, A. K. (2018). Sustainable agriculture—Enhancing environmental benefits, food nutritional quality and building crop resilience to abiotic and biotic stresses. *Agriculture*, *8*(1), 8. doi:10.3390/agriculture8010008

Rosegrant, M. W., Ringler, C., & Zhu, T. (2009). Water for agriculture: Maintaining food security under growing scarcity. *Annual Review of Environment and Resources*, *34*(1), 205–222. doi:10.1146/annurev.environ.030308.090351

Sahle, A., & Potting, J. (2013). Environmental life cycle assessment of Ethiopian rose cultivation. *The Science of the Total Environment*, *443*, 163–172. doi:10.1016/j.scitotenv.2012.10.048 PMID:23183227

Sajid, M., Basit, A., Ullah, I., Shah, S., Haleema, B., Nawab, K., Ahmad, I., Ahmad, N., Khan, A., & Muhammad, M. (2021). Screening of citrus rootstocks for salinity tolerance based on initial growth attributes and leaves proline content. *Element. Educ. Online*, *19*, 3128–3140.

Sajyan, T. K., Alturki, S. M., & Sassine, Y. N. (2020). Nano-fertilizers and their impact on vegetables: Contribution of Nano-chelate Super Plus ZFM and Lithovit®-standard to improve salt-tolerance of pepper. *Annals of Agricultural Science*, *65*(2), 200–208. doi:10.1016/j.aoas.2020.11.001

Samir, K., & Lutz, W. (2017). The human core of the shared socioeconomic pathways: Population scenarios by age, sex and level of education for all countries to 2100. *Global Environmental Change*, *42*, 181–192. doi:10.1016/j.gloenvcha.2014.06.004 PMID:28239237

Sanna, V., Pala, N., & Sechi, M. (2014). Targeted therapy using nanotechnology: Focus on cancer. *International Journal of Nanomedicine*, *9*, 467. PMID:24531078

Saravi, S. S. S., & Shokrzadeh, M. (2011). Role of pesticides in human life in the modern age: a review. *Pesticides in the modern world-risks and benefits, 5*, 3-12.

Savci, S. (2012). An agricultural pollutant: Chemical fertilizer. *International Journal of Environmental Sciences and Development*, *3*(1), 73–80. doi:10.7763/IJESD.2012.V3.191

Sekhon, B. S. (2014). Nanotechnology in agri-food production: An overview. *Nanotechnology, Science and Applications*, 31–53. doi:10.2147/NSA.S39406 PMID:24966671

Shang, Y., Hasan, M. K., Ahammed, G. J., Li, M., Yin, H., & Zhou, J. (2019). Applications of nanotechnology in plant growth and crop protection: A review. *Molecules (Basel, Switzerland)*, *24*(14), 2558. doi:10.3390/molecules24142558 PMID:31337070

Shelar, A., Nile, S. H., Singh, A. V., Rothenstein, D., Bill, J., Xiao, J., Chaskar, M., Kai, G., & Patil, R. (2023). Recent advances in nano-enabled seed treatment strategies for sustainable agriculture: Challenges, risk assessment, and future perspectives. *Nano-Micro Letters*, *15*(1), 54. doi:10.100740820-023-01025-5 PMID:36795339

Shiferaw, B., Prasanna, B. M., Hellin, J., & Bänziger, M. (2011). Crops that feed the world 6. Past successes and future challenges to the role played by maize in global food security. *Food Security*, *3*(3), 307–327. doi:10.100712571-011-0140-5

Shimada, D. (2015). Multi-level natural resources governance based on local community: A case study on semi-natural grassland in Tarōji, Nara, Japan. *The International Journal of the Commons*, *9*(2), 486. doi:10.18352/ijc.510

Shukla, A. K., Behera, S. K., Pakhre, A., & Chaudhari, S. (2018). Micronutrients in soils, plants, animals and humans. *Indian Journal of Fertilisers*, *14*(3), 30–54.

Singh, A., Rajput, V., Singh, A. K., Sengar, R. S., Singh, R. K., & Minkina, T. (2021). Transformation Techniques and Their Role in Crop Improvements: A Global Scenario of GM Crops. *Policy Issues in Genetically Modified Crops: A Global Perspective*, 515–542. doi:10.1016/B978-0-12-820780-2.00023-6

Singh, A., Sharma, R., Rawat, S., Singh, A. K., Rajput, V. D., Fedorov, Y., Minkina, T., & Chaplygin, V. (2022). Nanomaterial-plant interaction: Views on the pros and cons. *Toxicity of Nanoparticles in Plants: An Evaluation of Cyto/Morpho-Physiological, Biochemical and Molecular Responses*, 47–68. doi:10.1016/B978-0-323-90774-3.00015-5

Singh, A., Singh, N., Hussain, I., Singh, H., & Singh, S. (2015). Plant-nanoparticle interaction: An approach to improve agricultural practices and plant productivity. *International Journal of Pharmaceutical Science Invention*, *4*(8), 25–40.

Singh, H., Sharma, A., Bhardwaj, S. K., Arya, S. K., Bhardwaj, N., & Khatri, M. (2021). Recent advances in the applications of nano-agrochemicals for sustainable agricultural development. *Environmental Science. Processes & Impacts*, *23*(2), 213–239. doi:10.1039/D0EM00404A PMID:33447834

Singh, K. U., Kumar, A., Raja, L., Kumar, V., Vashney, N., & Chhetri, M. (2022). An artificial neural network-based pest identification and control in smart agriculture using wireless sensor networks. *Journal of Food Quality*, *2022*, 2022. doi:10.1155/2022/5801206

Singh, S., Kumar, R., Panchal, R., & Tiwari, M. K. (2021). Impact of COVID-19 on logistics systems and disruptions in food supply chain. *International Journal of Production Research*, *59*(7), 1993–2008. doi:10.1080/00207543.2020.1792000

Sinha, B. B., & Dhanalakshmi, R. (2022). Recent advancements and challenges of Internet of Things in smart agriculture: A survey. *Future Generation Computer Systems*, *126*, 169–184. doi:10.1016/j.future.2021.08.006

Smith, M. T., & Berjak, P. (2017). Deteriorative changes associated with the loss of viability of stored desiccation-tolerant and desiccation-sensitive seeds. In *Seed development and germination* (pp. 701–746). Routledge. doi:10.1201/9780203740071-26

Smith, P., Haberl, H., Popp, A., Erb, K., Lauk, C., Harper, R., Tubiello, F. N., de Siqueira Pinto, A., Jafari, M., Sohi, S., Masera, O., Böttcher, H., Berndes, G., Bustamante, M., Ahammad, H., Clark, H., Dong, H., Elsiddig, E. A., Mbow, C., & Rose, S. (2013). How much land-based greenhouse gas mitigation can be achieved without compromising food security and environmental goals? *Global Change Biology*, *19*(8), 2285–2302. doi:10.1111/gcb.12160 PMID:23505220

Soetan, K., Olaiya, C., & Oyewole, O. (2010). The importance of mineral elements for humans, domestic animals and plants: A review. *African Journal of Food Science*, *4*(5), 200–222.

Teh, C. Y., Budiman, P. M., Shak, K. P. Y., & Wu, T. Y. (2016). Recent advancement of coagulation–flocculation and its application in wastewater treatment. *Industrial & Engineering Chemistry Research*, *55*(16), 4363–4389. doi:10.1021/acs.iecr.5b04703

Tudi, M., Daniel Ruan, H., Wang, L., Lyu, J., Sadler, R., Connell, D., Chu, C., & Phung, D. T. (2021). Agriculture development, pesticide application and its impact on the environment. *International Journal of Environmental Research and Public Health*, *18*(3), 1112. doi:10.3390/ijerph18031112 PMID:33513796

Ukwattage, N. L., Ranjith, P., & Bouazza, M. (2013). The use of coal combustion fly ash as a soil amendment in agricultural lands (with comments on its potential to improve food security and sequester carbon). *Fuel, 109*, 400–408. doi:10.1016/j.fuel.2013.02.016

Van der Elst, K., & Williams, A. (2019). Environment and natural resources. In Agriculture & Food Systems to 2050: Global Trends, Challenges and Opportunities (pp. 193-214). World Scientific.

van Dijk, M., Morley, T., Rau, M. L., & Saghai, Y. (2021). A meta-analysis of projected global food demand and population at risk of hunger for the period 2010–2050. *Nature Food, 2*(7), 494–501. doi:10.103843016-021-00322-9 PMID:37117684

Van Wesenbeeck, C., Keyzer, M., Van Veen, W., & Qiu, H. (2021). Can China's overuse of fertilizer be reduced without threatening food security and farm incomes? *Agricultural Systems, 190*, 103093. doi:10.1016/j.agsy.2021.103093

Vejan, P., Khadiran, T., Abdullah, R., & Ahmad, N. (2021). Controlled release fertilizer: A review on developments, applications and potential in agriculture. *Journal of Controlled Release, 339*, 321–334. doi:10.1016/j.jconrel.2021.10.003 PMID:34626724

Veneklaas, E. J., Lambers, H., Bragg, J., Finnegan, P. M., Lovelock, C. E., Plaxton, W. C., Price, C. A., Scheible, W. R., Shane, M. W., White, P. J., & Raven, J. A. (2012). Opportunities for improving phosphorus-use efficiency in crop plants. *The New Phytologist, 195*(2), 306–320. doi:10.1111/j.1469-8137.2012.04190.x PMID:22691045

Verma, K. K., Song, X.-P., Joshi, A., Tian, D.-D., Rajput, V. D., Singh, M., Arora, J., Minkina, T., & Li, Y.-R. (2022). Recent trends in nano-fertilizers for sustainable agriculture under climate change for global food security. *Nanomaterials (Basel, Switzerland), 12*(1), 173. doi:10.3390/nano12010173 PMID:35010126

Verma, S. K., Das, A. K., Patel, M. K., Shah, A., Kumar, V., & Gantait, S. (2018). Engineered nanomaterials for plant growth and development: A perspective analysis. *The Science of the Total Environment, 630*, 1413–1435. doi:10.1016/j.scitotenv.2018.02.313 PMID:29554761

Weeks, J. J. Jr, & Hettiarachchi, G. M. (2019). A review of the latest in phosphorus fertilizer technology: Possibilities and pragmatism. *Journal of Environmental Quality, 48*(5), 1300–1313. doi:10.2134/jeq2019.02.0067 PMID:31589737

Weidner, T., Yang, A., & Hamm, M. W. (2021). Energy optimisation of plant factories and greenhouses for different climatic conditions. *Energy Conversion and Management*, *243*, 114336. doi:10.1016/j.enconman.2021.114336

WHO. (2019). *Microplastics in drinking-water*. WHO.

WHO. (2020). *The state of food security and nutrition in the world 2020: transforming food systems for affordable healthy diets* (Vol. 2020). Food & Agriculture Org.

Willett, W., Rockström, J., Loken, B., Springmann, M., Lang, T., Vermeulen, S., Garnett, T., Tilman, D., DeClerck, F., Wood, A., Jonell, M., Clark, M., Gordon, L. J., Fanzo, J., Hawkes, C., Zurayk, R., Rivera, J. A., De Vries, W., Majele Sibanda, L., ... Murray, C. J. L. (2019). Food in the Anthropocene: The EAT–Lancet Commission on healthy diets from sustainable food systems. *Lancet*, *393*(10170), 447–492. doi:10.1016/S0140-6736(18)31788-4 PMID:30660336

Woyessa, D. (2022). Weed control methods used in agriculture. *American Journal of Life Science and Innovation*, *1*(1), 19–26. doi:10.54536/ajlsi.v1i1.413

Wright, B. (2014). Global biofuels: Key to the puzzle of grain market behavior. *The Journal of Economic Perspectives*, *28*(1), 73–98. doi:10.1257/jep.28.1.73

Xi, L., Zhang, M., Zhang, L., Lew, T. T., & Lam, Y. M. (2022). Novel materials for urban farming. *Advanced Materials*, *34*(25), 2105009. doi:10.1002/adma.202105009 PMID:34668260

Yang, D., Tu, Y., Wang, X., Cao, C., Hu, Y., Shao, J., Weng, L., Mou, X., & Dong, X. (2021). A photo-triggered antifungal nanoplatform with efflux pump and heat shock protein reversal activity for enhanced chemo-photothermal synergistic therapy. *Biomaterials Science*, *9*(9), 3293–3299. doi:10.1039/D1BM00457C PMID:33861264

Zhang, W., Musante, C., White, J. C., Schwab, P., Wang, Q., Ebbs, S. D., & Ma, X. (2017). Bioavailability of cerium oxide nanoparticles to Raphanus sativus L. in two soils. *Plant Physiology and Biochemistry*, *110*, 185–193. doi:10.1016/j.plaphy.2015.12.013 PMID:26754029

Zhu, J.-K. (2016). Abiotic stress signaling and responses in plants. *Cell*, *167*(2), 313–324. doi:10.1016/j.cell.2016.08.029 PMID:27716505

Chapter 6
Nanoparticle-Based Priming-Enhanced Agricultural Practices:
Promoting Sustainable Plant Growth and Development

Abhishek Singh
https://orcid.org/0000-0002-0049-6910
Faculty of Biology, Yerevan State University, Armenia

Sapna Rawat
University of Delhi, India

Vishnu D. Rajput
https://orcid.org/0000-0002-6802-4805
Academy of Biology and Biotechnology, Southern Federal University, Rostov-on-Don, Russia

Karen Ghazaryan
https://orcid.org/0000-0002-8973-2595
Faculty of Biology, Yerevan State University, Armenia

Tatiana Minkina
Academy of Biology and Biotechnology, Southern Federal University, Rostov-on-Don, Russia

Abdel Rahman Mohammad Al Tawaha
Al Hussein bin Talal University, Jordan

Athanasios Alexiou
https://orcid.org/0000-0002-2206-7236
Novel Global Community Educational Foundation, Australia

Karansher Singh Sandhu
Bayer Crop Sciences, USA

ABSTRACT

The challenges of climate change, resource depletion, and biodiversity loss pose significant risks to agriculture. To ensure sustainable food production and tackle

DOI: 10.4018/978-1-6684-7232-3.ch006

these issues, an agricultural revolution is necessary. Nanotechnology offers promising solutions for enhancing agricultural sustainability, and one such approach is seed nano-priming. This technique effectively alters seed metabolism and signaling pathways, influencing not only germination and seedling establishment but also the entire plant life cycle. Extensive research has demonstrated numerous benefits of seed nano-priming, including improved plant growth, increased productivity, and enhanced nutritional quality of food. This chapter provides an overview of recent advancements in nanotechnology for seed nano-priming and explores the challenges and possibilities associated with its implementation in sustainable agriculture.

INTRODUCTION

Nanotechnology has gained considerable attention in recent years for its potential applications in various industries, including agriculture. Nanoparticles, tiny particles in the nanometer range, show promise in enhancing plant growth and development, making them valuable for sustainable agriculture. Acting as priming agents, nanoparticles can improve seed germination, increase nutrient absorption, boost plant resilience to environmental stressors, and enhance overall plant health (Nile SH et al., 2022; Rhaman MS et al., 2022; Sharma B et al., 2023; Singh VD et al., 2021; Singh A et al., 2022).

One significant advantage of nanoparticle use in agriculture is their ability to improve seed germination rates. Applying nanoparticles to seeds or seedlings enhances their ability to absorb water, resulting in faster and more consistent germination. This, in turn, leads to higher crop yields and improved plant establishment (Singh A et al., 2022). Moreover, nanoparticles can act as carriers of essential nutrients and micronutrients. Their small size and large surface area enable efficient delivery of nutrients to plants, ensuring optimal uptake. This targeted nutrient delivery system reduces nutrient losses and enhances nutrient utilization efficiency, minimizing the environmental impact of fertilizers while maximizing crop productivity (Zhang et al., 2011). Nanoparticles have also shown potential in mitigating abiotic stresses, such as drought, salinity, and heavy metal toxicity. When applied to plants, nanoparticles enhance their tolerance to these adverse conditions, enabling them to thrive. Nanoparticles function as antioxidants, protecting plant cells from oxidative damage by scavenging reactive oxygen species (Jiang HS et al, 2017; Sharma S et al., 2019). Additionally, nanoparticles improve photosynthesis efficiency and regulate stomatal behavior, allowing plants to conserve water during drought periods (Wu H et al., 2017; Jiang HS et al, 2017; Sharma S et al., 2019). In addition to their role in abiotic stress management, nanoparticles offer promise in enhancing plant resistance to pests and diseases. Some nanoparticles possess antimicrobial properties, inhibiting the growth of pathogens and reducing reliance on chemical pesticides (Hernández-Díaz JA et al., 2021; Al-Samarrai AM et al., 2012). Furthermore, nanoparticles can stimulate the plant's defense mechanisms, triggering a systemic acquired resistance response that enhances the plant's ability to defend against pests and diseases (Al-Samarrai AM et al., 2012).

The integration of nanoparticles in sustainable agriculture holds potential for addressing pressing challenges faced by the global farming community. By improving seed germination, nutrient uptake, stress tolerance, and disease resistance, nanoparticles can contribute to increased crop yields, reduced environmental impact, and enhanced food security (Wang D et al., 2022). However, further research is necessary to fully comprehend the long-term effects and potential risks associated with nanoparticle use in agriculture (Singh A et al., 2022).

Nanopriming Technology

When a seed is fertilized, it develops into an ovule containing the germinating embryo and sufficient food for plant growth. Priming seeds is a crucial step to improve their quality and potential yield before planting (Ashraf M et al., 2005). It involves soaking seeds in a solution that allows water absorption for germination without radicle protrusion through the seed coat. This method ensures uniform germination rates and maximizes seed yield. Nanoparticles have emerged as a novel technological advancement for seed priming (Dhal P et al., 2022). Unlike other priming methods using substances adsorbed on the seed or seed coating, seed nano-priming involves using suspensions or nano-formulations where nanoparticles may or may not be absorbed by the seeds (Garg R et al., 2023). In most cases, nanoparticles remain as a coating on the seed's exterior. This coating can include fungicides or bactericides for storage or planting purposes. Research has shown that nanoparticles, such as carbon nanotubes, can enhance seed germination by improving water uptake (Safdar M et al., 2022; Xin X et al., 2022). Carbon nanotubes have been found to increase the expression of water channel proteins in various plant species, leading to improved water absorption and increased blossom production in tomato plants (Khodakovskaya MV et al., 2013). Seed nano-priming, using metallic, biogenic metallic, and polymeric nanoparticles, has demonstrated the stimulation of plant growth, morphological and metabolic improvements, and enhanced resistance to biotic and abiotic stresses (Safdar M et al., 2022). By influencing the expression of genes involved in metabolic processes and phytohormone production, seed nano-priming accelerates root and shoot development. It also enhances antioxidant levels and enzyme activities, making plants more resilient to pests and environmental stresses in the field. The process of nanopriming initiates specific metabolic reactions similar to early germination stages, including rapid water uptake, starch hydrolysis, cell wall loosening, weakened endosperm, and rapid embryo and root-shoot growth (Nile SH et al., 2022; Imtiaz H et al., 2023). This results in higher germination rates, improved seedling development, increased yield, and enhanced quality. Seed nanopriming also impacts the molecular and cellular mechanisms of plant-environment interactions, promoting improved plant performance and soil sustainability.

Factors such as soil moisture, priming treatment, duration of exposure, and storage conditions influence the success of priming. Several studies have demonstrated the ability of nanoparticles to penetrate seed coats, increase water uptake, and positively affect germination and flowering (Imtiaz H et al., 2023; Hassanisaadi, M et al., 2022). Various nanoparticles, including metal-metal oxide and carbon-based nanoparticles, have been used as seed pretreatments to enhance germination, seedling growth, and stress tolerance in different crops (Nile SH et al., 2022). While priming seeds through rehydration before planting has seen limited research, seed nanopriming offers an alternative mechanism that does not require drying. However, there are still many unanswered questions regarding the physiological and molecular mechanisms of nanopriming-induced seed germination, as comprehensive studies in this area are lacking. This chapter can provide an overview of different nanoparticles used for seed priming and their potential stimulatory or protective effects against biotic and abiotic stress.

APPLICATION OF NANOPRIMING GROWTH AND DEVELOPMENT OF PLANTS FOR SUSTAINABLE AGRICULTURE

Enhanced Germination and Seedling Establishment

Seed nanopriming encompasses the utilization of nanoparticles to treat seeds, resulting in notable enhancements in seed germination and early seedling growth. Nanoparticles possess a distinctive capability to permeate the seed coats, facilitating the regulated release of advantageous compounds and enzymes (Rai-Kalal P et al., 2021). This regulated release mechanism aids in the alleviation of seed dormancy, fostering prompt and consistent germination. Furthermore, the presence of nanoparticles stimulates the seedlings, amplifying their vitality and overall development. Through the improvement of germination speed and uniformity, seed nanopriming fosters improved stand establishment, thereby ensuring a more resilient and fruitful crop yield (Amna Ali B et al., 2022).

Stress Tolerance

Nanoprimed seeds display improved resilience to diverse environmental stresses, rendering them more adaptable to demanding growth conditions (Alabdallah NM et al., 2021; Zhao L et al, 2020). Nanoparticles play a vital role in balancing plant hormones, mitigating oxidative damage induced by stress factors, and enhancing antioxidant defense mechanisms. This augmented stress tolerance empowers plants to endure unfavorable circumstances like drought, salinity, and extreme temperatures, thus supporting their sustained growth and productivity (Zhao L et al., 2020). By strengthening stress tolerance, seed nanopriming promotes agricultural sustainability by safeguarding crop yields against unpredictable environmental challenges (Nile SH et al., 2022).

Nutrient Uptake and Utilization

Seed nanopriming plays a crucial role in maximizing nutrient uptake and utilization efficiency in plants. Through the application of nanomaterials during nanopriming, root development and architecture are enhanced, leading to increased surface area for nutrient absorption. Additionally, nanoparticles aid in the release of enzymes that facilitate nutrient mobilization, thereby improving nutrient availability for plant growth. By enhancing the plants' capacity to absorb and utilize nutrients efficiently, seed nanopriming contributes to enhanced crop productivity and supports effective nutrient management in agriculture (Chen N et al., 2022). Hence, they can be used to enhance the bioavailability and uptake of essential micronutrients in plants, leading to improved nutritional quality. This approach, known as biofortification, can help address nutrient deficiencies in crops and contribute to improved human health and nutrition (Elemike EE et al., 2019; Singh A et al., 2022).

Disease and Pest Resistance

Nanopriming presents a promising approach to combat plant diseases and pests, which pose substantial risks to crop health and productivity. Nanomaterials have the ability to act as physical barriers, impeding pathogen invasion and limiting the spread of diseases within plants. Additionally, nanoparticles can serve as carriers for biopesticides or antimicrobial agents, delivering targeted protection against pests and diseases (Rai M et al., 2012). By bolstering the plant's immune response and enabling effective pest control, seed nanopriming plays a crucial role in minimizing yield losses and enhancing the overall health and quality of crops. Nanoparticles can be employed as carriers for beneficial microorganisms, such as biocontrol agents, which can help manage plant diseases and pests. The nanoparticles facilitate the targeted delivery of these biocontrol agents, increasing their effectiveness in protecting crops while minimizing the use of chemical pesticides (Kumar S et al., 2020).

Quality Preservation

Seed nanopriming is of paramount importance in maintaining seed quality during storage, guaranteeing the availability of superior seeds for optimal crop production. Through the application of nanoparticles during seed priming, seeds are safeguarded against deterioration induced by aging, environmental elements, or microbial attacks. The regulated release of bioactive compounds from the nanomaterials mitigates oxidative stress, preserving seed viability and vigor throughout extended storage durations (Dos Santos et al., 2020). By upholding seed quality, seed nanopriming fosters seed longevity and augments the success of subsequent plantings, thereby contributing to the promotion of sustainable agricultural practices (Jaghav R et al., 2023).

Improved Plant Root Colonization

Nanoparticles offer a promising avenue for enhancing the colonization of beneficial microorganisms, particularly mycorrhizal fungi, in plant roots. Mycorrhizal fungi form a mutualistic symbiotic relationship with plant roots, where they establish connections with the plant's root system. This symbiosis leads to various benefits for both the fungi and the plants.

By applying nanoparticles, the colonization and establishment of mycorrhizal fungi in plant roots can be improved (Dimpka CO et al., 2015). Nanoparticles can facilitate the attachment and growth of fungal hyphae, enhancing their ability to penetrate the root tissues and establish a strong symbiotic association. This improved colonization results in a larger surface area for nutrient exchange between the fungi and the plant roots. The presence of mycorrhizal fungi in the root system enhances nutrient uptake for the plant. The fungal hyphae extend into the soil, effectively exploring a larger soil volume and accessing nutrients that are otherwise inaccessible to plant roots. The fungi can extract nutrients, particularly phosphorus, more efficiently and deliver them to the plant. In return, the plant provides carbohydrates to the fungi, supporting their growth and survival. The enhanced nutrient uptake facilitated by mycorrhizal colonization contributes to improved plant growth, development, and overall performance. Plants with well-established mycorrhizal associations exhibit increased biomass production, enhanced tolerance to environmental stresses such as drought and nutrient deficiencies, and improved nutrient use efficiency. The symbiotic relationship between the plant and mycorrhizal fungi also contributes to sustainable nutrient cycling in the soil, reducing the need for external fertilizers and promoting efficient nutrient utilization.

By utilizing nanoparticles to enhance mycorrhizal colonization, farmers and agricultural practitioners can harness the benefits of this symbiotic relationship to optimize plant nutrition, reduce fertilizer dependency, and improve overall crop productivity in a sustainable manner (Timmusk S et al., 2017; Nile SH et al., 2022). However, it is crucial to conduct further research to ensure the safety and efficacy of nanoparticle applications in promoting mycorrhizal colonization and to better understand the potential long-term impacts on the soil ecosystem.

Water and Irrigation Management

Nanoparticles offer potential solutions for water and irrigation management in agriculture, particularly in areas where water scarcity or drought conditions pose challenges. Through their unique properties, nanoparticles can improve water retention in the soil and enhance water uptake efficiency by plants. Nanoparticles can be applied to the soil or incorporated into irrigation systems to enhance water retention (Wang X et al., 2021). Due to their small size and large surface area, nanoparticles can modify the soil structure, creating a more porous and water-absorbent environment. This increased water-holding capacity helps to reduce water loss through evaporation and runoff, ensuring a more efficient use of irrigation water.

Moreover, nanoparticles can enhance the water uptake efficiency of plants (AI-Selwey WA et al., 2022). By applying nanoparticles to plant roots or incorporating them into irrigation water, the nanoparticles can improve the root's ability to absorb water from the soil. The nanoparticles can facilitate better water penetration into the root system, allowing for enhanced hydration of the plant. This improved water uptake efficiency reduces water wastage and increases the plant's ability to withstand drought conditions. Additionally, nanoparticles can aid in the regulation of plant stomatal behavior. Stomata are small openings on the plant's leaves that control water loss through transpiration. Nanoparticles can influence the opening and closing of stomata, allowing for better control over water loss. This regulation helps plants conserve water during dry periods and reduces the overall water requirements for plant growth.

By utilizing nanoparticles for water and irrigation management, farmers can optimize water usage, conserve water resources, and mitigate the impact of water scarcity or drought conditions on crop production. It is important to note that further research is needed to fully understand the long-term effects, optimal application methods, and potential environmental impacts of nanoparticle-based water management strategies in agriculture (Singh VD et al., 2021).

Soil Remediation

Nanoparticles offer a promising approach for soil remediation by assisting in the removal of contaminants, including heavy metals and organic pollutants (Rajput VD et al., 2022). Their unique properties and reactivity enable them to facilitate the breakdown, immobilization, or removal of these contaminants, thereby improving soil quality and reducing environmental risks. In the case of heavy metal contamination, nanoparticles can interact with the metal ions present in the soil. They can form complexes or precipitates with the contaminants, effectively immobilizing them and reducing their bioavailability. This immobilization prevents the uptake of heavy metals by plants, reducing the potential for toxicity and environmental contamination. Nanoparticles can also aid in the degradation of organic pollutants in the soil (Trivedi M et al., 2022). They can serve as catalysts or carriers for enzymes that break down organic compounds, accelerating their degradation into non-toxic byproducts. Additionally, nanoparticles can enhance microbial activity in the soil, promoting the growth of pollutant-degrading microorganisms and facilitating the natural remediation processes. Furthermore, nanoparticles can assist in the removal of contaminants through processes such as adsorption or filtration. Their high surface area and reactivity allow them to adsorb or trap contaminants, effectively removing them from the soil matrix. This can be particularly beneficial for the removal of organic pollutants or emerging contaminants from agricultural soils (Mathur S et al., 2022).

By utilizing nanoparticles for soil remediation, contaminated soils can be treated, leading to improved soil quality and reduced environmental risks. However, it is important to consider the specific characteristics of the contaminants, soil conditions, and potential interactions with the nanoparticles to ensure their effectiveness and minimize any unintended environmental consequences. Robust research and proper monitoring are necessary to evaluate the long-term impacts and safety of nanoparticle-based soil remediation strategies.

Controlled Release of Agrochemicals

Nanoparticles offer a valuable method for the controlled release of agrochemicals, such as fertilizers, pesticides, and growth regulators (Machado TO et al., 2022). By acting as carriers, nanoparticles enable a targeted delivery system that enhances the effectiveness of these agrochemicals while minimizing their environmental impact and reducing resource wastage. In the case of fertilizers, nanoparticles can be loaded with nutrients and released gradually over time. This controlled release mechanism ensures a steady and sustained supply of nutrients to plants, promoting optimal growth and minimizing nutrient losses through leaching or runoff. The nanoparticles can protect the nutrients from being immobilized or washed away, enhancing their availability for plant uptake and maximizing nutrient use efficiency. Similarly, nanoparticles can be utilized to deliver pesticides or biopesticides in a controlled manner. The nanoparticles can encapsulate or adsorb the active ingredients, allowing for a slow and sustained release (Tao R et al., 2023). This targeted delivery ensures that the pesticides are released at the desired location, such as the plant surface or specific tissues, improving their efficacy while reducing the overall amount of pesticide needed. By minimizing the amount of pesticide released into the environment, nanoparticle-based controlled release systems help mitigate potential ecological risks associated with excessive pesticide use (Tao R et al., 2023).

Moreover, nanoparticles can serve as carriers for growth regulators, such as hormones or signaling molecules. The controlled release of these regulators can provide precise timing and dosage, ensuring optimal effects on plant growth, development, and yield. This targeted delivery system allows for better synchronization between the growth regulator application and the specific growth stages of the plant. By employing nanoparticles for the controlled release of agrochemicals, farmers and agricultural practitioners can optimize the effectiveness of these inputs while minimizing their environmental impact. This approach promotes more sustainable agricultural practices by reducing chemical usage, minimizing nutrient losses, and mitigating potential risks to ecosystems and human health. However, it is important to consider the specific characteristics of the nanoparticles, the agrochemicals involved, and their interactions to ensure the desired release profiles and minimize any unintended consequences. Thorough research and appropriate testing are essential to evaluate the efficacy, safety, and long-term impacts of nanoparticle-based controlled release systems in agricultural applications.

Enhanced Crop Quality

Seed nanopriming plays a role in enhancing crop quality attributes, including nutritional composition, shelf life, and sensory characteristics of harvested produce. Nanoparticles, through their interactions with plants, can modify biochemical pathways and metabolic processes, leading to the development of desirable traits in crops (Azeem A et al., 2023). Nanopriming can influence the synthesis and accumulation of various bioactive compounds in plants, such as antioxidants, vitamins, and phytochemicals (Nile SH et al., 2022). By modulating these pathways, nanoparticles can enhance the nutritional composition of crops, increasing their levels of essential nutrients and beneficial compounds. This can result in improved nutritional value and health benefits for consumers. Additionally, nanoparticles can influence post-harvest characteristics of crops, including shelf life and quality retention. They can regulate enzymatic activity and biochemical processes involved in post-harvest deterioration, such as the breakdown of cell membranes, oxidative stress, and microbial growth. This can extend the shelf life of harvested produce, reduce spoilage, and maintain desirable sensory attributes, such as color, flavor, and texture. Furthermore, nanopriming can affect the expression of genes related to crop quality attributes. By modifying gene expression patterns, nanoparticles can induce changes in the biosynthesis of flavor compounds, pigments, and volatile compounds that contribute to sensory characteristics. This can lead to improved taste, aroma, and overall sensory appeal of the harvested crops. By enhancing crop quality attributes, seed nanopriming contributes to the production of higher-value crops with improved nutritional profiles, extended shelf life, and enhanced sensory attributes (Nile SH et al., 2022). This can benefit both farmers and consumers by increasing the market value of the produce, reducing food waste, and providing consumers with more nutritious and appealing food options.

However, it is important to conduct further research to understand the specific effects of different nanoparticles on crop quality attributes, evaluate the safety of nanoparticle applications, and ensure regulatory compliance in food production. Proper monitoring and testing are essential to ensure the quality, safety, and consumer acceptance of nanoprimed crops in the market.

Concern

It is important to address the potential nanotoxicity concerns associated with the use of nanoparticles in agriculture. While nanoparticles offer promising benefits for sustainable agriculture, their potential impact on human health and the environment should be carefully evaluated. Nanoparticles can exhibit unique physicochemical properties that may pose risks when released into the environment (Behl T et al., 2022). Their small size and large surface area can facilitate interactions with living organisms, potentially leading to unintended consequences. Some nanoparticles have been found to have toxic effects on organisms, including humans, animals, and beneficial soil microorganisms. To ensure the safe use of nanoparticles in agriculture, it is crucial to conduct comprehensive research on their toxicity and potential environmental impacts (Behl T et al., 2022). This research should include studies on the bioaccumulation, persistence, and potential transfer of nanoparticles in the food chain. It is also important to assess their effects on soil microbial communities, non-target organisms, and ecosystems.

Regulatory bodies and scientific organizations are actively working to establish guidelines and regulations for the safe use of nanoparticles in agriculture. These guidelines aim to ensure the responsible and sustainable implementation of nanoparticle-based technologies while minimizing potential risks. Furthermore, the development and use of biodegradable or environmentally friendly nanoparticles can help mitigate concerns related to nanotoxicity. By designing nanoparticles that degrade over time or using materials that have minimal adverse effects, the potential risks associated with nanoparticle use can be reduced. Ultimately, it is essential to strike a balance between harnessing the benefits of nanoparticle-based technologies in agriculture and addressing potential nanotoxicity concerns. Ongoing research, risk assessment, and regulatory measures are crucial in ensuring the safe and sustainable application of nanoparticles in agriculture, thereby maximizing their benefits while minimizing potential risks (Nile SH et al., 2022).

CONCLUSION

In conclusion, the application of seed nanopriming using nanoparticles holds great potential for promoting sustainable agriculture practices. Throughout this study, we have explored the various benefits and applications of nanopriming in plant growth and development, including enhanced germination and seedling establishment, stress tolerance, nutrient uptake, disease resistance, and seed quality preservation. Nanopriming has been shown to improve seed germination rates, leading to more uniform and robust crop establishment. It enhances stress tolerance in plants, allowing them to withstand challenging environmental conditions and maintain productivity. Moreover, nanopriming optimizes nutrient uptake and utilization efficiency, contributing to improved crop productivity and efficient nutrient management. Furthermore, nanopriming offers effective solutions for combatting plant diseases and pests, reducing yield losses and promoting crop health. It also plays a crucial role in preserving seed quality during storage, ensuring the availability of high-quality seeds for optimal crop production. However, it is important to address concerns regarding nanotoxicity and potential environmental impacts associated with nanoparticle use (Bhattacharjee R et al., 2022). Ongoing research, risk assessment, and regulatory measures are necessary to ensure the safe and responsible implementation of nanopriming technologies in agriculture. By harnessing the benefits of nanopriming while mitigating potential risks, we can pave the way for a more sustainable and productive agricultural system. The continued exploration of nanopriming techniques, along with further research and development, will help unlock the full potential of nanoparticles in promoting sustainable agriculture and addressing the challenges faced by the global farming community.

Overall, seed nanopriming using nanoparticles offers a promising avenue for sustainable agriculture, contributing to increased crop yields, improved crop quality, reduced environmental impact, and enhanced food security. By embracing this technology responsibly, we can work towards a more resilient, efficient, and environmentally conscious agricultural system that meets the needs of present and future generations.

ACKNOWLEDGEMENT

The research was financially supported by the Ministry of Science and Higher Education of the Russian Federation (no. FENW-2023-0008) and the Strategic Academic Leadership Program of the Southern Federal University ("Priority 2030"). The work was supported by Science Committee of RA, research project № 21AG-4C075.

REFERENCES

Al-Samarrai, A. M. (2012). Nanoparticles as alternative to pesticides in management plant diseases-a review. *International Journal of Scientific and Research Publications*, *2*(4), 1–4.

Al-Selwey, W. A., Alsadon, A. A., Ibrahim, A. A., Labis, J. P., & Seleiman, M. F. (2023). Effects of zinc oxide and silicon dioxide nanoparticles on physiological, yield, and water use efficiency traits of potato grown under water deficit. *Plants*, *12*(1), 218. doi:10.3390/plants12010218 PMID:36616345

Alabdallah, N. M., & Hasan, M. M. (2021). Plant-based green synthesis of silver nanoparticles and its effective role in abiotic stress tolerance in crop plants. *Saudi Journal of Biological Sciences*, *28*(10), 5631–5639. doi:10.1016/j.sjbs.2021.05.081 PMID:34588874

Amna, A. B., Azeem, M. A., Qayyum, A., Mustafa, G., Ahmad, M. A., & Chaudhary, H. J. (2022). Bio-fabricated silver nanoparticles: A sustainable approach for augmentation of plant growth and pathogen control. In Sustainable agriculture reviews 53: nanoparticles: a new tool to enhance stress tolerance (pp. 345-371). Cham: Springer International Publishing.

Ashraf, M., & Foolad, M. R. (2005). Pre-sowing seed treatment—A shotgun approach to improve germination, plant growth, and crop yield under saline and non-saline conditions. *Advances in Agronomy*, *88*, 223–271. doi:10.1016/S0065-2113(05)88006-X

Azeem, A., Abbas, N., Azeem, S., Iqbal, Z., & Ul-Allah, S. (2023). Physiological and Molecular Mechanism of Nanoparticles Induced Tolerance in Plants. In *Emerging Contaminants and Plants: Interactions, Adaptations and Remediation Technologies* (pp. 233–248). Springer International Publishing. doi:10.1007/978-3-031-22269-6_9

Behl, T., Kaur, I., Sehgal, A., Singh, S., Sharma, N., Bhatia, S., Al-Harrasi, A., & Bungau, S. (2022). The dichotomy of nanotechnology as the cutting edge of agriculture: Nano-farming as an asset versus nanotoxicity. *Chemosphere, 288*, 132533. doi:10.1016/j.chemosphere.2021.132533 PMID:34655646

Bhattacharjee, R., Kumar, L., Mukerjee, N., Anand, U., Dhasmana, A., Preetam, S., Bhaumik, S., Sihi, S., Pal, S., Khare, T., Chattopadhyay, S., El-Zahaby, S. A., Alexiou, A., Koshy, E. P., Kumar, V., Malik, S., Dey, A., & Proćków, J. (2022). The emergence of metal oxide nanoparticles (NPs) as a phytomedicine: A two-facet role in plant growth, nano-toxicity and anti-phyto-microbial activity. *Biomedicine and Pharmacotherapy, 155*, 113658. doi:10.1016/j.biopha.2022.113658 PMID:36162370

Chen, N., Yao, P., Zhang, W., Zhang, Y., Xin, N., Wei, H., & Zhao, C. (2022). Selenium nanoparticles: Enhanced nutrition and beyond. *Critical Reviews in Food Science and Nutrition*, 1–12. PMID:35848122

Dhal, P., Sahu, G., Dhal, A., Mohanty, S., & Dash, S. K. (2022). Priming of vegetable seeds: A review. *The Pharma Innovation Journal, 11*(2), 519–525.

Dimkpa, C. O., McLean, J. E., Britt, D. W., & Anderson, A. J. (2015). Nano-CuO and interaction with nano-ZnO or soil bacterium provide evidence for the interference of nanoparticles in metal nutrition of plants. *Ecotoxicology (London, England), 24*(1), 119–129. doi:10.100710646-014-1364-x PMID:25297564

Dos Santos, C. A., Ingle, A. P., & Rai, M. (2020). The emerging role of metallic nanoparticles in food. *Applied Microbiology and Biotechnology, 104*(6), 2373–2383. doi:10.100700253-020-10372-x PMID:31989225

Elemike, E. E., Uzoh, I. M., Onwudiwe, D. C., & Babalola, O. O. (2019). The role of nanotechnology in the fortification of plant nutrients and improvement of crop production. *Applied Sciences (Basel, Switzerland), 9*(3), 499. doi:10.3390/app9030499

Garg, R., & Maheshwari, S. (2023). SEED PRIMING AND FORTIFICATION OF SEEDS USING NANOTECHNOLOGY: A REVIEW. *EPH-International Journal of Agriculture and Environmental Research, 9*(1), 11–17. doi:10.53555/eijaer.v9i1.68

Hassanisaadi, M., Barani, M., Rahdar, A., Heidary, M., Thysiadou, A., & Kyzas, G. Z. (2022). Role of agrochemical-based nanomaterials in plants: Biotic and abiotic stress with germination improvement of seeds. *Plant Growth Regulation, 97*(2), 375–418. doi:10.100710725-021-00782-w

Hernández-Díaz, J. A., Garza-García, J. J., Zamudio-Ojeda, A., León-Morales, J. M., López-Velázquez, J. C., & García-Morales, S. (2021). Plant-mediated synthesis of nanoparticles and their antimicrobial activity against phytopathogens. *Journal of the Science of Food and Agriculture, 101*(4), 1270–1287. doi:10.1002/jsfa.10767 PMID:32869290

Imtiaz, H., Shiraz, M., Mir, A. R., Siddiqui, H., & Hayat, S. (2023). Nano-Priming Techniques for Plant Physio-Biochemistry and Stress Tolerance. *Journal of Plant Growth Regulation*, 1–21. doi:10.100700344-023-10981-6

Jadhav, R., Pawar, P., Choudhari, V., Topare, N., Raut-Jadhav, S., Bokil, S., & Khan, A. (2023). An overview of antimicrobial nanoparticles for food preservation. *Materials Today: Proceedings, 72*, 204–216. doi:10.1016/j.matpr.2022.07.045

Jiang, H. S., Qiu, X. N., Li, G. B., Li, W., & Yin, L. Y. (2014). Silver nanoparticles induced accumulation of reactive oxygen species and alteration of antioxidant systems in the aquatic plant Spirodela polyrhiza. *Environmental Toxicology and Chemistry, 33*(6), 1398–1405. doi:10.1002/etc.2577 PMID:24619507

Khodakovskaya, M. V., Kim, B. S., Kim, J. N., Alimohammadi, M., Dervishi, E., Mustafa, T., & Cernigla, C. E. (2013). Carbon nanotubes as plant growth regulators: Effects on tomato growth, reproductive system, and soil microbial community. *Small, 9*(1), 115–123. doi:10.1002mll.201201225 PMID:23019062

Kumar, S., Nehra, M., Dilbaghi, N., Marrazza, G., Tuteja, S. K., & Kim, K. H. (2020). Nanovehicles for plant modifications towards pest-and disease-resistance traits. *Trends in Plant Science, 25*(2), 198–212. doi:10.1016/j.tplants.2019.10.007 PMID:31780333

Machado, T. O., Grabow, J., Sayer, C., de Araújo, P. H., Ehrenhard, M. L., & Wurm, F. R. (2022). Biopolymer-based nanocarriers for sustained release of agrochemicals: A review on materials and social science perspectives for a sustainable future of agri-and horticulture. *Advances in Colloid and Interface Science, 303*, 102645. doi:10.1016/j.cis.2022.102645 PMID:35358807

Mathur, S., Singh, D., & Ranjan, R. (2022). Remediation of heavy metal (loid) contaminated soil through green nanotechnology. *Frontiers in Sustainable Food Systems, 6*, 932424. doi:10.3389/fsufs.2022.932424

Nile, S. H., Thiruvengadam, M., Wang, Y., Samynathan, R., Shariati, M. A., Rebezov, M., Nile, A., Sun, M., Venkidasamy, B., Xiao, J., & Kai, G. (2022). Nano-priming as emerging seed priming technology for sustainable agriculture—Recent developments and future perspectives. *Journal of Nanobiotechnology*, *20*(1), 1–31. doi:10.118612951-022-01423-8 PMID:35659295

Rai, M., & Ingle, A. (2012). Role of nanotechnology in agriculture with special reference to management of insect pests. *Applied Microbiology and Biotechnology*, *94*(2), 287–293. doi:10.100700253-012-3969-4 PMID:22388570

Rai-Kalal, P., & Jajoo, A. (2021). Priming with zinc oxide nanoparticles improve germination and photosynthetic performance in wheat. *Plant Physiology and Biochemistry*, *160*, 341–351. doi:10.1016/j.plaphy.2021.01.032 PMID:33548801

Rajput, V. D., Singh, A., Minkina, T., Rawat, S., Mandzhieva, S., Sushkova, S., Shuvaeva, V., Nazarenko, O., Rajput, P., Komariah, Verma, K. K., Singh, A. K., Rao, M., & Upadhyay, S. K. (2021). Nano-enabled products: Challenges and opportunities for sustainable agriculture. *Plants*, *10*(12), 2727. doi:10.3390/plants10122727 PMID:34961197

Rhaman, M. S., Tania, S. S., Imran, S., Rauf, F., Kibria, M. G., Ye, W., Hasanuzzaman, M., & Murata, Y. (2022). Seed priming with nanoparticles: An emerging technique for improving plant growth, development, and abiotic stress tolerance. *Journal of Soil Science and Plant Nutrition*, *22*(4), 4047–4062. doi:10.100742729-022-01007-3

Safdar, M., Kim, W., Park, S., Gwon, Y., Kim, Y. O., & Kim, J. (2022). Engineering plants with carbon nanotubes: A sustainable agriculture approach. *Journal of Nanobiotechnology*, *20*(1), 1–30. doi:10.118612951-022-01483-w PMID:35701848

Sharma, B., Tiwari, S., Kumawat, K. C., & Cardinale, M. (2023). Nano-biofertilizers as bio-emerging strategies for sustainable agriculture development: Potentiality and their limitations. *The Science of the Total Environment*, *860*, 160476. doi:10.1016/j.scitotenv.2022.160476 PMID:36436627

Sharma, S., Singh, V. K., Kumar, A., & Mallubhotla, S. (2019). Effect of nanoparticles on oxidative damage and antioxidant defense system in plants. *Molecular Plant Abiotic Stress: Biology and Biotechnology*, 315-333.

Singh, A., Sharma, R., Rawat, S., Singh, A. K., Rajput, V. D., Fedorov, Y., & Chaplygin, V. (2022). Nanomaterial-plant interaction: Views on the pros and cons. In *Toxicity of Nanoparticles in Plants* (pp. 47–68). Academic Press. doi:10.1016/B978-0-323-90774-3.00015-5

Tao, R., You, C., Qu, Q., Zhang, X., Deng, Y., Ma, W., & Huang, C. (2023). Recent advances in the design of controlled-and sustained-release micro/nanocarriers of pesticide. *Environmental Science. Nano*, *10*(2), 351–371. doi:10.1039/D2EN00446A

Timmusk, S., Seisenbaeva, G., & Behers, L. (2018). Titania (TiO2) nanoparticles enhance the performance of growth-promoting rhizobacteria. *Scientific Reports*, *8*(1), 617. doi:10.103841598-017-18939-x PMID:29330479

Trivedi, M., Kedari, S., & Nikalje, G. C. (2022). Role of nanoparticles in remediation of contaminated soil. In *The Role of Nanoparticles in Plant Nutrition under Soil Pollution: Nanoscience in Nutrient Use Efficiency* (pp. 353–370). Springer International Publishing. doi:10.1007/978-3-030-97389-6_15

Wang, D., Saleh, N. B., Byro, A., Zepp, R., Sahle-Demessie, E., Luxton, T. P., Ho, K. T., Burgess, R. M., Flury, M., White, J. C., & Su, C. (2022). Nano-enabled pesticides for sustainable agriculture and global food security. *Nature Nanotechnology*, *17*(4), 347–360. doi:10.103841565-022-01082-8 PMID:35332293

Wang, X., Jiang, J., Dou, F., Sun, W., & Ma, X. (2021). Simultaneous mitigation of arsenic and cadmium accumulation in rice (Oryza sativa L.) seedlings by silicon oxide nanoparticles under different water management schemes. *Paddy and Water Environment*, *19*(4), 569–584. doi:10.100710333-021-00855-6

Wu, H., Tito, N., & Giraldo, J. P. (2017). Anionic cerium oxide nanoparticles protect plant photosynthesis from abiotic stress by scavenging reactive oxygen species. *ACS Nano*, *11*(11), 11283–11297. doi:10.1021/acsnano.7b05723 PMID:29099581

Xin, X., Zhao, F., Judy, J. D., & He, Z. (2022). Copper stress alleviation in corn (Zea mays L.): Comparative efficiency of carbon nanotubes and carbon nanoparticles. *NanoImpact*, *25*, 100381. doi:10.1016/j.impact.2022.100381 PMID:35559887

Zhang, F., Cui, Z., Fan, M., Zhang, W., Chen, X., & Jiang, R. (2011). Integrated soil–crop system management: Reducing environmental risk while increasing crop productivity and improving nutrient use efficiency in China. *Journal of Environmental Quality*, *40*(4), 1051–1057. doi:10.2134/jeq2010.0292 PMID:21712573

Zhao, L., Lu, L., Wang, A., Zhang, H., Huang, M., Wu, H., Xing, B., Wang, Z., & Ji, R. (2020). Nano-biotechnology in agriculture: Use of nanomaterials to promote plant growth and stress tolerance. *Journal of Agricultural and Food Chemistry*, *68*(7), 1935–1947. doi:10.1021/acs.jafc.9b06615 PMID:32003987

Section 4

Nanopriming-Based Plant Breeding and Omics Response for Sustainable Agriculture

Chapter 7
Emerging Frontiers in Plant Breeding Molecular Markers and Nanopriming for Enhanced Crop Improvement

Sapna Rawat
Delhi University, India

ABSTRACT

Plant breeding plays a crucial role in improving plant genotypes and ensuring global food security. However, traditional breeding methods have limitations in effectively selecting traits, particularly quantitative traits, and can be time-consuming and unreliable. Breeders have introduced molecular markers into plant breeding practices to address these problems, resulting in the fields of "smart breeding," "molecular breeding," and "marker-assisted breeding." To pinpoint the genetic areas that regulate desired features, a number of DNA-based molecular marker systems, including RFLP, RAPD, AFLP, SSR, and SNP, have been created. These markers eliminate the need for phenotype-based selection by streamlining the identification of target genomic areas based on genotypes.

INTRODUCTION

In the realm of plant breeding, the integration of molecular markers has revolutionized breeding strategies and opened new avenues for improvement. Molecular markers serve as valuable tools for identifying genomic regions associated with traits of interest and accelerating selection based on marker genotypes rather than relying solely on phenotypic observations (Collard and Mackill, 2008; Varshney et al., 2009).

DOI: 10.4018/978-1-6684-7232-3.ch007

The mapping of these loci to particular genomic regions has been made easier by the creation of high-density linkage maps and the discovery of QTLs using molecular markers. Linkage maps can be created by analysing the relationships between genotype and phenotype in populations, and marker-trait correlations can direct indirect selection of genes or QTLs using marker genotypes, or marker-assisted selection (MAS) (Ribaut & Hoisington, 1998). In comparison to conventional breeding techniques, MAS provides a more efficient, effective, dependable, and economical strategy. In a variety of crop species, such as rice, wheat, maize, barley, tuber crops, pulses, oilseeds, and horticultural crops, molecular markers tightly linked to agronomically significant genes have emerged as molecular tools for MAS. This has increased global food production and enhanced the effectiveness of traditional plant breeding (Mackill et al., 1999; McCouch and Doerge, 1995; Eagles et al., 2001; Van Sanford et al., 2001; Stuber et al., 1999; Tuberosa et al., 2003; Thomas, 2003; Barone, 2004; Fregene et al., 2001; Gebhardt & Valkonen, 2001; Kelly et al., 2003; Svetleva et al., 2003).

Nano priming using nanoparticles is a valuable technique in marker-based plant breeding Radhakrishnan, R et al., 2013). It improves the breeding process by enhancing efficiency and effectiveness. Nano priming offers several benefits, including enhanced germination rates and uniformity, increased seedling vigor, improved stress tolerance, and efficient delivery of molecular markers (Khodakovskaya, M et al., 2009, Pattemore, J. A et al., 2008). These advantages enable breeders to obtain a uniform population of seedlings, select healthier and more robust plants, enhance stress tolerance, and efficiently incorporate molecular markers into breeding programs (Shukla P et al., 2019). Nano priming, also referred to as nanoparticle seed priming, has garnered considerable interest in the agricultural and plant science domains. This technique entails the utilization of nanoparticles on seeds before sowing to augment seed germination, seedling growth, and overall plant development. Nano priming shows promise as an effective strategy to enhance crop productivity, bolster stress tolerance, and optimize nutrient utilization in an environmentally sustainable and efficient manner. However, further research is needed to optimize nanoparticle types, concentrations, and application methods for specific crops and desired traits in marker-based plant breeding.

What Are Genetic Markers?

Genetic markers play a crucial role in identifying genetic variation among species or organisms and are not directly involved in controlling specific traits (Winter & Kahl, 1995). These markers can be categorized into three main types: morphological markers, biochemical markers, and DNA markers.

Morphological markers are visually identifiable phenotypic characteristics, while biochemical markers involve allelic variants of enzymes that can be detected through specific staining methods. However, these markers have limitations as their number may be limited and their expression can be influenced by environmental factors and developmental stages of the plant. On the other hand, DNA markers, also known as molecular markers, are widely utilized in plant breeding due to their abundance and selective neutrality. They result from various types of DNA mutations, including substitutions, rearrangements, or errors during DNA replication of tandemly repeated DNA. Unlike the other marker types, DNA markers are typically located in non-coding regions of DNA and are not affected by environmental factors or the developmental stage of the plant (Winter & Kahl, 1995). The applications of DNA markers in plant breeding are diverse and include assessing genetic diversity within germplasm and determining cultivar identity (Baird et al., 1997; Jahufer et al., 2003; Winter & Kahl, 1995).

DNA markers can be broadly classified into three categories based on their detection methods:

1. Hybridization-based markers: An example is Restriction Fragment Length Polymorphism (RFLP), which involves fragmenting DNA using a restriction enzyme, separating the fragments based on length through agarose gel electrophoresis, and detecting them using labeled DNA probes. RFLP markers find common use in genome mapping and variation analysis, such as genotyping, forensics, paternity tests, and hereditary disease diagnostics (Winter & Kahl, 1995).
2. Polymerase chain reaction (PCR)-based markers: This category includes Random Amplification of Polymorphic DNA (RAPD), Amplified Fragment Length Polymorphism (AFLP), and Simple Sequence Repeats (SSR). RAPD utilizes arbitrary primers to amplify DNA segments, which are then analyzed for differences in band patterns. AFLP is a PCR-based fingerprinting technique that selectively amplifies DNA fragments and compares unique fingerprints for the genomes of interest. SSR markers consist of stretches of DNA with repeated nucleotide motifs, and their amplification products exhibit length polymorphisms due to variation in the number of repeat motifs. SSR markers are codominant, meaning they indicate differences in size and can be used to analyze genetic variation (Winter & Kahl, 1995).
3. DNA sequence-based markers: This class encompasses Single Nucleotide Polymorphisms (SNPs), which are variations in single nucleotides within a locus resulting from point mutations. SNPs are abundant in genomes and can have bi- or multi-allelic forms (Gupta et al., 1999; Joshi et al., 1999; Winter & Kahl, 1995).

Visualization of genetic differences revealed by DNA markers often involves techniques such as gel electrophoresis and staining with chemicals like ethidium bromide or silver, or detection using radioactive or colorimetric probes. Polymorphic markers can distinguish variations between individuals or species, while monomorphic markers cannot. Polymorphic markers can be further classified as either codominant, indicating differences in size, or dominant, where they are either present or absent (Winter & Kahl, 1995).

How to Construct a Linkage Map?

The construction of a linkage map involves determining the relative distances between genetic markers along chromosomes, providing valuable information about the chromosomal locations of genes and Quantitative Trait Loci (QTLs) (Paterson, 1996a). QTL mapping utilizes recombination during meiosis to analyze genes and markers in progeny, with markers closely linked being transmitted together more frequently (Paterson, 1996a). The construction of a linkage map typically involves three steps. Firstly, a mapping population is created, consisting of individuals that differ in the traits of interest (McCouch & Doerge, 1995; Paterson, 1996a). Secondly, polymorphic markers that exhibit differences between the parents are identified, and their segregation ratios are analyzed (Xu et al., 1997; Sayed et al., 2002). Lastly, linkage analysis is performed using genotypic data to construct the linkage map, with linkage between markers determined by LOD scores and software programs such as Mapmaker/EXP, MapManager QTX, and JoinMap (Lander et al., 1987; Lincoln et al., 1993a; Manly et al., 2001; Stam, 1993).

Mapping functions, such as the Kosambi and Haldane mapping functions, are used to convert recombination fractions into genetic distances, with the relationship between genetic and physical distances influenced by recombination hotspots and cold spots (Hartl & Jones, 2001; Kearsey & Pooni, 1996; Paterson, 1996a).

QTL Analysis

The process of identifying genes or Quantitative Trait Loci (QTL) involves the use of markers to divide the mapping population into genotypic groups, and significant differences in phenotypic means among these groups indicate the presence of a linked QTL (Tanksley, 1993). The proximity of a marker to a QTL influences the likelihood of recombination events between them, with a significant p-value indicating close linkage (Tanksley, 1993). Conversely, when a marker is loosely linked or unlinked to a QTL, no significant differences in means are observed (Tanksley, 1993).

How to Detect QTLs?

Three widely used methods for detecting QTLs are as follows:

1. Single-marker analysis, also referred to as single-point analysis, is a straightforward approach that examines the association between QTLs and individual markers using statistical techniques such as t-tests, ANOVA, and linear regression (Manly et al., 2001; Nelson, 1997). This method does not require a complete linkage map and can be performed using basic statistical software programs. However, its effectiveness decreases as the distance between a QTL and a marker increases.
2. Simple interval mapping (SIM) utilizes linkage maps and simultaneously assesses intervals between adjacent pairs of linked markers along chromosomes, offering greater statistical power compared to single-point analysis (Lander & Botstein, 1989; Liu, 1998). Software packages such as MapMaker/QTL and QGene are commonly employed for SIM (Lincoln et al., 1993b; Nelson, 1997).
3. Composite interval mapping (CIM) is the most commonly used method for QTL mapping. It combines interval mapping with linear regression and incorporates additional genetic markers into the statistical model, providing improved precision and effectiveness in QTL mapping compared to single-point analysis and interval mapping alone. Software programs such as QTL Cartographer, MapManager QTX, and PLABQTL are commonly used for CIM (Basten et al., 1994, 2001; Manly et al., 2001; Utz & Melchinger, 1996).

These three methods vary in their statistical power and precision for detecting QTLs, with CIM being widely preferred due to its superior performance compared to single-point analysis and interval mapping.

Analysis of Interval Mapping Results

Interval mapping methods, including SIM and CIM, are utilized to identify potential QTL positions between adjacent linked markers. The statistical measure commonly employed to represent the results is the logarithmic of odds (LOD) score or likelihood ratio statistic (LRS), with the LRS calculated as LRS = 4.6 × LOD (Liu, 1998). The QTL position with the highest LOD value is considered the most probable location of the QTL.

In graphical representations, the output from interval mapping displays linkage groups formed by markers on the x-axis, while the y-axis represents the test statistic (LOD score or LRS). To establish a QTL as statistically significant or "real," the peak or maximum of the test statistic must surpass a predetermined significance threshold. These thresholds are determined through permutation tests (Churchill & Doerge, 1994). Additionally, bootstrapping, a resampling method, can be employed to determine the confidence interval of QTL positions (Visscher et al., 1996). Mapping software programs like MapManager QTX (Manly et al., 2001) readily incorporate bootstrapping for QTL analysis.

Factors Impacting Detection of QTL

Interval mapping methods, such as Simple Interval Mapping (SIM) and Composite Interval Mapping (CIM), are employed to identify potential positions of Quantitative Trait Loci (QTL) between adjacent linked markers. The outcomes of these methods are typically represented using a statistical measure known as the logarithm of odds (LOD) score or likelihood ratio statistic (LRS) (Liu, 1998). The LRS is calculated as LRS = 4.6 × LOD. The QTL position with the highest LOD value is considered the most probable location of the QTL.

In graphical representations, the output of interval mapping is typically depicted with markers forming linkage groups along the x-axis and the test statistic (LOD score or LRS) along the y-axis. To establish a QTL as statistically significant or "real," the peak or maximum of the test statistic must exceed a predetermined significance threshold. These thresholds are determined using permutation tests (Churchill & Doerge, 1994). Another method called "bootstrapping," which involves resampling, can be utilized to determine the confidence interval of QTL positions (Visscher et al., 1996). Notably, certain mapping software programs like MapManager QTX (Manly et al., 2001) readily incorporate bootstrapping for QTL analysis.

Confirmation of QTL Maps

It is highly advisable to independently replicate or confirm QTL mapping studies, as emphasized by Lander and Kruglyak (1995). Replication studies involving separate populations play a crucial role in validating the outcomes of the primary QTL mapping study. These independent populations can be generated using the same parental genotypes or closely related genotypes utilized in the original study.

The replication of QTL mapping studies holds significance for several reasons. Firstly, QTL mapping studies conducted with limited population sizes may exhibit low power in detecting QTLs and a notable bias in estimating QTL effects. Independent replication studies employing larger populations can provide more precise and reliable estimates of QTL positions and effects (Melchinger et al., 1998). Despite the importance of replication studies, they are often lacking in QTL mapping research. This could be attributed to various factors, including limited research funding and time constraints, as well as a potential lack of awareness regarding the necessity of confirming results. However, it is crucial to emphasize the need for independent confirmation of QTL mapping studies to ensure the robustness and generalizability of the findings.

By conducting replication studies, researchers can reinforce the validity of their QTL mapping results and enhance confidence in the identified QTLs. Replication studies contribute to the overall comprehension of the genetic architecture of the investigated traits and foster the advancement of knowledge in the field of QTL mapping.

Marker-Assisted Selection (MAS)

Marker-assisted selection (MAS) is a powerful tool in plant breeding that utilizes genetic markers to select plants carrying specific genes or quantitative trait loci (QTLs) associated with desirable traits. The development and application of markers for MAS involve several important steps:

1. High-resolution mapping: By employing larger population sizes and a greater number of markers, high-resolution mapping aims to identify tightly-linked markers. This fine mapping process helps in pinpointing markers closely linked to the target QTLs, which can be used as reliable markers for MAS. The required population size for high-resolution mapping may vary depending on the specific study.
2. Marker validation: It is essential to validate markers by testing their effectiveness in independent populations and diverse genetic backgrounds. Marker validation ensures that the markers reliably determine the presence of the target phenotype. Validation studies are crucial for assessing marker performance and their applicability in different breeding programs and genetic backgrounds (Collins et al., 2003; Sharp et al., 2001).
3. Marker conversion: In certain cases, markers may need to be converted into other types to address issues of reproducibility or to overcome the complexity, time-consuming nature, or high cost associated with certain marker techniques. Marker conversion allows for the utilization of markers that are more suitable for the specific breeding objectives (e.g., converting RAPDs to more reliable marker types like RFLPs or AFLPs) (Collins et al., 2003).

MAS offers several advantages over conventional breeding methods, including:

1. Time-saving: MAS eliminates the need for extensive field trials by enabling selection at the seedling stage based on marker-assisted evaluation.

2. Early selection: MAS allows for the selection of desired genotypes at an early stage, reducing the time required for trait evaluation.
3. Elimination of phenotypic evaluation: MAS bypasses the laborious and subjective process of phenotypic evaluation, making the selection process more efficient and objective.
4. Gene pyramiding: MAS enables the simultaneous selection and combination of multiple genes or QTLs, facilitating the development of superior genotypes.
5. Avoidance of linkage drag: By utilizing tightly-linked markers, MAS helps in avoiding the transfer of undesirable or deleterious genes located near the target QTLs.
6. Selection for traits with low heritability: MAS allows for the selection of traits with low heritability, which may be challenging through conventional breeding methods.
7. Testing for traits with limited phenotypic evaluation: In cases where phenotypic evaluation is impractical or difficult, MAS provides a means to test for specific traits using genetic markers.

When selecting markers for MAS, any markers that are tightly linked to the target QTLs can be used. Even selecting a single QTL through MAS can be beneficial in plant breeding, provided that the selected QTL explains a substantial proportion of the phenotypic variance for the trait (Tanksley, 1993). It is also important to consider the stability of the selected QTL across different environments (Hittalmani et al., 2002). Some studies have shown that the development of MAS is cost-effective compared to conventional breeding methods, although there may be an initial high cost associated with marker development, subsequent marker usage is often more economical (Yu et al., 2000).

Overall, MAS offers significant advantages in plant breeding, accelerating the selection process and facilitating the development of improved cultivars with desired traits.

What Is Nano Priming?

Nano priming, also known as seed priming using nanoparticles, has gained significant attention in agriculture and plant science. It involves applying nanoparticles to seeds before planting to improve seed germination, seedling growth, and overall plant performance. Nano priming shows promise in enhancing crop productivity, stress tolerance, and nutrient utilization in a sustainable manner. Nanoparticles used in nano priming can vary in composition, including metallic nanoparticles (e.g., silver, iron oxide), carbon-based nanoparticles (e.g., carbon nanotubes), and other types (e.g., silica, zinc oxide). These nanoparticles possess unique properties that make them suitable for plant-based applications.

It enhances seed germination by providing optimal hydration and activating metabolic processes, leading to faster and more uniform germination. This aids in improving crop establishment and achieving higher stand densities. Additionally, nano priming improves seedling vigor, nutrient uptake, enzymatic activities, and plant hormone levels, resulting in enhanced root and shoot development, increased biomass accumulation, and improved plant architecture (Hussain HA et al., 2018; Mahakham W et al., 2019; Dehghani Bidgoli, A et al., 2019)

Nano priming also enhances plants' ability to tolerate abiotic and biotic stresses. It activates stress-responsive genes, induces antioxidant defense mechanisms, and regulates hormonal balance, making plants more resilient to factors like drought, salinity, heat, and pathogens. This aspect is crucial for sustainable agriculture and the development of stress-tolerant crop varieties (Kaushik P et al., 2019; Khan MN et al., 2021).

Furthermore, nano priming enables targeted and controlled delivery of bioactive compounds, growth regulators, and genetic material to seeds. This facilitates the incorporation of molecular markers, genes, or other biologically active substances during priming, enabling marker-based breeding and advanced plant improvement strategies.

Role of Nano Priming in Marker-Based Plant Breeding

Nano priming using nanoparticles plays a crucial role in marker-based plant breeding as it improves the efficiency and effectiveness of the breeding process. Marker-based breeding involves using molecular markers to select and breed plants with desired traits. Nano priming offers several advantages in this context:

1. **Enhanced Germination**: Nano priming improves seed germination rates and uniformity by providing optimal hydration, activating metabolic processes, and promoting early growth. This results in faster and more synchronized germination, enabling breeders to obtain a uniform population of seedlings for marker analysis and selection (Sharma P et al., 2020).
2. **Increased Seedling Vigor**: Nano priming enhances the vigor and health of seedlings by enhancing nutrient uptake, activating beneficial enzymes, and stimulating physiological processes. This leads to healthier and more robust seedlings, which is advantageous for accurate marker analysis and selection as healthier seedlings are more likely to express their genetic potential (Kalita P et al., 2021).

3. **Stress Tolerance**: Nano priming improves the stress tolerance of plants, which is particularly important for marker-based breeding focused on enhancing resistance to environmental stresses. Nanoparticles activate stress-responsive genes, enhance antioxidant systems, and regulate hormone levels, improving plants' ability to cope with abiotic and biotic stresses (Singh S et al., 2021; Ghosh P et al., 2019). This enables breeders to select plants with improved stress tolerance based on molecular markers.
4. **Efficient Marker Delivery**: Nano priming facilitates the delivery of molecular markers into seeds. Nanoparticles can serve as carriers for DNA markers, allowing their targeted delivery and uptake by seeds (Lal M et al., 2020). This simplifies the incorporation of molecular markers into breeding programs, enabling breeders to efficiently select plants based on desired genetic traits.

By combining nano priming with marker-based breeding approaches, breeders can enhance the accuracy, efficiency, and success of their breeding programs. Nano priming optimizes seed germination and seedling development, improves stress tolerance, and enables targeted delivery of molecular markers. However, further research is needed to optimize nanoparticle types, concentrations, and application methods for specific crop species and desired traits in marker-based plant breeding.

CONCERNS

Nano priming in plant breeding include limited understanding of its long-term effects and potential risks, variability and optimization challenges, high cost and scalability concerns, potential environmental impact, regulatory considerations, compatibility and integration issues, and ethical and social considerations. Ethical considerations surrounding nano priming in plant breeding extend beyond technological aspects and encompass concerns related to intellectual property rights, ownership, and equitable access to improved varieties. These ethical implications necessitate careful deliberation and proactive measures to address them effectively. The economic viability of nano priming in plant breeding programs, particularly on a large scale, requires a thorough assessment of the production and application costs associated with nanoparticles. Evaluating the scalability and cost-effectiveness of nano priming is crucial to determine its practicality and feasibility for widespread adoption in commercial agriculture. Addressing these limitations through further research, risk assessment, and responsible implementation is important to ensure the safe and effective use of nano priming in plant breeding programs.

CONCLUSION

The integration of molecular markers in plant breeding, along with the application of nano priming techniques, offers a powerful approach for crop improvement. Molecular markers enable precise trait selection and breeding program optimization, while nano priming enhances seed germination, stress tolerance, and overall crop productivity. By combining these strategies, breeders can develop improved crop varieties with enhanced agronomic traits and resilience. This integrated approach holds great promise for addressing global agricultural challenges and achieving a more sustainable and secure food supply. Despite the promising potential of nano priming, further research is needed to optimize nanoparticle types, concentrations, application methods, and their effects on different crop species and environmental conditions. Additionally, evaluating the safety and potential environmental impacts of nanoparticles in agriculture is essential for their responsible use.

ACKNOWLEDGMENTS

The authors are thankful to Dr Abhishek Singh and Dr Vishnu Rajput for helping during the preparation of this manuscript.

REFERENCES

Baird, V., Abbott, A., Ballard, R., Sosinski, B., & Rajapakse, S. (1997). Current Topics in Plant Molecular Biology: Technology Transfer of Plant Biotechnology. CRC Press, Boca Raton. P. Gresshoff (Ed.), DNA Diagnostics in Horticulture, 111–130.

Barone, A. (2004). Molecular marker-assisted selection for potato breeding. *American Journal of Potato Research*, *81*(2), 111–117. doi:10.1007/BF02853608

Basten, C., Weir, B., & Zeng, Z.-B. (2001). *QTL Cartographer*. Department of Statistics, North Carolina State University.

Basten, C. J., Weir, B. S., & Zeng, Z.-B. (1994). Zmap-a QTL cartographer. J.S.G.C. Smith.

Beattie, A., Larsen, J., Michaels, T., & Pauls, K. (2003). Mapping quantitative trait loci for a common bean (*Phaseolus vulgaris L.*) ideotype. *Genome*, *46*(3), 411–422. doi:10.1139/g03-015 PMID:12834057

B. J. Benkel, W. F. Chesnais, J. P. Gibson, B. W. Kennedy, & E. B. Burnside (Eds.). *Proceedings of the 5th World Congress on Genetics Applied to Livestock Production: Computing Strategies and Software*. Published by the Organizing Committee.

Churchill, G. A., & Doerge, R. W. (1994). Empirical threshold values for quantitative trait mapping. *Genetics*, *138*(3), 963–971. doi:10.1093/genetics/138.3.963 PMID:7851788

Collins, H. M., Panozzo, J. F., Logue, S. J., Jefferies, S. P., & Barr, A. R. (2003). Mapping and validation of chromosome regions associated with high malt extract in barley (*Hordeum vulgare L.). Australian Journal of Agricultural Research*, *54*(12), 1223–1240. doi:10.1071/AR02201

Eagles, H., Bariana, H., Ogbonnaya, F., Rebetzke, G., Hollamby, G., Henry, R., Henschke, P., & Carter, M. (2001). Implementation of markers in Australian wheat breeding. *Australian Journal of Agricultural Research*, *52*(12), 1349–1356. doi:10.1071/AR01067

Gebhardt, C., & Valkonen, J. P. T. (2001). Organization of genes controlling disease resistance in the potato genome. *Annual Review of Phytopathology*, *39*(1), 79–102. doi:10.1146/annurev.phyto.39.1.79 PMID:11701860

George, M. L. C., Prasanna, B. M., Rathore, R. S., Setty, T. A. S., Kasim, F., Azrai, M., Vasal, S., Balla, O., Hautea, D., Canama, A., Regalado, E., Vargas, M., Khairallah, M., Jeffers, D., & Hoisington, D. (2003). Identification of QTLs conferring resistance to downy mildews of maize in Asia. *Theoretical and Applied Genetics*, *107*(3), 544–551. doi:10.100700122-003-1280-6 PMID:12759731

Gupta, P., Varshney, R., Sharma, P., & Ramesh, B. (1999). Molecular markers and their applications in wheat breeding. *Plant Breeding*, *118*(5), 369–390. doi:10.1046/j.1439-0523.1999.00401.x

Hackett, C. (2002). Statistical methods for QTL mapping in cereals. *Plant Molecular Biology*, *48*(5/6), 585–599. doi:10.1023/A:1014896712447 PMID:11999836

Haley, C., & Andersson, L. (1997). Linkage mapping of quantitative trait loci in plants and animals, P. Dear (Ed.), Genome mapping–A practical approach. Oxford University Press.

Hartl, D., & Jones, E. (2001). *Genetics: Analysis of Genes and Genomes*. Jones and Bartlett Publishers.

Hittalmani, S., Shashidhar, H. E., Bagali, P. G., Huang, N., Sidhu, J. S., Singh, V. P., & Khush, G. S. (2002). Molecular mapping of quantitative trait loci for plant growth, yield and yield related traits across three diverse locations in a doubled haploid rice population. *Euphytica, 125*(2), 207–214. doi:10.1023/A:1015890125247

Jahufer, M., Barret, B., Griffiths, A., & Woodfield, D. (2003). DNA fingerprinting and genetic relationships among white clover cultivars. J. Morton (Ed.), *Proceedings of the New Zealand Grassland Association*. Taieri Print Limited, Dunedin.

Jansen, R., & Stam, P. (1994). High resolution of quantitative traits into multiple loci via interval mapping. *Genetics, 136*(4), 1447–1455. doi:10.1093/genetics/136.4.1447 PMID:8013917

Jansen, R., & Stam, P. (1994). High resolution of quantitative traits into multiple loci via interval mapping. *Genetics, 136*(4), 1447–1455. doi:10.1093/genetics/136.4.1447 PMID:8013917

Joshi, S., Ranjekar, P., & Gupta, V. (1999). Molecular markers in plant genome analysis. *Current Science, 77*, 230–240.

Kearsey, M., & Pooni, H. (1996). *The genetical analysis of quantitative traits*. Chapman & Hall. doi:10.1007/978-1-4899-4441-2

Kelly, J. D., Gepts, P., Miklas, P. N., & Coyne, D. P. (2003). Tagging and mapping of genes and QTL and molecular marker-assisted selection for traits of economic importance in bean and cowpea. *Field Crops Research, 82*(2-3), 135–154. doi:10.1016/S0378-4290(03)00034-0

Lander, E., & Botstein, D. (1989). Mapping Mendelian factors underlying quantitative traits using RFLP linkage maps. *Genetics, 121*(1), 185–199. doi:10.1093/genetics/121.1.185 PMID:2563713

Lander, E. S., Green, P., Abrahamson, J., Barlow, A., Daly, M. J., Lincoln, S. E., & Newburg, L. (1987). Mapmaker an interactive computer package for constructing primary genetic linkage maps of experimental and natural populations. *Genomics, 1*(2), 174–181. doi:10.1016/0888-7543(87)90010-3 PMID:3692487

Lander, E. S., & Kruglyak, L. (1995). Genetic dissection of complex traits: Guidelines for interpreting and reporting linkage results. *Nature Genetics, 11*(3), 241–247. doi:10.1038/ng1195-241 PMID:7581446

Lincoln, S., Daly, M., & Lander, E. (1993a). *Constructing genetic linkage maps with MAPMAKER/EXP. Version 3.0* (3rd ed.). Whitehead Institute for Biomedical Research Technical Report.

Lincoln, S., Daly, M., & Lander, E. (1993b). *Mapping genes controlling quantitative traits using MAPMAKER/QTL. Version 1.1* (2nd ed.). Whitehead Institute for Biomedical Research Technical Report.

Liu, B. (1998). *Statistical Genomics: Linkage*. Mapping and QTL Analysis CRC Press.

Mackill, D. J., Nguyen, H. T., & Zhan, J. (1999). Use of molecular markers in plant improvement programs for rain fed lowland rice. *Field Crops Research*, *64*(1-2), 177–185. doi:10.1016/S0378-4290(99)00058-1

Manly, K. F., Cudmore Robert, H. Jr, & Meer, J. M. (2001). Map Manager QTX, cross-platform software for genetic mapping. *Mammalian Genome*, *12*(12), 930–932. doi:10.100700335-001-1016-3 PMID:11707780

McCouch, S. R., & Doerge, R. W. (1995). QTL mapping in rice. *Trends*, *11*, 482–487. PMID:8533164

McCouch, S. R., & Doerge, R. W. (1995). QTL mapping in rice. *Trends in Genetics*, *11*(12), 482–487. doi:10.1016/S0168-9525(00)89157-X PMID:8533164

Mehlenbacher, S. A. (1995). Classical and molecular approaches to breeding fruit and nut crops for disease resistance. *HortScience*, *30*(3), 466–477. doi:10.21273/HORTSCI.30.3.466

Melchinger, A. E., Utz, H. F., & Schon, C. C. (1998). Quantitative trait locus (QTL) mapping using different testers and independent population samples in maize reveals low power of QTL detection and large bias in estimates of QTL effects. *Genetics*, *149*(1), 383–403. doi:10.1093/genetics/149.1.383 PMID:9584111

Michelmore, R. (1995). Molecular approaches to manipulation of disease resistance genes. *Annual Review of Phytopathology*, *33*(1), 393–427. doi:10.1146/annurev.py.33.090195.002141 PMID:18999967

Nelson, J. C. (1997). Qgene—Software for marker-based genomic analysis and breeding. *Molecular Breeding*, *3*(3), 239–245. doi:10.1023/A:1009604312050

Paterson, A., Tanksley, S., & Sorrels, M. E. (1991a). DNA markers in plant improvement. *Advances in Agronomy*, *44*, 39–90. doi:10.1016/S0065-2113(08)60578-7

Paterson, A. H. (1996a). Making genetic maps. In A. H. Paterson (Ed.), *Genome Mapping in Plants, R. G. Landes Company, San Diego, California* (pp. 23–39). Academic Press.

Ribaut, J.-M., & Hoisington, D. (1998). Marker-assisted selection: New tools and strategies. *Trends in Plant Science*, *3*(6), 236–239. doi:10.1016/S1360-1385(98)01240-0

Sayed, H., Kayyal, H., Ramsey, L., Ceccarelli, S., & Baum, M. (2002). Segregation distortion in doubled haploid lines of barley (*Hordeum vulgare L.*) detected by simple sequence repeat markers. *Euphytica*, *225*(2), 265–272. doi:10.1023/A:1015861610226

Sharp, P. J., Johnston, S., Brown, G., McIntosh, R. A., Pallotta, M., Carter, M., Bariana, H. S., Khartkar, S., Lagudah, E. S., Singh, R. P., Khairallah, M., Potter, R., & Jones, M. G. K. (2001). Validation of molecular markers for wheat breeding. *Australian Journal of Agricultural Research*, *52*(12), 1357–1366. doi:10.1071/AR01052

Snowdon, R., & Friedt, W. (2004). Molecular markers in Brassica oilseeds breeding: Current status and future possibilities. *Plant Breeding*, *123*(1), 1–8. doi:10.1111/j.1439-0523.2003.00968.x

Stam, P. (1993). Construction of integrated genetic linkage maps by means of a new computer package: JoinMap. *The Plant Journal*, *3*(5), 739–744. doi:10.1111/j.1365-313X.1993.00739.x

Stuber, C. W., Polacco, M., & Senior, M. L. (1999). Synergy of empirical breeding, marker-assisted selection, and genomics to increase crop yield potential. *Crop Science*, *39*(6), 1571–1583. doi:10.2135/cropsci1999.3961571x

Svetleva, D., Velcheva, M., & Bhowmik, G. (2003). Biotechnology as a useful tool in common bean (*Phaseolus vulgaris L.*) improvement: A review. *Euphytica*, *131*(2), 189–200. doi:10.1023/A:1023983831582

Tanksley, S. D. (1993). Mapping polygenes. *Annual Review of Genetics*, *27*(1), 205–233. doi:10.1146/annurev.ge.27.120193.001225 PMID:8122902

Tuberosa, R., Salvi, S., Sanguineti, M. C., Maccaferri, M., Giuliani, S., & Landi, P. (2003). Searching for quantitative trait loci controlling root traits in maize: A critical appraisal. *Plant and Soil*, *255*(1), 35–54. doi:10.1023/A:1026146615248

Utz, H., & Melchinger, A. (1996). PLABQTL: A program for composite interval mapping of QTL. *J Quant Trait Loci*, *2*(1).

Van Sanford, D., Anderson, J., Campbell, K., Costa, J., Cregan, P., Griffey, C., Hayes, P., & Ward, R. (2001). Discovery and deployment of molecular markers linked to *Fusarium* head blight resistance: An integrated system for wheat and barley. *Crop Science*, *41*(3), 638–644. doi:10.2135/cropsci2001.413638x

Visscher, P., Thompson, R., & Haley, C. (1996). Confidence intervals in QTL mapping by bootstrapping. *Genetics*, *143*(2), 1013–1020. doi:10.1093/genetics/143.2.1013 PMID:8725246

Vos, P., Hogers, R., Bleeker, M., Reijans, M., van de Lee, T., Hoernes, M., Frijters, A., Pot, J., Peleman, J., Kuiper, M., & Zabeau, M. (1995). AFLP: A new technique for DNA fingerprinting. *Nucleic Acids Research*, *23*(21), 4407–4414. doi:10.1093/nar/23.21.4407 PMID:7501463

Williams, K. J. (2003). The molecular genetics of disease resistance in barley. *Australian Journal of Agricultural Research*, *54*(12), 1065–1079. doi:10.1071/AR02219

Winter, P., & Kahl, G. (1995). Molecular marker technologies for plant improvement. *World Journal of Microbiology & Biotechnology*, *11*(4), 438–448. doi:10.1007/BF00364619 PMID:24414752

Xu, Y., Zhu, L., Xiao, J., Huang, N., & McCouch, S. R. (1997). Chromosomal regions associated with segregation distortion of molecular markers in F2, backcross, doubled haploid, and recombinant inbred populations in rice (*Oryza sativa L.*). *Molecular & General Genetics*, *253*(5), 535–545. doi:10.1007004380050355 PMID:9065686

Young, N. D. (1996). QTL mapping and quantitative disease resistance in plants. *Annual Review of Phytopathology*, *34*(1), 479–501. doi:10.1146/annurev.phyto.34.1.479 PMID:15012553

Yu, K., Park, S., & Poysa, V. (2000). Marker-assisted selection of common beans for resistance to common bacterial blight: Efficacy and economics. *Plant Breeding*, *119*(5), 411–415. doi:10.1046/j.1439-0523.2000.00514.x

Zeng, Z.-B. (1993). Theoretical basis for separation of multiple linked gene effects in mapping quantitative trait loci. *Proceedings of the National Academy of Sciences of the United States of America*, *90*(23), 10972–10976. doi:10.1073/pnas.90.23.10972 PMID:8248199

Chapter 8
Impact of Nanopriming and Omics-Based Applications for Sustainable Agriculture

Himani Sharma
Uttaranchal University, India

Atin Kumar
https://orcid.org/0000-0002-1653-2146
Uttaranchal University, India

Rajat Singh
Uttaranchal University, India

Satendra Kumar
Sardar Vallabhbhai Patel University of Agriculture and Technology, India

ABSTRACT

The concept of omics technologies in the agriculture sector is gaining importance nowadays due to its diverse applications like genomics which is based on genome sequences, transcriptomic which is based on gene expression and profiling, proteomics which is based on protein content, and metabolomics which involves the study of the metabolism of an organism. A revolutionary transformation in the last decade in the domain of omics has made analytical and scientific procedures cost-effective, easier, and faster and has increased accuracy. With the introduction of technologies involving omics in the field of agriculture, it has now become easier to identify and recognize the presence of different genes and genomic sequences liable for various agronomic traits like genes responsible for increasing yield, and genes for inducing resistance against biotic and abiotic stresses. Utilization of such technologies has resulted in obtaining accurate gene interpretation and identifying gene markers, thus acting as a guiding hand in the development of new crop varieties with some desirable traits.

DOI: 10.4018/978-1-6684-7232-3.ch008

INTRODUCTION

In order to supply the increasing global population with the food they need, crop plants must urgently increase their yield and nutritional value. The ability to enhance the quantitative and qualitative characteristics of crops has been made feasible by the development of high-capacity omics technologies. Omics technologies refer to a comprehensive method for characterizing and quantifying a biological organism's genes, transcripts, proteins, and metabolites, which are respectively known to be genomics, transcriptomics, proteomics, and metabolomics. (Varshney et al., 2013)

Most crop characteristics are complicated computable traits that are managed by various genes (Core et al., 2008). The economic characteristics of crop plants have been effectively studied and modified using omics technologies to increase yield and nutritional value. New genes have been found and gene expression profiles have been studied thanks to the evolution of next-generation sequencing technologies (NGS Technologies) and the availability of genetic data of ideal agricultural plants (Mochida and Shinozaki, 2010). It takes a lot of decoding and analysis of gene expression to find the important genes that connect to the required phenotypic traits. The discovery of marked loci connected with some phenotypic features has permitted to investigate of various intricate natural biological processes and phases such as plant growth, development, ability to tolerate biotic and abiotic stresses, etc.

The need to develop agricultural approaches that are environmentally friendly and considerate of the environment's assets (water, soil, and biodiversity) is growing in the face of climate change and an expanding world population. This is necessary to ensure that future generations' survival is not jeopardised and to increase productivity so that the current population's food needs can be met. Therefore, sustainable agriculture can be seen as a way to build the resilience of regions and communities.

Sustainable agriculture aims to increase productivity while using fewer resources such as energy and water, to protect the ecosystem by reducing the underlying causes of soil deterioration and environmental damage, and to preserve natural ecosystems. More crucially, we require an agricultural system that can meet social demands and support long-term growth at the same time. Since the Green Revolution of the 1960s, agriculture has become incredibly reliant on nitrogen and pesticide input. Environmental pollution results from this, which is not sustainable on a long-term basis.

Despite the difficulties we face, there are many opportunities as well, particularly in light of the rapidly advancing omics technologies. Thousands of species now have access to high-quality reference genomes, and some species even have multiple reference genomes. The recognition of core- and pan-genomes is made possible by genome-wide resequencing of many kinds. Our knowledge of the domestication of crops and advancement is increased by this. Genomes, epigenomes, transcriptomes, epi transcriptomes, and proteomes are only a few of the layers at which omics data are produced. (Salgotra et al., 2014)

At this time, it is possible to produce omics information regarding any animal at a fair price. It is significant to highlight that numerous informatics methods for matching data are being created quickly in order to extract biological significance from omics data.

Novel technologies like targeted gene editing and the knowledge gathered from omics data can be used to develop novel types and plants for sustainable agriculture. There are currently enormous volumes of omics knowledge. The combined collection of omics knowledge will allow for the quick and high quantities identification of multiple genes in succession for the right attribute. This will significantly alter how this is done in which we do what we're currently studying by moving from analysis of networks or specific genes to networks. Upcoming plants can be developed utilizing an approach with three stages that combines the deep understanding of hybridization and modification gained from omics data with cutting-edge gene editing strategies.

Genomic, proteomic, transcriptomic, and metabolomic technologies have increased the uniformity and reliability of plant breeding, cutting down on time and expense while developing superior agricultural crops that are stress-resistant yet still have an abundance of nutrients. In order to better manage pests, omics has shed light on the genetic mechanisms underlying insect susceptibility to pesticides as well as plant herbicide tolerance. Enhanced genotypes and greater comprehension of the processes underlying insect and weed resilience can result from correlating genes to characteristics since it increases the credibility of science. Through the use of omics, it is now feasible to understand the complex interactions involving the proteins, genes, and compounds that lead to the phenotype. This holistic method mainly depends on computer analysis, bioinformatics, and numerous additional disciplines of biology in order to safeguard and enhance crops. (Rhee et al., 2006)

Omics has the possibility of helping increase studies on agriculture in the fields of nutrition, wellness, vitality, chemical raw materials, and specialty chemicals, as well as safeguard, optimise, and even enhance the health of the planet. Specific traits that are intriguing are the primary objective of omics technology. Omics can boost the dietary content of food to the benefit of customers, especially lycopene-rich tomatoes, slowly ripening fruits, and produce with potent antioxidant capacities Ahmad et al., (2012).

With the use of genetics, we can develop nutritious foods that go beyond the requirements of basic dietary intake, like seed oils that lack trans fats but instead incorporate omega-3 fatty acids, which are beneficial to the heart, or cassava melons with more protein in order to combat starvation. Assessing alterations to protein nutritional value and quantity for optimal nourishment for humans and animals, increasing vitamin and mineral concentrations to cure dietary shortages, and reducing allergens and ant nutritional chemicals that degrade the nutritional value of foods are all feasible through the application of omics. Food comics is another phrase used in the omics field to describe research on food and nutrition that uses integrated omic approaches (Ibanez, 2013).

Omic technologies enable visualisation or tracking of all the changes that occur when an organism's genetic makeup, nutritional status, or environment are changed, demonstrating a comprehension of the changes in plant metabolism brought on by interactions with the environment. Omics can shed light on animals about which we previously believed we knew everything. For a long time, it was believed that the domesticated apple's primary ancestor was the wild apple of central Asia. But utilising genomic markers, it was discovered that the domestic apple Malus domestica's genome received an unusually high contribution from the European crab apple. Genetic research on the existence of plants can be very successful. Studies on domesticated tomatoes have shown that early domestication and current plant breeding have caused genetic alterations throughout time that have resulted in cultivars that are substantially larger than their wild counterparts.

The genes in charge of the proteins that bestow or prevent the desired features can be identified using omics. After being found, these genes can be silenced in plants, transferred from a single plant to another, or even introduced from another species, creating transgenic plants. So, rather than getting a gene or genes through pollination, a transgenic plant has one or more purposefully inserted genes (i.e., transgenes). There are numerous beneficial agronomic features present in transgenic crop cultivars (Ahmad et al., 2012).

By transferring one or two specific genes that correspond for those attributes into agricultural crops in vitro, novel varieties of plants with desirable characteristics can be developed with greater precision than using traditional plant breeding (Chokheli et al., 2022).

Genomic techniques for agricultural biotechnology allow for accurate, regulated genetic engineering and selection with marker assistance, accelerating the development of new crop varieties. Not only will you save time, but; plants can also be given new traits that were unimaginable before the omics era, such the ability to produce industrial chemicals and biopharmaceuticals (Ahmad et al., 2012). Gene expression research identifies the useful gene products that result in the phenotype, which is genetic knowledge that can be applied to the improvement of plants. In contrast to conventional plant breeding, the desired gene or genes can be added to a plant, and RNAi (Ribonucleic acid interference) can be used to silence a gene, to more quickly establish a phenotype (Kumar et al., 2022).

An effective technique for comprehending biological systems is transcriptomics. Next-generation sequencing (NGS), a transcriptomic technology, gives researchers the opportunity to better comprehend the functional components of the genome. To shed further light on how genes work, proteomics can be used to ascertain if the production of mRNA (Messenger Ribonucleic acid) results in protein synthesis. The texture, yield, flavour, and nutritional value of almost all food products are significantly influenced by the hundreds of thousands of different proteins that plants produce (Roberts et al., 2002).

Additionally, proteomics can advance knowledge of pesticide biodegradation, mode of behaviour, and resistance mechanisms, facilitating the development of innovative, risk-free, and efficient pesticides. If there have been any unfavourable changes to the overall metabolite composition, metabolomics can show whether the composition of changed and standard plants is equivalent or identical. Using mass spectrometry (MS) and nuclear magnetic resonance (NMR) analyses, metabolic profiling has been used to look at how metabolism is regulated and how metabolites fluctuate in accordance with environmental influences such light, humidity, temperature, soil type, salt, pests and chemical pesticides, fertilizers, and genetic alterations (Aliferis and Jabaji, 2011).

Genomics in Agriculture

When studying an organism's entire gene pool, scientists use collective and holistic methods known as genomics. Crop plant genomic studies have shed light on the total amount of genes present or their genetic makeup, and the role of genes in different metabolic activities. Genomic tools use DNA (deoxyribonucleic acid) sequencing and gene mapping to investigate the genomes. Automated Sanger's technique of genome sequencing was a component of first-generation sequencing (Sanger et al., 1977). During DNA synthesis in artificial conditions by enzyme DNA polymerase, through the addition of di-deoxy nucleotides / chain terminators, popularly called Sanger's method of sequencing. However, this method was very expensive, time-consuming, and labor-intensive for whole genome sequencing. The capabilities and opportunities for researching gene sequences have increased thanks to NGS techniques. According to Table 1, the second-generation approaches (Illumina and Roche 454) were incredibly effective, affordable, and offered a thorough understanding of genome sequences. The Helicos method, a single-molecule synthesis process using reverse terminators, is one of the third-generation sequencing techniques. It was found in Stephen

Table 1. Genomic technologies used for sequencing of crop plants

Genomics techniques	Discovery	Analysis	Software Tools
Shotgun Sequencing	Staden, 1979	DNA fragments are determined on the basis of overlapping in genetic codes	Sequencing Software
Computerized Chemical DNA Sequencing	Applied Biosystems, 1986	Use of Laser beam light to observe the presence of fluorescence dye	Assembler
SoLid	Barnes, 2002	Sequencing by ligation	Sequencing software
Illumina	Landegren, 1988	Solid phase bridge amplification	Sequencing software
Helicos	Braslavsky, 2003	Sequencing system using single molecule	Sequencing software

Quake's laboratory and thus was subsequently made available for purchase by Helicos Braslavsky et al., (2003).

Applications of Genomic Technologies

a) Genome sequencing and Gene prediction

The studies on genome-wide associations offer a more comprehensive knowledge about the functioning of all genes and their interaction with each other. The evolution of genomic technologies has made it possible for us to create agricultural crops with highly advantageous economic characteristics. Crop biotechnology has made use of genome sequencing, following functional interpretation, and molecular research.

The expansion of next-generation sequencing technologies (NGS) has aided comparative genomics studies' ability to identify genes. In 2000, the first plant itself, Arabidopsis thaliana, had its entire genome sequenced. It was then annotated to determine nearly 25,000 functional genes that were used as model/ideal crop plants for other crops' gene interpretation. In the comparative genomic study, newly sequenced genomes are compared to model crops, which ultimately led to the generation of certain novel genes (Table 2).

b) Genetic Improvement of Crop plants

All related aspects of omics have advanced agricultural research for bettering food crops, feedstock, and environmental preservation (Ahmad et al., 2012). The functional genes correlated with a particular character/trait have been recognized thanks to genome sequencing and gene expression studies, and this knowledge could be used to enhance crop plants through the introduction of genes or post-transcriptional gene silencing. (Table 3). Genomic technologies have made it possible to create

Table 2. Number of genes predicted in different agricultural crops by using NGS technologies

Crops	Haploid number of chromosomes	Number of genes predicted	Reference
Wheat/Triticum spp.	X=21	94,000 genes	Brenchley et al., 2012
Cabbage/ Brassica spp.	X=9	45,758 genes	Liu et al., 2014
Moong/Vigna radiata	X=11	22,427 genes	Kang et al., 2014
Bajra/Pennisetum glaucum	X=7	38,579 genes	Varshney et al., 2017
Rice/Oryza sativa	X=12	37,544 genes	Yu et al., 2002

functional foods like maize that can withstand drought, longer-lasting bananas, and rice with higher production of grains. For the purpose of creating modified crops with desired economic characteristics, a process referred to as mutation breeding exposes plants to mutagenic agents. Marker-assisted breeding has been used to select offspring with the intended traits. Simple sequence repeats (SSR) and single nucleotide polymorphisms (SNP), two molecular markers identified by genome sequencing methods, have been used to improve agricultural crops. (Salgotra et al. 2014)

Challenges of Genomic Technologies

The assembly of brief genomic sequence fragments would be complicated by the existence of highly frequently occurring DNA in plant genome sequences. The final sequencing manuscript would contain errors because of the gaps in genome sequences. Agriculture products' polyploidy and heterozygosity present challenges

Table 3. Functional food crops developed by genomic techniques

Crops	Techniques employed	Functional food crops developed	Benefits
Rice	Transgenic	Golden rice	More vitamin content
Plum and Apricot	Cross Hybridization	Plumcot	Vitamin A fortified
Tobacco	Transgenic	Edible vaccine	Monoclonal antibodies against West Nile Virus infection
Tomato and Banana	Transgenic	Plant bodies	Vaccine against Diphtheria, Tuberculosis and Hepatitis B
Mustard	Transgenic	Low concentration of saturated fatty acid	Reduced amount of saturated fatty acid

when assembling their sequences. Numerous discovered genes have not yet had their functional annotation completed. (Singh et al., 2022)

Transcriptomic in Agriculture

The perception of expressed gene sequences in a particular tissue/organ at a particular moment is provided by the transcriptome. The cell's transcriptome is constantly changing under various circumstances, and a single gene can produce multiple transcripts due to the alternative splicing mechanism. Control of alternative splicing can be investigated by the arrangement of transcripts, thus making transcriptome a dynamic process in nature. Knowledge of gene functioning, transcription levels, and molecular mechanisms underlying cellular metabolism is made possible by the research of the transcriptome (Borovitz and Chory 2004). The recognition of a particular gene for a particular biological trait and the identification of key enzymes in a metabolic pathway are both made possible by the expression analysis of genes (Table 4).

The development of NGS technologies made it possible to acquire inexpensive transcriptome assemblies that help in the recognition of genes (Mochida and Shinozaki, 2010). Information about various functional markers, such as simple sequence repeats (SSR) and single nucleotide polymorphisms (SNP), connected with stress-resistant responses is provided by the analysis of transcriptome assemblies. By standardizing the data with the help of various statistical models after getting the quantitative computation of each transcript, different gene expression patterns need to be examined. (Lowe et al., 2017)

Applications of Transcriptomics in Agriculture

The differences in transcriptome information during phase of seed germination, growth, development, and various types of stresses have been explained positively

Table 4. List of transcriptomics techniques with their applications

Procedure Used	Technique for analysis	Uses	Reference
Expressed Sequence Tag (EST)	Short nucleotide sequence created from a single transcript of RNA	Sequence information for pre microarray design	Marra et al., 1998
Serial Analysis of Gene Expression (SAGE)	Sequence Expression profiling	Expression analysis of plants with less characterized genome	Velculescu et al., 2000
Massive Parallel Signature Sequencing (MPSS)	Create sequences via complex series of hybridization	Identify and quantify the RNA transcript	Velculescu et al., 2000

by microarray technology. In a combinatorial method known as functional genomics, putative genes associated with a particular biological feature are found by using genomic data and expression analysis. By using this method, the region in the wheat genome important for the development of seeds has been identified (Jordan et al., 2007). Crop genomes have been used to map QTLs pertained to grain development as well as counteraction to stresses that are either abiotic or biotic in nature, and these QTLs have been effectively used to enhance crop varieties (Saha et al., 2013). RNA interference (RNAi), mutagenesis, and epigenetics are examples of functional methods that can be used as gene-silencing instruments to improve agricultural crops. A genome mapping method called bulk segregate analysis (BSA) is used to locate the chromosomal region associated with the expression of profitable significant crop characteristics. This method was effectively used to develop pigeon peas that are resistant to various disease. Molecular switches are DNA transcriptional factors that are used to bind to particular DNA sequences and control biological processes. By comparing recently sequenced plant species' genomes to the TF sequences of model plants, it was possible to identify different transcriptional factors (TF) of those species (*Arabidopsis thaliana*). (Table 5)

Challenges in Transcriptomics Technologies

The ability to functionally analyze target traits has been hindered by the restricted availability of molecular markers. The obstacles to whole transcriptome analysis include a dearth of appropriate infrastructure and restricted availability of high efficiency and affordable phenotyping programmes. Since the transcriptome data are extremely tissue and time-dependent, careful sample collection and evaluation are required. (Singh et al., 2022)

Table 5. Traits of plants identified by transcriptomic techniques

Crop	Resistant against	Approach used	Reference
Pea/Pisum sativum	Resistant to Stemphylium blight	Quantitative Trait Analysis	Saha et al., 2013
Groundnut/Arachis hypogea	Rust resistant	Back cross progenies	Varshney et al., 2013
Alfalfa/Medicago sativa	Aluminium toxicity	Microarray approach	Zhao et al., 2013
Pigeon pea	Resistant to fusarium wilt	Bulk Segregation Analysis	Kotresh et al., 2006

Proteomics in Agriculture

Proteomics is the study of the identification and assessment of all proteins generated from a genome. The genetic code of the protein is essential for regulating its shape and also performing various crucial functions. Protein sequences are examined in proteomics to determine the proportional concentrations of different post-translational modifications. Unlike genomics, proteomics has dynamic and vigorous nature and are subject to translational and post-translational changes. The knowledge of proteomics facilitates comprehension of challenging biological events and intracellular reactions to environmental stress (Renaut et al. 2006). Because proteomics explains how proteins are involved in maintaining cell homeostasis and in cell signalling pathways, it has emerged as an essential instrument for genetic improvement of crops and is required for structural maintenance.

Applications of Proteomics in Agriculture

Crop plants have been developed with improved photosynthesis capacities and resistance to abiotic stresses. It has been noted that C4 plants have two different classes of chloroplasts, making them better and more effective in converting energy. To discover various proteins accountable for more productive light fixation, a similar proteomic study of C3 and C4 plant chloroplasts had been conducted (Zhao et al., 2013). The proteins basically necessary for hybrid selection which were responsible for male sterility traits. Ability of crop plants to develop and produce male sterility traits are impacted by heat stress conditions. The organ-specific proteomic research was carried out on a wide range of crop plants to identify the proteins that accumulate specifically during abiotic stress (Hossain and Komatsu, 2013). In order to create transgenic plants with desired traits, the information from proteomic studies has been successfully applied. (Gong and Wang, 2013)

Challenges in Proteomics Technologies

Proteins are extracted in large quantities from the samples, which would make it difficult to analyze the desired protein. Right now, there are no established procedures for proteomic analysis and data interpretation. Nowadays, more sophisticated technologies are preferred for the precise characterization of proteins; earlier technologies, like 2-DE (2-dimensional gel electrophoresis), are not assumed to be trustworthy. Less reproducibility of proteomics data is caused by biological differences in proteins. As a result, the examination should be meticulously carried out in a controlled environment. For the purpose of predicting protein activity, proteomics also uses in silico tools and software. When using datasets that are currently accessible, homology search is frequently used to predict the functions of proteins, which can result in inaccurate predictions. (Fernie and Yan, 2019)

METABOLOMICS IN AGRICULTURE

The concentration of metabolites is what gives vegetables their taste, aroma, storage ability, and sustainability.

All constituents in a biological material must be identified and measured as part of metabolomics. On the basis of their availability at a particular phase of development, metabolites are divided into primary and secondary metabolites. Modern methods like mass spectrometry (MS), nuclear magnetic resonance (NMR), and Fourier-transform infrared (FTIR) spectroscopy are improving, making it simpler to analyse compounds. (Memelink, 2005)

The extraction of the sample at a specific moment is necessary for the metabolomics analysis (Hall, 2006). Our understanding of plant metabolism and the variety of different biological reactions which takes part in plant development, differentiation, resistance to stress, and resistance in plants against various disease causing organisms is made possible by the use of metabolomics techniques. Bioinformatics tools are used to create large databases that are then further processed in order to examine the metabolic network's component parts. The study of metabolomics improves our knowledge of intricate molecular networks and complex biosynthetic pathways. Modern molecular breeding programs have been developed with the aid of computational methods such as molecular dynamics simulations and three-dimensional structural modelling. (Kusano et al., 2007)

Applications of Metabolomics in Agriculture

For the sake of creating metabolic profiles in order to study different stress mechanisms, enhance economically important traits, and determine the influence of heredity, metabolomics studies have been carried out in the agricultural sector. The use of metabolic profiles to assess the effects of variations in the seasons, geographic locations, and natural variation has been effective. The classification of transgenic plants has involved the evaluation of metabolic profiles (Table 6). Orthologous enzymes with related catalytic characteristics have been found thanks to metabolic flux analysis. Additionally, metabolic tests for chemotaxis analysis, developmental stages, and growth profiles have been carried out. (Singh et al., 2022)

Table 6. Metabolic profiles of crop plants identified by omics technologies

Crop	Metabolic profile/Biochemical studied	Reference
Tomato/Solanum lycopersicum	Study of carotenoids and isoprenoids	Fraser et al., 2000
Tomato/Solanum lycopersicum	Presence of carbohydrates	Al-Babili et al., 2001
Maize/Zea mays	Provitamin A, lycopene £-cyclase	Fraser et al., 2000
Rice/Oryza sativa	Glycine betaine as markers for stress tolerance	Al-Babili et al., 2001

FUTURE PERSPECTIVE

Nowadays, there is a huge availability of omics data. Prompt and high-input detection of numerous genes simultaneously for an appropriate trait will be made practicable through the combination of different types of omics data. Redirecting from single gene analysis to various pathways or network analysis will fundamentally improve the way we conduct our current study. It is possible to grow future crops using a three-step process with a huge knowledge of adaptation and enhancement learned from omics data in addition with new gene modification mechanisms. In this way, the genetic breeding of crops in the future will be able to fulfil a variety of human needs at the same time reforming agricultural systems too.

CONCLUSION

It is necessary to create innovative crop varieties with greater yields, greater thermo-tolerance, and less insecticide use in order to combat the issues of the expanding population, shifting climatic patterns, and environmental stress. High-dimensional biological information generated by advanced technologies has been analysed statistically to show the relationships between various omics studies. The basis for the evaluation of metabolic routes and system biological studies has been laid by the analysis of omics studies, which have improved the precision of the biological evaluation. The characterization of genome function, including gene ontology, pathway analysis, and system modelling, has greatly benefited from the use of omics technologies for the study of different agricultural crop expression. However, the restoration of environmentally beneficial traits in wild crop plant types has been made possible by functional assessment of molecular markers and QTL mapping, which has increased their sustainability. The development of novel resistant varieties has benefited from research on how the metabolism of crops reacts to different biotic and abiotic stress conditions. The sample collection and analysis must be done very cautiously because the transcriptome, proteome, and metabolome data show high variations across periods, stages, and environmental circumstances.

REFERENCES

Ahmad, Z. A., Yeap, S. K., Ali, A. M., Ho, W. Y., Alitheen, N. B., & Hamid, M. (2012). scFv antibody: Principles and clinical application. *Clinical & Developmental Immunology*, *2012*, 980250. doi:10.1155/2012/980250 PMID:22474489

Al-Babili, S., Ye, X., Lucca, P., Potrykus, I., & Beyer, P. (2001). Biosynthesis of beta-carotene (provitamin A) in rice endosperm achieved by genetic engineering. *Novartis Foundation Symposium*, *236*, 219–228. PMID:11387982

Aliferis, K. A., & Jabaji, S. (2011). Metabolomics – A robust bioanalytical approach for the discovery of the modes-of-action of pesticides: A review. *Pesticide Biochemistry and Physiology*, *100*(2), 105–117. doi:10.1016/j.pestbp.2011.03.004

Borovitz, J. O., & Chory, J. (2004). Genomics tools for QTL analysis and gene discovery. *Current Opinion in Plant Biology*, *7*(2), 132–136. doi:10.1016/j.pbi.2004.01.011 PMID:15003212

Braslavsky, I., Hebert, B., Kartalov, E., & Quake, S. R. (2003). Sequence information can be obtained from single DNA molecules. *Proceedings of the National Academy of Sciences of the United States of America*, *100*(7), 3960–396. doi:10.1073/pnas.0230489100 PMID:12651960

Brenchley, R., Spannag, M., Pfeifer, M., Barker, G. L. A., D'Amore, R., Allen, A. M., & McKenzie, N. (2012). Analysis of the bread wheat genome using whole-genome shotgun sequencing. *Nature*, *491*(7426), 705–710. doi:10.1038/nature11650 PMID:23192148

Chokheli, V. A., Shmaraeva, A. N., Kornienko, I. V., Stepanenko, V. V., Rajput, V. D., Singh, A., Kumari, A., Dmitriev, P. A., Azarov, A. S., Minkina, T. M., & Varduni, T. V. (2022). Assessment of genetic diversity and DNA fingerprinting of rare species of the genus Crambe (Brassicaceae). *Crop & Pasture Science*. doi:10.1071/CP22027

Core, L. J., Waterfall, J. J., & Lis, J. T. (2008). Nascent RNA sequencing reveals widespread pausing and divergent initiation at human promoters. *Science*, *322*(5909), 1845–1848. doi:10.1126cience.1162228 PMID:19056941

Fernie, A. R., & Yan, J. (2019). *De novo* domestication: An alternative route toward new crops for the future. *Molecular Plant*, *12*(5), 615–631. doi:10.1016/j.molp.2019.03.016 PMID:30999078

Fraser, P. D., Pinto, M. E., Holloway, D. E., & Bramley, P. M. (2000). Application of high-performance liquid chromatography with photodiode array detection to the metabolic profiling of plant isoprenoids. *The Plant Journal*, *24*(4), 551–558. doi:10.1046/j.1365-313x.2000.00896.x PMID:11115136

Gong, C. Y., & Wang, T. (2013). Proteomic evaluation of genetically modified crops: Current status and challenges. *Frontiers in Plant Science*, *4*, 41. doi:10.3389/fpls.2013.00041 PMID:23471542

Hall, R. D. (2006). Plant metabolomics: From holistic hope to hype to hot topic. *The New Phytologist*, *169*(3), 453–468. doi:10.1111/j.1469-8137.2005.01632.x PMID:16411949

Hossain, Z., & Komatsu, S. (2013). Contribution of proteomic studies towards understanding plant heavy metal stress responses. *Frontiers in Plant Science*, *3*, 310. doi:10.3389/fpls.2012.00310 PMID:23355841

Ibanez, I., Gornish, E. S., Buckley, L., Debinski, D. M., Hellmann, J., Helmuth, B., HilleRisLambers, J., Latimer, A. M., Miller-Rushing, A. J., & Uriarte, M. (2013). Moving forward in global-change ecology: Capitalizing on natural variability. *Ecology and Evolution*, *3*(1), 170–181. doi:10.1002/ece3.433 PMID:23404535

Jordan, M. C., Somers, D. J., & Banks, T. W. (2007). Identifying regions of the wheat genome controlling seed development by mapping expression quantitative trait loci. *Plant Biotechnology Journal*, *5*(3), 442–453. doi:10.1111/j.1467-7652.2007.00253.x PMID:17388792

Kang, Y. J., Kim, S. K., Kim, M. Y., Lestari, P., Kim, K. H., Ha, B. K., Jun, T. H., Hwang, W. J., Lee, T., Lee, J., Shim, S., Yoon, M. Y., Jang, Y. E., Han, K. S., Taeprayoon, P., Yoon, N., Somta, P., Tanya, P., Kim, K. S., & Lee, S.-H. (2014). Genome sequence of mung bean and insights into evolution within Vigna species. *Nature Communications*, *5*(1), 5443. doi:10.1038/ncomms6443 PMID:25384727

Kotresh, H., Fakrudin, B., Punnuri, S., Rajkumar, B., Thudi, M., Paramesh, H., Lohithswa, H., & Kuruvinashei, M. S. (2006). Identification of two RAPD markers genetically linked to a recessive allele of a Fusarium wilt resistance gene in pigeon pea (Cajanus cajan L.) Millsp.). *Euphytica*, *149*(1-2), 113–120. doi:10.100710681-005-9059-2

Kumar, P., Singh, A., Rajput, V. D., Yadav, A. K. S., Kumar, P., Singh, A. K., & Minkina, T. (2022). *Role of artificial intelligence, sensor technology, big data in agriculture: next-generation farming*. Bioinformatics in Agriculture. doi:10.1016/B978-0-323-89778-5.00035-0

Kusano, M., Fukushima, A., Kobayashi, M., Hayashi, N., Jonson, P., & Mortiz, T. (2007). Application of a metabolomics method combining one-dimensional and two-dimensional gas chromatography time of light/ mass spectrometry to metabolic phenotyping of natural variants in rice. *Journal of Chromatography. B, Analytical Technologies in the Biomedical and Life Sciences*, *855*(1), 71–79. doi:10.1016/j.jchromb.2007.05.002 PMID:17556050

Liu, S., Liu, Y., Yang, X., Tong, C., Edwards, D., Parkin, I.A., & Zhao, M. (2014). The Brassica oleracea genome reveals the asymmetrical evolution of polyploidy genomes. *Nature Communications*.

Lowe, R., Shirley, N., Bleackley, M., Dolan, S., & Shafee, T. (2017). Transcriptomics technologies. *PLoS Computational Biology*, *13*(5), 1005457. doi:10.1371/journal.pcbi.1005457 PMID:28545146

Marra, M. A., Hillier, L., & Waterston, R. H. (1998). Expressed sequence tag-EST abolishing bridges between genomes. *Trends in Genetics*, *14*(1), 4–7. doi:10.1016/S0168-9525(97)01355-3 PMID:9448457

Memelink, J. (2005). Tailoring the plant metabolome without a loose stitch. *Trends in Plant Science*, *10*(7), 305–307. doi:10.1016/j.tplants.2005.05.006 PMID:15950519

Mochida, K., & Shinozaki, K. (2010). Genomics and Bioinformatics Resources for Crop Improvement. *Plant & Cell Physiology*, *51*(4), 497–523. doi:10.1093/pcp/pcq027 PMID:20208064

Renaut, J., Hausman, J. F., & Wisniewski, M. E. (2006). Proteomics and low-temperature studies: Bridging the gap between gene expression and metabolism. *Physiologia Plantarum*, *126*(1), 97–109. doi:10.1111/j.1399-3054.2006.00617.x

Rhee, S. Y., Dickerson, J., & Xu, D. (2006). Bioinformatics and its applications in plant biology. *Annual Review of Plant Biology*, *57*(1), 335–360. doi:10.1146/annurev.arplant.56.032604.144103 PMID:16669765

Roberts, C. M., Mcclean, C. J., Veron, J. E. N., Hawkins, J. P., Allen, G. R., McAllister, D. E., Mittermeier, C. G., Schueler, F. W., Spalding, M., Wells, F., Vynne, C., & Werner, T. B. (2002). Marine Biodiversity Hotspots and Conservation Priorities for Tropical Reefs. *Science*, *295*(5558), 1280–1284. doi:10.1126cience.1067728 PMID:11847338

Saha, G.C., Sarker, A., Chen, W., Vandemark, G.J., & Muehlbauer, F.J. (2013). Inheritance and linkage map positions of genes conferring agromorphological traits in *Lens culinaris* Medic. *International Journal of Agronomy*. doi:10.1155/2013/618926

Salgotra, R. K., Gupta, B. B., & Stewart, C. N. Jr. (2014). From genomics to functional markers in the era of next-generation sequencing. *Biotechnology Letters*, *36*(3), 417–426. doi:10.100710529-013-1377-1 PMID:24129954

Sanger, F., Nicklen, S., & Coulson, A. R. (1977). DNA sequencing with chain-terminating inhibitors. *Proceedings of the National Academy of Sciences of the United States of America*, *74*(12), 5463–5467. doi:10.1073/pnas.74.12.5463 PMID:271968

Singh, A., Rajput, V. D., Rawat, S., Sharma, R., Singh, A. K., Kumar, P., Singh, A. K., Minkina, T., Singh, R. P., & Singh, S. (2022). *Geoinformatics and Nanotechnological Approaches for Coping Up Abiotic and Biotic Stress in Crop Plants*. Sustainable Agriculture Systems and Technologies., doi:10.1002/9781119808565.ch17

Singh, A., Sengar, R., Singh, R., Shahi, U. P., & Yadav, M. K. (2022). Vaishali, Gangwar LK, Rajput VD. Effects of zinc oxide nanoparticles for promoting seed germination of rice (Oryza sativa L.) under salinity stress. Ecology. *Environmental Conservation*, *28*, 254–259. doi:10.53550/EEC.2022.v28i07s.042

Singh, A., Sengar, R. S., Rajput, V. D., Minkina, T., & Singh, R. K. (2022). Zinc Oxide Nanoparticles Improve Salt Tolerance in Rice Seedlings by Improving Physiological and Biochemical Indices. *Agriculture*, *12*(7), 1014. doi:10.3390/agriculture12071014

Singh, A., Sengar, R. S., Shahi, U. P., Rajput, V. D., Minkina, T., & Ghazaryan, K. A. (2022). Prominent Effects of Zinc Oxide Nanoparticles on Roots of Rice (Oryza sativa L.) Grown under Salinity Stress. *Stresses*, *3*(1), 33–46. doi:10.3390tresses3010004

Singh, A., Singh, A. K., Rawat, S., Pal, N., Rajput, V. D., Minkina, T., Sharma, R., Singh, N. P., & Tripathi, J. N. (2022). Satellite-Based Quantification of Methane Emissions from Wetlands and Rice Paddies Ecosystems in North and Northeast India. *Hydrobiology*, *1*(3), 317–330. doi:10.3390/hydrobiology1030023

Varshney, R. K., Shi, C., Thudi, M., Mariac, C., Wallace, J., Qi, P., Zhang, H., Zhao, Y., Wang, X., Rathore, A., Srivastava, R. K., Chitikineni, A., Fan, G., Bajaj, P., Punnuri, S., Gupta, S. K., Wang, H., Jiang, Y., Couderc, M., & Xu, X. (2017). Pearl millet genome sequence provides a resource to improve agronomic traits in arid environments. *Nature Biotechnology*, *35*(10), 969–976. Advance online publication. doi:10.1038/nbt.3943 PMID:28922347

Varshney, R. K., Song, C., Saxena, R. K., Azam, S., Yu, S., Sharpe, A. G., Cannon, S., Baek, J., Rosen, B. D., Tar'an, B., Millan, T., Zhang, X., Ramsay, L. D., Iwata, A., Wang, Y., Nelson, W., Farmer, A. D., Gaur, P. M., Soderlund, C., & Cook, D. R. (2013). The draft genome sequence of chickpea (Cicer arietinum) provides a resource for trait improvement. *Nature Biotechnology*, *31*(3), 240–248. doi:10.1038/nbt.2491 PMID:23354103

Velculescu, V. E., Vogelstein, B., & Kinzler, K. W. (2000). Analysing uncharted transcriptomes with SAGE. *Trends in Genetics*, *16*(10), 423–425. doi:10.1016/S0168-9525(00)02114-4 PMID:11050322

Yu, J., Hu, S., Wang, J., Wong, K. S., Li, S., Liu, B., Deng, Y., Dai, L., Zhou, Y., Zhang, X., Cao, M., Liu, J., Sun, J., Tang, J., Chen, Y., Huang, X., Lin, W., Ye, C., Tong, W., & Yang, H. (2002). A draft sequence of the rice genome (Oryza sativa L. ssp. indica). *Science*, *296*(5565), 79–92. doi:10.1126cience.1068037 PMID:11935017

Zhao, Q., Chen, S., & Dai, S. (2013). C4 photosynthetic machinery: Insights from maize chloroplast proteomics. *Frontiers in Plant Science*, *4*, 85. doi:10.3389/fpls.2013.00085 PMID:23596450

Section 5

Nanopriming and Geoinformatics: Combined Approaches for Crop Production and Soil Health Monitoring

Chapter 9 Nanopriming and Geoinformatics: A Combined Approach for Crop Production for Monitoring of Soil Health

P. Smriti Rao
Integral University, India

Setu Ratnam
Integral University, India

Ashish David
Sam Higginbottom University of Agricultural Science and Technology, India

Raghunandan Khatana
Krishi Vigyan Kendra, Abusar, India

Akshita Barthwal
R&D Bayer Crop Science, India

ABSTRACT

A soil survey is a component of an improved agricultural research and advisory programme. It provides in-depth knowledge on soils and a full assessment of the region's soil resources. It offers the information needed to organize activities for managing soil and land use. With the help of the soil survey report and field samples, a database of soil characteristics is created. Although the soil survey report is mostly based on a low-intensity survey, it can be used to achieve a number of planning goals. Geographical information systems (GIS), Photogrammetry, remote sensing, global positioning systems (GPS), and other technologies are useful for analysing soil, creating various vegetation indices, creating digital elevation models, determining land use and land cover, and many other tasks. These computational models can then be used by land managers, soil scientists, researchers, and many other professionals to make well-informed suggestions.

DOI: 10.4018/978-1-6684-7232-3.ch009

INTRODUCTION

The information about soils that has been accumulated over many years is now integrated into soil science. Globally, different soil kinds and qualities exist in different locations. The goal of soil scientists is to identify and comprehend natural soils. Good cultivable sites have tendency of variation in their output potential and are also quantitively challenged. The variety of soils present in a village, district, or state affects how well they may be used for various purposes. due to such scarce supply, we need thorough understanding of the characteristics, nature, and distribution of soil resources on the terrain. This data is gathered via soil mapping and surveying (Atkinson and Tate, 2000). The variability of soils is a significant constant. The soil is uniform. Geographic knowledge on soils and soil qualities can be used in any endeavor involving soil, from building a park to farm crops fertigation. The soil pit continues to be the foundation of soil mapping despite all the recent technological improvements. However, for obvious reasons, a soil scientist cannot collect numerous soil samples from the landscape to determine the types of soil available and their limits. Even though soil mapping and investigations still rely heavily on it, soil sugaring is invasive and labor-intensive. Ground penetrating radar, geopositioning systems, electromagnetic induction, and remote sensing of surface features are some of the non-intrusive processes of soil appraisal that are being used more frequently to help locate soil boundaries and identify subsurface characteristics (Brady and Weil, 2002). For the analysis and reporting of soil surveys for challenging land projection evaluations, a digital geographic information system may be able to make the greatest use of the enormous amount of data present in a nation (GIS). The significance of geoinformatics in monitoring soil health is discussed in this chapter.

GEOINFORMATICS

The term “geoinformatics” refers to the intersection of science and technology that deals with the collecting, qualification, and classification of spatial data as well as its processing, dissemination, and storage. It is a crucial tool for gathering, processing, and generating data from both spatial and non-spatial sources. Remote sensing, GPS, geographic information systems, and relational database management systems are some of the components that make up geoinformatics.

TOOLS AND TECHNIQUES USED IN GEOINFORMATICS

Global Positioning System (GPS)

Positions on the earth can be found using the Global Positioning System (GPS), it is a satellite-based navigation system. In any weather, GPS offers worldwide, constant (24 hours/day), real-time endeavor, three-dimensional locationing, timing, and navigation. Global Positioning System satellites transmit various signals that enable GPS receivers to determine their location as a tool for precision agriculture. The real time information accessed makes it possible to receive continuous position updates while the satellite it rotating. Mapping of crop, soil, and water measurements is made possible by having accurate location data at all times. Farm equipment may be located using GPS within three meters of a field location by using a network of satellites. The most popular applications of GPS in agriculture are for variable rate fertilizer/pesticide applicators and yield mapping. In order to evaluate the geographical variability and site-specific input application, GPS are crucial for pinpointing the precise location in the field. The GPS's has a positional (horizontal) accuracy from 10 to 20 metres. For precision farming, necessary accuracies are crucial, which is attained by a differentially corrected positioning mode, in such cases the GPS techniques can be harnessed.

Remote Sensing Technique

The science of remote sensing (RS) involves drawing conclusions about physical objects without actually touching them from data taken at a distance. Various implements like satellite, balloon, rocket, aircraft, or even a ground-based sensor-supporting stand can serve as a stand for a sensor to be attached as part of a remote sensing system. Remote sensing sensors are being used in a number of imaging systems on aircraft and spacecraft. Current picture systems from spacecraft platforms include the French National Earth Observation Satellite (SPOT), Indian Remote Sensing Satellites, and others. On the basis of the supposition that agricultural landscape elements which have been consistently recognizable signatures on the kind of remote sensing data, the recognition, identification, measurement, and monitoring of agricultural phenomena which are predicted. These recognisable signatures indicate crop type, maturity level, density, shape, vigour, wetness, and other characteristics. The type of sensor employed and the area of the electromagnetic spectrum employed for sensing have a significant impact on the detection of features.

Geographical Information System (GIS)

A computerized system for storing and retrieving data called a geographic information system (GIS) this can be used to analyse and manage the real time geographical data regarding crop productivity and agronomic parameters. A greater understanding of the interactions between yield, fertility, insects, pests, weeds, and other productivity related elements can be gained by using GIS, as can decision-making based on these spatial linkages amongst them. There are numerous farm information systems (FIS) that are used as straight forward programmes to build databases at the farm level. LORIS is one illustration of such a farm information systems FIS. The LORIS (Local Resources Information System) is constitutes of modules allowing for the data import, construction of raster files using multiple gridding techniques, storage of raster data in databases, creation of digitalized agro-resource maps, and more. Base maps for topography, soil type, N, P, K, and other nutrient levels, soil moisture, pH, etc. are all included in a full farm GIS. It is also possible to keep information on crop rotation, tillage, pesticide applications, nutrient, yields, etc (Bolstad, 2002).

Using GIS, it is possible to produce fertility, pest, and weed intensity maps, which may subsequently be used to produce a roadmap that depict the optimumization of rates at which fertilizers, nutrients or pesticides should be applied (Burrough, 1998).

GEOINFORMATICS AND PRECISION AGRICULTURE

1) Precision farming is a method of crop production that uses system engineering and inputs that are provided "as needed." with recent revolutionary developments in precision farming, with the implementation of GIS, plug and play microcomputers, Global Positioning System, and automated farm machinery control, crop production can be modernized.
2) This strategy uses integrated soil and plant information, engineering, and economic management technologies to regulate temporal variability and spatial variability in agricultural areas at the micro level.
3) It is well knowledge that weather and soil conditions vary throughout time and place. Given this inherent uncertainty, management choices should be location- and time-specific rather than predetermined and consistent.
4) By better matching inputs applied to crop needs, precision agriculture offers methods for customizing production inputs to particular plots within a field, potentially lowering input costs, boosting yields, and minimizing environmental consequences.

5) The three production-related components of data collection, analysis, and application are covered by modern technologies utilised in precision agriculture: (a) input of data or information; (b) precision information processing; and (c) final recommendations as feedback.
6) The main goal of soil data gathering is to comprehend the intrinsic variable elements influencing crop growth and output.
7) The traditional method for mapping soil variability often involves manual grid sampling and interpolation using geostatistical methods over the entire agricultural landscape.
8) To establish causal connections, the inherent variability and the yield's variability map can be connected. By doing this, reasonably uniform management zones might be created. Such samples take a lot of effort and money.
9) The remote sensing data range has been extensively explored, and further research is being done to determine the spectral response of the soil which might be connected to different soil traits and attributes.
10) The non-destructive Hyperspectral remote sensing is affordable, and capable of spatial prediction, it has been examined for surface soil characterization as an alternative to traditional laboratory soil analysis.
11) The capacity to gather non-visible information over a spatial perspective on vast scales is provided by hyperspectral image data or various other form of imaging spectrometry technology, which has numerous narrow and contiguous bands, provides information on reflectance that is nearly laboratory-quality.

SOIL MAPPING

To address the needs of planning at various levels, soil maps at various scales, ranging from 1:1 million to 1:4,000, are necessary. Small scale soil maps of 1:1 million are required for macro-level planning at the national level since the scale of a soil map directly correlates with the frequent information content and intensive field investigations that are conducted. With a comprehensive interpretation of the soil related information to determine the appropriateness and constraints for various agricultural and land related uses, soil maps at 1:250,000 scale give information for planning at the regional, state national level and need less intensive soil observations and time. Soil maps at a scale of 1:50,000, where associations of soil series are shown, are used to plan resource conservation and optimal land use at the district and state level and necessitate only moderately intensive fieldwork. Extremely high resolution satellite data or large-scale aerial images helps to created maps at 1:50,000 scales and at 1:8,000 or 1:4,000 scales which are Large scale soil maps, are specific purpose maps that can be created through intensive field observations. To create strategies for the soil conservation and reclamation, information on degraded lands such as issues with soil salinity, eroded soils, waterlogged and sodden areas, jhum cultivated lands (shifting agriculture), etc. is necessary at various rationales (Sehgal et al., 1987).

Remote Sensing Intensively Used for soil Mapping

The use of remote sensing for soil and land dependence and degradation mapping has sped up formerly arduous, subjective, and time-consuming soil survey programmes. Due to the synoptic picture provided by aerial photography and satellite data, determining soil types and boundaries required less field labour (Soil Survey Staff, 1951).

a) Soil Mapping method
   - soil mapping is done by using Multispectral satellite data up to family association level, with topographic variations serving as the foundation for displaying soil variability (1:50000).

b) Visual Image Interpretation
   - Shape, size, shades, tone, shadow, texture, quality, pattern, location, union and association are the key factors.
   - This is reasonably easy and cheap.
   - A number of researchers have found that remote sensing technology offers superior efficiency at mapping scales of 1:50000 and 1:10000.

c) Digital technique
   - The development of numerical analysis of remote sensing data using computers was driven by necessitate to analyze vast qualtity of data more quickly and extract insights.
   - The spectrum variations are used by computer assisted approaches for classification.
   - The aspect of Pattern recognition in remote sensing aids in the identification of homogeneous areas, which can be utilised as a platform for doing thorough fieldwork and creating models between field characteristics and remote sensing data.
   - Accurate boundary delineation is a major issue with traditional soil survey and soil cartography. It is laborious and time-consuming to conduct field observations using a traditional soil survey.
   - The best option, with a clearer definition of soil mapping units, is remote sensing data combined with auxiliary data.

SITE SPECIFIC NUTRIENT MANAGEMENT (SSNM)

Site specific nutrient management (SSNM), a comparatively modern technique for nutrient recommendations, is mostly based on the native nutrient delivery from the soil and the crop's nutrient requirement to achieve the desired yield. The SSNM recommendations could be based simply on plant analysis or based on soil combined with plant analysis. Because it offers a method for crops like rice, maize, wheat, pulses, etc. with nutrients as per requirement, the SSNM aids in enhancing NUE.Profitability growth is the main advantage for farmers of enhanced nutrient management strategies.By reducing excessive fertilisation rates and eliminating fertiliser application when the crop doesn't need it, the SSNM prevents fertiliser waste.

Importance of SSNM

Making the best use possible of the native sources of nutrients already in place, including crop waste.

the triple nutrients N, P, and K applications are tailored to the crop's location and seasonal requirements.

- Utilizing the leaf colour chart guarantees that nitrogen is supplied when and in the quantity that the crop requires, preventing fertiliser waste.
- Determine the P and K fertilizers needed to meet the crop needs using nitrogen omission plots.

This guarantees that P and K are applied in the proportion that the rice crop requires.

Zn, S, and micronutrient applications follow local randomization. choosing the most cost-effective fertiliser source combo.

Integration with other ICM techniques, such as the utilization of high-quality seeds, ideal plant density, Integrated Nutrient Management, Integrated Pest Management (IPM), and efficient water resource management in the plants.

SSNM Strategy

The recent approach to nutrient recommendations is based mostly on the crop's need for nutrients to achieve the desired production and the native nutrient supply from the soil. The SSNM recommendations could be based simply on plant analysis or based on soil combined with plant analysis.

Plant Based SSNM

The crop's nutritional state is the best indicator of soil nutrient supplies and crop nutrient demands. As a result, plant analysis is strategised. SSNM was initially tested for low-land rice, but it later showed promise for a number of modern approaches of fertilizer recommendations in rice, wheat, and other mono cropping, rice-based production systems common in Asian nations.

The following are the top five steps for creating field-specific fertiliser NPK recommendations:

1) Yield goal Selection
    - A yield target that is greater than 70–80% of the variety-specific maximum yield (Ymax) must be chosen.
    - "Ymax" is the highest grain yield that can be produced, with crop growth unaffected by any other factors and only the local climate to contend with.
    - The reasoning behind choosing a yield goal that is between 70 and 80 percent of Ymax is because internal NUEs start to decline at extremely high yield levels close to Ymax.
    - To determine the Ymax of a particular crop variety under a specific climatic condition, crop growth models (such as DSSAT) can be utilised.
2) Assessment of Crop nutrient requirement
    - The amount of nutrients that a crop must absorb depends on both the yield target and "Y-max".
    - In SSNM, models for the quantitative evaluation of tropical soil fertility (QUEFTS) are used to calculate nutrient requirements.
    - A crop variety's nutritional needs may be lower in a high-yielding season than in a low-yielding one for a given yield goal.
3) Estimation of Indigenous nutrient supplies
    - The entire amount of a certain nutrient that is available to the crop from the soil during the cropping cycle, while others are non-limiting, is known as the indigenous nutrient supply (INS).
    - water, air deposition, and crop wastes integrated incorporated into the soil are the main component of INS
    - It is calculated by monitoring plant nutrient absorption rate in an omission plot buried in the farmer's field, where all other minor and major nutrients (N, P, or K) are applied in sufficient amounts but not the one in question.
4) Computation of Fertilizer nutrient rates

  - The preceding processes (1-3) and the anticipated fertilizers recovery efficiency (i.e. RE, Kg of fertilizer nutrient taken up by the crop per Kg of applied nutrient) are used to calculate field-specific fertilizer N, P, or K recommendations.
  - According to studies, rice has RE values of 40–60% for Nitrogen, 20–30% for Phosphorus, and 40–50% for potash when the nutrients are added as water-soluble fertilizer sources.

5) Dynamic Adjustments of N application Rates
  - In contrast to the P and K fertilizer rates and application schedules, which were calculated above, which are basally applied, which means at the time of sowing or planting, the N schedules and rates can be further adjusted as a chlorophyll metre (commonly referred to as SPAD) reads or leaf colour chart (LCC) signals.
  - Recent on-farm studies in India and elsewhere have shown that SPAD/LCC-based N management schedules in rice and wheat have a significant advantage over the conventionally advised N application involving 2 or 3 splits during crop growth regardless of N supplying capacity of the soils in terms of yield grain, N use efficiency, and economic returns.

Soil Sum Plant Based SSNM

- Plant nutrient requirements for a better higher target yield (at least 80% of "Ymax"), and RE of applied nutrients are considered while creating a fertiliser use schedule to produce the highest possible economic yield of a crop variety.
- Using documented data for comparable crop growing situations, the total amount of nutrients needed for the intended yield and RE are estimated.
- The crop's (variety's) nutrient need is then met with recommendations for field-specific fertiliser rates that don't deplete the soil's reserves.
- These recommendations, which are based on crop responses to soil tests, are currently being used to help numerous field crops meet their yield goals.
- Recent research on intensive cropping systems has demonstrated that the above method to fertiliser recommendations offers larger economic improvements than the NPK fertiliserprogrammes that are typically recommended by soil testing laboratories.

SOIL TEST CROP RESPONSE (STCR)

For farm profitability and environmental conservation, fertilizer applications must be manageable and judicious because it is one of the most expensive agricultural inputs. Farmers that utilize fertilizers inefficiently harm the quality of various resources such as the soil and water as well as the production of their crops. Knowledge of the ideal fertiliser doses for each crop is required to increase farm profitability in a variety of soil-climate situations. The best method for figuring out the best fertiliser dosages is called Soil Test Based Integrated Fertilizer Recommendation (STBIFR), which is maneuvered based on the data of soil test and crop response studies. Due to hefty fertilizer costs and high yielding, fertilizer-responsive crop varieties, a quantitative basis for fertilizer recommendations has been developed. To build a quantitative foundation for determining the profit-maximizing dose of fertilisers based on soil test for any crop, soil test crop response (STCR) study should be carried out. The STCR technique bases fertiliser dose recommendations on soil contribution and production level. The STCR notion is more specific, meaningful, and quantifiable. The soil test crop response (STCR) approach for targeted yield, which is different from other methods of fertiliser recommendation such as soil test-based recommendation, general recommended doses (GRD), critical value approachs, etc., indicates both soil test based fertiliser doses and the level of yield that can be optimized with better agronomic practices.

Soil testing is a quick chemical examination that determines the amount of accessible nutrients in the soil and involves evaluation, interpretation, and fertiliser recommendations based on the findings. a chemical technique for calculating a soil's ability to supply nutrients. Prior to the planting of a crop (field, vegetable, or ornamental), it can ascertain the nutrient state of the soil.

Objectives of STCR

- To investigate the connection between various soil test results for N, P, and K availability and crop yield response.
- To generate yield targeting equations for significant crops so that fertiliser advice can be given.
- To assess various soil testing techniques for their applicability in real-world settings.
- To assess the potential reduction in crop fertiliser requirements associated with the concurrent application of organic manure.

Targeted Yield Approach

A strategy that falls under STCR is the targeted yield idea. The crop must absorb a specific amount of nutrients in order to produce a given yield. Once this is established for a particular yield, the effectiveness of the soil's accessible nutrients' pool and the necessity for fertiliser can be used to determine the amount of fertilizer needed.

Concepts of STCR

The STCR idea was proposed by Ramamoorthy in 1987, and the ICAR formed the AICRP on STCR in 1967.The STCR technique seeks to establish a foundation for exact quantitative adjustment of fertiliser doses for desired levels of crop output under varied soil test values and response. These are examined in follow-up verification by field trials to support soil testing laboratories for their advisory purpose under specific soil, crop, and agro-climatic conditions. STCR offers the link between a soil test value and crop production.

Methods of STCR

- **Gradient Experiment**

At the experiment location, an artificial gradient in soil fertility is generated in this phase by performing the following steps:

- Depending on the size of the field, the experiment area was divided into three or four equal strips.
- In the first strip, no fertilisation was applied; in the second, only one application; and in the third, two applications were made, presumably increasing the fertiliser dose as the number of strips increased.
- Grow a variety of crops, such as maize and sorghum, for use as feed. Pre-sowing and post-harvest soil samples were taken from each strip and analysed.
- Plant analysis following a thorough crop harvest.
    - **Test Crop Experiment**

This phase of the field experiment was carried out using the following approach when it was confirmed that fertility gradients had been established in the experiment field:

- Each strip is divided into the same number of plots and treatments.
- Each plot's initial soil sample is taken and examined.

- The experiment is set up using statistical design, and test crops are used in an experiment with various fertilizer levels.
- After the crop has reached maturity, the yield from each plot is estimated. Additionally, samples of plant and soil are taken from each plot and evaluated.
- The essential information, such as nutrient need (kg/q), fertiliser, soil, water, and organic manure efficiency (%) for producing fertilizer recommendations, can be worked out using soil test results, the yield and nutrient uptake data, and applied fertiliser doses of the treated and control plots.

Target Yield Equation

The underlying principle of this idea is Liebig's law of minimum. Ramamoorthy and colleagues in 1967 developed the theoretical foundation and experimental evidence to support Truog's initial assertion that Libig's law of the minimum applies equally to N, P, and K.

A method based on yield targeting is distinctive among the other methods of fertiliser advice in that it not only provides the degree of yield but also the soil test-based fertiliser dose. The targeted yield idea strikes a balance between nutrients that are applied and those that are already present in the soil.

STCR Approach

An intricate interaction of soil, water, seed, fertilisers, and other agro-chemicals result in the agricultural production system. Therefore, in order to sustain such a complex system, careful management of all the inputs is necessary. Under typical circumstances, it is advised to collect soil samples from fields that are less than 20 acres in size. Grid soil sampling intensifies sampling while adhering to the same principles as soil sampling. Grid soil sampling's objective is to create an application map, which is a nutrient demand map. The computer mounted on the variable rate fertiliser spreader loads an application map, and the computer uses the application map and GPS receiver to control a product delivery controller that modifies the amount and type of fertiliser product in accordance with the application map.

NANOTECHNOLOGY

The Greek term "nano" means "Dwarf," and 1nm is equal to 10-9 metres. The study of events and the manipulation of substances at the atomic, molecular, and macromolecular scaleswhere properties differ noticeably from those at larger scalesis known as **nanoscience**.

In nanotechnologies, manipulation of shape and sizes at nano-scale can directly impacts the design, manufacture, characters and application of technical structure, devises and system.

Nanotechnology is a technique used to create structures and controls that are only a few atoms and molecules in size.

A **nanoparticle** is a tiny item that functions as a single entity for the purposes of properties and transport.

Concept of Nanotechnology

The physicist Richard Feynman first proposed the concept of nanotechnology in 1959.Norio Taniguchi had framed the phrase "nanotechnology" for the first time in 1974. An ultrafine unit with dimensions measured in nanometers is a nanoparticle. Nanoparticles can be produced artificially and occur in nature. Nanoparticles exhibit the following characteristics. They have vast surface areas, are highly mobile in the free state, and could display quantum phenomena.

The foundation of nanotechnology is two major strategies.

1) Bottom-up approach: devices and materials are constructed from molecular building blocks and brought together chemically through molecular recognition.
2) Top-down approach: Without atomic level control, nano things are built from huge entities.

NANOFERTILIZERS

Nano-fertilizers are nutrient carriers or holder of Nano-dimensions capablilities of holding a bounty of nutrient ions due to their enlarged surface area and releasing them steadily and slowly that are proportionate with the crop demand. Efficiencies of Nutrient use of conventional fertilizers rarely exceed 30-35%, 18-20%, and 35-40% for most essential elements like N, P, and K, respectively. According to research that hasn't changed in over a decade, nanoparticles with large surfaces may store large amounts of nutrients and release them gradually and consistently, facilitating nutrient uptake that is in line with crop needs without the negative impacts of specialized fertiliser inputs.

One of these new facilities, which works in one of three ways, encapsulates nutrients inside a nano particle. The nutrient may be enclosed in nanoporous substances with a thin polymer film coating.delivered in the form of nanoscale-sized particles or emulsions. the fertilizer relewase from nanofertilizers can be optimized and efficiently maneuvered by nanodevices. This will prevent unfavorable nutrient losses to soil, water, and air through direct internalization by crops and prevent the interaction of nutrients with soil, microorganisms, water, and air.Among the different kinds of nano devices, there are nano-porous zeolites and nanofertilizers with controlled release.

NANO BIOSENSERS

A modified biosensor known as a nanobiosensor is a small analytical instrument or unit that includes a biologically derived sensitive element connected to a physico-chemical transducer at nano level. Nanobiosensors are nanosensors containing immobilised bioreceptor probes that are selective for target analytic molecules. A nanobiosensor is typically constructed at the nanoscale to collect, process, and analyse data at the atomic level.Their uses include monitoring metabolites, detecting various microbes and diseases, and detecting analytes including urea, glucose, insecticides, etc.

Characteristics of Ideal Biosensers

- Highly specific, i.e., a sensor should have discriminatory properties of distinguishing between analyte and any "other" material, for the purposes of the analysis.
- Under typical storage circumstances, stable.
- The specific way analytes interact should be unaffected by any physical factors, including, temperature, pH, and stirring.
- Reaction time should be minimum.
- The resulting answers must be exact, repeatable, linear over the appropriate analytical range, and electrical noise-free.
- The nanobiosensor have to be minute,, harmless, biocompatible and non-antigenic.
- It must be affordable, transportable, and suitable for usage by operators with some level of ability.

Constitution of Nano Biosenser

Three pieces typically make up a nanobiosensor: a physiologically sensitised probe, a transducer, and a detector.

1) The biologically sensitive components (probe), which may include tissue, microorganisms, organelles receptors, enzymes, antibodies, nucleic acids, molecular imprints, lectins, etc., are either bio-derivative materials or bio-

mimetic components that has capabilities to access signals from sample (the target analytes) and then transmit them to the transducer. The development of future nanobiosensors may be greatly aided by such a nano-receptor.

2) The transducer serves as an interface, converting the energy from the reaction at the bioreceptor/sensitive biological element into a measured electrical output.
3) The detector elements are designed to capture the transducers' signals, which are subsequently sent to a device called microprocessor for amplification and analysis.

Role of Nano Biosensor

Nanomaterial-based biosensors currently provide exciting advantages over conventional biosensors. Nanobiosensors offer notable benefits like improved detection sensitivity/specificity and have a wide range of possible uses, such as environmental and bioprocess management, food quality control, agriculture, biodefense, and, particularly, medicinal applications. However, in this case, we are interested in the function of nanobiosensors in agriculture and agro-products. The following is a list of some possible uses for nanobiosensors:

- As a Contaminants and Other Molecules Detection Device.
- As a for Soil Quality and Disease Assessment Diagnostic Tool
- As an Effective DNA and Protein Detection Tool
- As Sustainable Agriculture promoter.

USE OF NANOTECHNOLOGY IN SOIL

The poor world has long relied heavily on agriculture. Currently, nanotechnology is a rapidly developing branch of science that is being used in a variety of scientific fields, including agriculture. Nanotechnology seems to be a more shrewd approach to sustainable agriculture. Utilize nanopesticides and nanofertilizers to boost productivity. Utilize hydrogels and nanozeolites to improve soil quality. Use nanoparticles to stimulate crop growth. utilisingnanosensors and wireless communication devices, provide intelligent monitoring.

Use of Nanotechnology in Agronomical Processes Like Tillage

Mechanical tillage techniques enhance soil texture and structure, boost porosity, better distribute soil particles, and ultimately change the physical characteristics of soil. The use of nanomaterials improves the pH and structure of the soil. Along with lowering soil erosion, it also lessens the mobility, availability, and heavy metal toxicity. In addition to lowering the soil's shear strength, nanoparticles in the soil also reduce cohesion and internal friction. Reduced soil particle adhesion makes it easier and more energy-efficient to smash lumps.

Nanotechnology in Fertilizers

Fertilizers have been essential in increasing the yield of food grains in India.Despite the overwhelming success in grain outputnon judicious use of fertilizers had led to diminish the loss in organic loads in the soil and thus yield in various prominent cash as well regular crops have been slacked.The groundwater is impacted by excessive nitrogen fertiliser use, and eutrophication in aquatic environments results.The fact that nitrogen fertiliser use efficiency is 20–50% and phosphorus fertiliser use efficiency is 10–25% is alarming. Only by using nano fertilisers, an emerging alternative to conventional fertilisers, is it feasible to eliminate eutrophication and drinking water with a potential buildup of nutrients in soil. Additionally, nanotechnology has reduced the cost of environmental protection and increased nutrient use efficiency in the fertilizer application can be created, which will ensure good substitutes for multiple soluble fertilisers. Throughout the growth of the crop, slow release of nutrient must be ensured to maximum absorption of the fertilizers by the plant and no time lag in between these processes. Zeolites, being a class of naturally occurring minerals with a layered, honeycomb-like crystal structure, can be maneuvered to gradually release nutrients into the environment. In addition to other slowly dissolving substances providing essential elements like calcium and phosphorous, and a full range of minor and trace nutrients, it is possible to load nitrogen and potassium into its network of interconnecting tunnels and cages. there is "on demand" release of nutrients from the pockets of zeolites, which have capacity to held various nutrients. Nanomembranes can be applied to fertiliser particles to enable the gradual release of nutrients. It's also important to carefully consider the Nano-composites that are being considered to give all the nutrients in the proper amounts through "Smart" delivery systems. as it is well known fact that more than 50 to 70 percent of the nitrogen leaches by water of into the air due to rapid degeneration of the nitrogenous fertilizers when applied to the soil resulting into low efficiency thus economic losses. One of these new facilities, which work in one of three ways, encapsulates nutrients inside a nanoparticle. The nutrient supplemented to the soil in the form of nano-emultions ort nano—particles which are thinly layered in polymer coating to ensure their stability in the soil or they may be applied in form of some nonporous nano-materials which can sustain the irrigation and can stay stable in the soil.

Nanotechnology in Water Use

Alumina filters, carbon nanotubes filters etc are some of the microscopic components which are extensively used in the process of water filtration and purification. the tiny pores in magnetic nanoparticles, nanocatalysts, and zeolite filtration membranes helps in such endeavors. Nearly all types of water impurities, such as water turbidity, soil bacteria, pathogens, viruses, and other organic contaminants, may be eliminated by carbon nanotube membranes and nanofibrous alumina filters.

Nanopriming Importance for Soil Health

Nano-priming involves the use of nanoparticles (NPs) smaller than 100 nm, and "priming" refers to the growth of stress tolerance under regular, mild stress. An all-inclusive solution for enhancing crops, the surrounding habitat, and animal health is the controlled delivery of ingredients (pesticides & fertilisers) in nano-encapsulation on-site. To maintain soil fertility and health, efforts are being made to create nano-agrochemicals that can regulate the expulsion of certain nutrients. Utilising carriers that can entrap, encapsulate, incorporate, or bind active chemicals, agricultural nano formulations were created. For increased crop output, various nano-fertilizer administration methods have been used often, including soil, foliar spray, seed priming, seedling root dipping, fertigation drip tape, and aerosol dusting (Abbasi et al., 2021).

Drought is a significant ecological issue that has an impact on crop productivity and nutrient content, which eventually has an impact on human nutrition. According to the investigations, a drought delayed the appearance of grain heads and fag leaves by 6–17 days, whereas for ZnO NPs, the delays were only four days. Furthermore, drought reduced grain yield by 76%, but using ZnO NPs increased yield by 22–83% after the drought. ZnO NPs enhance N uptake during drought stress, reinforce key food grains with Zn minerals, and encourage plant growth. They also increase yield. Through Zn-mediated improvements in photosynthetic pigment, reactive oxygen scavengers, and lipid peroxidation reduction, Zn-based fertilisers have been demonstrated to be essential in reducing wheat plant drought stress. There are several methods for fertilising with Zn, including foliar spraying and soil mixing. This offers significant promise for increasing the adaptability of agricultural production systems, maintaining food safety for humans and animals, minimising nutrient loss, and lowering environmental pollution brought on by N-fertilizers.

CONCLUSION

The technology for soil testing and fertilisation recommendations can stop ineffective fertilisation and low fertiliser use rates. Precision farming is a useful tool for increasing farm productivity while minimising the impact on soil fertility and other crucial inputs. Precision farming's key element is geospatial technology. Nutrient expert, SSNM, and other geospatial technology-based agricultural systems are frequently utilised to determine the recommended nutrient levels in fallow or standing crop fields. They are helpful for determining the site-specific fertiliser rate requirements and for developing foliar nutrient maps using data from foliar analysis and decision support systems. The use of all these technologies helps farmers increase their income by using less input.

REFERENCES

Abbasi Khalaki, M., Moameri, M., Asgari Lajayer, B., & Astatkie, T. (2021). Influence of nano-priming on seed germination and plant growth of forage and medicinal plants. *Plant Growth Regulation*, *93*(1), 13–28. doi:10.100710725-020-00670-9

Adio, S. O., Omar, M. H., Asif, M., & Saleh, T. A. (2017). Arsenic and selenium removal from water using biosynthesized nanoscale zero-valent iron: A factorial design analysis. *Process Safety and Environmental Protection*, *107*, 518–527. doi:10.1016/j.psep.2017.03.004

Ahmad, R., Mohsin, M., Ahmad, T., & Sardar, M. (2015). Alpha amylase assisted synthesis of TiO_2 nanoparticles: Structural characterization and application as antibacterial agents. *Journal of Hazardous Materials*, *283*, 171–177. doi:10.1016/j.jhazmat.2014.08.073 PMID:25270329

Akl, B. A., Nader, M. M., & El-Saadony, M. T. (2020). Biosynthesis of silver nanoparticles by Serratia marcescens ssp sakuensis and its antibacterial application against some pathogenic bacteria. *Journal of Agricultural Chemistry and Biotechnology*, *11*(11), 1–8. doi:10.21608/jacb.2020.76656

Ambhorkar, P., Wang, Z., Ko, H., Lee, S., Koo, K. I., Kim, K., & Cho, D. I. D. (2018). Nanowire-based biosensors: From growth to applications. *Micromachines*, *2018*(9), 679. doi:10.3390/mi9120679 PMID:30572645

Amooaghaie, R., Saeri, M. R., & Azizi, M. (2015). Synthesis, characterization and biocompatibility of silver nanoparticles synthesized from Nigella sativa leaf extract in comparison with chemical silver nanoparticles. *Ecotoxicology and Environmental Safety*, *120*, 400–408. doi:10.1016/j.ecoenv.2015.06.025 PMID:26122733

Annamalai, J., Ummalyma, S. B., Pandey, A., & Bhaskar, T. (2021). Recent trends in microbial nanoparticle synthesis and potential application in environmental technology: A comprehensive review. *Environmental Science and Pollution Research International*, *28*(36), 49362–49382. doi:10.100711356-021-15680-x PMID:34331227

Araya-Castro, K., Chao, T. C., Durán-Vinet, B., Cisternas, C., Ciudad, G., & Rubilar, O. (2020). Green synthesis of copper oxide nanoparticles using protein fractions from an aqueous extract of Brown Algae Macrocystis pyrifera. *Processes (Basel, Switzerland)*, *9*(9), 78. doi:10.3390/pr9010078

Arif, R., & Uddin, R. (2021). A review on recent developments in the biosynthesis of silver nanoparticles and its biomedical applications. *Medical Devices & Sensors*, *4*(4), 10158. doi:10.1002/mds3.10158

Atkinson, P. M., & Tate, J. J. (2000). *Advances in Remote Sensing and GIS Analysis*. Wiley.

Babu, P. J. (2021). Nanotechnology mediated intelligent and improved food packaging. *International Nano Letters*, (12), 1–14.

Baghaienezhad, M., Boroghani, M., & Anabestani, R. (2020). Silver nanoparticles synthesis by coffee residues extract and their antibacterial activity. *Nanomedicine Research Journal*, (5), 29–34.

Bartolucci, C., Antonacci, A., Arduini, F., Moscone, D., Fraceto, L., Campos, E., Attaallah, R., Amine, A., Zanardi, C., Cubillana-Aguilera, L. M., Palacios Santander, J. M., & Scognamiglio, V. (2020). Green nanomaterials fostering agrifood sustainability. *Trends in Analytical Chemistry*, *125*, 115840. doi:10.1016/j.trac.2020.115840

Bhatt, D., Bhatt, M.D., Nath, M., Dudhat, R., Sharma, M., & Bisht, D.S. (2020). Application of nanoparticles in overcoming different environmental stresses. Protective chemical agents in the amelioration of plant abiotic stress. *Biochemistry and Molcular Perspect.* 635–654.

Bhavyasree, P. G., & Xavier, T. S. (2020). Green synthesis of Copper Oxide/Carbon nanocomposites using the leaf extract of Adhatoda vasica Nees, their characterization and antimicrobial activity. *Heliyon*, *6*(6), 03323. doi:10.1016/j.heliyon.2020.e03323 PMID:32072042

Bolstad, P. (2002). GIS Fundaments. Eider Press.

Brady, N. C., & Well, R. R. (2002). *The Nature and Properties of Soils (Thirteen Edition). Peerson Prentice Hall, Peerson Education, Inc. Published in India by Dorling Kindersley*. Pvt. Ltd.

Burrough, P. A. (1998). *Principles of Geographical Information Systems for Land Resources Assessment* (2nd ed.). Oxford University Press.

Carbone, M., Donia, D. T., Sabbatella, G., & Antiochia, R. (2016). Silver nanoparticles in polymeric matrices for fresh food packaging. *Journal of King Saud University. Science*, *28*(4), 273–279. doi:10.1016/j.jksus.2016.05.004

Chand, K., Jiao, C., Lakhan, M. N., Shah, A. H., Kumar, V., Fouad, D. E., Chandio, M. B., Maitlo, A. A., Ahmed, M., & Cao, D. (2021). Green synthesis, characterization and photocatalytic activity of silver nanoparticles synthesized with Nigella sativa seed extract. *Chemical Physics Letters*, *763*, 138218. doi:10.1016/j.cplett.2020.138218

Chaturvedi, V., & Verma, P. (2015). Fabrication of silver nanoparticles from leaf extract of Butea monosperma (Flame of Forest) and their inhibitory effect on bloom-forming cyanobacteria. *Bioresources and Bioprocessing*, *2*(2), 1–8. doi:10.118640643-015-0048-6

Chaudhary, P., Fatima, F., & Kumar, A. (2020). Relevance of nanomaterials in food packaging and its advanced future prospects. *Journal of Inorganic and Organometallic Polymers and Materials*, *30*(12), 5180–5192. doi:10.100710904-020-01674-8 PMID:32837459

Chauhan, R., Reddy, A., & Abraham, J. (2015). Biosynthesis of silver and zinc oxide nanoparticles using Pichia fermentans JA2 and their antimicrobial property. *Applied Nanoscience*, *5*(5), 63–71. doi:10.100713204-014-0292-7

Daphedar, A. B., Kakkalameli, S. B., Melappa, G., Taranath, T. C., Srinivasa, C., Shivamallu, C., Syed, A., Marraiki, N., Elgorban, A. M., & Veerapur, R. (2021). Genotoxic assay of silver and zinc oxide nanoparticles synthesized by leaf extract of Garcinia livingstonei T. Anderson: A comparative study. *Pharmacognosy Magazine*, (17), 114.

Das, M., & Smita, S. S. (2018). Biosynthesis of silver nanoparticles using bark extracts of Butea monosperma (Lam.) Taub. and study of their antimicrobial activity. *Applied Nanoscience*, *8*(8), 1059–1067. doi:10.100713204-018-0721-0

Dekkers, S., Krystek, P., Peters, R. J., Lankveld, D. P., Bokkers, B. G., van Hoeven-Arentzen, P. H., Bouwmeester, H., & Oomen, A. G. (2011). Presence and risks of nanosilica in food products. *Nanotoxicology*, *5*(5), 393–405. doi:10.3109/17435390.2010.519836 PMID:20868236

Demirer, G. S., Silva, T. N., Jackson, C. T., Thomas, J. B., Ehrhardt, D. W., Rhee, S. Y., Mortimer, J. C., & Landry, M. P. (2021). Nanotechnology to advance CRISPR–Cas genetic engineering of plants. *Nanotechnology*, (16), 243–250. PMID:33712738

Dera, M. W., & Teseme, W. B. (2020). Review on the Application of Food Nanotechnology in Food Processing. *American Journal of Engineering and Technology Management*, *5*(5), 41–47. doi:10.11648/j.ajetm.20200502.12

Dhand, V., Soumya, L., Bharadwaj, S., Chakra, S., Bhatt, D., & Sreedhar, B. (2016). Green synthesis of silver nanoparticles using Coffea arabica seed extract and its antibacterial activity. *Materials Science and Engineering C*, *58*, 36–43. doi:10.1016/j.msec.2015.08.018 PMID:26478284

Dhivahar, J., Khusro, A., Elancheran, L., Agastian, P., Al-Dhabi, N. A., Esmail, G. A., Arasu, M. V., Kim, Y. O., Kim, H., & Kim, H. J. (2020). Photo-mediated biosynthesis and characterization of silver nanoparticles using bacterial xylanases as reductant: Role of synthesized product (Xyl-AgNPs) in fruits juice clarification. *Surfaces and Interfaces*, (21), 100747.

Durán, N., Nakazato, G., & Seabra, A. B. (2016). Antimicrobial activity of biogenic silver nanoparticles, and silver chloride nanoparticles: An overview and comments. *Applied Microbiology and Biotechnology*, *100*(15), 6555–6570. doi:10.100700253-016-7657-7 PMID:27289481

Ealia, S. A. M., & Saravanakumar, M. P. (2017) A Review on the Classification, Characterisation, Synthesis of Nanoparticles and Their Application. In *IOP Conference Series: Materials Science and Engineering*. IOP Publishing: Bristol, UK.

Echeverría, I., López-Caballero, M. E., Gómez-Guillén, M. C., Mauri, A. N., & Montero, M. P. (2018). Active nanocomposite films based on soy proteins-montmorillonite-clove essential oil for the preservation of refrigerated bluefin tuna (Thunnus thynnus) fillets. *International Journal of Food Microbiology*, *266*, 142–149. doi:10.1016/j.ijfoodmicro.2017.10.003 PMID:29216554

Elsakhawy, T., Omara, A. E. D., Alshaal, T., El-Ramady, H., Ghazi, A., El-Nahrawy, S., & Elhawat, N. (2018). Nanomaterials and plant abiotic stress in agroecosystems. *Environmental Biodiversity and Soil Security*, *2*(2), 73–94. doi:10.21608/jenvbs.2018.3897.1030

Fatemi, M., Mollania, N., Momeni-Moghaddam, M., & Sadeghifar, F. (2018). Extracellular biosynthesis of magnetic iron oxide nanoparticles by Bacillus cereus strain HMH1: Characterization and in vitro cytotoxicity analysis on MCF-7 and 3T3 cell lines. *Journal of Biotechnology*, *270*, 1–11. doi:10.1016/j.jbiotec.2018.01.021 PMID:29407416

Feitshans, I. L., & Sabatier, P. (2022). Global health impacts of nanotechnology law: Advances in safernano regulation. *Materials Today: Proceedings*, *985*(67), 994. doi:10.1016/j.matpr.2022.08.377

Hamad, A. F., Han, J. H., Kim, B. C., & Rather, I. A. (2018). The intertwine of nanotechnology with the food industry. *Saudi Journal of Biological Sciences*, *25*(1), 27–30. doi:10.1016/j.sjbs.2017.09.004 PMID:29379352

Handford, C. E., Dean, M., Henchion, M., Spence, M., Elliott, C. T., & Campbell, K. (2014). Implications of nanotechnology for the agri-food industry: Opportunities, benefits and risks. *Trends in Food Science & Technology*, *40*(2), 226–241. doi:10.1016/j.tifs.2014.09.007

He, Y., Du, Z., Ma, S., Cheng, S., Jiang, S., Liu, Y., Li, D., Huang, H., Zhang, K., & Zheng, X. (2016). Biosynthesis, antibacterial activity and anticancer effects against prostate cancer (PC-3) cells of silver nanoparticles using Dimocarpus Longan Lour. Peel extract. *Nanoscale Research Letters*, *11*(11), 1–10. doi:10.118611671-016-1511-9 PMID:27316741

Hyllested, J. Æ., Palanco, M. E., Hagen, N., Mogensen, K. B., & Kneipp, K. (2015). Green preparation and spectroscopic characterization of plasmonic silver nanoparticles using fruits as reducing agents. *Beilstein Journal of Nanotechnology*, *6*(6), 293–299. doi:10.3762/bjnano.6.27 PMID:25821667

Ioannou, A., Gohari, G., Papaphilippou, P., Panahirad, S., Akbari, A., Dadpour, M. R., Krasia-Christoforou, T., & Fotopoulos, V. (2020). Advanced nanomaterials in agriculture under a changing climate: The way to the future? *Environmental and Experimental Botany*, *176*, 104048. doi:10.1016/j.envexpbot.2020.104048

Jadoun, S., Arif, R., Jangid, N. K., & Meena, R. K. (2021). Green synthesis of nanoparticles using plant extracts: A review. *Environmental Chemistry Letters*, *19*(1), 355–374. doi:10.100710311-020-01074-x

Kamran, K., Kemmerling, B., Shutaywi, M., & Mashwani, Z. U. R. (2020). Nano zinc elicited biochemical characterization, nutritional assessment, antioxidant enzymes and fatty acid profiling of rapeseed. *PLoS One*, (15), 0241568. PMID:33170873

Kasote, D. M., Lee, J. H., Jayaprakasha, G. K., & Patil, B. S. (2021). Manganese oxide nanoparticles as safer seed priming agent to improve chlorophyll and antioxidant profiles in watermelon seedlings. *Nanomaterials (Basel, Switzerland)*, *11*(11), 1016. doi:10.3390/nano11041016 PMID:33921180

Korbekandi, H., Mohseni, S., Mardani Jouneghani, R., Pourhossein, M., & Iravani, S. (2016). Biosynthesis of silver nanoparticles using Saccharomyces cerevisiae. *Artificial Cells, Nanomedicine, and Biotechnology*, *44*(1), 235–239. doi:10.3109/21691401.2014.937870 PMID:25101816

Kumar, P., Singh, A., Rajput, V. D., Yadav, A. K. S., Kumar, P., Singh, A. K., & Minkina, T. (2022) Role of artificial intelligence, sensor technology, big data in agriculture: next-generation farming, Bioinformatics in Agriculture. Academic Press.

Kumar, Y., Singh, K. T. T., & Raliya, R. (2021). Nanofertilizers and their role in sustainable agriculture. *Annals of Plant and Soil Reserach*, *23*(3), 238–255. doi:10.47815/apsr.2021.10067

Mallakpour, S., & Azadi, E. (2021). Nanofiltration membranes for food and pharmaceutical industries. *Emergent Materials*, 1–15.

Musial, J., Krakowiak, R., Mlynarczyk, D. T., Goslinski, T., & Stanisz, B. J. (2020). Titanium dioxide nanoparticles in food and personal care products—What do we know about their safety? *Nanomaterials (Basel, Switzerland)*, *10*(10), 1110. doi:10.3390/nano10061110 PMID:32512703

Neme, K., Nafady, A., Uddin, S., & Tola, Y. B. (2021). Application of nanotechnology in agriculture, postharvest loss reduction and food processing: Food security implication and challenges. *Heliyon*, *7*(7), 08539. doi:10.1016/j.heliyon.2021.e08539 PMID:34934845

Okey-Onyesolu, C. F., Hassanisaadi, M., Bilal, M., Barani, M., Rahdar, A., Iqbal, J., & Kyzas, G. Z. (2021). Nanomaterials as nanofertilizers and nanopesticides: An overview. *ChemistrySelect*, *6*(6), 8645–8663. doi:10.1002lct.202102379

Pandey, G. (2018). Challenges and future prospects of agri-nanotechnology for sustainable agriculture in India. *Environmental Technology & Innovation*, *11*(11), 299–307. doi:10.1016/j.eti.2018.06.012

Pattanayak, S., Mollick, M. M. R., Maity, D., Chakraborty, S., Dash, S. K., Chattopadhyay, S., Roy, S., Chattopadhyay, D., & Chakraborty, M. (2017). Butea monosperma bark extract mediated green synthesis of silver nanoparticles: Characterization and biomedical applications. *Journal of Saudi Chemical Society*, *21*(6), 673–684. doi:10.1016/j.jscs.2015.11.004

Prasad, R., Kumar, V., & Prasad, K. S. (2014). Nanotechnology in sustainable agriculture: Present concerns and future aspects. *African Journal of Biotechnology*, (13), 705–713.

Qu, Y., You, S., Zhang, X., Pei, X., Shen, W., Li, Z., Li, S., & Zhang, Z. (2018). Biosynthesis of gold nanoparticles using cell-free extracts of Magnusiomyces ingens LH-F1 for nitrophenols reduction. *Bioprocess and Biosystems Engineering*, *41*(3), 359–367. doi:10.100700449-017-1869-9 PMID:29188359

Rajam, K. S., Rani, M. E., Gunaseeli, R., & Munavar, M. H. (2017). Extracellular synthesis of silver nanoparticles by the fungus Emericella nidulans EV4 and its application. *Indian Journal of Experimental Biology*, (55), 262–265.

Rajput, V. D., Minkina, T., Feizi, M., Kumari, A., Khan, M., Mandzhieva, S., Sushkova, S., El-Ramady, H., Verma, K. K., Singh, A., Hullebusch, E. D., Singh, R. K., Jatav, H. S., & Choudhary, R. (2021). Effects of Silicon and Silicon-Based Nanoparticles on Rhizosphere Microbiome, Plant Stress and Growth. *Biology (Basel)*, *10*(10), 791. doi:10.3390/biology10080791 PMID:34440021

Raliya, R., Biswas, P., & Tarafdar, J. C. (2015). TiO_2 nanoparticle biosynthesis and its physiological effect on mung bean (Vigna radiata L.). *Biotechnology Reports (Amsterdam, Netherlands)*, *5*(5), 22–26. doi:10.1016/j.btre.2014.10.009 PMID:28626678

Raliya, R., Saharan, V., Dimkpa, C., & Biswas, P. (2017). Nanofertilizer for precision and sustainable agriculture: Current state and future perspectives. *Journal of Agricultural and Food Chemistry*, (66), 6487–6503. PMID:28835103

Ramachandraiah, K., Han, S. G., & Chin, K. B. (2015). Nanotechnology in meat processing and packaging: Potential applications—A review. *Asian-Australasian Journal of Animal Sciences*, *28*(2), 290–302. doi:10.5713/ajas.14.0607 PMID:25557827

Ratan, Z. A., Haidere, M. F., Nurunnabi, M. D., Shahriar, S. M., Ahammad, A. J., Shim, Y. Y., Reaney, M. J., & Cho, J. Y. (2020). Green chemistry synthesis of silver nanoparticles and their potential anticancer effects. *Cancers (Basel)*, *12*(12), 855. doi:10.3390/cancers12040855 PMID:32244822

Ravichandran, V., Vasanthi, S., Shalini, S., Shah, S. A. A., & Harish, R. (2016). Green synthesis of silver nanoparticles using Atrocarpus altilis leaf extract and the study of their antimicrobial and antioxidant activity. *Materials Letters*, *180*, 264–267. doi:10.1016/j.matlet.2016.05.172

Sahani, S., & Sharma, Y. C. (2021). Advancements in applications of nanotechnology in global food industry. *Food Chemistry, 342*, 128318. doi:10.1016/j.foodchem.2020.128318 PMID:33189478

Sajid, M., & Płotka-Wasylka, J. (2020). Nanoparticles: Synthesis, characteristics, and applications in analytical and other sciences. *Microchemical Journal, 154*, 104623. doi:10.1016/j.microc.2020.104623

Saminathan, K. (2015). Biosynthesis of silver nanoparticles using soil Actinomycetes Streptomyces sp. *International Journal of Current Microbiology and Applied Sciences*, (4), 1073–1083.

Saravanakumar, A., Ganesh, M., Jayaprakash, J., & Jang, H. T. (2015). Biosynthesis of silver nanoparticles using Cassia tora leaf extract and its antioxidant and antibacterial activities. *Journal of Industrial and Engineering Chemistry, 28*, 277–281. doi:10.1016/j.jiec.2015.03.003

Saxena, R., Tomar, R. S., & Kumar, M. (2016). Exploring nanobiotechnology to mitigate abiotic stress in crop plants. *Journal of Pharmaceutical Sciences and Research*, (8), 974.

Sehgal, J. L., Saxena, R. K., & Vadivelu, S. (1987). Field. Manual- Soil Resource Mapping of Different States of India, *Technical Bulletin No. 13, NBSS&LUP, Nagpur.*

Seleiman, M. F., Almutairi, K. F., Alotaibi, M., Shami, A., Alhammad, B. A., & Battaglia, M. L. (2021). Nano-Fertilization as an Emerging Fertilization Technique: Why Can Modern Agriculture Benefit from Its Use? *Plants*, (10), 2. PMID:33375026

Singh, A., Rajput, V. D., Rawat, S., Sharma, R., Singh, A. K., Kumar, P., Singh, A. K., Minkina, T., Singh, R. P., & Singh, S. (2022). *Geoinformatics and Nanotechnological Approaches for Coping Up Abiotic and Biotic Stress in Crop Plants. Sustainable Agriculture Systems and Technologies*. Wiley.

Singh, A., Sengar, R. S., Rajput, V. D., Minkina, T., & Singh, R. K. (2022). Zinc Oxide Nanoparticles Improve Salt Tolerance in Rice Seedlings by Improving Physiological and Biochemical Indices. *Agriculture, 12*(7), 1014. doi:10.3390/agriculture12071014

Singh, A., Sengar, R. S., Shahi, U. P., Rajput, V. D., Minkina, T., & Ghazaryan, K. A. (2023). Prominent Effects of Zinc Oxide Nanoparticles on Roots of Rice (Oryza sativa L.) Grown under Salinity Stress. *Stresses, 3*(1), 33–46. doi:10.3390tresses3010004

Singh, A., Sengar, R. S., Singh, R., Shahi, U. P., & Yadav, M. K. (2022). Effects of zinc oxide nanoparticles for promoting seed germination of rice (Oryza sativa L.) under salinity stress. *Ecology Environmental & Conservation*, *28*, S254–S259. doi:10.53550/EEC.2022.v28i07s.042

Singh, A., Singh, A. K., Rawat, S., Pal, N., Rajput, V. D., Minkina, T., Sharma, R., Singh, N. P., & Tripathi, J. N. (2022). Satellite-Based Quantification of Methane Emissions from Wetlands and Rice Paddies Ecosystems in North and Northeast India. *Hydrobiology*, *1*(3), 317–330. doi:10.3390/hydrobiology1030023

Singh, H., Sharma, A., Bhardwaj, S. K., Arya, S. K., Bhardwaj, N., & Khatri, M. (2021). Recent advances in the applications of nano-agrochemicals for sustainable agricultural development. *Environmental Science. Processes & Impacts*, *23*(2), 213–239. doi:10.1039/D0EM00404A PMID:33447834

Singh, P., Kim, Y. J., Wang, C., Mathiyalagan, R., El-Agamy Farh, M., & Yang, D. C. (2016). Biogenic silver and gold nanoparticles synthesized using red ginseng root extract, and their applications. *Artificial Cells, Nanomedicine, and Biotechnology*, (44), 811–816. PMID:25706249

Singh, S., Sangwan, S., Sharma, P., Devi, P., & Moond, M. (2021). Nanotechnology for sustainable agriculture: An emerging perspective. *Journal of Nanoscience and Nanotechnology*, *21*(6), 3453–3465. doi:10.1166/jnn.2021.19012 PMID:34739798

Soil Survey Staff. (1951). *Soil Survey Manual. U.S. Department of Agriculture. Handbook No. 18*. Oxford & IBH Publishing Co.

Sun, Q., Li, J., & Le, T. (2018). Zinc oxide nanoparticle as a novel class of antifungal agents: Current advances and future perspectives. *Journal of Agricultural and Food Chemistry*, *66*(43), 11209–11220. doi:10.1021/acs.jafc.8b03210 PMID:30299956

Usman, M., Farooq, M., Wakeel, A., Nawaz, A., Cheema, S. A., Rehman, H., Ashraf, I., & Sanaullah, M. (2020). Nanotechnology in agriculture: Current status, challenges and future opportunities. *The Science of the Total Environment*, *721*, 137778. doi:10.1016/j.scitotenv.2020.137778 PMID:32179352

Verma, K. K., Song, X. P., Joshi, A., Tian, D. D., Rajput, V. D., Singh, M., Arora, J., Minkina, T., & Li, Y. R. (2022). Recent Trends in Nano-Fertilizers for Sustainable Agriculture under Climate Change for Global Food Security. *Nanomaterials (Basel, Switzerland)*, *12*(12), 173. doi:10.3390/nano12010173 PMID:35010126

Vishwakarma, K., Upadhyay, N., Kumar, N., Tripathi, D. K., Chauhan, D. K., Sharma, S., & Sahi, S. (2018). Potential applications and avenues of nanotechnology in sustainable agriculture. In *Nanomaterials in Plants, Algae, and Microorganisms*. Academic Press.

Wang, Y., Deng, C., Rawat, S., Cota-Ruiz, K., Medina-Velo, I., & Gardea-Torresdey, J. L. (2021). Evaluation of the effects of nanomaterials on rice (Oryza sativa L.) responses: Underlining the benefits of nanotechnology for agricultural applications. *ACS Agricultural Science & Technology*, *1*(1), 44–54. doi:10.1021/acsagscitech.1c00030

Yadu, B., Xalxo, R., Chandra, J., Kumar, M., Chandrakar, V., & Keshavkant, S. (2021). Applications of Nanomaterials to Enhance Plant Health and Agricultural Production. In *Plant Responses to Nanomaterials*. Springer. doi:10.1007/978-3-030-36740-4_1

Yu, H., Park, J. Y., Kwon, C. W., Hong, S. C., Park, K. M., & Chang, P. S. (2018). An overview of nanotechnology in food science: Preparative methods, practical applications, and safety. *Journal of Chemistry*, *2018*, 1–10. doi:10.1155/2018/5427978

Zhao, L., Lu, L., Wang, A., Zhang, H., Huang, M., Wu, H., Xing, B., Wang, Z., & Ji, R. (2020). Nano-biotechnology in agriculture: Use of nanomaterials to promote plant growth and stress tolerance. *Journal of Agricultural and Food Chemistry*, *68*(7), 1935–1947. doi:10.1021/acs.jafc.9b06615 PMID:32003987

Section 6

Nanopriming-Based Biotic and Abiotic Stress Management

Chapter 10
Nanopriming:
A Potential Solution for Heavy Metal Toxicity Alleviation in Crops

Sachin Kumar
Himachal Pradesh Agriculture University, Palampur, India

Chiranjeev Kumar
Himachal Pradesh Agriculture University, Palampur, India

Shilpa Manhas
Himachal Pradesh Agriculture University, Palampur, India

Ankita Sharma
Punjab Agricultural University, India

Richa Jaswal
Himachal Pradesh Agriculture University, Palampur, India

ABSTRACT

Sustainable crop production required steady and balanced means of nutrients and other growth factors during the crop growth period. Heavy metal accumulation in plant and soil ecosystems along with its toxicity effects is much more pronounced in crops. Heavy metals interfere with crop germination, plant nutrient uptake, photosynthesis, and response to fertilizer and organic inputs which in turn make crops vulnerable to various adverse conditions. Priming seeds and other plant propagating materials with suitable nano-priming materials may enhance the overall plant photosynthetic rate, enhance vigor, and increase plant biomass with ensuring food security and productivity in a long-term approach. Nano priming with nano-materials has higher efficiency due to its smaller size and large surface area coverage which then perfectly adsorbed on the surface of plant growing materials such as seed and other materials to remove the attachment and concentration of heavy metals in plants.

DOI: 10.4018/978-1-6684-7232-3.ch010

INTRODUCTION

Agriculture output is being reduced by numerous biotic and abiotic variables, including limited freshwater supplies, climate change and ineffective agrochemicals use (Singh et al., 2022). Millions of dollars in agriculture lost due to salinity and drought (Salam et al., 2022). Halophytes are salt-tolerant plants while glycophytes are less salt-tolerant plants that are not adapted to salinity. Modern agriculture uses agrochemicals more and more frequently. The pesticides account of 2.5×10^9 kg applied each year are vanished to the atmosphere and runoff, or both as they fail to reach their targets (Anwar et al., 2022). The other serious issue for the atmosphere, crop efficiency, and nutrition safety is heavy metals (HMs) contamination. As per reported, about 70% of HMs and combinations move in a person's body through foodstuff. The continual discharge of metal remaining into cultivated top soil has an antagonistic effect on the soil reaction, saltiness, and productiveness of the soil. Worldwide climate situation has made the issues worse by placing further constraints on produce and nutrition security. Initial advance phases of plants, e.g., seed propagation and development, are very vulnerable to ecological restrictions, which results in poor growth and production limitations. Modern day agriculture requires advanced technologies to produce crops with high yields and sustainability. An overview of developing pollutants is shown in figure 1. To protect seeds during storage, enhance germination and germination synchrony, raise crops' ability to resist the pathogens or environmental stress conditions, and enhance growth, nano-priming can be applied to seeds. The use of nano-priming can be helpful in lowering the amounts of various chemicals like insecticides, herbicides or inorganic manures needed. According to recent research, seed nano-priming can trigger a variety of genetic factor during seed sprouting, particularly the factors which are involved to resist plant stress conditions. The studies have already produced promising results, the use of nanotechnology in seed priming is a relatively fresh arena of study (Ye et al., 2020). Since nanoparticles have antibacterial properties and ability to carry microbicide mediums, seed nano-priming could be employed to protect kernels. Plants also have capacity to tackle biotic and abiotic stress through various mechanisms such as avoidance mechanism that prevent contact to stress, tolerance mechanisms that let the crop to hold out against stress, Acclimation means plants alter their physiology and response stress. The Abscises acid (ABA), Jasmonic acid, Ethylene, and calcium are helpful to control plant stress reactions. Many plants have the capability to withstand specific stress and are therefore regarded as stress-resistant. The ability of the organism to adapt or acclimatise to the stress is a prerequisite for stress resistance. Additionally, bio-fortification of seeds by nano-priming can be employed to encourage an improvement in food quality and output. Figure 1 depicts the possible uses of seed nano-priming in crop cultivation.

Abiotic Stress

Figure 1. A schematic diagram shows different abiotic stress factors inducing stress and the potential of NPs to alleviate stress and promote plant growth
(Salam et al., 2022)

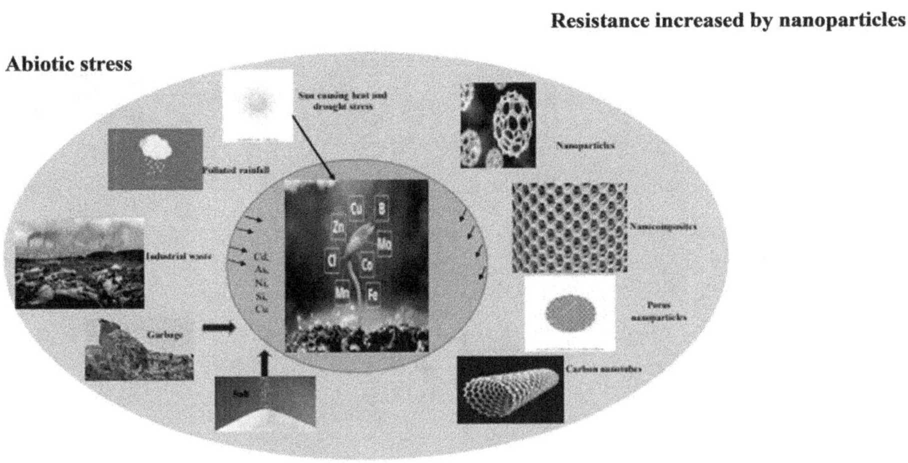

Nano priming

Pre-treating seeds before planting are known as priming. Examples of priming techniques that have advantages include better seed propagation, growth, and development, augmented tolerance against abiotic and biotic stress, better produce, and higher quantity of micronutrients in crops. To lessen the dormancy of seeds, nano-priming, matri-conditioning, magneto-priming, bio-priming and hormonal priming can all be used. For achieving workable agricultural yields, nano-priming is believed to be superior to old seed treatment procedures. In nano-priming, nanoparticles (NPs) smaller than 0.1 micrometer are used and "priming" is the process of building stress toleration through mild, ongoing stress. Figure 2 illustrates various forms of nano-priming and their effects. According to reports, nano-priming may possibly stimulate kernel propagation and strength in various crops (Maharramov et al., 2019). Additionally, bio fortification of seeds by nano-priming can be employed to encourage an improvement in food quality and production (Singh et al., 2021). Table 1 depicts the different types of nano-priming and its influences on various crops.

Heavy metals are biological materials that weathering processes have produced and are found in the soil. They have a specific gravity of roughly 5 g cm^{-3}. Heavy metals are divided into those that are necessary (essential) and those that are not (non-essential). Zn, Fe, Co, Mo, Ni, and other important metals are required in

Figure 2. Different types of nano-priming and its influences on various crops (Nile et al., 2022)

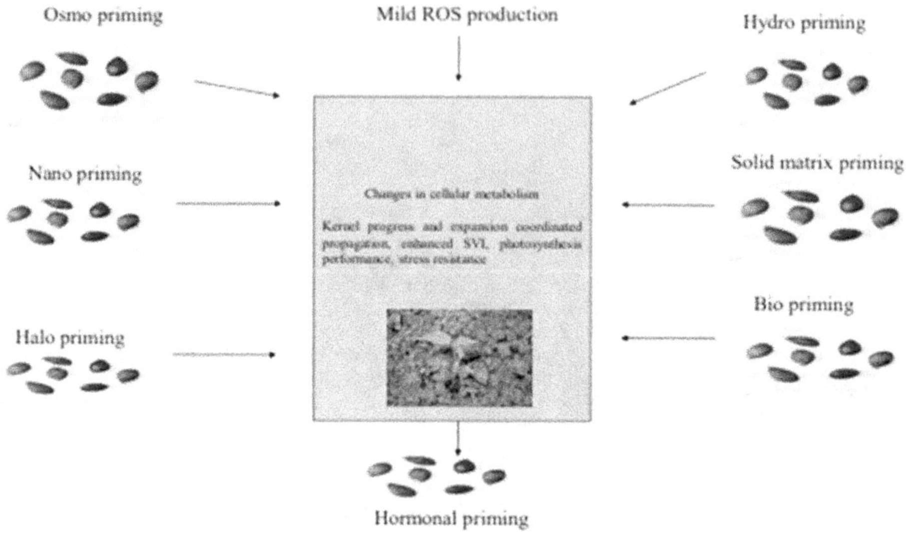

minute amounts by living beings for the best operation of their biological processes. Co, Fe, Cu, Mo, and Zn are recognized as being necessary for humans. As, Cd, Ag, Pb, and Hg are among non-essential elements. At very low quantities, they might be tolerable, but at higher concentrations, they become poisonous. According to Young and Blevin, (1981), at some degrees of exposure and absorption, all heavy metals may be hazardous to the majority of organisms.

Oxides as well as hydroxides of Cr, Mn, and Fe are frequently found in soils and sediments can be seen as surface coatings and floatable particles on clay. When exposed to anoxic and/or acidic conditions, Mn (IV) and Mn (III) oxides and hydroxides change into the more soluble manganous form (Moore and Patrick, 1989). Nearly all of the reducible Mn is converted into water-soluble and exchangeable form below pH 5.

Sources of Heavy Metals in Soil

Synthetic Compounds: Some soils lack the heavy metals (such as Cu, Co, Mo, Fe, Mn, Zn and Ni,) which are essential for healthy plant growth, but in intensive agricultural systems, huge quantities of fertilizers are frequently added to the soil to supply enough N, P, and K for crop growth. Heavy metals (such as Pb and Cd) are present in trace amounts in the compounds used to supply these elements as impurities, which, following repeated fertilizer application, may considerably increase the amount of these metals in the soil.

Pesticide: Although intensive agricultural systems routinely add massive amounts of fertilizer to the soil to offer enough N, P, and K for crop growth, some soils lack the heavy metals (such as Fe, Cu, Mn, Co, Mo, Zn, and Ni) necessary for healthy plant growth. The compounds used to deliver these elements contain trace levels of heavy metals (including Pb and Cd) as impurities; nevertheless, after repeated fertilizer applications, the amount of these metals in the soil may significantly increase.

Manures: Heavy metals including Cd, As, Cr, Pb, Cu, Ni, Hg, Mo, Se, Zn, and others end up accumulating in the soil after the application of numerous bio-solids (municipal sewage sludge, such as cattle manures, and composts) to land. High levels of Zn, As and Cu are present in the manures produced by animals fed such diets, and if these manures are frequently spread over confined areas of land, a significant amount of these metals may eventually accumulate in the soil.

Sewage sludge, also known as bio-solids, is a solid byproduct of wastewater treatment that is mostly made up of organic materials that can be recovered. Cu, Cd, Pb, Ni, Cr, and Zn are the heavy metals that are most frequently found in bio-solids. In rare instances, metals added to soils in the form of bio-solid applications could seep through the soil profile and contaminate groundwater. Recent studies on some bio-solids-treated New Zealand soils have revealed increased concentrations of Cd, Ni, and Zn in drainage leachates.

Table 1. Organic manures and inorganic amendments, a source for heavy metals immobilization (Raymond et al, 2011)

Organic amendments		Inorganic amendments	
Material	**Heavy metals immobilized**	**Material**	**Heavy metals immobilized**
Bark sawdust	Pb, Cd, Cu, Hg	Lime	Cd, Zn, Ni, Pb, Cu
Xylogen	Hg, Zn, Pb	Phosphate salt	Zn, Pb, Cu, Cd
Chitosan	Hg, Cd, Cr	Hydroxyapatite	Pb, Zn, Cu, Cd
Bagasse	Pb	Fly ash	Cr, Cd, Pb, Cu, Zn,
Poultry manure	Zn, Cu, Pb, Cd	Slag	Pb, Cd, Cr, Zn
Cattle manure	Cd	Ca-montmorillonite	Pb, Zn
Rice hulls	Pb, Cd, Cr	Portland cement	Pb, Cr, Cu, Zn
Sewage sludge		Bentonite	Pb
Leaves	Cr, Cd	–	–
Straw	Pb, Cd, Cr	-	–

Industrial Wastes: When mining, heavier and larger particles that have gathered at the bottom of the flotation cell are immediately released into natural depressions, such as onsite wetlands, where they can build up to significant concentrations. Zn and Pb ore have been extensively mined and smelted, contaminating the soil and jeopardizing both human and ecological health. Many businesses, such as those in the production of textiles, tanning, petrochemicals (from unintentional oil spills or the use of products derived from petroleum), insecticides, and pharmaceutical facilities, produce additional molecules with incredibly diverse compositions. Many are also rarely, if ever, applied to the ground and could be hazardous due to the presence of heavy metals (Zn, Pb and Cr) or toxic organic compounds. Others have very little plant nourishment or lack soil conditioning abilities.

Toxicity in Soil

Soil microbes: The soil's toxicity from heavy metals is a serious issue. Heavy metal pollution modifies the size, composition, and activity of the microbial population in addition to having detrimental impacts on several elements that have an impact on the quality and production of plants. The main causes of soil contamination with heavy metals are a variety of metals, including Cu, Ni, Cd, Zn, Cr, and Pb. The characteristics of the soil, including its organic matter, clay content, and pH, have a significant impact on how much of an impact metals have on biological and biochemical characteristics. Heavy metals affect soil enzymatic activity indirectly by altering the microbial community that generates enzymes. Heavy metals have detrimental impacts on the soil biota by obstructing crucial microbial processes and lowering the number and activity of soil microorganisms. The long-term effects of heavy metals can both increase the tolerance of bacterial communities and fungi, such as arbuscular mycorrhiza (AM) fungi, which can be essential in the regeneration of damaged ecosystems. Heavy metals caused bacterial community biomass and diversity to decline in contaminated soils, while also causing a relative increase in soil actinomycetes. Cd is more harmful to enzymes than Pb because of its greater mobility and weaker affinities for soil colloids.

Soil Enzyme: Heavy metals especially Pb reduce the availability of several soil enzymes such as catalase, Urease, invertase and acid phosphatase. The urease remained unaffected by arsenic, which inhibits phosphatase and sulfatase. Cd contamination has a negative effect on the arylsulfatase, protease, urease and alkaline phosphatase but has no impact on invertase. Each soil enzyme has a different level of sensitivity to heavy metals. Mn, Cr, Cd, Pb and Zn were the metals that inhibited urease action in decreasing order. The diversity and activity of soil microbes are important for the recycling of plant nutrients, preservation of soil structure, detoxification of toxins, and control of plant pests. Plant growth communities are an important indicator of soil quality. On soil microorganisms, heavy metals have detrimental impacts that alter the variety, population density, and general activity of these organisms. Increases in metal content often have a detrimental effect on soil microbes, including respiration rate and enzyme activity, which appear to be excellent indicators of soil pollution. Lead-contaminated soil showed a slight modification in its microbial composition.

Toxicity in Humans

Cadmium: Humans are adversely affected by chronic consumption of hazardous metals, and the adverse effects only become apparent after a number of years of exposure. Cadmium is a well-known heavy metal with an 8.65-times-higher specific gravity than water (Cd). The organs that are most vulnerable to cadmium toxicity are the liver, placenta, kidneys, lungs, brain, and bones. Depending on the level of exposure, symptoms of effects can include everything from nausea and vomiting to cramping in the stomach, dyspnea, and muscle weakness.

A fatal outcome from severe exposure is pulmonary oedema. Sub-chronic inhalation exposure to cadmium and its compounds may cause renal consequences as well as pulmonary symptoms (emphysema, bronchiolitis, and alveolitis). Cd has been linked to a variety of clinical problems, including cancer, anosmia, cerebrovascular infarction, heart failure, proteinuria, emphysema and the development of eye cataracts, to varying degrees.

Zinc: Clinical manifestations of zinc toxicosis includes anaemia, liver failure, renal failure, icterus (yellow mucus membrane), vomiting, diarrhoea, and bloody urine.

Copper: Due to its role as an electron giver or acceptor in metallo enzymes, copper (Cu) is a crucial component of mammalian diet. On the other hand, contact to extreme amounts of Cu can have harmful impacts on health. Humans are largely exposed to copper through food and water consumption. Humans who consume too much copper may experience severe corrosion and mucosal irritability, liver and kidney damage, broad capillary damage, and depression after stimulation of the central nervous system. Additionally, there may be severe gastrointestinal discomfort and potential liver and kidney necrosis.

Nickel: Skin irritation and harm to the lungs, neurological system, and mucous membranes are only a few of the side effects of Ni exposure.

Lead: Humans are physiologically and neurologically hazardous to lead (Pb). Acute Pb poisoning may cause illness and death by disrupting the kidney, reproductive system, liver, and brain. Even in really tiny amounts, Pb outranks the dangers. Lead toxicity's teratogenic effect is a particularly harmful side effect. Additionally, lead poisoning impairs the production of haemoglobin and damages the cardiovascular system, central nervous system and peripheral nervous system both acutely and chronically. Anoxia, anaemia, exhaustion, digestive issues, and weariness are further

Table 2. Regulatory guidelines for heavy metals (Raymond et al, 2011)

Metals	Soil concentration range	Regulatory limits
Cd	0.10–345	100
Cr	0.05–3 950	100
Hg	<0.01-1800	270
Pb	1.00–69 000	600
Zn	150-5000	1500

chronic consequences. High blood pressure, muscular and joint pain, and pregnancy problems are all possible effects of lead exposure.

Other side effects include neurological problems, severe and permanent brain damage, and damage to the urinary system. Lead in its inorganic form often affects the central nervous system and peripheral nervous system and other bio systems; however, lead in its organic form primarily affects the CNS. Children are affected by lead, especially those between the ages of 2 and 3, since it causes the brain's grey matter to develop poorly, which lowers their intelligence quotient (IQ). The body absorbs it more readily when Ca and Zn levels are low.

Chromium: chromium (Cr), is the tenth prevalent element in the earth's mantle, which can be found as Cr (III) or Cr (VI). Cr (VI) is dangerous to both plants and animals due to its properties as a strong oxidizing agent, corrosive, toxic substance that is soluble in both alkaline and slightly acidic water. Cr (VI) can pass cell membranes and oxidase biological molecules, making it dangerous.

Mercury: Mercury is poisonous, and its role in human biochemistry and physiology is unknown. Congenital malformations, gastrointestinal problems, and spontaneous abortion are all brought on by inorganic forms of mercury (like corrosive esophagitis and hematochezia). Acrodynia (pink disease) causes a rash and the desquamation of the feet and hands. Other signs of mercury poisoning by its organic forms, such as monomethyl and dimethyl mercury, include erythrism (an abnormal irritation or sensitivity of an organ or body part to stimulation), gingivitis, and stomatitis, neurological disorders, complete damage to the brain and CNS, and are also connected to congenital malformation. Arsenic interferes with respiration by inhibiting the synthesis of adenosine triphosphate (ATP), coagulating protein, and forming complexes with coenzymes. In all of its oxidation states, it is potentially carcinogenic in compounds, and excessive exposure can be fatal. Arsenic poisoning is a disease that can mimic Guillain-Barre syndrome and is occasionally mistaken for it. Guillain-Barre syndrome is an immunological sickness that arises when the body's immune system incorrectly attacks a part of the PNS, causing nerve inflammation that causes muscle paralysis.

Nano-Remediation of HMs Contaminated Soils

Using NPs to detoxify Heavy Metals pollutants present in the soil and other settings, nano-remediation is an economical and environmentally benign method. By absorbing the heavy metals, stimulating the reaction and decreasing the hazardous valence to stabilize the metallic state, the cutting-edge remediation method has demonstrated its efficacy in the elimination of heavy metals. To successfully eliminate HMs, a variety of NPs (metallic, carbonaceous, metallic oxide, nanocomposites, and polymeric,) have been utilized.

The efficiency of phytoremediation is increased by nanotechnology-based phytoremediation because it removes heavy metal-contaminated soil, water, and organic matter with the use of nanoparticles (Awasthi et al., 2022). The targeted main pollutants are adsorbed, degraded or changed as a result of the NPs' unique physicochemical properties, which function as a catalyst and lower the activation energy needed to break down the molecules. Additionally, polymeric nanoparticles such as nano-spheres and nano-capsules can efficiently remove pesticides and hydrocarbons (Chauhan et al., 2020).

Varied carbon-based nanomaterials have different structural arrangements, including fullerenes, carbon nanotubes, and graphene. Cd and harmful organic solvent dichlorobenzene were removed from the soil by carbon nanotube (CNT) nano-

Figure 3. List of nano-materials used for the removal of heavy metals (HMs)

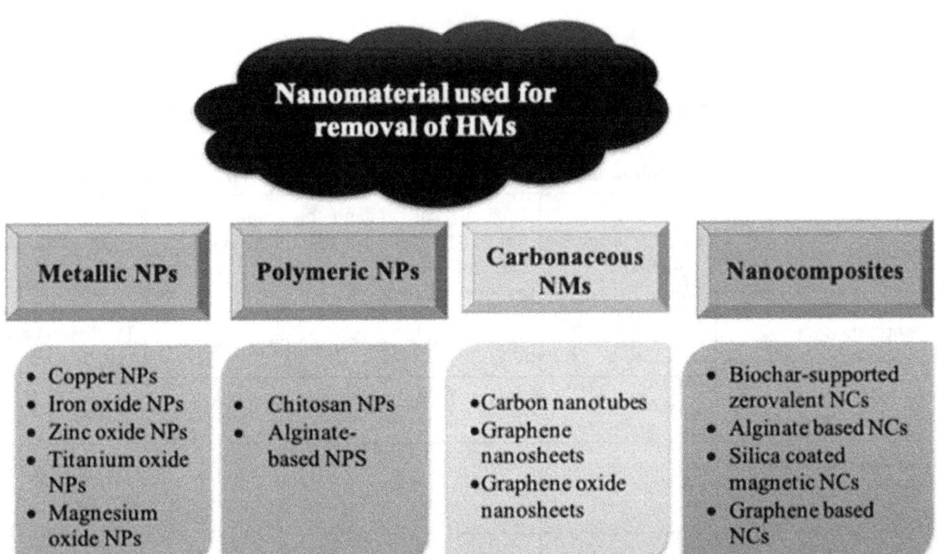

sponges. The removal of pollutants using environmentally friendly nanomaterials and nanoparticles is known as nano-bioremediation. It is a low-cost technique for using plants and bacteria to break down hazardous substances in order to lessen pollution and enhance soil quality. The term "nano-bioremediation" refers to a wide range of methods and tools, including nano-biosensors, nano-sorbents, nano-biosurfactants, and nano-phytoremediation. However, the function of several nanomaterials in reducing the stress caused by heavy metal contamination of soil.

Impact of Seed Priming on Pre-Germinative Based Metabolism

Seeds face adverse environmental impacts whether they are in a germinating stage or in post-harvest storehouse or in the maturity stage. During the time of these stresses, all the seeds are subordinated to extreme oxidative damage of structures of nucleic acid, protein, and lipids (Ventura et al., 2012). The seed coats are protected due to the release of secondary metabolites.

The DNA repairing process must be planned in a way to maintain vigor and enhance the germination process of seeds. DNA after conservation after the forming process the durability of cell cycle is preceded inside the embryo that ultimately leads to the replication of DNA. Oxidative injury as in the form of imperfect form process contributes to the death of cells (Ventura et al., 2012).

In the process of DNA restoration, the pre-germination based metabolism starts as and when the seed starts to consume water and reactive oxygen species store or buildup due to imbibition. These imbibition also conserve the integrity of genome (Cordoba-Canero et al., 2014) of original stage of seed are important DNA restoration

Table 3. Function of several nanomaterials in reducing stress caused by HMs

Types of nanomaterial	Pollutant	Efficiency of remediation	References
Nanotubes	Organophosphate and HM	Unzipped carbon nanotube, multi walled CNT and Single- walled CNT helped in the removal of HM contaminants by 22% removal efficiency	Fosso- Kankeu et al., (2014)
Biosorption *Saccharomyces cervisiae*	Cd, Pb	76.9 and 70.7% for Cd and Pb respectively	De Rossi et al., (2020)
Nanobiosensors *P. aeruginosa 9027*	HM	Rhamnolipids bioremediation of heavy metal	Ibrahim et al., (2013)
Microbial remediation *Bacillus sp. KL1, KL4 and KL6*	As	77% As in KL1 at 24 h	Taran et al., (2019)
Phytoremediation *Euphorbia macroclada*	Cd, Ni, Pb, Zn and Cu	92% Pb, 31.50% Ni, 74.66% Cd, 76.05% Zn and 74.66% Cu	Mohsenzadeh and Rad (2011)

pathways. The sub-pathway of Recap-coupled nucleotide form clearly recognizes and neutralizes the unwanted transcribed genomic materials and other active genes that have been examined in rice seeds during germination by Macovei et al., (2014) who also focused on the activity of DNA helicase that helps in unwinding of DNA during its replication. The relation of microRNAs that opt helicase in the process of DNA impairment-related responses as well as its restoration in rice seeds also has been examined under oxidative stress conditions confirmed by γ- shaft (Macovei and Tuteja, 2013). Ionizing radiations and their response towards seed growth were experimented and accordingly treatment adjustments were done (Grover and Khan, 2014).

Impact of Seed Priming on Oxidative Metabolism

Seed physiology plays a key role to mitigate the extreme oxidative damage due to Reactive Oxygen Species in plants. Enzymes similar to NADPH oxidases, oxalate oxidases, peroxidases enhance mitochondrial as well as respiration activities. The reactive oxygen species directly affect the metabolism and oxidation processes in plants and among these H_2O_2 species play an important role as a signaling patch. There is a need for reactive oxygen species detoxifying system to be equipped for providing safety towards oxidative damage. There are certain enzymes that are efficient to neutralize oxidative damage (i.e. *superoxide dismutase*, *glutathione reductase*, and *catalase*) as well as non-enzymatic antioxidant compounds such as reduced glutathione and ascorbic acid. At lower concentrations, the ROS products also play as signaling motes similar to hydrogen peroxide, hydrogen sulfide, and nitric oxide affecting dormancy to the germination phase (Wojtyla et al., 2016). An antioxidant protective system establishment is very much necessary to protect the seedlings from the oxidative damage caused by catalase during the last phase of germination. Priming is related to the induction of the antioxidant system as it has been verified through an integrated approach of proteomic and Transcriptomic studies. The primed genes are responsible for the decomposition of triacylglycerol compounds into nano fractions as in case of *B. napus* by through the process of gluconeogenesis. The special type of genes are responsible for water-holding priming of rape seed that encodes for recap factors and is associated with after hormo priming with ABA. It has been verified that primed seeds contain particular genes that may help in the germination of unprimed seeds (Kubala et al., 2015). Gel-free shotgun is another indispensable proteomic tool that is solely used to detect and quantify protein at large.

Impact of Seed Priming on Alteration in Metabolomics

Metabolome consist of entire metabolite sets that have lower molecular mass and also act as a quantitative dimension that influences sample metabolism. Functional genomics has gained special significance in case of plants. There are a combination of several ways and tools like gas chromatography, HPLC, atmospheric pressure chemical ionization, electrospray ionization etc. used to study maximum number as well as volume of metabolite. Priming also helps in separating composites that are mixed with nanoseconds and also helps in germinating the plants at any adverse conditions. Metabolites are produced in the particular organelle of the plant cell and show chemical responses during the birth process. Gibberellins play an important role in the case of smoothening of germination and have interactions with phenols or sugars and this makes interference in detecting metabolites (Wu et al., 2014).

Impact of Seed Priming on Seed Germination and Seedling Growth

Priming of seeds shows enhancement in imbibition as well as in metabolic processes that boost seed germination, and better seedling growth which leads to proper development in both normal and stressful conditions (Nayban et al., 2017). Seed priming also enhances the time of germination as early and supports better growth even in stress conditions that compared to undressed or non-primed seeds (Chen et al., 2012). In one study by Ahmadvand et al., (2012) showed that KNO_3^- primed seeds of soybean produced increased fresh weight, dry weight, root length, and other physiological parameters and also germination percent increased.

Impact of Seed Priming on Crop Nutrition and Yield

Crop nutrient deficiencies and incapability to uptake nutrients cause low yields and poor quality produce that cannot mitigate the global hunger at a large stake. Seed priming is a wonderful and advanced pathway to not only enhance seed emergence but also increase nutrient uptake by plants that are necessary for smooth biochemical processes and this helps in enrichment of nutrients in grains thus produced (Singh et al., 2015). Bhowmick et al., (2010) reported that seed priming was a suitable strategy to increase crop yields. Srivastava and Bose (2012) took ($Mg(NO_3)_2$ and KNO_3) mariners as a suitable priming agent to apply in rice crop and in their study reported that these priming agents increased physiological parameters as well as responses as compared to non-primed seeds.

Abiotic and Biotic Stress

The degradation of resources will create more difficult situations. Salinity, drought, floods, temperature extremes, heavy metals, radiation, etc. are examples of abiotic stressors. It has a substantial part in the loss of important species across globe (Singh et al., 2023). The occurrence of diseases or invasions from fungi, bacteria, oomycetes, nematodes, and herbivores are examples of biotic stress. Globally, diseases brought on by these pathogens are a major cause of yield loss. Sessile plants are unable to ignore these environmental stimuli since they are sessile. Successful completion of the lifetime requires expertise in handling these stressors. Numerous studies have shown that seed priming reduces the time it takes for seedlings to emerge and results in healthy plants that can withstand a variety of biotic and abiotic challenges. Additionally, researchers discovered that priming with TiO2 NPs boosted the chlorophyll content and anti-oxidative enzymes in maize and tomato plants, resulting in resistance to salt stress.

Distribution of Minerals

Mineral imbalances in soils can have a negative impact on plant health either incidentally by altering plant nutrition or water uptake or directly by having a lethal effect on plant cells. Plant stress can be caused by a number of abnormalities in the elemental composition of soils, e.g. large accumulation of Na^+ and Cl^- ions, poisonous particles like As and Cd and also little accumulation of key minerals like calcium, magnesium, nitrogen and phosphorus. Due to gratin's effective nutrition and water uptake, priming seeds with NPs promotes germination. To increase its effects, nutritional priming can be paired with nano-priming. Increased nutrient availability and improved plant stress tolerance are both benefits of nutrient priming. Numerous studies have produced favorable findings.

The *Zea mays* seedlings could benefitted in terms of the photosynthesis, growth, and kernel propagation by using magnesium hydroxide NP (macronutrient). Silver nanoparticles increased potassium content while lowering sulfur, phosphorus, and magnesium levels in oriental lilies. Fe NPs (500 ppm) foliar sprays can improve the Fe content and seedling vigor of southern pea (*Vigna unguiculata*) plants (Delfani et al., 2014). The use of inorganic manures in Oryza sativa could be effectively replaced with nano-pyrite. Nano-priming using zinc solutions improved the biological yield of *Triticum aestivum* L. and chickpeas. Another method involves "nutrient priming," which involves soaking seeds in a specific concentration of nutrients prior to planting. The water uptake efficacy, nutritional obtainability, propagation, all are improved if kernels are treated with essential or beneficial nutrients e.g. macronutrients and micronutrients. Micronutrient seed priming increases osmosis for controlling water

Table 4. Different types of nano-priming and its influences on various crops

No.	Crops	Priming NPs	NPs concentration	Physiological/Biochemical/ molecular changes	Reference
1.	*Solanum lycopersicum* L.	Nanosilicon dioxide nSiO2, NiO NPs	0.8% 0.2%	Affect kernel propagation Increased CAT, GR, and SOD	Faisal et al., (2013)
2.	*Capsicum annum* L.	Anatase nanoparticle (nTiO2)	7.52 g/L nTiO	Enhanced propagation, mass, radicle, plumule length and Germination rate index, radicle and plumule length and potency	Dehkourdi et al., (2014)
3.	*Pisum sativum* L.	ZnO NPs AgNPs	125, 250, and 500 ppm 60 mg/L AgNPs	Enhanced root length and kernel propagation	Mukherjee et al., (2014)
4.	*Triticum aestivum* L.	Phyllanthus emblica fruit extract with AgNPs (B-AgNPs) FeNPs Ag NPs	10 ppm B-AgNPs 2.0 mg/L 10 ppm	Enhanced seed propagation, decreased ROS injuries Increased the leaf and root growth Some protiens associated with metabolism and protection of cell in plant were activated	Kannaujia et al., (2019)
5.	Sugarcane (*Saccharum* spp.)	AgNPs	25 and 50 ppm	Production of ROS that induces the antioxidant response of the plant	Bello-bello et al., (2017)
6.	*Glycine max*	CeO2 NPs	100 mg kg/L CeO_2 NPs	Photosynthetic rate was enhanced	Cao et al., (2018)
7.	*Allium cepa*	AuNPs	0.1, 0.5, 1 mm	Increase in development and mitotic index	Alharbi et al., (2016)
8.	*Zea mays* L	AuNPs, gold nanoparticles GNPs (Galanga rhizome extracts (GRE) 2 mL GRE: 10 mL HAuCl4)	5–15 mg/L 5–15 ppm GNPs at ppm, 10 ppm GNPs at 15 ppm	Increased kernel propagation Enhanced total chlorophyll content i.e. 35–53% in GNP primed lines	Arnott et al., (2021)
9.	*Cucumis sativus*	CeO2 and CuO NPs	CeO_2 and CuO NPs	Increased fruit production per plant	Hong et al., (2016)
10.	*Oryza sativa*	CeO2 NPs	125 mg/L	Enhanced lipid peroxidation in rice roots	Rico et al., (2013)

during germination. For instance, priming mung beans with sodium molybdate dehydration for 5 hours increased the yield. Similarly, using macronutrients like potassium increases plants' ability to withstand environmental stress.

CONCLUSION AND FUTURE PERSPECTIVES

Recently, agricultural soils have been contaminated with heavy metals (HMs), posing a threat to global food security and human health. The accumulation of HMs in plants and soil ecosystems, as well as their toxicity effects, are more pronounced nowadays. The application of nanomaterials is specifically the transformation into nano fertilizers and/or nanoparticles in agriculture. In crops like wheat, maize, rice, soybeans, and other staple foods, heavy metal accumulation is a serious threat to food safety. It is possible to greatly reduce metal ion translocation between roots and uppermost parts by using nano-fertilizers. However, the translocation of vital nutrients like K, Mg, and Fe is significantly improved from higher portions of the plant to the rachises with the use of nano-fertilizers. The use of NPs may mimic plant defense mechanisms such as peroxidases, superoxide dismutase, and catalase, which continually reduce the ROS in the plants. Furthermore, NPs can accumulate in plant and microbial systems, thus entering the food chain which poses a threat to human health. Therefore, future research should focus on creating NPs that are biodegradable and biocompatible in order to reduce their buildup in agricultural systems.

REFERENCES

Ahmadvand, G., Soleimani, F., Saadatian, B., & Pouya, M. (2012). Effect of seed priming with potassium nitrate on germination and emergence traits of two soybean cultivars under salinity stress conditions. *American-Eurasian Journal of Agricultural & Environmental Sciences*, *12*, 769–777.

Alharbi, N. S., Bhakyaraj, K., Gopinath, K., Govindarajan, M., Kumuraguru, S., Mohan, S., & Benelli, G. (2016). Gum-mediated fabrication of eco-friendly gold nanoparticles promoting cell division and pollen germination in plant cells. *Journal of Cluster Science*, *28*(1), 507–517. doi:10.100710876-016-1130-8

Anwar, T. A., Qureshi, H., Parveen, N., Mahmood, S., Haider, M. Z., Mumtaz, S., & Ahmed, A. E. (2022). Herbicidal effectiveness of wild poisonous plant Rhazya stricta using different media by the sandwich method. *Pakistan Journal of Botany*, *55*(10.30848).

Arnott, A., Galagedara, L., Thomas, R., Cheema, M., & Sobze, J. M. (2021). The potential of rock dust nanoparticles to improve seed germination and seedling vigor of native species: A review. *The Science of the Total Environment*, *775*, 145139. doi:10.1016/j.scitotenv.2021.145139

Awasthi, G., Nagar, V., Mandzhieva, S., Minkina, T., Sankhla, M. S., Pandit, P. P., & Srivastava, S. (2022). Sustainable amelioration of heavy metals in soil ecosystem: Existing developments to emerging trends. *Minerals (Basel)*, *12*(1), 85. doi:10.3390/min12010085

Bello-Bello, J. J., Chavez-Santoscoy, R. A., Lecona-Guzman, C. A., Bogdanchikova, N., Salinas-Ruíz, J., Gomez-Merino, F. C., & Pestryakov, A. (2017). Hormetic response by silver nanoparticles on in vitro multiplication of sugarcane (Saccharum spp. Cv. Mex 69-290) using a temporary immersion system. *Dose-Response*, *15*(4), 1559325817744945. doi:10.1177/1559325817744945 PMID:29238274

Bhowmick, M. K., Biswas, P. K., Sen, P., & Bhattacharyya, P. (2010). Studies on seed priming, row spacing and foliar nutrition in chickpea under rainfed conditions in West Bengal, India. In: *Proceedings of the international seminar on climate change and environmental challenges of 21st century, December* (pp 7–9). Research Gate.

Bhowmick, M. K., Biswas, P. K., Sen, P., & Bhattacharyya, P. (2010, December). Studies on seed priming, row spacing and foliar nutrition in chickpea under rainfed conditions in West Bengal, India. In *Proceedings of the international seminar on climate change and environmental challenges of 21st century, December* (pp. 7-9). Research Gate.

Cao, Z., Rossi, L., Stowers, C., Zhang, W., Lombardini, L., & Ma, X. (2018). The impact of cerium oxide nanoparticles on the physiology of soybean (Glycine max (L.) Merr.) under different soil moisture conditions. *Environmental Science and Pollution Research International*, *25*(1), 930–939. doi:10.100711356-017-0501-5 PMID:29076022

Chauhan, R., Yadav, H. O., & Sehrawat, N. (2020). Nanobioremediation: A new and a versatile tool for sustainable environmental cleanup-Overview. *Journal of Materials & Environmental Sciences*, *11*(4), 564–573.

Chen, K., Fessehaie, A., & Arora, R. (2012). Dehydrin metabolism is altered during seed osmopriming and subsequent germination under chilling and desiccation in Spinacia oleracea L. cv. Bloomsdale: Possible role in stress tolerance. *Plant Science*, *183*, 27–36. doi:10.1016/j.plantsci.2011.11.002 PMID:22195574

Córdoba-Cañero, D., Roldán-Arjona, T., & Ariza, R. R. (2014). Arabidopsis ZDP DNA 3′-phosphatase and ARP endonuclease function in 8-oxoG repair initiated by FPG and OGG 1 DNA glycosylases. *The Plant Journal*, *79*(5), 824–834. doi:10.1111/tpj.12588 PMID:24934622

De Rossi, A., Rigueto, C. V., Dettmer, A., Colla, L. M., & Piccin, J. S. (2020). Synthesis, characterization, and application of Saccharomyces cerevisiae/alginate composites beads for adsorption of heavy metals. *Journal of Environmental Chemical Engineering*, *8*(4), 104009. doi:10.1016/j.jece.2020.104009

Dehkourdi, E. H., Chehrazi, M., Hosseini, H., & Hosseini, M. (2014). The effect of anatase nanoparticles (TiO2) on pepper seed germination (Capsicum annum L.). [IJB]. *International Journal of Biosciences*, *4*(5), 141–145.

Delfani, M., Baradarn Firouzabadi, M., Farrokhi, N., & Makarian, H. (2014). Some physiological responses of black-eyed pea to iron and magnesium nanofertilizers. *Communications in Soil Science and Plant Analysis*, *45*(4), 530–540. doi:10.1080/00103624.2013.863911

Faisal, M., Saquib, Q., Alatar, A. A., Al-Khedhairy, A. A., Hegazy, A. K., & Musarrat, J. (2013). Phytotoxic hazards of NiO-nanoparticles in tomato: A study on mechanism of cell death. *Journal of Hazardous Materials*, *250*, 318–332. doi:10.1016/j.jhazmat.2013.01.063 PMID:23474406

Fosso-Kankeu, E., Mulaba-Bafubiandi, A. F., & Mishra, A. K. (2014). Prospects for immobilization of microbial sorbents on carbon nanotubes for biosorption: bioremediation of heavy metals polluted water. *Application of nanotechnology in water research, 39*.

Grover, S., & Khan, A. S. (2014). Effect of ionizing radiation on some characteristics of seeds of wheat. *International Journal of Science and Technology Resources*, *3*, 2277–8616.

Hong, J., Wang, L., Sun, Y., Zhao, L., Niu, G., Tan, W., & Gardea-Torresdey, J. L. (2016). Foliar applied nanoscale and microscale CeO2 and CuO alter cucumber (Cucumis sativus) fruit quality. *The Science of the Total Environment*, *563*, 904–911. doi:10.1016/j.scitotenv.2015.08.029 PMID:26351199

Ibrahim, M. L., Ijah, U. J. J., Manga, S. B., Bilbis, L. S., & Umar, S. (2013). Production and partial characterization of biosurfactant produced by crude oil degrading bacteria. *International Biodeterioration & Biodegradation*, *81*, 28–34. doi:10.1016/j.ibiod.2012.11.012

Kannaujia, R., Srivastava, C. M., Prasad, V., Singh, B. N., & Pandey, V. (2019). Phyllanthus emblica fruit extract stabilized biogenic silver nanoparticles as a growth promoter of wheat varieties by reducing ROS toxicity. *Plant Physiology and Biochemistry*, *142*, 460–471. doi:10.1016/j.plaphy.2019.08.008 PMID:31425972

Kubala, S., Garnczarska, M., Wojtyla, Ł., Clippe, A., Kosmala, A., Żmieńko, A., & Quinet, M. (2015). Deciphering priming-induced improvement of rapeseed (Brassica napus L.) germination through an integrated transcriptomic and proteomic approach. *Plant Science*, *231*, 94–113. doi:10.1016/j.plantsci.2014.11.008 PMID:25575995

Macovei, A., Garg, B., Raikwar, S., Balestrazzi, A., Carbonera, D., Buttafava, A., & Tuteja, N. (2014). Synergistic exposure of rice seeds to different doses of-ray and salinity stress resulted in increased antioxidant enzyme activities and gene-specific modulation of tc-ner pathway. *BioMed Research International*, *2014*, 2014. doi:10.1155/2014/676934 PMID:24551849

Macovei, A., & Tuteja, N. (2013). Different expression of miRNAs targeting helicases in rice in response to low and high dose rate γ-ray treatments. *Plant Signaling & Behavior*, *8*(8), 25128. doi:10.4161/psb.25128 PMID:23733055

Maharramov, A. M., Hasanova, U. A., Suleymanova, I. A., Osmanova, G. E., & Hajiyeva, N. E. (2019). The engineered nanoparticles in food chain: Potential toxicity and effects. *SN Applied Sciences*, *1*(11), 1–25. doi:10.100742452-019-1412-5

Mohsenzadeh, F., & Rad, A. C. (2011). Application of nano-particles of Euphorbia Macroclada for bioremediation of heavy metal polluted environments. In *International Conference on Nanotechnology and Biosensors IPCBEE* (Vol. 25). Academia.

Moore, P. A. Jr, & Patrick, W. H. Jr. (1989). Manganese availability and uptake by rice in acid sulfate soils. *Soil Science Society of America Journal*, *53*(1), 104–109. doi:10.2136ssaj1989.03615995005300010019x

Mukherjee, A., Peralta-Videa, J. R., Bandyopadhyay, S., Rico, C. M., Zhao, L., & Gardea-Torresdey, J. L. (2014). Physiological effects of nanoparticulate ZnO in green peas (Pisum sativum L.) cultivated in soil. *Metallomics*, *6*(1), 132–138. doi:10.1039/C3MT00064H PMID:24190632

Nayban, G., Mandal, A. K., & De, B. K. (2017). Seed priming: A low-cost climate-resilient tool for improving germination, growth and productivity of mungbean. *SATSA Mukhaptra Annual Technical*, (21), 162–172.

Nile, S. H., Thiruvengadam, M., Wang, Y., Samynathan, R., Shariati, M. A., Rebezov, M., & Kai, G. (2022). Nano-priming as emerging seed priming technology for sustainable agriculture recent developments and future perspectives. *Journal of Nanobiotechnology*, *20*(1), 1–31. doi:10.118612951-022-01423-8 PMID:35659295

Raymond, A. W., & Felix, E. O. (2011). Heavy metals in contaminated soils: A review of sources, chemistry, risks and best available strategies for remediation. *International Scholarly Research Network ISRN Ecology. Article ID*, *402647*, 1–20.

Rico, C. M., Morales, M. I., Barrios, A. C., McCreary, R., Hong, J., Lee, W. Y., & Gardea-Torresdey, J. L. (2013). Effect of cerium oxide nanoparticles on the quality of rice (Oryza sativa L.) grains. *Journal of Agricultural and Food Chemistry, 61*(47), 11278–11285. doi:10.1021/jf404046v PMID:24188281

Salam, A., Afridi, M. S., Javed, M. A., Saleem, A., Hafeez, A., Khan, A. R., & Gan, Y. (2022). Nano-priming against abiotic stress: A way forward towards sustainable agriculture. *Sustainability (Basel), 14*(22), 14880. doi:10.3390u142214880

Singh, A., Sengar, R. S., Rajput, V. D., Minkina, T., & Singh, R. K. (2022). Zinc oxide nanoparticles improve salt tolerance in rice seedlings by improving physiological and biochemical indices. *Agriculture, 12*(7), 1014. doi:10.3390/agriculture12071014

Singh, A., Sengar, R. S., Shahi, U. P., Rajput, V. D., Minkina, T., & Ghazaryan, K. A. (2023). Prominent effects of zinc oxide nanoparticles on roots of rice (*Oryza sativa* L.) grown under salinity stress. *Stresses, 3*(1), 33–46. doi:10.3390tresses3010004

Singh, A., Sengar, R. S., Sharma, R., Rajput, P., & Singh, A. K. (2021). Nano-priming technology for sustainable agriculture. *Biogeosystem Technique, 8*(2), 79–92.

Singh, H., Jassal, R. K., Kang, J. S., Sandhu, S. S., Kang, H., & Grewal, K. (2015). Seed priming techniques in field crops-A review. *Agricultural Reviews (Karnal), 36*(4), 251–264. doi:10.18805/ag.v36i4.6662

Srivastava, A. K., & Bose, B. (2012). Effect of nitrate seed priming on phenology, growth rate and yield attributes in rice (*Oryza sativa L.*). *Vegetos Int J Plant Res, 25*(2), 174–181.

Taran, M., Fateh, R., Rezaei, S., & Gholi, M. K. (2019). Isolation of arsenic accumulating bacteria from garbage leachates for possible application in bioremediation. *Iranian Journal of Microbiology, 11*(1), 60. doi:10.18502/ijm.v11i1.707 PMID:30996833

Ventura, L., Donà, M., Macovei, A., Carbonera, D., Buttafava, A., Mondoni, A., & Balestrazzi, A. (2012). Understanding the molecular pathways associated with seed vigor. *Plant Physiology and Biochemistry, 60*, 196–206. doi:10.1016/j.plaphy.2012.07.031 PMID:22995217

Wojtyla, Ł., Lechowska, K., Kubala, S., & Garnczarska, M. (2016). Different modes of hydrogen peroxide action during seed germination. *Frontiers in Plant Science, 7*, 66. doi:10.3389/fpls.2016.00066 PMID:26870076

Wu, X., Li, N., Li, H., & Tang, H. (2014). An optimized method for NMR-based plant seed metabolomic analysis with maximized polar metabolite extraction efficiency, signal-to-noise ratio, and chemical shift consistency. *Analyst (London)*, *139*(7), 1769–1778. doi:10.1039/C3AN02100A PMID:24519438

Ye, Y., Cota-Ruiz, K., Hernandez-Viezcas, J. A., Valdes, C., Medina-Velo, I. A., Turley, R. S., & Gardea-Torresdey, J. L. (2020). Manganese nanoparticles control salinity-modulated molecular responses in Capsicum annuum L. through priming: A sustainable approach for agriculture. *ACS Sustainable Chemistry & Engineering*, *8*(3), 1427–1436. doi:10.1021/acssuschemeng.9b05615

Young, G. J., & Blevins, R. D. (1981). Heavy metal concentrations in the Holston river basin (Tennessee). *Archives of Environmental Contamination and Toxicology*, *10*(5), 541–560. doi:10.1007/BF01054878 PMID:7305447

Chapter 11
Nanopriming Effective Tools for the Field of Horticultural Crops to Copping Up With Biotic and Abiotic Stresses

Shipra Singh Parmar
Dr. Y.S. Parmar University of Horticulture and Forestry, India

Divya Pandey
Dr. Y.S. Parmar University of Horticulture and Forestry, India

ABSTRACT

Horticulture crops are continually impacted by severe biotic and abiotic stresses, which induce various impacts leading to global food scarcity scenarios. In light of this, establishing a balance between horticultural output and food demand necessitates the development of strategies to acclimate plants to stressful environments. In recent years, seed priming has been of significant interest to scientists. It leads to induced changes in seed metabolism promoting successful establishment of crops through accelerated seed germination and enhanced growth of growing seedlings. The application of nanotechnology in the current era holds the promise in combating the environmental stressors through the use of nanomaterials such as nanoseed priming technologies. Seeds are nano primed to hasten germination, which is followed by improved germination, biotic and abiotic stress tolerance, and to increase plant production.

INTRODUCTION

Seeds, the key component in the field of agriculture and offers the basis for a crop to develop naturally. The genetic basis of variability is inherent in seeds, which

DOI: 10.4018/978-1-6684-7232-3.ch011

significantly contributes to the final production. Therefore, seed is an essential component of every nation's food security plan. For the purpose of ensuring uniform seed germination, high horticultural productivity, and financial security in commercial agriculture, the most effective use of production resources is essential (Acharya et al., 2020). Germination of seed and its subsequent establishment as a seedling is considered as one of the foundation phases of plant life cycle because they determine any agricultural plant's field productivity by establishing a perfect plant stand.

Most horticultutal species on fruit trees reproduce vegetatively, but some fruits, like jamun, citrusand phalsa, are still propagated by seeds. The rigid or semihard seed coverings seen in the majority of commercial fruits may hinder water permeability and, as a result, seed germination.As a result, there may be inadequate oxygen levels, which could further cause damage. This could lead to an even greater lack of oxygen, impairing development. Many crop species have semipermeable seed coat layers to limit solute leakage, water absorption, and gas exchange (Abdelmigid et al., 2022). Nicking, scarification including removal of seed coat are among the few techniques used to improve and hasten the seed coat's permeability to allow better water and air adsorption; but these techniques has varied efficiencies. Therefore to enhance seed germination and improve seedling vigour, newer seed technologies are necessary. To increase crop productivity and seed vigour without harming ecosystems, seed priming is a cutting-edge ecological seed technology. Dutta (2018) in his work opined that, when applied to various crops, seed priming as evolved as a technique that add value in terms of quick seed germination and growth. In order to boost metabolic activity and raise germination rate without radicle emergence through the seed coat, An effective pre-sowing method or therapy called "seed priming" involves consuming seeds along with a particular solution (Paparella et al., 2015). Various ways in which priming technique is essential for the crop plants are as follows:

(i) Improve seedling germination and establishment
(ii) Accelerated and coordinated germination
(iii) Higher absorption of nutrients
(iv) Enhanced resistance to biotic and abiotic stress and
(v) Eventually, rise in yield and production.

TYPES OF SEED PRIMING

In the past, priming materials and techniques have been utilised to improve seed germination and seedling performance in numerous plant species. The types of priming currently being used are halo-priming, hydro-priming, solid matrix-priming, bio-priming and osmo-priming. The agents used in the different types of priming for example water in hydro-priming, osmotic agents in osmo-priming, salt solution in halo-priming, beneficial microbes in bio-priming and solution containing solid particle in solid matrix-priming. Seed priming is known to affect the enzymes that actively break down large sized molecules into smaller size for the overall development of the embryo, eventually resulting in faster seedling growth and establishment (Song and He, 2021). Farmers would benefit from these biological impacts of priming since they would cut down on the expenses and time invested in resowing and added irrigations. The fundamental drawback of priming, however, and the main obstacle to its widespread use is reduced storability as well as viability of primed seeds as drying to initial moisture is essential (Abdelmigid et al., 2022). Researchers have recently looked at the resilience of seeds primed with nutrient-rich nanoparticles. Abbasi et al. (2021) in his experiments has shown that the use of primers in the form of nano particles has proved to be more effective for achieving workable yields compared to the traditional priming methods. Figure 1 represents various type of seed primimg.

PRINCIPLES OF SEED PRIMING

In agriculture germination is a vital stage for plant development and crop quality (Abbasi et al., 2020). Rapid seedling growth causes rapid root and leaf elongation, which favours transpiration flow and nutrient uptake as well as biomass production (Mahakham et al., 2016). Early seedlings are among the most vulnerable times in a plant's life cycle, and slow germination exposes them to a number of environmental stressors or illnesses, reducing their vigour and crop output and causing economic losses to farmers (Acharya et al., 2019). Auxins regulate the dormancy and germination of seeds (Wu et al., 2020). Excessive ROS levels result in significant oxidative damage, which hinders seed germination (Bailly, 2019). In order to be included in the so-called oxidative window, which enables proper completion of germination, ROS level must be controlled (Bailly, 2019).

Preparing seeds for planting is a common agricultural practise known as "seed priming," which aims to encourage seed germination and plant growth (Carrillo-Reche et al., 2018). Typically, seeds are dried after being soaked in water, or they may be physically primed using ultraviolet light (Lemmens et al., 2019). In order to proceed with phase I and II sufficient water absorption should occur to initiate pre-germination metabolic pathways. The reverse mobilisation capacity, transcriptome reprogramming, cell wall thinning, and translation and posttranslational modification predispositions are all altered as a result of this process, which influences seed metabolism at the molecular and cellular levels. On absorption, it creates a particular

Figure 1. Different types of seed priming

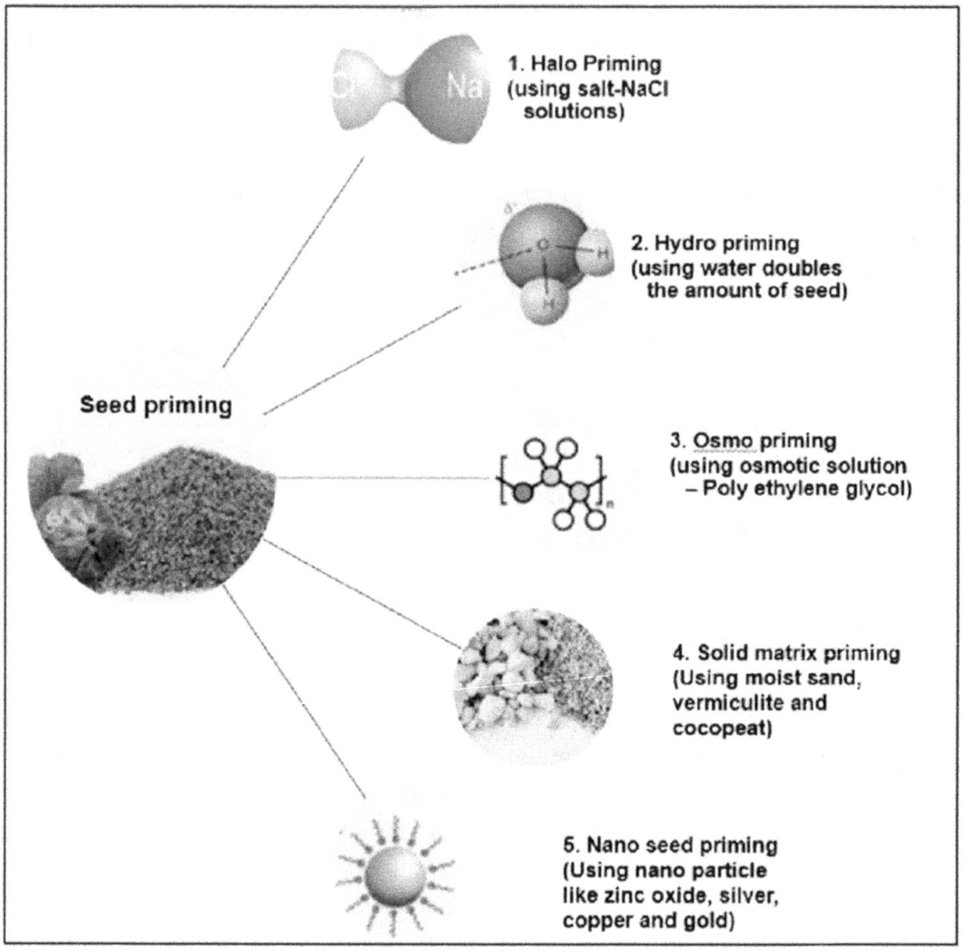

physiological state that strengthens and improves the germination and vitality of seeds that have been primed (Carrillo-Reche et al., 2018). Soaking and subsequent drying may lead to development of cross resistance mainly triggered with the production of HSPs (heat shock proteins), antioxidant mechanisms as well as other reactions related to stress. Therefore, seed priming has been used to improve the quality of produce and increase productivity by encouraging seedling vigour, strengthening plants' resistance to stress, boosting and coordinating germination, etc (Carrillo-Reche et al., 2018; Lemmens et al., 2019). There are several methods for priming seeds, including hydro-priming and hydrothermo-priming, which involves treating seeds with water for a period of time, usually between 7 - 14 hours, to keep them hydrated and facilitate the onset of phase II of germination (Carrillo-Reche et al.,

2018). This technique can be used in both cold and hot climates (Noorhosseini et al., 2017). Osmo-priming makes use of solutions with lower water potential to control hydration (roughly 10–20%) and change seed metabolism via an abiotic stress. Other pre-sowing procedures include the use of plant growth regulators for hormo-priming (Sytar et al., 2018), halopriming for salt-based solutions (Saddiq et al., 2019), and bio-priming for microorganisms (Lemmens et al., 2019).

SEED NANO-PRIMING

Scientists have studied how nanomaterials affect plant germination and growth in order to promote their use in horticulture, as nanotechnology has advanced across many fields. Undoubtedly, nanotechnology is a major factor in the transformation of agriculture and food production, and it has the potential to advance the farming sector (Fraceto et al., 2016). The germination rate, seedling vigour and dry weight of crops are all improved by the new technology known as nano-priming, which uses nanoparticles smaller than 100 nm. It has been proposed that seed priming with micronutrient nanoparticles might be a novel way to increase the rate of germination and the vigour and growth of seedlings (Singh et al., 2021). Nano priming enhances the permeability of the seed coat which in turn enables greater imbibitions of water, oxygen and nutrients thus, causing higher germination seedling growth rate (Dutta,2018). There are two types of nano particles that have been used as seed priming agents: metal-based and carbon-based nanotube (Zuverza-Mena et al., 2017). In significant crops, carbon nanotube priming has improved germination, growth, uptake of water and nutrients, and yield. Silicon, iron, copper, copper oxide, and magnesium nanoparticles can be rapidly absorbed, transported, and stored in plant tissues, thereby regulating physiological and biochemical processes that promote plant growth and development (Rastogi et al., 2017). Additional nanopriming substances, including AgNPs (nanosilver) and TiO2 (nanotitanium dioxide), may affect various plant growth parameters like leaf chlorophyll content, plant germination and establishment, shoot and root length, plant's fresh and dry weight, water uptake, total biomass and other antioxidant contents (Dadarwal et al., 2022). In order to create secure nanoscale biomaterials for agricultural purposes, green nanoparticle synthesis has been added to the precision of nanotechnology. Plant extracts can be used to make metal or metal oxide nanoparticles by acting as reducing stabilisers of the corresponding precursor molecules (Mohamm et al., 2016). Green nano particle synthesis' central tenet is that the organic components in plant extract stabilise nanoparticles, which then shrink and encase the metal ions in an aqueous solution (Husaain et al., 2019). Because traditional seed priming typically involves hydropriming with water or solutions in which the chemicals are adsorbed on the seed and the seed gets coated with these nutrients, biopolymers, or hormones, the terms "seed priming" and "seed nano priming" are not interchangeable. In seed nano-priming, suspensions and nanoformulations are used, and the seeds may or may not absorb the nanoparticles (Acharya et al., 2019). Even after being absorbed, the majority of nanoparticles remain as a coating on the seed surface (Acharya et al., 2019; Montanha et al., 2020). When used in the field or for storage, these seed coverings are combined with fungicides or bactericides to prevent disease (Gross et al., 2020). Khodakovskaya et al (2009)'s study was one of the first to demonstrate how nanomaterials can impact seed germination. These researchers found that even without seed priming, seeds can absorb carbon nanotubes. It has been shown that a variety of nanomaterials, including metallic, biogenic metallic, and polymeric nanoparticles, are effective for seed nano-priming (Siddaiah et al., 2018). This causes altered gene expression, which accelerates the growth of shoots and roots and interferes with metabolic processes like hormone synthesis. As a result of an increase in antioxidant activity and enzymatic activities in the defence system, plants exhibit greater resistance to pests and other biotic and abiotic stimuli in the environment after seed priming (Itroutwar et al., 2019; Siddaiah et al., 2018).

APPLICATION OF VARIOUS NANOPARTICLES IN SEED PRIMING

Copper Nanoparticles (Cu-NPs)

Copper (Cu) is a crucial element, for growth and adaptation related to photosynthesis. It takes part in protein and hydrocarbon exchanges. Ascorbic acid oxidase and polyphenol oxidase are two oxidising enzymes that contain copper as a component (Leng et al., 2015). Cu is necessary for plant metabolism and overall development. Copper deficiency is indicated by curled leaves, however excess copper can be hazardous (Leng et al., 2015). Cu-NPs that are released from a variety of products may have an impact on people and natural systems (Chen et al., 2012). However, further research is necessary to determine the impact of Cu-NPs on plants, so some previous findings are uncertain (Leng et al., 2015).

Iron Nanoparticles (Fe-NPs)

Fourth most prevalent element in the crust of the planet is iron (Fe), which is nontoxic (Li et al., 2006). All living things require iron as a micronutrient. It is required for the synthesis of chlorophyll, the process of photosynthesis, and respiration (Najafi Disfani et al., 2017). Nanoparticles of iron oxide (FeO), play a key role in the germination, growth, and increased productivity of plants. Since iron is the most common component of chlorophyll, increasing FeO-NPs reduces iron deficiency while raising levels of chlorophyll a and b (Ghafariyan et al., 2013). Due to their crucial function in plants, iron oxide NPs have a substantial impact on the development of agriculture and other fields (Li et al., 2006). FeO-NPs are also employed as nano-fertilizers to enhance the availability of iron to the plants, boost height, biomass, and root length, and modify the activity of hormones and antioxidant enzymes.

Silica Nanoparticles (Si-NPs)

The most common element on earth after oxygen is silicon (Si), which makes up about 31% of the planet's crust (Dietz, Herth, 2011). Biology and electronics are just two examples of the many scientific, technological, and engineering fields in which nanosilica is used. As per Wang et al.(2011) Si may aid plants in overcoming salt stress. Si contributes to increased leaf mechanical strength, photosynthetic activity, plant resistance to biotic or abiotic stress, organ lifetime, and improved leaf light absorption. The epidermal tissue of the plant's secretory organs contains si (Rastogi et al., 2019). Improvements in crop quantity and quality have been seen when silicon fertilisers have been used in a variety of soil types and climates. Because of their distinct physiological characteristics, silicon NPs have the ability to affect plant metabolism (Rastogi et al., 2019). Silica NPs may potentially carry DNA and other substances into the cells of plants and animals (Wang et al., 2011). All aspects of germination, including growth speed, seed radicle height and dry weight are improved by nanosilica (Rastogi et al., 2019; Wang et al., 2011).

Silver Nanoparticles (Ag-NPs)

Currently, second only to carbon nanotubes in terms of usage, Ag-NPs are one of the most popular nanoparticles. These NPs' antimicrobial properties are widely applied to a variety of products, such as detergents, textiles, and polymers (Awasthi et al., 2017). Ag-NPs are highly valued for their specific features such as optical, electrical, and magnetic capabilities. They can therefore be combined with cryogenic superconducting components, electronic parts, biosensor materials, cosmetics, composite fibres, ethylene action monitoring, and antimicrobial plant applications. Ag-NPs are believed to be helpful in gardening and other agricultural activities due to their capacity to shield seeds against bacterial and fungal infection (Parveen, Rao, 2015).

Zinc Oxide Nanoparticles (ZnO-NPs)

Zinc's role in the synthesis of carbohydrates and chlorophyll makes it a crucial mineral for the growth of plants. When zinc levels in plants are increased, detrimental heavy metal absorption is reduced, and therefore their toxic effects are minimised. Zinc nanoparticles are mostly used in agriculture because they affect plant anatomy and physiological responses (Awasthi et al., 2017). ZnO is required to regulate the actions of numerous other enzymes, including superoxide dismutase and dehydrogenases, as well as the metabolism of phytohormones (Rajput et al., 2021a). Recent studies have shifted their attention to how ZnO-NPs affect the germination of seeds in a variety of plant species (Awasthi et al., 2017; Gaafar et al., 2020; Sharma et al., 2021). The biological activity of metal-based NPs has been noted to have a negative impact on plants when ZnO-NPs nanofertilizer is used, though (Itroutwar et al., 2020). Numerous studies claim that nano-ZnO is so toxic that it can prevent the growth of plant roots.

Biogenic Nanoparticles

It has been demonstrated that microbes are crucial nano factories with huge opportunities as eco-friendly and economical instruments, avoiding poisonous, nasty chemicals and the high intensity demand necessary for physiochemical synthesis (Singh et al., 2015). Microorganisms are capable of absorbing and detoxifying heavy metals because they have a variety of reductase enzymes that can change metal salts into metallic nanoparticles with a narrow size distribution and reduced polydispersity. Recent research has looked at a variety of bacteria for the production of silver and gold nanoparticles, including *Pseudomonas deceptionensis* (Jo et al., 2015), *Weissella oryzae* (Singh et al., 2015), *Bacillus methylotrophicus* (Wang et al., 2015), and *Bhargavaea indica* (Singh et al., 2015). Titanium dioxide (TiO_2) and silver nanoparticles are two examples of the numerous types of nanoparticles made from plant extracts such pomegranate, coffee, mango, and banana (Abdelmigid et al., 2021).

EFFECT OF NANO PRIMING

Seed Germination

One of the most important phases of plant establishment in agriculture is germination, which is crucial for crop quality. There are three phases to the germination of seeds. Imbibition is the first stage of it; during this phase, the seed rapidly absorbs water, activating protein synthesis, transcription, and mitochondrial activity in its basal metabolism. Restricted water absorption is a characteristic of phase II, also referred to as the activation or lag phase but hyperactive metabolism due to the formation of amylases, endoxylanases, and phytases, which are essential for mobilising reserves and embryo growth. During the phase III, the seeds once more show rapid water intake, and the embryo grows to the point where the radicle protrudes. Due to the general hydration and rehydration mechanisms, phase II of the nanopriming process lasts longer than typical seed germination. There are two hydration mechanisms in primed seed, but only one under non primed seed. Seedling development is inhibited throughout the drying process, however, metabolic activity is still present, which is essential for the production of metabolites, antioxidant capacity, solubility of stored materials, priming, stress tolerance, and repair mechanisms. As a result, stress tolerance, climatic adaptation, and the quick and reliable establishment of seedlings are all improved (Dadarwal et al., 2022). Figure 1 depicts the mechanism of nano seed priming in relation to the regulated seed germination process.When it reaches the seed coat, nano-particles produces ROS (reactive oxygen species), which activates a number of subsequent actions. Theoretically, it was suggested that nanoparticles play a part in promoting seed germination via.

Formation of nanopores on the seed coat

(i) Mild ROS-induced stressor, and
(ii) Functioning as a nano catalyst for increased starch-hydrolyzing enzyme activity (Singh et al., 2020).

The production of ROS is necessary to break seed dormancy and promote germination, and this can be done by upping the production of gibberellin and shifting storage proteins (Nile et al., 2022). The production of ROS is necessary to break seed dormancy and promote germination, and this can be done by upping the production of gibberellin and shifting storage proteins. This dosage was proven to be efficient for pomegranate germination (Abdelmigid et al., 2022). The optimal nano titanium dioxide concentration on strawberry seeds was reported to be 7.5% for high germination percentage, fresh weight of seedling, and vigour index (Hosseini et al., 2013). The findings suggested that nanosized TiO_2 in the correct proportion might encourage strawberry seed germination and seedling development. Ahmed et al. (2018) found that soaking acid lime seeds in Nano-selenium solution at 50 ppm could help shorten the time needed for nursery growth and produce more robust, healthy plants. In water melon, seed priming using silver nano particles (AgNPs) can improve seed emergence and stand establishment (Acharya et al., 2020).

Vegetative Stage

A revolutionary technique for enhancing plant development is seed priming with metal nanoparticles or carbon-based nanomaterials. Its use can increase horticultural productivity and product quality. When plants are exposed to nano-ZnO particles, their height, biomass, level of chlorophyll, and yield are all improved. Additionally, the primed treated seed has a greater Zn content in the leaf and seed.

Increases in biomass, stomatal density, epidermal cell number, vascular tissue size, and crop water uptake are all seen in seeds that have been primed with multiwall carbon nanotubes (Joshi et al., 2018). Total indoles and total phenols content increased in acid lime seedling leaves, and the number of leaves/seedling and leaf area were both significantly increased by nano-selenium solution primed seed at 50 ppm (Ahmed et al., 2018). There was also an increase in the pigments carotenoid, chlorophyll a, and b. The absorption, translocation, and accumulation of nitrogen, phosphorus, potassium, iron, zinc, and manganese in leaves were encouraged by nano primed selenium seeds. At 20 mg/l, manganese oxide nano particles for seed priming significantly alter the chlorophyll and antioxidant profiles of seedlings of water melon. However, at a concentration of 40 mg/l, priming had a noticeable effect on the leaf phenolic acid and phytohormone profiles in watermelon seedlings (Kasote et al., 2021).

EFFECT OF NANO PRIMING ON BIOTIC STRESS

Stress causes diminished output and financial damage. Insects and microbial pathogens (bacteria, viruses, nematodes, and fungus) are the primary causes of biotic stress. Since plants are sessile by nature, they interact with various biotic agents during their whole life cycle. These biotic agents could be microorganisms such as viruses, fungus, bacteria, insects and nematodes. Presence of these biotic agents may affect the microenvironment of the plant and also modifies the internal physiological and biochemical condition of the system; accordingly, the yield potential of horticultural cropping systems decreases. Nonetheless, with the invention of current nano technology, the agriculture industry is continuously expanding, surpassing various yield limits and boosting crop output despite in the face of numerous biotic agents that cause biotic stressors. Seed nano priming technology is utilised to combat biotic stressors, and growth regulators are employed for this reason (Bose et al., 2018). Nanoparticles have antibacterial and anti fungal activity which improves germination and development. Citrus seedling damping off percentage and the percentage of albino seedlings were significantly decreased by a nano-selenium solution at 50 ppm for seed priming

(Ahmed et al., 2018). Metal sulphide nanoparticles applied to seeds significantly reduced a seedling blight in crops (Ahuja et al., 2019). Since silver is well recognized for being particularly efficient against microbial activity, its impact on certain phytopathogens known to cause illness has been recorded. According to a study, nano silver particles have an inhibitory effect on phytopathogenic fungi, which slows their proliferation and the spread of illness (Dadarwal et al., 2022). Bacterial wilt, which is caused by *Ralstonia solanacearum*, is one of the most prevalent diseases of *Solanum* spp (RS). On solanaceous crops like potato and tomato plants, 100 micro gram per mili litre of chitosan nanoparticles applied as soil-amended curative therapies resulted in a decrease in the incidence and severity of disease (Khairy et al., 2022).

EFFECT OF NANO PRIMING ON ABIOTIC STRESS

Throughout the course of their lifecycle, plants frequently experience a variety of abiotic stresses. Various biochemical, genetic, physiological, and metabolic changes in plants are brought on by oxidative stress, metabolic changes, and decreased yields. Abiotic stress is brought on by environmental elements like cold, heat, salinity, and drought. In horticulture, seed nano-priming can improve seed quality and increase tolerance to challenging circumstances. Nanoparticles have the ability to directly combat infections and modify seed and plant metabolism, which strengthens the innate immune system, changes hormone synthesis, and increases plant resistance to diseases and abiotic stress (Panyuta et al., 2016). Due to early vigour, seed priming with nanoparticles reduces the overall period of seed emergence, encourages uniform germination, and enhances crop stand (Thakur et al., 2019). The effect of carbon nano material seed priming on the metabolic reactions and content of bioactive chemicals in tomato plants growing under salinity stress conditions was investigated. The results demonstrated that CNMs can work as elicitors and activate tomato plants' antioxidant defence mechanism. Since the application of CNMs enhanced the antioxidant potential of the fruit by boosting lycopene and glutathione, the rise in antioxidant chemicals seemed to be consistent. While the combination of CNMs and salinity stress increased lycopene, ascorbic acid, phenols, glutathione, and flavonoids (Garcia et al., 2022). By absorbing water more quickly and recovering the seed metabolism through the repair of vital biological components like nucleic acids and proteins, nanoprimed seeds are ecofriendly and useful alternatives that offer greater plant protection against a variety of abiotic challenges. Application of nanopriming promotes high root and shoot biomass at higher levels of abiotic stress conditions because the buildup of a high concentration of dehydrogenase enzymes in cells improves cell elongation in the rhizome and tends to speed up the growth of the root system, ultimately producing high biomass (Yousefi et al., 2017). Nanoparticles like ZnO and Fe also enhance the development of plant characteristics like shoot and root dry biomass due to their optimal nutrient absorption and restricted usage at specific locations and stages of plants under various plant stressors (Rizwan et al., 2019). Under a hazardous concentration of heavy metals, seed priming with nano particles improves plant photosynthesis and decreases super oxide dismutase, electrolyte leakageand POX activity in leaves (Rizwan et al., 2019). Increased germination and seedling vigour index were seen in plants under salt stress when seeds primed with nanosilica particles due to higher levels of ROS, gibberellin content, amylase, and lipase enzyme activities during respiration (Naguib and Abdalla, 2019). Aldolase and isocitrate lyase, two enzymes important for respiration, both experienced an increase in activity. In the end, under a variety of abiotic conditions, nanopriming enhances the physiological and metabolic state of agricultural plant seeds. Under drought circumstances, a modest dosage of multiwalled carbon nanotubes exhibited favourable impacts on germination and root development (Rahimi et al., 2016).To maintain a high net photosynthetic rate under water deficit condition, nano-ZnO (100/mgL) was applied to the plant. It also increased stomatal mobility and water consumption efficiency (Sun et al., 2020). Furthermore, under drought conditions, cytoplasmic invertase, phosphoglucoisomerase, and UDP-glucopyrophosphorylase activity increased, leading to an increase in glycolysis metabolism as well as sucrose and starch production in leaves. Pavithra et al. (2016) discovered that ZnO-treated plants had higher Zn levels, biomass, and increased chlorophyll content in their leaves and seeds when compared to control. When seeds were primed with silver and silicon nanoparticles, germination was also boosted under saline circumstances (Almutairi, 2016). Singh et al. (2023) reported that ZnO-NPs application reduced oxidative stress and boosted antioxidant buildup in the roots of sustainably grown plants, thus also minimising the adverse consequences of salinity stress. In order to investigate the impact of seed priming with zinc oxide and iron nano particles on plant development under the influence of cadmium toxicity, An experiment by Rizwan et al. (2019) revealed that nano primed seeds increased plant height and dry weight of shoot and root. It was also noted that nanopriming had a substantial impact on photosynthesis. The nanopriming reduced the leakage of peroxidase, superoxide, and electrolytes in leaves under cadmium toxicity. Additionally, it is hypothesised that silver particle nanopriming increases -amylase activity and is responsible for an increase in the accumulation of soluble sugar, both of which help seedling establishment (Mahakham et al., 2017).

Table 1. Application of nano seed priming for various abiotic stresses

Plant	Nano material	Abiotic stress	References
Basella sp	Zinc oxide and Iron nano particles	Heavy metal stress	Gupta et al., 2022
Brassica sp	Cerium nano particles	Salinity	Khan et al., 2022
Calendula officinalis	Silica nano particles	Drought	Rahimi et al., 2021
Linum usitatissimum	Iron nano particles	Drought	Mazhar et al., 2022
Solanum lycopericum	Carbon nano tubes	Salinity	Garcia et al., 2022
Solanum lycopericum	Silver and titanium oxide nano particles	Drought	Sonawane et al., 2021
Spinacia oleracea	Zinc oxide nano particles	Salinity	Zafar et al., 2022

NANOPRIMING TO ENHANCE SUSTAINABILITY

As was already mentioned, nanopriming is crucial for seed germination, seedling establishment, and yield improvement. Aside from them, nano primed seeds plays a vital function in enhancing fertiliser efficiency and lowering both input costs and soil contamination. In soils with low nutrient levels, the use of nanopyrite priming in various crops increases crop productivity and nutrient use efficiency. For biofortifying crops, micronutrient seed priming, which raises the amount of micronutrients Zn and Fe in horticultutalcrops, has produced encouraging results (Kumar et al., 2020). It has been demonstrated that using seed nano-priming improves root growth and increases the synthesis of enzymes necessary for nutrient uptake, allowing for a reduction in the quantity of fertiliser used in the field.

CONCLUSION

Nanotechnology has the potential to transform traditional agriculture from one that depends on the use of agrochemicals to one that is more sustainable by treating seeds. When those systems can support plant establishment while also protecting against biotic and abiotic challenges, productivity and food quality will increase. Although using nanoparticles for seed priming has a number of advantages. Comparing seed treatments to foliar sprays and soil inoculations, the exposure to nanomaterials is reduced. Another benefit is the small amounts of controlled nanomaterials used for seed priming that can be applied in factories, preventing a significant release of these substances into the environment. Another emerging technology for the future that can support environmentally conscious farming and sustainable agricultural development is green nanopriming. Green nanomaterials have the potential to enhance farmer socioeconomic condition, create safe and superior horticultural productsand conserve the environment. The use of nanoparticle seed priming can alter crop management, resulting in lower chemical application rates, a reduction in contamination risks, and more environmentally friendly horticultural practises for growers, consumers, and the environment.

REFERENCES

Abbasi Khalaki, M., Moameri, M., Asgari Lajayer, B., & Astatkie, T. (2020). Influence of nano-priming on seed germination and plant growth of forage and medicinal plants. *Plant Growth Regulation*, *93*(93), 13–28.

Abdelmigid, H. M., Alyamani, A. A., Hussien, N. A., Morsi, M. M., & Alhumaidi, A. (2022). Integrated approaches for adsorption and incorporation testing of green-synthesized TiO2NPs mediated by seed-priming technology in *Punica granatum* L. *Agronomy (Basel)*, *12*(7), 1601. doi:10.3390/agronomy12071601

Acharya, P., Jayaprakasha, G. K., Crosby, K. M., Jifon, J. L., & Patil, B. S. (2019). Green-synthesized nanoparticles enhanced seedling growth, yield, and quality of onion (*Allium cepa* L.). *ACS Sustainable Chemistry & Engineering*, *7*(17), 14580–14590. doi:10.1021/acssuschemeng.9b02180

Acharya, P., Jayaprakasha, G. K., Crosby, K. M., Jifon, J. L., & Patil, B. S. (2020). Nanoparticle-mediated seed priming improves germination, growth, yield, and quality of watermelons (*Citrullus lanatus*) at multilocations in Texas. *Scientific Reports*, *10*(1), 5037. doi:10.103841598-020-61696-7 PMID:32193449

Ahuja, R., Sidhu, A., & Bala, A. (2019). Synthesis and evaluation of iron (ii). *European Journal of Plant Pathology*, *155*(1), 163–171. doi:10.100710658-019-01758-3

Almutairi, Z. M. (2016). Influence of silver nano-particles on the salt resistance of tomato (*Solanum lycopersicum*) during germination. *International Journal of Agriculture and Biology*, *18*, 125–129. doi:10.17957/IJAB/15.0114

Awasthi, A., Bansal, S., Jangir, L. K., Awasthi, G., Awasthi, K. K., & Awasthi, K. (2017). Effect of ZnO nanoparticles on germination of *Triticum aestivum* Seeds. *Macromolecular Symposia*, *376*(1). 10.1002/masy.201700043

Bailly, C. (2019). The signalling role of ROS in the regulation of seed germination and dormancy. *The Biochemical Journal*, *476*(20), 3019–3032. doi:10.1042/BCJ20190159 PMID:31657442

Bose, B., Kumar, M., Singhal, R. K., & Mondal, S. (2018). *Impact of seed priming on the modulation of physico-chemical and molecular processes during germination, growth, and development of crops. Advances in seed priming*. Springer. doi:10.1007/978-981-13-0032-5_2

Carrillo-Reche, J., Vallejo-Marín, M., & Quilliam, R. S. (2018). Quantifying the potential of "on-farm" seed priming to increase crop performance in developing countries. A meta-analysis. Agron. *Journal of Sustainable Development*, *38*, 1–14.

Chen, Y., Wang, D., Zhu, X., Zheng, X., & Feng, L. (2012). Long-term effects of copper nanoparticles on wastewater biological nutrient removal and N2O generation in the activated sludge process. *Environmental Science & Technology*, *46*(22), 12452–12458. doi:10.1021/es302646q PMID:23110389

Dietz, K. J., & Herth, S. (2011). Plant nanotoxicology. *Trends in Plant Science*, *16*(11), 582–589. doi:10.1016/j.tplants.2011.08.003 PMID:21906987

Dutta, P. (2018). Seed priming: New vistas and contemporary perspectives. In *Advances in seed priming* (pp. 3–22). Springer. doi:10.1007/978-981-13-0032-5_1

Fraceto, L. F., Grillo, R., de Medeiros, G. A., Scognamiglio, V., Rea, G., & Bartolucci, C. (2016). Nanotechnology in agriculture: Which innovation potential does it have? *Frontiers in Environmental Science*, *4*, 20. doi:10.3389/fenvs.2016.00020

Gaafar, R. M., Diab, R. H., Halawa, M. L., Elshanshory, A. R., El-Shaer, A., & Hamouda, M. M. (2020). Role of zinc oxide nanoparticles in ameliorating salt tolerance in soybean. *Egyptian Journal of Botany*, *60*(0), 733–747. doi:10.21608/ejbo.2020.26415.1475

Ghafariyan, M. H., Malakouti, M. J., Dadpour, M. R., Stroeve, P., & Mahmoudi, M. (2013). Effects of magnetite nanoparticles on soybean chlorophyll. *Environmental Science & Technology*, *47*(18), 10645–10652. doi:10.1021/es402249b PMID:23951999

Gross, M. S., Bean, T. G., Hladik, M. L., Rattner, B. A., & Kuivila, K. M. (2020). Uptake, metabolism, and elimination of fungicides from coated wheat seeds in Japanese quail (*Coturnix japonica*). *Journal of Agricultural and Food Chemistry*, *68*(6), 1514–1524. doi:10.1021/acs.jafc.9b05668 PMID:31977218

Gupta, N., Singh, P. M., Sagar, V., Pandya, A., Chinnappa, M., Kumar, R., & Bahadur, A. (2022). Seed priming with ZnOand Fe3O4Nanoparticles alleviate the lead toxicity in Basella Alba L. through reduced lead uptake and regulation of ROS. *Plants*, *11*(17), 2227. doi:10.3390/plants11172227 PMID:36079609

Gupta, R., & Xie, H. (2018). Nanoparticles in daily life: Applications, toxicity and regulations. *Journal of Environmental Pathology, Toxicology and Oncology*, *37*(3), 209–230. doi:10.1615/JEnvironPatholToxicolOncol.2018026009 PMID:30317972

Hussain, A., Rizwan, M., Ali, Q., & Ali, S. (2019). Seed priming with silicon nanoparticles improved the biomass and yield while reduced the oxidative stress and cadmium concentration in wheat grains. *Environmental Science and Pollution Research International*, *26*(8), 7579–7588. doi:10.100711356-019-04210-5 PMID:30661166

Itroutwar, P. D., Govindaraju, K., Tamilselvan, S., Kannan, M., Raja, K., & Subramanian, K. S. (2019). Seaweed-based biogenic ZnO nanoparticles for improving agromorphological characteristics of rice (Oryza sativa L.). *Plant Growth Regulation*, *392*(39), 717–728.

Jo, J. H., Singh, P., Kim, Y. J., Wang, C., Mathiyalagan, R., Jin, C. G., & Yang, D. C. (2015). Pseudomonas deceptionensis DC5-mediated synthesis of extracellular silver nanoparticles. *ArtifitialCells Nanomedicines and Biotechnology*, 1–6.

Joshi, A., Kaur, S., Dharamvir, K., Nayyar, H., & Verma, G. (2018). Multi-walled carbon nanotubes applied through seed-priming influence early germination, root hair, growth and yield of bread wheat (*Triticum aestivum* L.). *Journal of the Science of Food and Agriculture*, *98*(8), 3148–3160. doi:10.1002/jsfa.8818 PMID:29220088

Kasote, D. M., Lee, J. H. J., Jayaprakasha, G. K., & Patil, B. S. (2021). Manganese oxide nanoparticles as safer seed priming agent to improve chlorophyll and antioxidant profiles in watermelon seedlings. *Nanomaterials (Basel, Switzerland)*, *11*(4), 1016. doi:10.3390/nano11041016 PMID:33921180

Khairy, A. M., Tohamy, M. R. A., Zayed, M. A., Mahmoud, S. F., El-Tahan, A. M., El-Saadony, M. T., & Mesiha, P. K. (2022). Eco-friendly application of Nano-chitosan for controlling potato and tomato bacterial wilt. *Saudi Journal of Biological Sciences*, *29*(4), 2199–2209. doi:10.1016/j.sjbs.2021.11.041 PMID:35531227

Khan, M. N., Li, Y., Fu, C., Hu, J., Chen, L., Yan, J., Khan, Z., Wu, H., & Li, Z. (2022). CeO_2 nanoparticles seed priming increases salicylic acid level and ROS Scavenging Ability to improve rapeseed salt tolerance. *Global Challenges (Hoboken, NJ)*, *6*(7), 2200025. doi:10.1002/gch2.202200025 PMID:35860396

Khodakovskaya, M., Dervishi, E., Mahmood, M., Xu, Y., Li, Z., Watanabe, F., & Biris, A. S. (2009). Carbon nanotubes are able to penetrate plant seed coat and dramatically affect seed germination and plant growth. *ACS Nano*, *3*(10), 3221–3227. doi:10.1021/nn900887m PMID:19772305

Kumar, V., Singhal, R. K., Kumar, N., & Bose, B. (2020). *Micronutrient seed priming: A pragmatic approach towards abiotic stress management. New frontiers in stress management for durable agriculture*. Springer.

Lemmens, E., Deleu, L. J., De Brier, N., De Man, W. L., De Proft, M., Prinsen, E., & Delcour, J. A. (2019). The impact of hydro-priming and Osmo-priming on seedling characteristics, plant hormone concentrations, activity of selected hydrolytic enzymes, and cell wall and phytate hydrolysis in sprouted wheat (*Triticum aestivum* L.). *ACS Omega*, *4*(26), 22089–22100. doi:10.1021/acsomega.9b03210 PMID:31891089

Leng, X., Jia, H., Sun, X., Shangguan, L., Mu, Q., Wang, B., & Fang, J. (2015). Comparative transcriptome analysis of grapevine in response to copper stress. *Scientific Reports*, *5*(1), 17749. doi:10.1038rep17749 PMID:26673527

Li, L., Fan, M., Brown, R. C., Van Leeuwen, J., Wang, J., Wang, W., Song, Y., & Zhang, P. (2006). Synthesis, properties, and environmental applications of nanoscale iron-based materials: A review. *Critical Reviews in Environmental Science and Technology*, *36*(5), 405–431. doi:10.1080/10643380600620387

Mahakham, W., Sarmah, A. K., Maensiri, S., & Theerakulpisut, P. (2017). Nanopriming technology for enhancing germination and starch metabolism of aged rice seeds using phytosynthesized silver nanoparticles. *Scientific Reports*, *7*(1), 121. doi:10.103841598-017-08669-5 PMID:28811584

Mahakham, W., Theerakulpisut, P., Maensiri, S., Phumying, S., & Sarmah, A. K. (2016). Environmentally benign synthesis of phytochemicals-capped gold nanoparticles as nanopriming agent for promoting maize seed germination. *The Science of the Total Environment*, *573*, 1089–1102. doi:10.1016/j.scitotenv.2016.08.120 PMID:27639594

Mohammadinejad, R., Karimi, S., Iravani, S., & Varma, R. S. (2016). Plant-derived nanostructures: Types and applications. *Green Chemistry*, *18*(1), 20–52. doi:10.1039/C5GC01403D

Montanha, G. S., Rodrigues, E. S., Marques, J. P. R., de Almeida, E., Colzato, M., & Pereira de Carvalho, H. W. (2020). Zinc nanocoated seeds: An alternative to boost soybean seed germination and seedling development. *SN Applied Sciences*, *2*(5), 1–11. doi:10.100742452-020-2630-6

Naguib, D. M., & Abdalla, H. (2019). Metabolic status during germination of nano silica primed Zea mays seeds under salinity stress. *Journal of Crop Science and Biotechnology*, *22*(5), 415–423. doi:10.100712892-019-0168-0

Najafi Disfani, M., Mikhak, A., Kassaee, M. Z., & Maghari, A. (2017). Effects of Nano Fe/SiO2 fertilizers on germination and growth of barley and maize. *Archives of Agronomy and Soil Science*, *63*(6), 817–826. doi:10.1080/03650340.2016.1239016

Nonogaki, H. (2014). Seed dormancy and germination-emerging mechanisms and new hypotheses. *Frontiers in Plant Science*, *5*, 233. doi:10.3389/fpls.2014.00233 PMID:24904627

Nonogaki, H., Bassel, G. W., & Bewley, J. D. (2010). Germination—Still a mystery. *Plant Science*, *179*(6), 574–581. doi:10.1016/j.plantsci.2010.02.010

Noor, R., Yasmin, H., Ilyas, N., Nosheen, A., Hassan, M. N., Mumtaz, S., Khan, N., Ahmad, A., & Ahmad, P. (2022). Comparative analysis of iron oxide nanoparticles synthesized from ginger *(Zingiber officinale)* and cumin seeds (*Cuminum cyminum*) to induce resistance in wheat against drought stress. *Chemosphere*, *292*, 133201. doi:10.1016/j.chemosphere.2021.133201 PMID:34921860

Noorhosseini, S. A., Jokar, N. K., & Damalas, C. A. (2017). Improving seed germination and early growth of garden cress (Lepidium sativum) and basil (Ocimum basilicum) with hydro-priming. *Journal of Plant Growth Regulation*, *37*(37), 323–334.

Panyuta, O., Belava, V., Fomaidi, S., Kalinichenko, O., Volkogon, M., & Taran, N. (2016). The effect of pre-sowing seed treatment with metal nanoparticles on the formation of the defensive reaction of wheat seedlings infected with the eyespot causal agent. *Nanoscale Research Letters*, *11*(1), 92. doi:10.118611671-016-1305-0 PMID:26880729

Paparella, S., Araújo, S. S., Rossi, G., Wijayasinghe, M., Carbonera, D., & Balestrazzi, A. (2015). Seed priming: State of the art and new perspectives. *Plant Cell Reports*, *34*(8), 1281–1293. doi:10.100700299-015-1784-y PMID:25812837

Parveen, A., & Rao, S. (2015). Effect of nanosilver on seed germination and seedling growth in *Pennisetum glaucum*. *Journal of Cluster Science*, *26*(3), 693–701. doi:10.100710876-014-0728-y

Rahimi, D., Kartoolinejad, D., Nourmohammadi, K., & Naghdi, R. (2016). Increasing drought resistance of Alnus subcordata CA Mey. seeds using a Nano priming technique with multi-walled carbon nanotubes. *Journal of Forensic Sciences*, *62*, 269–278.

Rahimi, S., Hatami, M., & Ghorbanpour, M. (2021). Silicon-nanoparticle mediated changes in seed germination and vigor index ofMarigold (Calendula officinalis L.) compared to silicate under PEG-induced drought stress. *Gesunde Pflanzen*, *73*(4), 575–589. doi:10.100710343-021-00579-x

Rajput, V., Singh, A., Rawat, S., Sharma, R., Tatiana, M., Singh, A., Shuvaeva, V., & Nazarenko, O. (2022). Nanotechnology as the cutting edge for sustainable agriculture. In: Nanotechnology for Sustainable Agriculture An Innovative and Eco-Friendly Approach.

Rajput, V. D., Minkina, T., Fedorenko, A., Chernikova, N., Hassan, T., Mandzhieva, S., Sushkova, S., Lysenko, V., Soldatov, M. A., & Burachevskaya, M. (2021). Effects of zinc oxide nanoparticles on physiological and anatomical indices in spring barley tissues. *Nanomaterials (Basel, Switzerland)*, *11*(7), 1722. doi:10.3390/nano11071722 PMID:34208886

Rameshraddy, P., Pavithra, G. J., Rajashekar Reddy, B. H., Salimath, M., Geetha, K. N., & Shankar, A. G. (2017). Zinc oxide nanoparticles increases Zn uptake, translocation in rice with positive effect on growth, yield and moisture stress tolerance. *Indian Journal of Plant Physiology / Official Publication of the Indian Society for Plant Physiology*, *22*(3), 287–294. doi:10.100740502-017-0303-2

Rastogi, A., Tripathi, D. K., Yadav, S., Chauhan, D. K., Živčák, M., Ghorbanpour, M., El-Sheery, N. I., & Brestic, M. (2019). Application of silicon nanoparticles in agriculture. *3 Biotech*, *9*(3), 90. https://doi.org/ doi:10.1007/s13205-019-1626-7

Rastogi, A., Zivcak, M., Sytar, O., Kalaji, H. M., He, X., Mbarki, S., & Brestic, M. (2017). Impact of metal and metal oxide nanoparticles on plant: A critical review. *Frontiers in Chemistry*, *5*, 78. doi:10.3389/fchem.2017.00078 PMID:29075626

Rizwan, M., Ali, S., Ali, B., Adrees, M., Arshad, M., Hussain, A., Zia Ur Rehman, M., & Waris, A. A. (2019). Zinc and iron oxide nanoparticles improved the plant growth and reduced the oxidative stress and cadmium concentration in wheat. *Chemosphere*, *214*, 269–277. doi:10.1016/j.chemosphere.2018.09.120 PMID:30265934

Saddiq, M. S., Iqbal, S., Afzal, I., Ibrahim, A. M. H., Bakhtavar, M. A., Hafeez, M. B., Jahanzaib, M. M. M., & Maqbool, M. M. (2019). Mitigation of salinity stress in wheat (Triticum aestivum L.) seedlings through physiological seed enhancements. *Journal of Plant Nutrition*, *42*(10), 1192–1204. doi:10.1080/01904167.2019.1609509

Sharma, D., Afzal, S., & Singh, N. K. (2021). Nanopriming with phytosynthesized zinc oxide nanoparticles for promoting germination and starch metabolism in rice seeds. *Journal of Biotechnology*, *336*, 64–75. doi:10.1016/j.jbiotec.2021.06.014 PMID:34116127

Siddaiah, C. N., Prasanth, K. V. H., Satyanarayana, N. R., Mudili, V., Gupta, V. K., Kalagatur, N. K., Satyavati, T., Dai, X. F., Chen, J. Y., Mocan, A., Singh, B. P., & Srivastava, R. K. (2018). Chitosan nanoparticles having higher degree of acetylation induce resistance against pearl millet downy mildew through nitric oxide generation. *Scientific Reports*, *81*(8), 1–14. doi:10.103841598-017-19016-z PMID:29410438

Singh, A., Sengar, R. K., Sharma, R., Rajput, P., & Singh, A. K. (2021). Nano-priming technology for sustainable agriculture. *Biogeosystem Technique*, *8*(2), 79–92.

Singh, A., Sengar, R. S., Rajput, V. D., Minkina, T., & Singh, R. K. (2022). Zinc oxide nanoparticles improve salt tolerance in rice seedlings by improving physiological and biochemical indices. *Agriculture*, *12*(7), 1014. doi:10.3390/agriculture12071014

Singh, A., Sengar, R. S., Shahi, U. P., Rajput, V. D., Minkina, T., & Karen, A. (2023). Prominent Effects of Zinc Oxide Nanoparticles on Roots of Rice (*Oryza sativa* L.) Grown under Salinity Stress. *Stresses*, *3*(1), 33–46. doi:10.3390tresses3010004

Singh, P., Kim, Y. J., Singh, H., Mathiyalagan, R., Wang, C., & Yang, D. C. (2015). Biosynthesis of anisotropic silver nanoparticles by Bhargavaea indica and their synergistic effect with antibiotics against pathogenic microorganisms. *Journal of Nanomaterials*, 1–10. doi:10.1155/2015/234741

Singh, Y., Kaushal, S., & Sodhi, R. S. (2020). Biogenic synthesis of silver nanoparticles using *cyanobacterium Leptolyngbya* sp. WUC 59 cell-free extract and their effects on bacterial growth and seed germination. *Nanoscale Advances*, *2*(9), 3972–3982. doi:10.1039/D0NA00357C PMID:36132754

Sonawane, H., Arya, S., Math, S., & Shelke, D. (2021). Myco-synthesized silver and titanium oxide nanoparticles as seed priming agents to promote seed germination and seedling growth of *Solanum lycopersicum*: A comparative study. *International Nano Letters*, *11*(4), 371–379. doi:10.100740089-021-00346-w

Song, K., & He, X. (2021). How to improve seed germination with green nanopriming. *Seed Science and Technology*, *49*(2), 81–92. doi:10.15258st.2021.49.2.01

Sun, L., Song, F., Zhu, X., Liu, S., Liu, F., & Wang, Y. (2020). Nano-ZnO alleviates drought stress via modulating the plant water use and carbohydrate metabolism in maize. *Archives of Agronomy and Soil Science*, 115.

Sytar, O., Kumari, P., Yadav, S., Brestic, M., & Rastogi, A. (2018). Phytohormone priming: Regulator for heavy metal stress in plants. *Journal of Plant Growth Regulation*, *38*(38), 739–752.

Thakur, M., Sharma, P., & Anand, A. (2019). *Seed priming-induced early vigor in crops: An alternate strategy for abiotic stress tolerance. Priming and pretreatment of seeds and seedlings*. Springer.

Wang, C., Kim, Y. J., Singh, P., Mathiyalagan, R., Jin, Y., & Yang, D. C. (2016). Green synthesis of silver nanoparticles by *Bacillus methylotrophicus*, and their antimicrobial activity. *Artificial Cells, Nanomedicine, and Biotechnology*, *44*(4), 1127–1132. doi:10.3109/21691401.2015.1011805 PMID:25749281

Wang, H., Kou, X., Pei, Z., Xiao, J. Q., Shan, X., & Xing, B. (2011). Physiological effects of magnetite (Fe3O4) nanoparticles on perennial ryegrass (*Lolium perenne* L.) and pumpkin (*Cucurbita mixta*) plants. *Nanotoxicology*, *5*(1), 30–42. doi:10.3109/17435390.2010.489206 PMID:21417686

Waqas Mazhar, M., Ishtiaq, M., Maqbool, M., Akram, R., Shahid, A., Shokralla, S., Al-Ghobari, H., Alataway, A., Dewidar, A. Z., El-Sabrout, A. M., & Elansary, H. O. (2022). Seed priming with iron oxide nanoparticles raises biomass production and agronomic profile of water-stressed flax plants. *Agronomy (Basel)*, *12*(5), 982. doi:10.3390/agronomy12050982

Wu, M., Wu, J., & Gan, Y. (2020). The new insight of auxin functions: Transition from seed dormancy to germination and floral opening in plants. *Plant Growth Regulation*, *91*(91), 169–174. doi:10.100710725-020-00608-1

Yousefi, S., Kartoolinejad, D., & Naghdi, R. (2017). Effects of priming with multi-walled carbon nanotubes on seed physiological characteristics of Hopbush (*Dodonaea viscosa* L.) under drought stress. *The International Journal of Environmental Studies*, *74*(4), 528–539. doi:10.1080/00207233.2017.1325627

Zafar, S., Perveen, S., Kamran Khan, M., Shaheen, M. R., Hussain, R., Sarwar, N., Rashid, S., Nafees, M., Farid, G., Alamri, S., Shah, A. A., Javed, T., Irfan, M., & Siddiqui, M. H. (2022). Effect of zinc nanoparticles seed priming and foliar application on the growth and physio-biochemical indices of spinach (Spinacia oleracea L.) under salt stress. *PLoS One*, *17*(2), e0263194. doi:10.1371/journal.pone.0263194 PMID:35192615

Zhang, D., Ma, X. L., Gu, Y., Huang, H., & Zhang, G. W. (2020). Green synthesis of metallic nanoparticles and their potential applications to treat cancer. *Frontiers in Chemistry*, *8*, 799. doi:10.3389/fchem.2020.00799 PMID:33195027

Zuverza-Mena, N., Martínez-Fernández, D., Du, W., Hernandez-Viezcas, J. A., Bonilla-Bird, N., López-Moreno, M. L., Komárek, M., Peralta-Videa, J. R., & Gardea-Torresdey, J. L. (2017). Exposure of engineered nanomaterials to plants: Insights into the physiological and biochemical responses—A review. *Plant Physiology and Biochemistry*, *110*, 236–264. doi:10.1016/j.plaphy.2016.05.037 PMID:27289187

Chapter 12
Nanopriming-Based Management of Biotic Stresses for Sustainable Agriculture

Muhammad Majeed
University of Gujrat, Pakistan

Murad Muhammad
Chinese Academy of Sciences, China & University of Peshawar, Pakistan

Tanveer Hussain
Mirpur University of Science and Technology, Pakistan

Muqarrab Ali
Muhammad Nawaz Shareef University of Agriculture, Pakistan

Tayyaba Naz
Institute of Soil and Environmental Sciences, University of Agriculture, Faisalabad, Pakistan

Sehar Nawaz
University of Agriculture, Faisalabad, Pakistan

Muhammad Mazhar Iqbal
Government of Punjab, Pakistan

Ghulam Abbas
University of Agriculture, Dera Ismail Khan, Pakistan

ABSTRACT

An increasing number of agricultural uses for nanotechnology has also sparked interest in nano-priming of plants. Plant growth and seed production can be stunted by the physiological, psychological, and molecular alterations that are commonly associated with stressor events. Nano-priming is a novel technique for seed priming that enhances plant resilience in improved germination, growth, and yield in unfavorable environmental conditions. Unfavorable impacts of nano-priming upon seed germination, seedling, and plant growth features were detected despite the fact that the usage of nanoparticles increased environmental stress tolerance of these plants. The most crucial aspects of these plants' germination, morphology, and physiology that influence their establishment, development, and production are also

DOI: 10.4018/978-1-6684-7232-3.ch012

addressed, along with the impact of various nanoparticles. Using nanotechnology for agricultural purposes is to make growing crops easier and more efficient. Nanotechnology is a viable method for sustainability boosting agricultural output, which is essential to achieving this objective.

INTRODUCTION

Nanoparticles display peculiar physicochemical properties because of their out-of-the-ordinary size, surface roughness, chemical composition, stability, shape, and aggregation. (Saleh, 2020). Because nanozymes (nanomaterials that can mimic enzymes with antioxidant activities) may remove ROS, increasing plant stress tolerance, and nanotechnology can be developed into versatile places with extensive applications for plant science, it has tremendous potential as a tool in agriculture. (Lu et al., 2020). Researchers found that by coating plants with cerium oxide nanoparticles, they may improve their resistance to salt, drought, heat, and cold (Liu et al., 2021; Zhao et al., 2020). It may be utilized to make nanosensors for initial stress detection in plants, allowing for gradual and targeted administration of agrochemicals. Carbon nanotubes were used to detect reactive oxygen species (ROS), and calcium (Ca^{2+}), nitric oxide (NO), and they served as a vector for delivering genetic material (DNA or RNA) to facilitate genetic modification in non-model plant species (Yuan et al., 2018). In the sections to come, you'll learn about further ways that nanotechnology may be used in farming (Singh et al., 2021, 2022).

The effects of climate change on ecosystem processes lead to an increase in biotic and abiotic stressors. The quality of agricultural production is severely impacted by salinity, making it one of the most significant abiotic pressures (Yadav et al., 2020). Globally, food production is being disrupted by soil salinity, which interferes with physiochemical processes including the processes of photosynthesis capacity for antioxidants, and ion balancing. Excess sodium ion (Na^{+}) decreases the level of and potassium (K^{+}) nitrogen (N) in plant tissues by inhibiting the transporter protein molecules attributable to mineral ion uptake (Talukdar et al., 2020), thereby disrupting the physiological processes (Tang et al., 2015). Because of the ionic and osmotic strains, it places on plants, salinity stunts their development and, in extreme cases, may even kill them (Talukdar et al., 2020).

Nanotechnology, an interdisciplinary field that has shown promising results in several areas, has opened up new possibilities in the agricultural sector to tackle issues related to sustainable agriculture (Rai & Ingle, 2012). Nano-fertilizers, nano-pesticides, and nanosensors are all examples of cutting-edge technologies being developed to improve sustainable agriculture practices in resource- and waste-efficient ways. Using nanoparticles in agriculture and the environment could lessen the environmental toll of crop production without negatively impacting soil quality or biodiversity (Hofmann et al., 2020).

Agriculture and other plant-related disciplines (Haq et al., 2022; Tassadduq et al., 2022) have benefited from nanotechnology's potential for progress. Among the most important parts of every ecosystem is the plants inside it. Therefore, understanding how plants react to nanomaterials (NMs) is crucial to realizing sustainable agriculture's potential. The impact that NMs have on plants rely on their physicochemical and structural features (Holland, 1992). Furthermore, plant impacts may vary amongst plant species. Plant organs undergo changes in morphology and physiology as a result of their interaction with NMs (Oliveira et al., 2023).

Priming seeds is a potent approach for improving plant growth and output in the face of a variety of environmental conditions and obstacles. Many biotic and abiotic stresses may have substantial influence on the growth and development of plants (Hassanisaadi et al., 2022). When seeds germinate rapidly and without difficulty, there is a greater chance that the plant will flourish and build a robust root system. Because of its diminutive size and unique combination of physicochemical properties, nano-priming is a more effective method of seed priming than its counterparts (Abbasi Khalaki et al., 2021). Because of these physiological distinctions, various plant species (Majeed, Bhatti, & Amjad, 2021; Majeed, Mahmood, et al., 2022) respond to nano-priming with varied rates and patterns of growth. The majority of prior research has focused on one plant at a time, looking at how a nanomaterial affects its germination, growth, and development (Li et al., 2020).

Priming plant seeds with NPs has shown more success in affecting growth, quality, production, and reactive oxygen species (ROS) balance than other techniques of priming, including bulk material priming and hydro-priming (Kandhol et al., 2022). Toxic implications for seed germination or seedling development must be avoided by excluding NPs from purposeful priming until their potential impacts on plant and human health are well understood (Handy et al., 2012). Therefore, the purpose of this article is to offer the most up-to-date research on nano-priming of seeds from various plants, as well as make suggestions for further study in this area.

APPLICATION OF NANOPRIMING

Seed germinations, and by extension crop yields, are significantly impacted by biotic and abiotic stressors. Seeds may be treated in advance to combat the factors that hinder germination, plant development, and even dormancy. To combat biotic stressors, growers use a wide range of seed priming techniques (Zulfiqar, 2021). Numerous priming techniques exist, including but not limited to hydropriming, osmopriming, nutritional priming, chemical priming, biopriming, priming by plant growth regulators and plant extracts, nanopriming, and physical priming. Nanopesticides, nanosensors, nanofertilizers, and other nanopriming technologies can potentially enhance various crops' germination rate and resilience. (Hossain et al., 2020).

Food shortages (Majeed, Bhatti, Pieroni, et al., 2021) occur all around the world because agricultural output is hampered by a combination of abiotic and biotic stressors that stunt crop development (Onyekachi et al., 2019). The current focus of environmentalists is on enhancing agricultural productivity in hostile environments to guarantee secure food supplies (Brummer et al., 2011). Nanotechnology is a viable method for sustainingably boosting agricultural output, which is essential to achieving this objective. There is now evidence that seed nano-priming may stimulate not just germination and establishment of seedlings but also plant growth. Nanotechnology as an interdisciplinary field that studies materials with sizes between 1 and 100 nm, whether they are organic or inorganic (Satalkar et al., 2016). Green formulations of nanoparticles (NPs) have been shown to be good for the environment and farmers (Arif et al., 2021; Khan et al., 2022), stakeholders, and the agro-industrial sector (Omran & Baek, 2022). Seed quality, seedling establishment, crop output, and tolerance to environmental difficulties are all improved by the nanopriming method, which is based on a combination of seed priming and nanoparticles treatment, in comparison to unprimed or other agent-primed seeds (Chandrasekaran et al., 2020).

Increased soluble sugar levels for seedling development may arise from nanopriming treatment of seeds, which increases-amylase activity (Kumar et al., 2020). The mechanisms behind nano-priming-induced germination of seeds have been described. The production of hydroxyl radicals in cell wall thinning, the development of nanopores for enhanced nutrient-water absorption, the reactivation of internal ROS/antioxidant systems in seeds, and the design of a nanocatalyst to promote starch hydrolysis are all examples of such mechanisms. Another impact of nanoprimed seeds is elevated aquaporin gene expression (Upreti & Patro, 2022).

Global warming, salt stress, and water deficits are only a few of the modern environmental concerns that severely limit agricultural output. Saline conditions negatively affect agricultural output by up to 70% and affect more than 20% of the hosed land and 6% from the world's total acreage (Zaman et al., 2018). The buildup of reactive oxygen species (ROS) caused by salinity has a profound effect on the molecular makeup of enzymes, nucleic acids, and lipids, hence reducing agricultural output (Bano et al., 2021). It's also linked to a decline in photosynthesis due to suppression of normal electron transport chain operations and oxidative damage that compromises cell viability and chlorophyll structure, thereby wreaking havoc on plant metabolism (Lodeyro et al., 2012). In order to aid in homeostasis, regulate oxidative damages, and take part in detoxifying activities, plants must enhance their physiochemical reactions when exposed to salt. Using nano-priming exhibits superior efficacy compared to alternative seed priming techniques (Garg & Maheshwari, 2023).

Protecting plants through priming allows them to respond more quickly and efficiently to these stresses (Mauch-Mani et al., 2017). Seeds and young plants benefit from increased resistance to abiotic stressors via a cascade of metabolic processes triggered by priming (Nile et al., 2022). Therefore, plants that have been repeatedly subjected to stress demonstrate improved growth biomarkers and tolerance to stress when subjected to future stressors. Mild degrees of abiotic stressors and exposure to pathogens, pests, beneficial microbes, natural and synthetic substances, nanomaterials, and so on may set off a priming event (Gonzalez Guzman et al., 2022).

Improvements in the usage of nanoparticles (NPs) as insecticides, fertilizers, and sensors have led to a more sustainable agricultural sector (Mittal et al., 2020). Depending on their unique physicochemical properties, NPs can be tailored to enhance plant growth, development, and resilience. Increasing plant yields or nutritional value by using NPs as a priming element to boost seed germination and plant development is also being researched (Abbasi Khalaki et al., 2021). The very tiny size, large surface area, and gradual release rate of NPs all contribute to their usefulness in enhancing plant nutrient absorption. Priming with nanoparticles has several advantages, such as influencing metabolic pathways, physiology, activity of enzymes, and the interaction between phytohormones (Moghanloo et al., 2019).

This research is the first to suggest a room-temperature production of phytochemicals-capped GNPs utilizing rhizome extract from the galanga plant, which would be easy, cost-effective, and environmentally benign. Several methods were used to characterize the synthesized GNPs. As a means of advancing the agricultural uses of green nanotechnology, we nanoprimed old maize seedlings with a solution of GNPs at an ecologically feasible level (5 to 15 ppm).

Priming with 5 ppm GNPs boosted emergence by 83% compared to the unprimed controls (43%) and the hydroprimed group (56%). The vigor index of the seedlings produced from seeds primed with either 5 or 10 ppm GNPs was three times that of the control. At a dosage of 10 ppm, priming maize seedlings with GNPs improved both their physiology and biochemistry. Transmission electron microscopy (TEM) and inductively coupled plasma atomic emission spectroscopy (ICP-OES) have been used to show that GNPs can be taken up by seeds. However, the absence of GNPs in both the shoot and the root, as revealed by inductively coupled plasma optical emission spectroscopy (ICP-OES), suggests that the nanopriming technique effectively reduces the amount of Au taken up by the seedling. It was discovered that GNPs produced via photosynthesis were safer than those produced via chemical synthesis. This is the first study to indicate that GNPs coated with phytochemicals may serve as an effective nanopriming agent for reviving dormant agricultural plant seeds. Food shortages occur all around the world because agricultural output is hampered by biotic and abiotic stressors that inhibit crop development (Raza et al., 2019). The present attention of environmentalists is on enhancing agricultural productivity in hostile environments to guarantee secure food supplies. There is now evidence that seed nano-priming may stimulate not just germination and establishment of seedlings but also plant growth (Nile et al., 2022). Here, we'll look at how seed nano-priming may improve things like germination rates, plant health, and harvests.

The technique of seed priming, the process of seed germination, and the molecular, biochemical, and physiologic alterations brought on by nano-priming have all been explored (Nile et al., 2022). Nano-priming has been shown to reduce the negative effects of both abiotic and biotic challenges, making it a viable strategy for producing hardy crops (Mukherjee, 2019). The benefits of seed nano-priming for environmentally friendly farming have been discussed, as have some of the drawbacks. The present discourse delves into the potential of endophytes to enhance sustainable agriculture by virtue of their diverse advantages to host and non-host plants, seed germination, and abiotic stress tolerance through seed priming, employing a range of state-of-the-art seed coating methodologies (Jeger et al., 2021).

Given the scarcity of these resources, it is imperative that immediate action be taken to provide other means of dealing with their misuse and overexploitation (Olalekan et al., 2019). As a consequence of all the human interference, natural resources are under constant assault.

Nanomaterials may be created by either traditional methods or, more recently, environmentally friendly synthetic methods. The former often have a large negative effect on the environment or need a lot of energy, whereas the latter do not have these characteristics and are also cheaper (Olalekan et al., 2019).

Given the vast possibilities offered by nanomaterials, current studies have focused on natural biomasses, looking at various methods of valorizing these underutilized materials. There has also been some work on mixing nanomaterials to create hybrid, combined, or coated structures, which might allow for efficient modulation of the characteristics of the resultant nanoparticles (He et al., 2015). The potential to produce components that enhance, modify, or otherwise alter the characteristics of one or more chemicals contained in the produced systems makes this last point particularly noteworthy (Demil & Lecocq, 2010).

Stress reduction is greatly aided by the presence of some NMs. If successfully implemented, the notion of tailored nano-carriers has the potential to decrease the environmental effect of crop protection and the production of food while increasing crop yields (Toksha et al., 2021). The role of NMs in promoting seed germination and reducing the negative effects of abiotic and biotic stressors like drought, salt, temperatures, metal, UV-B radiation, and floods is examined in detail. Additionally, plant adaptations to NMs in extreme environments are discussed. We also evaluated NMs for their potential to improve germination as part of a seed priming strategy.

With the information presented here, nanoparticles may be used more effectively in farming. Eventually, this will cause nanotechnology to tailor inputs to specific farming requirements. Only a small number of research have shown the impact of nano-priming using different particles on the germination and growth of forage and medicinal plants (Afzal et al., 2021). As a result, this page provides of research into the effects of nano-priming with a variety of particles on the development of forage and medical plant seeds and seedlings.

Nano-priming improved the future performance of most medicinal and forage plants by hastening the germination of seeds, the growth and development of seedlings, the vigor of seedlings, the rate at which they emerged from the soil, and so on (Abbasi Khalaki et al., 2021). Priming can be described as the physiological process by which plants are stimulated to show improved growth biomarkers or stress tolerance (Kandhol et al., 2022).

Because of their vulnerability to environmental changes, plants have developed complex methods to cope with abiotic stress, which leads to decreased productivity. Such plant life disruptions, along with the continuously expanding global population, have led to a massive disparity between food supply and demand. It is vital to find suitable ways for acclimatizing plants to stressful settings in order to uncover adequate strategies for balancing crop output and food demand. This is necessary in order to strike a balance between crop yield and food demand (Dahal et al., 2019). Even at low concentrations and short times, abiotic stressors may trigger a number of signaling cascades and processes in plants (Vickers et al., 2009). As a consequence, plants (Arshad et al., 2022; Jamil et al., 2022) are better able to withstand future stress. Cross-tolerance refers to a plant's increased resistance to several stressors as a consequence of exposure to a single stress. Today, traditional breeding, mutant cultivation, genetic engineering, and genome modification are all employed to increase plant yield under stress (Rommens, 2007). Because of the constraints imposed by these technologies, there is an increasing need to identify an alternative that is both effective and economical for demonstrating plant resistance to a variety of stimuli.

Priming, often referred to as sensitization, is the process by which a plant's defensive system is engaged in response to an environmental stress event. This process makes the plant more resilient in the face of subsequent shocks (Mozgova et al., 2019). When plants are stressed, they enter a primed condition in which they respond to future pressures more quickly than they would if they were not primed. Chemical priming is gaining interest among scientists studying plant stress pathophysiology and stress management because it can increase plants' resistance to abiotic stressors through exposure to natural or synthetic chemical priming substances (including nanoparticles, NPs) (Johnson & Puthur, 2021).

Recently, nanotechnology has been created to enhance agricultural sustainability by supplying the agriculture industry with a number of useful instruments, including nanopesticides (Kim et al., 2018), Nanotechnology has a wide range of applications. Nanoparticles (NPs) are now being optimized for enhancing plant development, growth, and resistance under difficult circumstances based on their distinct physicochemical features. To increase plant production and food nutritional content, nanoparticles (NPs) are also being utilized as a priming agent (a method known as nano-priming) for seeds to improve germination (Rexlin et al., 2022). Nanoparticle exposure alters metabolic, physiological, pharmacological, and signaling pathways in seeds, which may account for some of the benefits of seed nano-priming. Seed germination, seedling viability, root and shoot length, and the antioxidant defense system are some of the characteristics of plant growth and development that are investigated in relation to nano-priming (Johnson & Puthur, 2021).

The application of nanotechnology to farming is a hot issue right now. Facilitating nano-enabled agriculture requires talking about how nanobiotechnology may be used to boost crop yields (Giraldo et al., 2019). There has not been an assessment of the research needs and gaps associated with nanobiotechnology in agriculture. This chapter focuses on how plants detect stress, identify it early, and adapt to it. In addition to seed nanopriming and the controlled release of agrochemicals, transgenic events have been promoted for usage in no model crop species (Gonzalez Guzman et al., 2022). Nanomaterials could be employed to convert ultraviolet and near-infrared radiation into visible light as a means of avoiding the capture of excess electrons. This phenomenon may account for the broadened wavelength range observed in photosynthesis. This chapter delves into the mechanisms by which plants are able to sustain photosynthesis in the absence of light. The present chapter aims to elucidate the potential of Nano biotechnology in agricultural settings for the purpose of optimizing crop yields.

There is a constant barrage of environmental pressures that jeopardize crop output. Beyond mitigating the effects of stressors, modern and effective strategies are needed to increase agricultural productivity. Many recent research have shown that nanotechnology may help plants grow and thrive in both stressful and non-stressful environments, making for a more sustainable agricultural system (Elsheery et al., 2020).

Priming seeds with NPs has been shown in a variety of plant studies to have positive effects on germination, nutrient absorption, antioxidant defense, and oxidative damage response, both in stressful and non-stressful environments (Zulfiqar et al., 2019). Several recent evaluations have examined the impact of nanotechnology's regulatory function on plant growth and development. However, there has not been a comprehensive and systematic evaluation of the regulatory functions of nano-seed priming.

There is a constant barrage of environmental pressures that jeopardize crop output. Beyond mitigating the effects of stressors, modern and effective strategies are needed to increase agricultural productivity. Many recent research have shown that nanotechnology may help plants grow and thrive in both stressful and non-stressful environments, making for a more sustainable agricultural system (Maggio et al., 2018). Improve germination, plant growth rate, productivity, and postharvest physiology with this cutting-edge method of priming seeds with NPs to control morphological and physio-biochemical parameters. Numerous plant studies have demonstrated that priming seeds with NPs improves germination, nutrient uptake, antioxidant defense, and oxidative damage response in both stressful and non-stressful conditions. Recent studies have analyzed how nanotechnology's regulatory function affects plant expansion and maturation (Bouwmeester et al., 2009). However, the regulatory roles of nano-seed priming have not been systematically evaluated. New information on nano-priming is compiled and analyzed for its application to plant growth in both normal and stressful conditions, as well as its potential for use in sustainable agriculture in the future.

Nano-priming is a powerful technique that may affect seed germination, seedling establishment, and ultimately the plant's longevity (Abbasi Khalaki et al., 2021). This is because of the process's potential to activate certain biochemical or signaling pathways in the seed. Due to the importance of proper germination and the timely emergence of seedlings for the subsequent development and harvest of a crop, priming is an absolute need. The use of nanoparticles with priming the seeds has been discovered to be more advantageous, suggesting the prospect of general adoption for nano-priming as opposed to other conventional priming procedures that in some cases contribute to either environmental or agricultural damage (Sonawane et al., 2021).

Intensive farming (including the use of farm machinery, high dosage fertilizer, and agrochemicals) is just one of many human activities that have increased agriculture's production and output. However, it alters the soil and ecosystem, decreasing the amount and quality of grain that can be harvested. Therefore, it is essential to explore innovative, efficient, ecologically (Khoja et al., 2022; Majeed, Tariq, et al., 2022) friendly technologies that may guarantee future generations have access to healthy food. Here, it's crucial to highlight the ways in which agricultural nanotechnology is contributing to a brighter, greener future. One innovative, cost-effective, and sustainable strategy for enhancing agricultural yields in the face of adverse environmental and climatic conditions is priming seeds with nano-chemicals or green chemicals like plant growth promoting rhizobacteria (PGPR) (Shelar et al., 2021).

The utilization of green chemicals such as chitosan, micronutrients including Fe, Zn, Cu, and organic and inorganic compounds such as PGPR hormones in the process of nanopriming (which involves the treatment of seeds with nanoparticles prior to sowing) has demonstrated remarkable and dynamic outcomes in various stages of seed growth, including germination, vegetative and reproductive phases, as well as in the quality attributes of the grains (Kumar et al., 2022). This method greatly lowers the amount of fertilizers, pesticides, and other agrochemicals that are needed to grow crops. It does this by making it easier for plants to absorb nutrients, move them to their reproductive parts, use water, and make use of other resources. Nanopriming increases stress proteins and genes, antioxidant compounds, and signaling substances to lower the crop's stress response. To help you understand how important nanopriming is, this chapter looks at recent advances and successes under unfavorable conditions, possible problems, and the chances of learning more about the NP mechanism so that farmers, researchers, and future generations can live in a sustainable environment. So, more research needs to be done on the physiological and chemical effects of nano-priming on plant growth and human health.

NANOPARTICLES IN AGRICULTURE

Nanomaterials of a size of 100 nm or less are considered to be of the one-dimensional variety. The high surface-to-volume ratio of nanomaterials gives them a distinct advantage over their conventional counterparts. Priming seeds, a process that alters seed metabolism to facilitate enhanced germination and quicker plant establishment in its environment, has attracted a lot of attention from scientists in recent years. The subsequent agricultural technological revolution may involve nanoscale (less than 100 nm) particles (Abobatta, 2018). NPs are well-suited for various uses in agriculture due to their exceptional physicochemical features, including their large surface area-to-volume rates and strong reactivity (Islam et al., 2022). Biomolecules (such as nucleotides, proteins, and activators), agrochemicals (such as fertilizers and insecticides), and sensors for monitoring plant health might all be loaded onto NPs and conveyed to the plant for controlled release. Soil quality, disease detection, pesticide residue analysis, plant growth, and agricultural yield are all areas where NPs could be used in agri-nano. Fertilizers and metal fertilizers made heavy use of NPs throughout the past decade

PRIMING SEEDS USING NANOPARTICLES (SEED NANOPRIMING)

Seed and plant pre-treatment with biological and chemical substances may induce a primed state (PS), which can deploy stronger and more rapid responses than those of a non-primed plant (Mhlongo et al., 2021). Priming seeds involves soaking them in various solutions before planting them. Various priming techniques are available for seed treatment, including halo-priming with salt solutions, hydro-conditioning with water, osmo-priming with osmotic agents, hormonal priming with solutions containing plant hormones, bio-priming with solutions containing valuable microbes, nano-priming with solutions containing NPs, and magneto-priming with the application of a magnetic field. During priming, biochemical alterations in the seed lead to enhanced germination rate, growth uniformity, abandonment, and seedling resilience to adverse environmental (Haq et al., n.d.; Majeed et al., 2023) conditions (Garcia et al., 2021). The seed priming process involves augmenting the seed's moisture content, thereby triggering the activation of certain chemicals that play a crucial role in the embryonic development of the seed and the exploration of the insipid endosperm. In addition, seed priming has the potential to initiate metabolic processes that enhance DNA replication, augment RNA levels, and expedite cellular restoration. Seed priming has enhanced the immune system by augmenting the synthesis of antioxidant enzymes such as catalase, superoxide dismutase, and glutathione reductase (Yadav et al., 2019). Priming enhances performance in various climatic circumstances, helps overcome dormancy, and boosts seed output, crop uniformity, and germination. This results in cell cycle regulation changes, ultra-structural development, and seed water content. The use of seed priming and foliar spray may mitigate the stress response during seed growth and seedling establishment.

PLANT RESPONSES TO NANOPRIMING SEEDS

Seed germination, development, stability, and physiology may all be enhanced with nanopriming by modifying absorption, metabolic processes, antioxidants, and photosynthesis. Many studies have shown that nanoprimed seeds have enhanced abilities in areas such as cellular balance and photosynthetic capacity, nutrient uptake and photosynthesis effectiveness, chlorophyll capacity and activity as an antioxidant, and defense mechanisms (such as osmotic pressure changes and stomatal movements) (Rai-Kalal et al., 2021b). It also stops the body from taking copper, cadmium, and zinc, which are all dangerous heavy metals.

SEED NANOPRIMING FOR THE REDUCTION OF ABIOTIC STRESS

Seeds are increasingly being subjected to both abiotic and biotic stress due to rising pollution levels and changes to the global environment (Rivero et al., 2022). In response to these stresses, cells may undergo various physicochemical changes. Cellular membranes, chloroplasts, mitochondria, ER, and peroxisomes are all potential sites for stress sensing (Das & Roychoudhury, 2014). Alterations in gene transcription, translation, or post-translational modifications (PTMs) of proteins are responsible for developing and altering proteins that perform specific roles in the stress response.

SMART FARMING WITH NANOTECHNOLOGY-ENABLED STRESS DETECTION AND EARLY DETECTION

Crop plants, rooted in the ground, have developed sophisticated defenses to withstand environmental stresses. Plants have a variety of methods to detect and respond to stress, and H_2O_2 is one such mechanism. Signaling molecules in plant stress responses include hydrogen peroxide, calcium ions, sugars (such as glucose and sucrose), gas molecules (such as Nitric oxide, Carbon monoxide, and Hydrogen sulphide), hormones produced by plants (Huang et al., 2021). Plants respond to salt stress by initiating signal transduction from the root to the shoot, which involves Ca^{2+} signaling processes. Under various pressures, the concentration profiles of these signaling molecules shift. Plants' Ca^{2+} signatures under osmotic and salinity stress have distinct maxima. Real-time, non-destructive identification of these signaling chemicals is only possible in a few model plant species (Kader & Lindberg, 2010).

DEVELOPING PLANT-BASED SENSORS FOR EARLY DETECTION OF STRESS

Reduced crop losses might be achieved by early diagnosis of stress in agricultural production. While remote sensing and hyperspectral imaging are now used to identify stress, features such as leaf chlorophyll fluorescence, morphological change, and hydration status represent the performance of plants after stress has accumulated (Li et al., 2014). It's helpful to watch for early warning signs of stress. Nanosensors' temporal and geographical resolution make them a promising addition to remote sensing for detecting crop stress via monitoring stress-signaling chemicals (Lew et al., 2020). In 2019, we suggested using nanosensors to transform plants into smart plant sensors with Professor Juan Pablo Giraldo of the University of California, Riverside, CA, USA (Wu & Li, 2022). Using fluorescent quenching or fluorescence shifting, nanosensors may detect analytes and report their detection. Transforming chemical signals into optical or radio frequencies detectable by agricultural equipment may be possible with nanosensors capable of detecting and reacting to particular stress-signaling chemicals like H_2O_2, glucose, and NO (Pal et al., 2023). Better management of farms and early diagnosis of plant stress might be possible with such a device. Not only can glucose be detected in real-time in vivo using a QD system, but recent breakthroughs in the field have shown that H_2O_2 can also be detected in plants in real-time in vivo under stress.

"Seed nanopriming" is the process of giving seeds nanoform nutrients at the seedling level to act as initial fertilizers for seeds and improve seed germination, plant growth, and yield." (Rai-Kalal et al., 2021a). Plant morphology and physiological and biochemical responses, may be altered by nanopriming. Nanopriming has yet to be tested in a controlled environment where its effects may be seen in plants under stress. Specifically, further research is needed to determine which genes are triggered and how they work together to create this impact. However, many unanswered questions about NPs need to be investigated, including their nontoxicity and the long-term consequences of exposure.

More work is needed to fully realize the suggested smart plant sensor, particularly in real-world settings. Heat and drought are only two of the many challenges crops face out in the field simultaneously (Mittler & Blumwald, 2010). A better precision is needed in the decoded signals provided by nanosensors in reaction to particular signaling molecules because combined stressors may confound chemical signaling processes. Nanosensors would benefit from increased sensitivity, selection, and accuracy to further their use in agriculture to allow greater decoded signal precision under field settings.

Nanoparticle (NP) toxicity to seed implantation and plant development has been the primary focus of these investigations. Germination of seeds is the initial stage in a plant's life cycle and is the determining factor in whether or not plants will develop in soils contaminated by NPs (Younis et al., 2021). Many kinds of plant show an increase or reduction in seed germination depending on the type of NP used, like metal or metal oxide. Plants' reactions to NPs will thus vary not just for plant species but also concerning NPs' chemical composition, molecular size, and concentration (Remédios et al., 2012). Consequently, several studies have focused on pinpointing the factors that determine the ultimate effect of NPs on the germination of seeds and plant development.

Nanopriming is a novel seed-priming process that uses nanoparticles to boost crop yield and spread awareness of the need of sustainable agriculture (Shelar et al., 2021). Because of their exceptional properties, such as their high surface area to mass ratio, NPs may effectively absorb and distribute compounds of interest. There is a lack of definitive evidence on the impact of NPs on seed germination, despite the many publications on the impact of NPs on plant growth parameters, such as many physiological and biochemical processes (Ali et al., 2021). Since it is currently unclear how NPs interact with seeds and plants, so their ultimate impact might be beneficial or harmful. Therefore, it is necessary to do an extensive study on the function of all critical components impacting the metal NPs/plant interaction, as this will offer a better knowledge of the ultimate consequences.

In addition, it is very adaptable to a wide range of soil and environmental factors, including salinization, which increases its relevance as a model crop for research into the processes behind stress tolerance (Koevoets et al., 2016). Identifying essential genes that regulate plant salinity tolerance may need knowledge of the physiological, biochemical, and molecular processes underlying these responses. Salinity has been shown to impair barley seed germination. Still, the comprehensive response mechanisms to salt stress tolerance created in the presence of AgNPs after nano priming has not been documented before (Cembrowska-Lech & Rybak, 2023). In light of these findings, we examined whether AgNPs affect antioxidant defense mechanisms in salt-stressed barley seed germination. Seed germination stimulation and salt phytotoxicity mitigation are two possible applications for AgNPs. AgNPs, which may increase germination for the presence of NaCl, were compared in their effect on non-dormant seeds across four imbibition time periods (before beginning the imbibition, during phase I, and during phases II).

By introducing helpful microbes or biological control elements into the root zone or seed early, biopriming may boost plant development and disease resistance in many ways—individuals belonging to the genus Trichoderma (Singh et al., 2018). Produce growth regulators that expedite seed germination and enhance the dry weight of stems and roots under conditions of drought stress. The formation of Trichoderma spp. colonies in the roots of plants. The augmentation of crop yield, enhancement of root growth and spread, fortification against non-living environmental factors, and amplification of nutrient accessibility are observed outcomes. The application of Trichoderma fungus has been found to enhance the uptake of phosphorus and other microelements by plants. The augmentation in question confers advantages to plants that are already endowed with resistance to infections and drought, as it enhances their growth, vitality, and disease resistance (Drobek et al., 2019). Numerous studies have demonstrated that the process of biopriming utilizing PGPR, which involves the combination of diverse bacteria and fungi, can potentially augment the water content, growth of seedlings, pigments, and protein content. This effect has been observed even in conditions of limited water and high salinity (Fazeli-Nasab et al., 2021).

Biopriming with PGPR has made it so that young seedlings don't have as many bad physiological and biochemical reactions to water stress. Biopriming is a pre-planting method that works well because it is a natural process that involves microorganisms in the soil. It is safe, doesn't pollute the environment, and helps keep people from getting sick (Rawat et al., 2021). The evaluation of bio-stabilization of nitrogen and solubilization of insoluble inorganic and organic phosphorus is necessary. This will be accompanied by a presentation of PGPR and priming. The role of bio-priming in the chlorophyll rate, operation, and function components of the germination process will also be examined. Additionally, the enhancement of plant development and growth under stress conditions, as well as the production of substances such as ACC deaminize, siderophore, phytohormones, and antioxidants, will be analyzed.

Nanobiotechnology in agriculture has yet to have its understanding gap and future research requirements assessed (Horton et al., 2017). The topics of this chapter are the sensing, prompt identification, and tolerance of stress in plants. Transgenic events have also been emphasized for use in nonmodel crop species, seed nanopriming and the controlled release of agrochemicals. Instead of trapping additional electrons, nanomaterials might convert UV and nIR radiation to visible light. An expanded spectrum for photosynthesis may result from these phenomena. The process through which plants may keep up photosynthesis even when darkness falls is explored. Thus, this chapter will pave the way for the effective use of nanobiotechnology in agriculture.

Recent research into calcium signaling has shown that calcium plays an important function in protecting plants from abiotic stressors and bolstering their own defenses. Under PEG-induced drought, canola plants were given calcium by seed nano-priming using calcium oxide nanoparticles (CaONPs) at doses of 25, 50, 75, and 100 ppm, respectively. PEG-6000 treatment-maintained water potential at -0.15 MPa and pH of 5.5 supplemented by Hoagland solutions at ambient temperature in order to replicate drought conditions for the seedlings throughout the germination experiment.

PEG-6000 wasn't administered to the control group. We used artificial climate chambers to test the pot following the germination test. Under drought stress, the germination rate and the seedlings' fresh weight were increased by 30% after being nano-primed with CaONPs at a concentration of 75 ppm. Under drought stress, seed nano-priming with 75 ppm CaONPs raised the overall number of leaves to 16%, total chlorophyll concentrations by 28.9%, the number of pods per branch by 33%, the number of seeds per pod by 35%, the weight of 100 seeds by 35%, and yield by 35.18%. Not only were stress signs reduced, but antioxidant enzyme levels also increased. Current studies suggest that CaONPs may be used as pre-sowing therapies for the seed to boost canola plants' productivity.

Nanobiotechnology in plants might play an important role in the future of environmentally responsible farming (Prasad et al., 2014). Agricultural application necessitates the development of nanomaterials with low metal content and high dispensability. To better understand the biological impact of nanozymes like Mn3O4 nanoparticles on plants under stress, we proposed that additional research is needed. More research is needed to understand the mechanisms behind seed nanopriming, such as nanoparticle absorption, distribution, and destiny, including their interactions with seeds. Using nanomaterials in plant engineering opens up new possibilities, such as producing chloroplasts in plants. In low-light environments, where plants need more electrons for photosynthesis, nanomaterials may convert nIR and UV radiation to visible light (Wu & Li, 2022).

CONCLUSION

The application of nanotechnology to farming is a hot issue right now. Facilitating nano-enabled agriculture requires talking about how nanobiotechnology may be used to boost crop yields. Priming is the physiological process by which plants are stimulated to show enhanced growth biomarkers or stress tolerance following treatment with biological/chemical stimulating agents or moderate stress. Priming seeds with NPs is a cutting-edge technique that regulates morphological and physio-biochemical traits to improve germination, plant growth development, productivity, and postharvest physiology. Seed nano priming, stress sensing and early detection, targeted distribution of agrochemicals, and transgenic events in non-model species of crops are all possible applications of plant nanobiotechnology. Nanotechnology is playing an increasingly important role in this area since it provides access to a plethora of novel nanomaterials, with a focus on the creation of novel bio-based solutions. Understanding how nanomaterials boost plant stress tolerance is crucial for tailoring nanomaterials to specific agricultural applications. Concerns about the use of nanoparticles in agriculture may be alleviated if policies and regulatory standards were developed to manage the associated biosafety issues. We think nanoparticles might be very useful in farming.

REFERENCES

Abbasi Khalaki, M., Moameri, M., Asgari Lajayer, B., & Astatkie, T. (2021). Influence of nano-priming on seed germination and plant growth of forage and medicinal plants. *Plant Growth Regulation*, *93*(1), 13–28. doi:10.100710725-020-00670-9

Abobatta, W. F. (2018). Nanotechnology application in agriculture. *Acta Scientific Agriculture*, *2*(6).

Afzal, S., Sharma, D., & Singh, N. K. (2021). Eco-friendly synthesis of phytochemical-capped iron oxide nanoparticles as nano-priming agent for boosting seed germination in rice (Oryza sativa L.). *Environmental Science and Pollution Research International*, *28*(30), 40275–40287. doi:10.100711356-020-12056-5 PMID:33447981

Ali, S., Mehmood, A., & Khan, N. (2021). Uptake, translocation, and consequences of nanomaterials on plant growth and stress adaptation. *Journal of Nanomaterials*, *2021*, 1–17. doi:10.1155/2021/6677616

Arif, U., Bhatti, K. H., Ajaib, M., Wagay, N. A., Majeed, M., Zeb, J., Hameed, A., & Kiani, J. (2021). Ethnobotanical indigenous knowledge of Tehsil Charhoi, District Kotli, Azad Jammu and Kashmir, Pakistan. *Ethnobotany Research and Applications*, *22*. doi:10.32859/era.22.50.1-24

Arshad, F., Waheed, M., Harun, N., Fatima, K., Khan, B. A., Fatima, K., Abbas, Z., Jabeen, S., & Majeed, M. (2022). Indigenous farmer's perception about fodder and foraging species of Semi-arid lowlands of Pakistan: A case study of District Kasur, Pakistan. *Taiwania*, *67*(4), 510–523. doi:10.6165/tai.2022.67.510

Bano, A., Gupta, A., Rai, S., Fatima, T., Sharma, S., & Pathak, N. (2021). Mechanistic role of reactive oxygen species and its regulation via the antioxidant system under environmental stress. *Plant Stress Physiology—Perspectives in Agriculture*, 1-18.

Bouwmeester, H., Dekkers, S., Noordam, M. Y., Hagens, W. I., Bulder, A. S., De Heer, C., Ten Voorde, S. E., Wijnhoven, S. W., Marvin, H. J., & Sips, A. J. (2009). Review of health safety aspects of nanotechnologies in food production. *Regulatory Toxicology and Pharmacology*, *53*(1), 52–62. doi:10.1016/j.yrtph.2008.10.008 PMID:19027049

Brummer, E. C., Barber, W. T., Collier, S. M., Cox, T. S., Johnson, R., Murray, S. C., Olsen, R. T., Pratt, R. C., & Thro, A. M. (2011). Plant breeding for harmony between agriculture and the environment. *Frontiers in Ecology and the Environment*, *9*(10), 561–568. doi:10.1890/100225

Cembrowska-Lech, D., & Rybak, K. (2023). Nanopriming of Barley Seeds—A Shotgun Approach to Improve Germination under Salt Stress Conditions by Regulating of Reactive Oxygen Species. *Plants*, *12*(2), 405. doi:10.3390/plants12020405 PMID:36679118

Chandrasekaran, U., Luo, X., Wang, Q., & Shu, K. (2020). Are there unidentified factors involved in the germination of nanoprimed seeds? *Frontiers in Plant Science*, *11*, 832. doi:10.3389/fpls.2020.00832 PMID:32587599

Dahal, K., Li, X.-Q., Tai, H., Creelman, A., & Bizimungu, B. (2019). Improving potato stress tolerance and tuber yield under a climate change scenario–a current overview. *Frontiers in Plant Science*, *10*, 563. doi:10.3389/fpls.2019.00563 PMID:31139199

Das, K., & Roychoudhury, A. (2014). Reactive oxygen species (ROS) and response of antioxidants as ROS-scavengers during environmental stress in plants. *Frontiers in Environmental Science*, *2*, 53. doi:10.3389/fenvs.2014.00053

Demil, B., & Lecocq, X. (2010). Business model evolution: In search of dynamic consistency. *Long Range Planning*, *43*(2-3), 227–246. doi:10.1016/j.lrp.2010.02.004

do Espirito Santo Pereira, A., Caixeta Oliveira, H., Fernandes Fraceto, L., & Santaella, C. (2021). Nanotechnology potential in seed priming for sustainable agriculture. *Nanomaterials (Basel, Switzerland)*, *11*(2), 267. doi:10.3390/nano11020267 PMID:33498531

Drobek, M., Frąc, M., & Cybulska, J. (2019). Plant biostimulants: Importance of the quality and yield of horticultural crops and the improvement of plant tolerance to abiotic stress—A review. *Agronomy (Basel)*, *9*(6), 335. doi:10.3390/agronomy9060335

El-Badri, A. M., Batool, M., Mohamed, I. A., Khatab, A., Sherif, A., Wang, Z., Salah, A., Nishawy, E., Ayaad, M., Kuai, J., Wang, B., & Zhou, G. (2021). Modulation of salinity impact on early seedling stage via nano-priming application of zinc oxide on rapeseed (Brassica napus L.). *Plant Physiology and Biochemistry*, *166*, 376–392. doi:10.1016/j.plaphy.2021.05.040 PMID:34153882

Elsheery, N. I., Helaly, M. N., El-Hoseiny, H. M., & Alam-Eldein, S. M. (2020). Zinc oxide and silicone nanoparticles to improve the resistance mechanism and annual productivity of salt-stressed mango trees. *Agronomy (Basel)*, *10*(4), 558. doi:10.3390/agronomy10040558

Fazeli-Nasab, B., Sayyed, R., Piri, R., & Rahmani, A. F. (2021). Biopriming and Nanopriming: green revolution wings to increase plant yield, growth, and development under stress condition and forward dimensions. *Antioxidants in Plant-Microbe Interaction*, 623-655.

Garcia, D., Arif, S., Zhao, Y., Zhao, S., Ming, L. C., & Huang, D. (2021). *Seed priming technology as a key strategy to increase crop plant production under adverse environmental conditions*. Research Gate.

Garg, R., & Maheshwari, S. (2023). SEED PRIMING AND FORTIFICATION OF SEEDS USING NANOTECHNOLOGY: A REVIEW. *EPH-International Journal of Agriculture and Environmental Research*, *9*(1), 11–17. doi:10.53555/eijaer.v9i1.68

Giraldo, J. P., Wu, H., Newkirk, G. M., & Kruss, S. (2019). Nanobiotechnology approaches for engineering smart plant sensors. *Nature Nanotechnology*, *14*(6), 541–553. doi:10.103841565-019-0470-6 PMID:31168083

Gonzalez Guzman, M., Cellini, F., Fotopoulos, V., Balestrini, R., & Arbona, V. (2022). New approaches to improve crop tolerance to biotic and abiotic stresses. *Physiologia Plantarum, 174*(1), e13547. doi:10.1111/ppl.13547 PMID:34480798

Handy, R. D., van den Brink, N., Chappell, M., Mühling, M., Behra, R., Dušinská, M., Simpson, P., Ahtiainen, J., Jha, A. N., Seiter, J., Bednar, A., Kennedy, A., Fernandes, T. F., & Riediker, M. (2012). Practical considerations for conducting ecotoxicity test methods with manufactured nanomaterials: What have we learnt so far? *Ecotoxicology (London, England), 21*(4), 933–972. doi:10.100710646-012-0862-y PMID:22422174

Haq, S. M., Tariq, A., Li, Q., Yaqoob, U., Majeed, M., Hassan, M., Fatima, S., Kumar, M., Bussmann, R. W., Farhan, M., Moazzam, U., & Aslam, M. (2022). *Influence of Edaphic Properties in Determining Forest Community Patterns of the Zabarwan Mountain Range in the Kashmir Himalayas*. MDPI, 1–15.

Haq, S. M., Yaqoob, U., Majeed, M., Amjad, M. S., Hassan, M., Ahmad, R., Waheed, M., Bussmann, R. W., Calixto, E. S., & Pro, J. (n.d.). *Quantitative ethnoveterinary study on plant resource utilization by indigenous communities in high-altitude regions*. Frontiers.

Hassanisaadi, M., Barani, M., Rahdar, A., Heidary, M., Thysiadou, A., & Kyzas, G. Z. (2022). Role of agrochemical-based nanomaterials in plants: Biotic and abiotic stress with germination improvement of seeds. *Plant Growth Regulation, 97*(2), 375–418. doi:10.100710725-021-00782-w

He, C., Liu, D., & Lin, W. (2015). Nanomedicine applications of hybrid nanomaterials built from metal–ligand coordination bonds: Nanoscale metal–organic frameworks and nanoscale coordination polymers. *Chemical Reviews, 115*(19), 11079–11108. doi:10.1021/acs.chemrev.5b00125 PMID:26312730

Hofmann, T., Lowry, G. V., Ghoshal, S., Tufenkji, N., Brambilla, D., Dutcher, J. R., Gilbertson, L. M., Giraldo, J. P., Kinsella, J. M., Landry, M. P., Lovell, W., Naccache, R., Paret, M., Pedersen, J. A., Unrine, J. M., White, J. C., & Wilkinson, K. J. (2020). Technology readiness and overcoming barriers to sustainably implement nanotechnology-enabled plant agriculture. *Nature Food, 1*(7), 416–425. doi:10.103843016-020-0110-1

Holland, J. H. (1992). Genetic algorithms. *Scientific American, 267*(1), 66–73. doi:10.1038cientificamerican0792-66

Horton, A. A., Walton, A., Spurgeon, D. J., Lahive, E., & Svendsen, C. (2017). Microplastics in freshwater and terrestrial environments: Evaluating the current understanding to identify the knowledge gaps and future research priorities. *The Science of the Total Environment, 586*, 127–141. doi:10.1016/j.scitotenv.2017.01.190 PMID:28169032

Hossain, A., Kerry, R. G., Farooq, M., Abdullah, N., & Tofazzal Islam, M. (2020). Application of nanotechnology for sustainable crop production systems. *Nanotechnology for food, agriculture, and environment*, 135-159.

Huang, D., Huo, J., & Liao, W. (2021). Hydrogen sulfide: Roles in plant abiotic stress response and crosstalk with other signals. *Plant Science, 302*, 110733. doi:10.1016/j.plantsci.2020.110733 PMID:33288031

Islam, F., Shohag, S., Uddin, M. J., Islam, M. R., Nafady, M. H., Akter, A., Mitra, S., Roy, A., Emran, T. B., & Cavalu, S. (2022). Exploring the journey of zinc oxide nanoparticles (ZnO-NPs) toward biomedical applications. *Materials (Basel), 15*(6), 2160. doi:10.3390/ma15062160 PMID:35329610

Jamil, M. D., Waheed, M., Akhtar, S., Bangash, N., Chaudhari, S. K., Majeed, M., Hussain, M., Ali, K., & Jones, D. A. (2022). Invasive Plants Diversity, Ecological Status, and Distribution Pattern in Relation to Edaphic Factors in Different Habitat Types of District Mandi Bahauddin, Punjab, Pakistan. *Sustainability (Basel), 14*(20), 13312. Advance online publication. doi:10.3390u142013312

Jeger, M., Beresford, R., Bock, C., Brown, N., Fox, A., Newton, A., Vicent, A., Xu, X., & Yuen, J. (2021). Global challenges facing plant pathology: Multidisciplinary approaches to meet the food security and environmental challenges in the mid-twenty-first century. *CABI Agriculture and Bioscience, 2*(1), 1–18. doi:10.118643170-021-00042-x

Johnson, R., & Puthur, J. T. (2021). Seed priming as a cost effective technique for developing plants with cross tolerance to salinity stress. *Plant Physiology and Biochemistry, 162*, 247–257. doi:10.1016/j.plaphy.2021.02.034 PMID:33711718

Kader, M. A., & Lindberg, S. (2010). Cytosolic calcium and pH signaling in plants under salinity stress. *Plant Signaling & Behavior, 5*(3), 233–238. doi:10.4161/psb.5.3.10740 PMID:20037468

Kandhol, N., Singh, V. P., Ramawat, N., Prasad, R., Chauhan, D. K., Sharma, S., Grillo, R., Sahi, S., Peralta-Videa, J., & Tripathi, D. K. (2022). Nano-priming: Impression on the beginner of plant life. *Plant Stress, 5*, 100091. doi:10.1016/j.stress.2022.100091

Khan, A. M., Li, Q., Saqib, Z., Khan, N., Habib, T., Khalid, N., Majeed, M., & Tariq, A. (2022). MaxEnt Modelling and Impact of Climate Change on Habitat Suitability Variations of Economically Important Chilgoza Pine (Pinus gerardiana Wall.) in South Asia. *Forests*, *13*(5), 715. doi:10.3390/f13050715

Khoja, A. A., Haq, S. M., Majeed, M., Hassan, M., Waheed, M., Yaqoob, U., Bussmann, R. W., Alataway, A., & Dewidar, A. Z. (2022). *Diversity*. Ecological and Traditional Knowledge of Pteridophytes in the Western Himalayas.

Kim, D. Y., Kadam, A., Shinde, S., Saratale, R. G., Patra, J., & Ghodake, G. (2018). Recent developments in nanotechnology transforming the agricultural sector: A transition replete with opportunities. *Journal of the Science of Food and Agriculture*, *98*(3), 849–864. doi:10.1002/jsfa.8749 PMID:29065236

Koevoets, I. T., Venema, J. H., Elzenga, J. T. M., & Testerink, C. (2016). Roots withstanding their environment: Exploiting root system architecture responses to abiotic stress to improve crop tolerance. *Frontiers in Plant Science*, *7*, 1335. doi:10.3389/fpls.2016.01335 PMID:27630659

Kumar, B., Singhal, R. K., Chand, S., Chauhan, J., Kumar, V., Mishra, U. N., Hidangmayum, A., Singh, A., & Bose, B. (2022). Nanopriming in sustainable agriculture: recent advances, emerging challenges and future prospective. *New and Future Developments in Microbial Biotechnology and Bioengineering*, 339-365.

Kumar, V. K., Muthukrishnan, S., & Rajalakshmi, R. (2020). Phytostimulatory effect of phytochemical fabricated nanosilver (AgNPs) on Psophocarpus tetragonolobus (L.) DC. seed germination: An insight from antioxidative enzyme activities and genetic similarity studies. *Current Plant Biology*, *23*, 100158. doi:10.1016/j.cpb.2020.100158

Lew, T. T. S., Sarojam, R., Jang, I.-C., Park, B. S., Naqvi, N. I., Wong, M. H., Singh, G. P., Ram, R. J., Shoseyov, O., Saito, K., Chua, N.-H., & Strano, M. S. (2020). Species-independent analytical tools for next-generation agriculture. *Nature Plants*, *6*(12), 1408–1417. doi:10.103841477-020-00808-7 PMID:33257857

Li, L., Zhang, Q., & Huang, D. (2014). A review of imaging techniques for plant phenotyping. *Sensors (Basel)*, *14*(11), 20078–20111. doi:10.3390141120078 PMID:25347588

Li, Y., Xu, X., Wu, Y., Zhuang, J., Zhang, X., Zhang, H., Lei, B., Hu, C., & Liu, Y. (2020). A review on the effects of carbon dots in plant systems. *Materials Chemistry Frontiers*, *4*(2), 437–448. doi:10.1039/C9QM00614A

Liu, J., Li, G., Chen, L., Gu, J., Wu, H., & Li, Z. (2021). Cerium oxide nanoparticles improve cotton salt tolerance by enabling better ability to maintain cytosolic K+/Na+ ratio. *Journal of Nanobiotechnology*, *19*(1), 153. doi:10.118612951-021-00892-7 PMID:34034767

Lodeyro, A. F., Ceccoli, R. D., Karlusich, J. J. P., & Carrillo, N. (2012). The importance of flavodoxin for environmental stress tolerance in photosynthetic microorganisms and transgenic plants. Mechanism, evolution and biotechnological potential. *FEBS Letters*, *586*(18), 2917–2924. doi:10.1016/j.febslet.2012.07.026 PMID:22819831

Lu, L., Huang, M., Huang, Y., Corvini, P. F.-X., Ji, R., & Zhao, L. (2020). Mn 3 O 4 nanozymes boost endogenous antioxidant metabolites in cucumber (Cucumis sativus) plant and enhance resistance to salinity stress. *Environmental Science. Nano*, *7*(6), 1692–1703. doi:10.1039/D0EN00214C

Maggio, A., Bressan, R. A., Zhao, Y., Park, J., & Yun, D.-J. (2018). It's hard to avoid avoidance: Uncoupling the evolutionary connection between plant growth, productivity and stress "tolerance". *International Journal of Molecular Sciences*, *19*(11), 3671. doi:10.3390/ijms19113671 PMID:30463352

Majeed, M., Bhatti, K. H., & Amjad, M. S. (2021). Impact of climatic variations on the flowering phenology of plant species in jhelum district, punjab, pakistan. *Applied Ecology and Environmental Research*, *19*(5), 3343–3376. doi:10.15666/aeer/1905_33433376

Majeed, M., Bhatti, K. H., Pieroni, A., Renata, S., Bussmann, R. W., Khan, A. M., Chaudhari, S. K., Aziz, M. A., & Amjad, M. S. (2021). *Gathered Wild Food Plants among Diverse Religious Groups in*. MDPI. http://scholar.google.com/scholar_url?url=https://www.mdpi.com/2304-8158/10/3/594/pdf&hl=sl&sa=X&d=16557763567181478619&ei=yrdPYOSADsbYmQHIqoSICA&scisig=AAGBfm3h_EDHFITvMVt55xuUBsvq3j_PUw&nossl=1&oi=scholaralrt&hist=RTqo9GEAAAAJ:3473121909733627244:AAGBfm

Majeed, M., Lu, L., Anwar, M. M., Tariq, A., Qin, S., El-Hefnawy, M. E., El-Sharnouby, M., Li, Q., & Alasmari, A. (2023). Prediction of flash flood susceptibility using integrating analytic hierarchy process (AHP) and frequency ratio (FR) algorithms. *Frontiers in Environmental Science*, *10*, 1037547. doi:10.3389/fenvs.2022.1037547

Majeed, M., Mahmood, A., Habib, T., Mushahid, M., Ali, H., Khan, N., & Ali, K. (2022). Vegetation analysis and environmental indicators of an arid tropical forest ecosystem of Pakistan. *Ecological Indicators*, *142*(May), 109291. doi:10.1016/j.ecolind.2022.109291

Majeed, M., Tariq, A., Haq, S. M., Waheed, M., Anwar, M. M., Li, Q., Aslam, M., Abbasi, S., Mousa, B. G., & Jamil, A. (2022). A Detailed Ecological Exploration of the Distribution Patterns of Wild Poaceae from the Jhelum District (Punjab), Pakistan. *Sustainability (Basel)*, *14*(7), 3786. doi:10.3390u14073786

Mauch-Mani, B., Baccelli, I., Luna, E., & Flors, V. (2017). Defense priming: An adaptive part of induced resistance. *Annual Review of Plant Biology*, *68*(1), 485–512. doi:10.1146/annurev-arplant-042916-041132 PMID:28226238

Mhlongo, M. I., Piater, L. A., Steenkamp, P. A., Labuschagne, N., & Dubery, I. A. (2021). Metabolomic evaluation of tissue-specific defense responses in tomato plants modulated by PGPR-priming against Phytophthora capsici infection. *Plants*, *10*(8), 1530. doi:10.3390/plants10081530 PMID:34451575

Mittal, D., Kaur, G., Singh, P., Yadav, K., & Ali, S. A. (2020). Nanoparticle-based sustainable agriculture and food science: Recent advances and future outlook. *Frontiers in Nanotechnology*, *2*, 579954. doi:10.3389/fnano.2020.579954

Mittler, R., & Blumwald, E. (2010). Genetic engineering for modern agriculture: Challenges and perspectives. *Annual Review of Plant Biology*, *61*(1), 443–462. doi:10.1146/annurev-arplant-042809-112116 PMID:20192746

Moghanloo, M., Iranbakhsh, A., Ebadi, M., & Oraghi Ardebili, Z. (2019). Differential physiology and expression of phenylalanine ammonia lyase (PAL) and universal stress protein (USP) in the endangered species Astragalus fridae following seed priming with cold plasma and manipulation of culture medium with silica nanoparticles. *3 Biotech*, *9*, 1-13.

Mozgova, I., Mikulski, P., Pecinka, A., & Farrona, S. (2019). Epigenetic mechanisms of abiotic stress response and memory in plants. *Epigenetics in plants of agronomic importance: Fundamentals and applications: Transcriptional regulation and chromatin remodelling in plants*, 1-64.

Mukherjee, D. (2019). Seed invigoration: An effective tool for conservation of endangered and valuable medicinal plant. *Advances in Plant Physiology*, *28*, 243–272.

Nile, S. H., Thiruvengadam, M., Wang, Y., Samynathan, R., Shariati, M. A., Rebezov, M., Nile, A., Sun, M., Venkidasamy, B., Xiao, J., & Kai, G. (2022). Nano-priming as emerging seed priming technology for sustainable agriculture—Recent developments and future perspectives. *Journal of Nanobiotechnology*, *20*(1), 1–31. doi:10.118612951-022-01423-8 PMID:35659295

Olalekan, R., Omidiji, A., Williams, E., Christianah, M., & Modupe, O. (2019). The roles of all tiers of government and development partners in environmental conservation of natural resource: A case study in Nigeria. *MOJ Ecology & Environmental Sciences*, *4*(3), 114–121. doi:10.15406/mojes.2019.04.00142

Oliveira, H. C., Seabra, A. B., Kondak, S., Adedokun, O. P., & Kolbert, Z. (2023). Multilevel approach of plant-nanomaterial relationship: From cells to living ecosystems. *Journal of Experimental Botany*, *74*(12), erad107. doi:10.1093/jxb/erad107

Omran, B. A., & Baek, K.-H. (2022). Valorization of agro-industrial biowaste to green nanomaterials for wastewater treatment: Approaching green chemistry and circular economy principles. *Journal of Environmental Management*, *311*, 114806. doi:10.1016/j.jenvman.2022.114806 PMID:35240500

Onyekachi, O. G., Boniface, O. O., Gemlack, N. F., & Nicholas, N. (2019). The effect of climate change on abiotic plant stress: a review. *Abiotic and biotic stress in plants, 17.*

Pal, R., Mondal, T., Dev, R., Parihar, M., Kumar, A., Sharma, D., & Kant, L. (2023). 5 Role of Enzyme-Mimicking Nanoparticles in Crop Plants. *Biotic Stress Management of Crop Plants using. Nanomaterials (Basel, Switzerland)*, 71–85.

Prasad, R., Kumar, V., & Prasad, K. S. (2014). Nanotechnology in sustainable agriculture: Present concerns and future aspects. *African Journal of Biotechnology*, *13*(6), 705–713. doi:10.5897/AJBX2013.13554

Rai, M., & Ingle, A. (2012). Role of nanotechnology in agriculture with special reference to management of insect pests. *Applied Microbiology and Biotechnology*, *94*(2), 287–293. doi:10.100700253-012-3969-4 PMID:22388570

Rai-Kalal, P., Tomar, R. S., & Jajoo, A. (2021a). H2O2 signaling regulates seed germination in ZnO nanoprimed wheat (Triticum aestivum L.) seeds for improving plant performance under drought stress. *Environmental and Experimental Botany*, *189*, 104561. doi:10.1016/j.envexpbot.2021.104561

Rai-Kalal, P., Tomar, R. S., & Jajoo, A. (2021b). Seed nanopriming by silicon oxide improves drought stress alleviation potential in wheat plants. *Functional Plant Biology*, *48*(9), 905–915. doi:10.1071/FP21079 PMID:34366002

Rawat, L., Bisht, T., & Naithani, D. C. (2021). Plant disease management in organic farming system: strategies and challenges. *Emerging trends in plant pathology*, 611-642. NIH.

Raza, A., Razzaq, A., Mehmood, S. S., Zou, X., Zhang, X., Lv, Y., & Xu, J. (2019). Impact of climate change on crops adaptation and strategies to tackle its outcome: A review. *Plants*, *8*(2), 34. doi:10.3390/plants8020034 PMID:30704089

Remédios, C., Rosário, F., & Bastos, V. (2012). Environmental nanoparticles interactions with plants: Morphological, physiological, and genotoxic aspects. *Le Journal de Botanique*.

Rexlin, J., Vijayakumar, S., Nilavukkarasi, M., Vidhya, E., Alharthi, N. S., Sajjad, M., Punitha, V., & Praseetha, P. (2022). Bioengineered ZnO nanoparticles as a nano priming agent in Cyamopsis tetragonoloba (L). Taub. to improve yield and disease resistance. *Applied Nanoscience*, 1–9. doi:10.100713204-022-02526-2

Rivero, R. M., Mittler, R., Blumwald, E., & Zandalinas, S. I. (2022). Developing climate-resilient crops: Improving plant tolerance to stress combination. *The Plant Journal*, *109*(2), 373–389. doi:10.1111/tpj.15483 PMID:34482588

Rommens, C. M. (2007). Intragenic crop improvement: Combining the benefits of traditional breeding and genetic engineering. *Journal of Agricultural and Food Chemistry*, *55*(11), 4281–4288. doi:10.1021/jf0706631 PMID:17488120

Saleh, T. A. (2020). Nanomaterials: Classification, properties, and environmental toxicities. *Environmental Technology & Innovation*, *20*, 101067. doi:10.1016/j.eti.2020.101067

Satalkar, P., Elger, B. S., & Shaw, D. M. (2016). Defining nano, nanotechnology and nanomedicine: Why should it matter? *Science and Engineering Ethics*, *22*(5), 1255–1276. doi:10.100711948-015-9705-6 PMID:26373718

Sharma, B., Tiwari, S., Kumawat, K. C., & Cardinale, M. (2023). Nano-biofertilizers as bio-emerging strategies for sustainable agriculture development: Potentiality and their limitations. *The Science of the Total Environment*, *860*, 160476. doi:10.1016/j.scitotenv.2022.160476 PMID:36436627

Shelar, A., Singh, A. V., Maharjan, R. S., Laux, P., Luch, A., Gemmati, D., Tisato, V., Singh, S. P., Santilli, M. F., Shelar, A., Chaskar, M., & Patil, R. (2021). Sustainable agriculture through multidisciplinary seed nanopriming: Prospects of opportunities and challenges. *Cells*, *10*(9), 2428. doi:10.3390/cells10092428 PMID:34572078

Singh, A., Rajput, V., Singh, A. K., Sengar, R. S., Singh, R. K., & Minkina, T. (2021). Transformation Techniques and Their Role in Crop Improvements: A Global Scenario of GM Crops. *Policy Issues in Genetically Modified Crops: A Global Perspective*, 515–542. doi:10.1016/B978-0-12-820780-2.00023-6

Singh, A., Sharma, R., Rawat, S., Singh, A. K., Rajput, V. D., Fedorov, Y., Minkina, T., & Chaplygin, V. (2022). Nanomaterial-plant interaction: Views on the pros and cons. *Toxicity of Nanoparticles in Plants: An Evaluation of Cyto/Morpho-Physiological, Biochemical and Molecular Responses*, 47–68. doi:10.1016/B978-0-323-90774-3.00015-5

Singh, A., Shukla, N., Kabadwal, B., Tewari, A., & Kumar, J. (2018). Review on plant-Trichoderma-pathogen interaction. *International Journal of Current Microbiology and Applied Sciences*, *7*(2), 2382–2397. doi:10.20546/ijcmas.2018.702.291

Sonawane, H., Arya, S., Math, S., & Shelke, D. (2021). Myco-synthesized silver and titanium oxide nanoparticles as seed priming agents to promote seed germination and seedling growth of Solanum lycopersicum: A comparative study. *International Nano Letters*, *11*(4), 371–379. doi:10.100740089-021-00346-w

Talukdar, S., Singha, P., Mahato, S., Shahfahad, Pal, S., Liou, Y.-A., & Rahman, A. (2020). Land-Use Land-Cover Classification by Machine Learning Classifiers for Satellite Observations—A Review. *Remote Sensing (Basel)*, *12*(7), 1135. doi:10.3390/rs12071135

Tang, X., Mu, X., Shao, H., Wang, H., & Brestic, M. (2015). Global plant-responding mechanisms to salt stress: Physiological and molecular levels and implications in biotechnology. *Critical Reviews in Biotechnology*, *35*(4), 425–437. doi:10.3109/07388551.2014.889080 PMID:24738851

Tassadduq, S. S., Akhtar, S., Waheed, M., Bangash, N., Nayab, D. E., Majeed, M., Abbasi, S., Muhammad, M., Alataway, A., Dewidar, A. Z., Elansary, H. O., & Yessoufou, K. (2022). Ecological Distribution Patterns of Wild Grasses and Abiotic Factors. *Sustainability (Basel)*, *14*(18), 11117. Advance online publication. doi:10.3390u141811117

Toksha, B., Sonawale, V. A. M., Vanarase, A., Bornare, D., Tonde, S., Hazra, C., Kundu, D., Satdive, A., Tayde, S., & Chatterjee, A. (2021). Nanofertilizers: A review on synthesis and impact of their use on crop yield and environment. *Environmental Technology & Innovation*, *24*, 101986. doi:10.1016/j.eti.2021.101986

Upreti, D., & Patro, T. U. (2022). Polymer Nanocomposite Membranes for Water Remediation. In *Advances in Nanostructured Materials* (pp. 227–279). Springer. doi:10.1007/978-981-16-8391-6_13

Vickers, C. E., Gershenzon, J., Lerdau, M. T., & Loreto, F. (2009). A unified mechanism of action for volatile isoprenoids in plant abiotic stress. *Nature Chemical Biology*, *5*(5), 283–291. doi:10.1038/nchembio.158 PMID:19377454

Wu, H., & Li, Z. (2022). Recent advances in nano-enabled agriculture for improving plant performance. *The Crop Journal, 10*(1), 1–12. doi:10.1016/j.cj.2021.06.002

Yadav, S., Modi, P., Dave, A., Vijapura, A., Patel, D., & Patel, M. (2020). Effect of abiotic stress on crops. *Sustainable crop production, 3.*

Yadav, S. K., Lal, S., Yadav, S., Laxman, J., Verma, B., Sushma, M., Choudhary, R., Singh, P., Singh, S., & Sharma, V. (2019). Use of nanotechnology in agri-food sectors and apprehensions: An overview. *Seed Res*, *47*(2), 99–149.

Younis, S. A., Kim, K.-H., Shaheen, S. M., Antoniadis, V., Tsang, Y. F., Rinklebe, J., Deep, A., & Brown, R. J. (2021). Advancements of nanotechnologies in crop promotion and soil fertility: Benefits, life cycle assessment, and legislation policies. *Renewable & Sustainable Energy Reviews*, *152*, 111686. doi:10.1016/j.rser.2021.111686

Yuan, P., Yang, T., & Poovaiah, B. (2018). Calcium signaling-mediated plant response to cold stress. *International Journal of Molecular Sciences*, *19*(12), 3896. doi:10.3390/ijms19123896 PMID:30563125

Zaman, M., Shahid, S. A., Heng, L., Shahid, S. A., Zaman, M., & Heng, L. (2018). Soil salinity: Historical perspectives and a world overview of the problem. *Guideline for salinity assessment, mitigation and adaptation using nuclear and related techniques*, 43-53.

Zhao, L., Lu, L., Wang, A., Zhang, H., Huang, M., Wu, H., Xing, B., Wang, Z., & Ji, R. (2020). Nanobiotechnology in agriculture: Use of nanomaterials to promote plant growth and stress tolerance. *Journal of Agricultural and Food Chemistry*, *68*(7), 1935–1947. doi:10.1021/acs.jafc.9b06615 PMID:32003987

Zulfiqar, F. (2021). Effect of seed priming on horticultural crops. *Scientia Horticulturae*, *286*, 110197. doi:10.1016/j.scienta.2021.110197

Zulfiqar, F., Navarro, M., Ashraf, M., Akram, N. A., & Munné-Bosch, S. (2019). Nanofertilizer use for sustainable agriculture: Advantages and limitations. *Plant Science*, *289*, 110270. doi:10.1016/j.plantsci.2019.110270 PMID:31623775

Chapter 13
Seed Nanopriming:
An Innovative Approach for Upregulating Seed Germination and Plant Growth Under Salinity Stress

Abhishek Singh
https://orcid.org/0000-0002-0049-6910
Faculty of Biology, Yerevan State University, Armenia

Shreni Agrawal
Parul Institute of Applied Science, Parul University, India

Vishnu D. Rajput
https://orcid.org/0000-0002-6802-4805
Academy of Biology and Biotechnology, Southern Federal University, Rostov-on-Don, Russia

Karen Ghazaryan
https://orcid.org/0000-0002-8973-2595
Faculty of Biology, Yerevan State University, Armenia

Hasmik S. Movsesyan
Yerevan State University, Armenia

Tatiana Minkina
Academy of Biology and Biotechnology, Southern Federal University, Rostov-on-Don, Russia

Abdel Rahman Mohammad Al Tawaha
Department of Science and Engineering, Al Hussein bin Talal University, Jordan

Athanasios Alexiou
https://orcid.org/0000-0002-2206-7236
Department of Science and Engineering, Novel Global Community Educational Foundation, Hebersham, Australia

Badal Singh
ICAR-National Bureau of Plant Genetic Resources, Pusa, India

Santosh Kumar Gupta
https://orcid.org/0000-0003-2245-8488
National Institute of Plant Genome Research, India

DOI: 10.4018/978-1-6684-7232-3.ch013

ABSTRACT

Soil salinity, one of the main abiotic risks to agricultural products, lowers agricultural yields, and as a result, the world's rising population confronts substantial food challenges. It's one of the world's most pressing abiotic issues, affecting 30% or more of irrigated land. Saline stress is caused by a number of different salt concentrations; however, NaCl is by far the most common salt in the ground. High salt stress is the reason for delay in seed germination, which may be caused by either the osmotic potential of the water or by the effects of certain toxic ions. This chapter covers seed priming and salt resistance approaches that generate biochemical, physiological, and structural processes in saline stress, and the interaction between seed priming and salt resistance. Priming with nanoparticles (NPs) needs more study to reduce the negative effects of salinity stress on a variety of crops grown in extreme conditions.

INTRODUCTION

Currently, agriculture is dealing with a number of issues, including output losses brought on by pests, the consequences of global warming, and the gradual dwindling of natural assets (Kah et al., 2019). Traditional farming methods have also contributed to environmental degradation through their reliance on the ever-present use of chemical fertilizers and pesticides (Kaivosoja et al., 2013). In order to meet the estimated 9–10 billion world population in 2050, crop productivity needs to increase by 25–70% (Scott et al., 2018). Therefore, the use of cutting-edge technology in agriculture is essential for the future of food production (A. Singh et al., 2021; Verma, Song, Joshi, Rajput, et al., 2022).

Abiotic plant stresses are the most common types of environmental challenges that plants face. Plants are subjected to significant amounts of abiotic stress in environments where salinity is the primary prevailing abiotic stressor. This causes a significant yearly crop decline and severely stunts the plant's growth and developmental processes. Salinity stress has a greater effect in the world's coastal, arid, and semi-arid regions (AbdElgawad et al., 2016). Inefficient irrigation practices, ineffective salinity reclamation techniques, and the use of marginal land for crop production are all contributing to the spread of salinity (Van Zelm et al., 2020). Over 800 million hectares, or about 6% of the world's land area, is covered by salty soils (Munns & Tester, 2008a). The FAO (2012) estimates that 60–80 million hectares of land have been impacted negatively by water-logging and related salinity (FAO WFP & IFAD, 2012). Salinity-related soil deterioration is a major issue that has had a detrimental influence on global agricultural production (Adhikary et al., 2022). Each year, more than 10 mha of agricultural soil is lost to salinity. More than half of the world's farmland will be lost by the end of the 21st century because of soil salinity (FAO WFP & IFAD, 2012). Plant growth is affected by soil salinity throughout all stages of development, though to varying degrees. High salinity may hinder seed germination, a crucial early growth stage. The most telling indicator of salt tolerance is whether or not a plant survives through the germination and emergence processes. As a result, the seedling and germination stages of development are particularly vulnerable to salt damage in the majority of crop species (Sharief, 2012). The inability of a germinating seed to absorb enough water is the underlying cause of poor germination (S. A. Ashraf et al., 2021).It is essential for plant stands to adapt to a salty environment during the early stages of growth (germination and seedling development). Salinity stress caused a delay in the germination of susceptible plant seeds (Ibrahim, 2016). Singh et al. (2022) found that slow germination of seeds and poor seedling establishment were both caused by soil salinity stress. Additionally, by restricting carbon uptake in the latter stages following stand formation, the Na+ lethal level present in leaves encourages a reduction in photosynthesis and early senility of the leaves (A. Hussain et al., 2018). Seeds germinate more slowly and have a lower success rate when exposed to high concentrations of salt (A. Singh et al., 2022). Seed germination rates and total amounts vary by variety and species at any given time. Osmotic stress or specific ion effects caused by salinity could have a negative impact on plant growth and seed germination. This triggers the closing of the plant's stomata and the intake of water and nutrients (such as potassium and calcium) as well as carbon dioxide (CO_2) (Parihar et al., 2015). Reduced early crop growth rates are a direct result of soil salinity stress. In crucial metabolic activities, particular ion action (high Na+ and Cl- ions) and osmosis activity result in a water shortage condition, causing plant growth to slow down in high-salinity environments (Hasanuzzaman et al., 2019; Munns & Tester, 2008a; Rahman et al., 2015).There are two causes of plant growth suppression caused by salt stress in soil water. An osmotic effect is produced, and secondly, a salt-specific ion effect is brought about. These two saline stress effects lead to an a two-phase developmental reaction to salt (Munns, 2005; Munns & Tester, 2008b).

A new technological agricultural revolution may be possible with the help of nanotechnology (D. Sharma et al., 2021). Nano-fertilizers and nano-pesticides have been developed to decrease the amount of fertilizers and pesticides used in agriculture, while simultaneously increasing crop yields and improving food quality (V. Rajput et al., 2020; Verma, Song, Joshi, Rajput, et al., 2022; Vishwakarma et al., 2018). Additionally, using nano-enabled plant protective technologies can reduce the damaging environmental consequences of agriculture, making it a more sustainable option (He et al., 2019). Applying nano-priming to seeds can protect them while they are stored, boost germination and germination synchronization, speed up plant growth, and increase crop resistance to abiotic and biotic stress, all of which can lead to lower pesticide and fertilizer usage (Manjunatha et al., 2016). Recent research has shown that nano-priming seeds can activate multiple genes during germination, including those involved in the plant's response to stress (De La Torre-Roche et al., 2020). Despite being a relatively new area of study, the application of nanomaterials for seed preparation has so far yielded promising results (D. Sharma et al., 2021). Seedling nano-priming is a helpful strategy for seed safety since several NPs have antibacterial activity and can load antimicrobial compounds. Additionally, seeds can be nutritionally fortified through nanopriming, which has been shown to improve both food quality and yield (De La Torre-Roche et al., 2020).

Nanopriming Technology for Seed Priming

Nanomaterials, especially NPs, are used for planting seeds in a novel method known as seed nano-priming (Shukla et al., 2019). Seed nano-priming is quite different from traditional seed priming since it generally employs water (hydropriming) or solutions comprising vitamins, minerals, hormones, or biological polymers that adhere on the surface of seed and may result in seed coating (or dressing). Seed nano-priming is done using solutions or nano-formulations, and the seeds can or cannot take the NPs in (Rai-Kalal & Jajoo, 2021). Even when NPs are taken up, the bulk is still left as a protective layer on the seed surface (M. R. Khan et al., 2019). Such seed covering can protect against infections in the growing environment or while being kept when combined with fungicides or antibacterial agents (Elmer et al., 2018). Tomato plants produced twice as many blossoms as usual because the carbon nanotubes enhanced water intake (Khodakovskaya et al., 2013). Similar findings from earlier research have demonstrated that carbon nanotubes can change the metabolism of seeds in plants including tomato, wheat, soybeans, and maize while enhancing the gene expression of several water channel proteins (Lahiani et al., 2013). Several nanomaterials have also shown promise for seed nano-priming, which would encourage plant development and improve structural and biochemical features. These materials include metallic, biogenic metallic, and polymeric NPs. This mechanism can stimulate quick development of roots and shoots by altering the functioning of genes that affect metabolic functions, such as the generation of phytohormones. The defense system is altered by seed nano-priming, which also increases antioxidant levels and enzyme activity enhancing crops' resistance to pathogens and other both abiotic and biotic stressors in the field (Abdel Latef et al., 2016; Siddaiah et al., 2018).

Impact of Nanopriming on Salinity Stress

Accelerated seed development and stand establishment play a critical role in determining agricultural yield in adverse conditions. Therefore, it is essential to increase the resistance of cultivated crops to high salinity stress. The ability of other cultivated crops to withstand salt is improved using a variety of techniques; one of the most effective methods is seed priming, that has been proven to be an effective way to reduce salt stress (Munns & Tester, 2008b). This approach reduces the ability to withstand abiotic stress while also bolstering crops' defenses. Pre-sowing treatments are used in this method to control the level of hydration in the seeds, allowing for the occurrence of certain metabolic activities before to germination (M. Ashraf & Foolad, 2007). Increased stimulation defense processes of a plant are the term used to describe this phenomenon (El-Badri et al., 2021). The amount of time between planting and a seed's emergence is shortened by seed priming. Target seeds are altered physiologically as a result, and during their pre-germination stage, they consume both natural and synthetic compounds. It is a practical method for improving the nutritional value of seeds since certain germination metabolic procedures, such hydraulic improvements, membrane reorganization and restructuring, and the avoidance of electrolyte seepage, are involved (Vishnu D. Rajput et al., 2021; V. K. Singh et al., 2020). Additionally, it promotes protein biosynthesis, repairs cellular bio-membranes, improves antioxidant enzymes, and aids DNA replication responses (Rastogi et al., 2019a). Compared to unprimed seedlings, which contain regular quantities of nitrate reductase and nitrogen in seedlings, priming of seeds increases the biochemical state of plants by enhancing hydrolyses' activity (such as amylase) and sugar content during germination (Filippou et al., 2014). Furthermore, the role of seed priming in enhancing crop germination and development in the face of various abiotic stresses has long been recognized (Farooq et al., 2020). With increased levels of superoxide dismutase (SOD), catalase (CAT), glutathione reductase (GR), peroxidase (POD), and ascorbic acid, primed seeds have an improved antioxidant enzyme defense system that controls stress proteins like aquaporin, dehydrins (DHNs), and late embryogenesis abundant proteins Seed priming increases seedling efficiency by promoting quick and uniform germination and strong seedling development, resulting in favorable physiological circumstances that promote more rapid and stronger crop germination (Abid et al., 2020; El-Zohri et al., 2021; A. K. S. H. Mohamed et al., 2017a; Sadat-Hosseini et al., 123 C.E.; D. Sharma et al., 2021; Verma, Song, Joshi, Tian, et al., 2022).

Mode of Action of Seed Nanopriming

Priming seeds is a process in which they are partially hydrated for a set amount of time in a controlled environment before being dried and planted. Priming seeds is a highly effective and crucial strategy that equips plants with a defense mechanism and resistance to abiotic stress. Priming seeds enables metabolic processes before germination in pretreated seedlings, maintains the ideal amount of moisture before planting, and restricts radicle development (Ali et al., 2021; Naz et al., 2016). Priming facilitates pre-germination processes vital to successful germination. Priming seeds in water aids in the initiation of germination-critical metabolic processes like dormancy breaking, enzyme activation, and water imbibition (H. Singh et al., 2021; N. Singh & Bhardwaj, 2016). Seed priming supplies moisture that is suitable for germination, which is a bit lower than what is really needed for germination but is optimal for the metabolic processes that occur during germination (Ali et al., 2021; Faisal et al., 2021; Liu et al., 2022). The lag phase, imbibition, and radicle protrusion over the test a are the three stages of germination triggered by watering the seed (Ibrahim, 2016). In seed priming, water is carefully metered to each individual seed in order to achieve optimal moisture for germination. Although this concentration is high enough to kick off the physiology behind the germination process, it is not enough to push the seed through to full germination (I. A. A. Mohamed et al., 2020; Usman et al., 2020). Primed seeds are able to bypass the dormant phase III and enter the active phases II and I of growth (Rastogi et al., 2019b). Primed seeds convert their reserves (nutrients and moisture) into food for the developing embryo before germination is complete. Once favorable circumstances are present, the prepared seeds go on to the third phase of germination, which is distinguished by the protrusion of the radicle (Lutts et al., 2016). After treatment, seeds are washed and dried to the proper moisture content for long-term storage. In order to preserve their quality and prevent them from deteriorating, seeds are dried after priming (Di Girolamo and Barbanti, 2012; Bhanuprakash, 2015; Pawar and Laware, 2018; Sano et al., 2020). Once radicle emergence has begun, seed drying has a negative impact on the seed's vitality and viability; however, by obtaining starting moisture in the third stage, this may be handled without any loss. (El-Badri et al., 2021). Seeds that have been treated or primed have improved desiccation tolerance and longevity, allowing them to be stored until they are ready to be planted (Kalteh et al., 2018; Shafiq et al., 2019a; Usman et al., 2020). The methods of storing and dehydrating an item are integral to its longevity or durability. These pretreated seeds' resistance to desiccation and their gradual evaporation increase the seed's longevity, while rapid evaporation could affect the soluble carbohydrate content (Abbasi Khalaki et al., 2020). While it's helpful to store primed seeds, they don't last as long as untreated seeds do in storage, so taking extra precautions to keep them in pristine condition is essential.

Priming is not merely a simulation of the initial stages of germination that occur after imbibition. By subjecting seeds to sufficient stress, it triggers the stress-responsive system and makes them more resilient to subsequent stresses (Jisha et al., 2012). Priming the seed to germinate involved metabolic processes such as RNA and protein synthesis, structural (membrane fortification during imbibition), genetic repair, and antioxidant mechanisms. Both healthy seedling development and germination depend on these mechanisms (Saddiq et al., 2019; Zahra et al., 2021).

Mitigation Effect of Seed Nanopriming at Physiological Level

Plants use a process called osmotic adjustment to ensure they have access to enough water for growth and germination (Van Zelm et al., 2020). Osmotic adjustment and ion homeostasis are two of the most important physiological processes that are aided by seed priming (Jia et al., 2020; Zhao et al., 2012). Seed germination is aided by osmolytes after priming. These include soluble sugars, proline, sugar alcohols, glycine betaine, organic acids, and trehalose (El-Badri et al., 2021; Munns & Tester, 2008b). Primed seeds produce substances that promote germination and boost resistance to upcoming stressful situations (Jiménez et al., 1997). Under stress, these osmolytes are crucial for defense and signaling processes (M. Ashraf & Foolad, 2007; Slama et al., 2015). These solutes and inorganic ions, which reduce the capacity of solutes of the cells and produce a possible gradient of water, cause fast seed germination and accelerate metabolic activities before to germination. According to Flowers and Colmer, (2015) a plant's ability to develop a tolerance to salty environments is facilitated by a physiological adaptation: the elimination of toxic ions. Plants' enhanced salinity tolerance is attributable, in part, to the accumulation of K+ and Ca^{2+} in primed seeds, which supports the maintenance of ionic adjustment and osmotic modification (Dvořák et al., 2021). Under saline stress conditions, primed seeds increase K^+ and Ca^{2+} and decrease Na^+ and Cl accumulation, according to multiple studies (Abdel Latef et al., 2016; Mahdy et al., 2020; Mazhar et al., 2022; Rai-Kalal & Jajoo, 2021; Shah et al., 2021; Wu et al., 2019). In addition, seed priming improves plant growth by fortifying the physiological foundation (germination) in the face of extreme adversity.

Mitigation Effect of Seed Nanopriming at Biochemical Level

Seed priming in high saline stress conditions results in biochemical changes such as altered enzyme activities and induced antioxidant defense. After priming, seeds exhibit enhanced activity of cell-metabolic enzymes like esterase, acid phosphatase, and glyoxysome enzymes, as well as reserve-consuming enzymes like isocitrate lyase and amylases (Di Girolamo & Barbanti, 2012). The insertion of non-enzymatic antioxidants such soluble sugars, GSH, proline, and ASA and the activity of antioxidant-producing enzymes are also increased by priming (El-Badri et al., 2021; Gaafar et al., 2020; Kar, 2015; V. D. Rajput et al., 2015; Shafiq et al., 2019b; D. Sharma et al., 2021). The formation of ROS such as H_2O_2 and O_2 is decreased and the scavenging of ROS is improved during seed germination thanks to these enhancements (P. Sharma et al., 2012). During the germination process the metabolic reactions generated by priming seeds are crucial for reducing dormancy, enabling enzymes, building proteins and nucleic acids like DNA and RNA, transcription, and translation. These functions are also crucial for repair mechanisms, germination metabolism, DNA replication, antioxidant enzyme activities, improving ATP content and mitochondrial quantity, fixing impaired seeds parts, and decreasing metabolites leakages. Priming typically stimulates cascading changes in biochemistry, metabolism, and physiology that boost a plant's ability to grow and germinate healthily, as well as increase its tolerance to abiotic stresses (Hosseinpour et al., 2020; Ibrahim, 2016; C. K. Lee et al., 2013). The advantages of seed priming over typical imbibition include upregulation of proteins and genes responsible for cell proliferation and division, translation, reserve mobilization, cell wall alterations, transcription, transport of water, stress oxidative response, membrane and DNA repair, and other processes (D. Sharma et al., 2021). According to several studies the antioxidant system is the most important defense mechanism in germinating primed seeds, as it controls the production of ROS when plants are subjected to abiotic stress (A. K. S. H. Mohamed et al., 2017b; Tanou et al., 2009; Verma, Song, Joshi, Tian, et al., 2022). Antioxidant enzymes like superoxide dismutase (SOD), glutathione reductase (GR), peroxidase (POX), and catalase (CAT) are all stimulated by seed priming, and levels of (i.e., ascorbic acid). Seed protection and the suppression of ROS are the results of these processes. Also, it promotes the incorporation of ROS-scavenging and crop-protecting photoprotective pigments (anthocyanin) (Jia et al., 2020). The insertion of MDA as an indicator of oxidative damage reduces with seed priming and rises with abiotic stress. Seed priming is associated with this decrease because of enhanced membrane repairing of antioxidant enzymes inductive reactions (Zulfiqar & Ashraf, 2021).

Mitigation Effect Seed Nanopriming at Molecular Level

Priming seeds improves germination consistency and efficiency by restoring metabolic function during osmotic adjustment, water uptake, and stimulation of metabolic development prior to germination. The synthesis of DNA, early mobilization of resources, and duplication repair mechanisms are all affected (Läuchli & Grattan, 2007). On the other hand, priming promotes sprouting activities like reserve mobilization, respiration, metabolizing energy, promoting the growth of embryonic cells, and weakening of the endosperm (Chen et al., 2005; Das & Roychoudhury, 2014; M. Hussain et al., 2018; Sadat-Hosseini et al., 123 C.E.).

A number of mechanisms exist for how seed priming improves germination, including the repair of DNA pathways, control of catalase and antioxidant enzymes, production of proteins and nucleic acids, and accumulation of steroids and phospholipids (Ozgur et al., 2013). Furthermore, it has been proposed that good seedling development from primed seeds is influenced by abiotic stress-responsive proteins, enhanced plasma membrane mobility, cell division, and elongation. In a similar vein, the priming of seeds can boost the activity of defense mechanisms that deal with abiotic stress. Increased HSPs, more protective proteins, and a burst of membrane efflux pumps are all results (Zhu, 2016). The enhanced protein synthesis and post-translational changes enhance abiotic stress resistance (Zhu, 2016).

Increased stress tolerance in germinating seeds is what causes the cross-tolerance manifestation caused by the priming process (De Villiers et al., 1996; Gapper & Dolan, 2006). Seed priming, which results in resistance to abiotic stress, is probably achieved in two ways. Priming is the first step in transforming dormant seeds into a germination-ready state (De Villiers et al., 1996; Gapper & Dolan, 2006). This is accomplished by stimulating the pre-germination metabolism processes that are accountable for transforming dormant seeds into a germination-ready state. In contrast to preventing radicle expansion, priming seeds with abiotic stressors encourages enzymatic stimulation, physiological responses to stress, osmotic regulation, and resistance to a variety of abiotic stresses. Some methods of stress tolerance create a "priming memory" in seeds (primed seeds) that can be used after the removal of stress to promote further tolerance (Mahakham et al., 2017).

The process of seed priming causes an upregulation of protein associated with cellular oxidative stress resilience, dormancy barrier elimination, membrane reconstruction directed the proteolysis process and post-translation processing ability, embryo development, tissue remodeling, and germination improvement (Kubala et al., 2015b; Chen et al., 2021). Gene expression leading to new RNAs and proteins as well as the DNA repair mechanism are examples of molecular responses triggered by the priming process (Varier et al., 2010). Maintaining adequate DNA repair rates within the embryo is necessary for successful germination. Imbibition progresses the growth of embryonic cells by replicating DNA and stimulates their cell cycle, increasing germination yield. However, faulty repair of DNA causes a large rise in cell mortality. Additionally, DNA repair is activated during seed priming, making it a key process in regulating any priming benefits. Genes involved in germination, plant growth, and antioxidant production are some of the many that can be influenced by seed priming to improve germination.

NP Accumulation and Nano-Phytotoxicity

Different phytotoxicity reports show that NPs can have impacts on root extension, proliferation of cells, growth, and metabolic processes can be both beneficial and detrimental. Gene transcription and translation, DNA repair, respiration, and energy metabolism are just a few of the many processes that occur during germination (C. W. Lee et al., 2010; P. Wang et al., 2016). Mitotic index, which measures how often cells divide, seems to agree with the rate at which roots expand. DNA double-strand breaks caused by micronuclei formation are believed to have a true mutagenesis impact, which might harm genetic information and cause spindle fiber malformation (Maity et al., 2020). Micronucleus formation, chromosome disruption, chromosome fragmentation, chromosome stickiness, chromosome laggardism, and diminution in the mitotic rate are the most noticeable abnormalities in plants exposed to Ag, Cu, TiO_2, Zn, ZnO, SeO, MWCNTs, tetra-methyl-ammonium-hydroxide (TMAH), and bismuth oxide (Bi_2O_3) NPs. The degree of abnormality varies with concentration, nanoparticle size, and length of exposure. Priming studies' cytotoxic and genotoxic effects are dependent on the concentration of NPs used, and this concentration varies between plant species. The discovery shows the fact that NPs with sizes in the range of 8 to 10 nm may reach the nucleus via nuclear pores demonstrates that there is a size-dependent interaction between NPs and cell components (S et al., 2018; Y. Singh et al., 2017; W. N. Wang et al., 2013). The interaction of the NPs with proteins may block proteins and cause flaws in cell division pathways, contact with antioxidant enzymes, or stimulation of ROS generation by cellular constituents. Antioxidant levels drop and gene expression shifts when DNA repair is interrupted or slowed down, and when oxidative stress rises due to ROS generated during cell organelle interaction (cell membrane, mitochondria). NPs toxicity results from three separate processes (M. N. Khan et al., 2017). To begin, NPs release their toxic materials into the contact media. DNA damage could be caused, for instance, by free Ag^+ ions released from Ag-NPs or by other toxic ions produced by soluble NPs. Through the Fenton-type reaction, transitional metal ions including Ag^+, Cu^+, Fe^{2+}, Cr^{5+}, Ni^{2+}, and Mn^{2+} can attach to DNA base pairs and contribute to the production of ROS (Bruce et al., 2007).

There have been numerous reports on cytotoxicity in mammals and bacteria, but much less on the role of NPs as a contributor to genotoxicity in plants. It has been found that exposure to Ag NPs causes cellular alterations at the ends of the roots of germinated barley and wheat seedlings (Vannini et al., 2014). The pretreatment of seeds with Ag NPs caused a variety of chromosomal abnormalities during metaphase and anaphase such as chromosomal loss, malformed nuclei, micronuclei, sticky chromosomes, binucleate cells, and aneuploidy. In addition, the mitotic index of NPs-exposed plants was significantly higher than that of controls. Mitotic abnormalities like nuclear dissolution, chromosomal disintegration, pole deformation during anaphase, trailing chromosome, chromosome separation, and C-metaphase are more common when exposed to Ag NPs because of their impacts on chromosomal configuration alterations and mitotic spindle actions at different cell cycle phases. When Ag NPs interact with the highly reactive tubulin-SH group, they disrupt the mitotic spindle's normal function (Kuriyama & Sakai, 1974). Wheat and barley plants treated with silver nano-priming performed better than control plants in terms of germinating seeds, length of shoots, dry and fresh biomass, and the related root characteristics. The effects of Ag NPs priming include aneuploid chromosomes, binucleate cells, deletion chromosomes, malformed nuclei, micronuclei, chromosomal breakage, and chromosome adhesion. This suggests that Ag NPs can enter plants and cause cellular damage and chromosomal disruptions (Abou-Zeid & Moustafa, 2014). DNA damage may have been induced when nano-Ag entered the cell or it may have resulted from the halting of DNA synthesis during the S-phase (Madhumathi et al., 2010). Deformed nuclei and structural chromosomal aberrations were found in all three plants exposed to Ag NPs. Onion root tips exposed to Ag and ZnO NPs had their cell division phases disrupted, metaphase interrupted, multiple chromosomes broken, and cells disintegrated (Ahmed et al., 2017). NPs regulate cell cycle processes including replication of DNA and division of cells, as well as may interact with protein kinases and their families to change how they operate in signal transduction (Magdolenova et al., 2014). The mitotic checkpoint protein PLK1 and its function were disrupted by TiO_2 NPs. Mitosis, including the formation of the contractile ring and cytokinesis, is regulated and controlled by these proteins.

Future Prospects

In the future, evaluating the capabilities of different crops under varying saline and environmental conditions will be of utmost importance. Researchers will continue to explore and investigate the use of different seed priming chemicals to enhance seed germination and growth in saline-stressed environments. They will investigate how diverse priming techniques, including osmo-priming, bio-priming, hydro-priming, halo-priming, hormonal priming, as well as different priming chemicals and their ratios, affect different plants grown in varying salt environments. Scientists will prioritize the rapid growth of seeds and crop production to achieve global food security goals, especially under adverse environmental conditions. Researchers will work to create cutting-edge seed priming techniques, such as using NPs or physical factors like microwaves, X-rays, gamma rays, or UV radiation. The ultimate goal of these techniques is to raise the total production of various crops while reducing the detrimental effects of salt stress. Crop-specific research will play a crucial role, as different crops may respond differently to seed priming techniques and salinity stress. Therefore, scientists will conduct tailored studies to understand the interactions between priming methods, priming agents, salinity conditions, and crop performance. By doing so, they will develop customized strategies to enhance the germination of seeds, growth, and total yield in areas subject to salt stress. By investing in these future prospects, researchers and scientists will contribute to ensuring food security for humanity. Their efforts will focus on optimizing seed priming techniques, exploring novel priming agents, and understanding crop-specific responses. This knowledge will empower the agricultural sector to overcome the challenges posed by saline-stressed environments and foster sustainable and productive agriculture in the years to come.

CONCLUSION

The use of seed nanopriming shows promise as a strategy to enhance seed germination, plant growth, and yield. The application of nanoform micronutrients at the seed level can lead to morphological, physiological, and biochemical changes in plants, improving their overall performance. However, further research is necessary to fully understand the molecular mechanisms behind these effects and identify the specific genes that are activated during nanopriming.

While Ag nanopriming has demonstrated positive effects on wheat and barley seeds, it is important to consider the potential risks associated with nanoparticle exposure. Studies have shown that Ag NPs can enter plant cells and disrupt chromosomes, causing various chromosomal abnormalities and cellular damage. NPs, including Ag and ZnO, have been found to affect cell division phases, leading to halted metaphase, chromosome breakage, and cell disintegration. Moreover, NPs have the ability to interact with and modify the function of protein kinases involved in signal transduction. This interaction can influence critical cell cycle events, such as DNA replication and cell division. For example, TiO_2NPs have been found to inhibit the function of proteins like PLK1, which regulate mitosis and cytokinesis. To fully harness the potential benefits of nanopriming while minimizing potential risks, it is crucial to conduct further research on nanotoxicity and long-term effects of nanoparticle exposure. This will ensure the development of safe and sustainable nanopriming techniques that can effectively enhance plant performance without compromising environmental and human health.

Summary

In summary, seed nanopriming holds great promise as a technique for improving seed germination, plant growth, and yield. However, a comprehensive understanding of the underlying molecular mechanisms, the specific genes involved, and the potential risks associated with nanoparticle exposure is necessary for its successful implementation. Continued research in these areas will contribute to the development of safe and effective nanopriming strategies for sustainable agriculture.

ACKNOWLEDGEMENT

The research was supported by the Strategic Academic Leadership Program of the Southern Federal University ("Priority 2030") and the Science Committee of RA, in the frames of the research project No 21AG-4C075.The work was supported by Science Committee of RA, research project № 21AG-4C075.

REFERENCES

Abbasi Khalaki, M., Moameri, M., Asgari Lajayer, B., & Astatkie, T. (2020). Influence of nano-priming on seed germination and plant growth of forage and medicinal plants. *Plant Growth Regulation, 93*(1), 13–28. doi:10.1007/s10725-020-00670-9

Abdel Latef, A. A. H., Abu Alhmad, M. F., & Abdelfattah, K. E. (2016). The Possible Roles of Priming with ZnO NPs in Mitigation of Salinity Stress in Lupine (Lupinus termis) Plants. *Journal of Plant Growth Regulation, 36*(1), 60–70. doi:10.1007/s00344-016-9618-x

AbdElgawad, H., Zinta, G., Hegab, M. M., Pandey, R., Asard, H., & Abuelsoud, W. (2016). High salinity induces different oxidative stress and antioxidant responses in maize seedlings organs. *Frontiers in Plant Science, 7*(MAR2016), 276. https://doi.org/ doi:10.3389/FPLS.2016.00276/BIBTEX

Abid, M., Zhang, Y. J., Li, Z., Bai, D. F., Zhong, Y. P., & Fang, J. B. (2020). Effect of Salt stress on growth, physiological and biochemical characters of Four kiwifruit genotypes. *Scientia Horticulturae, 271*, 109473. doi:10.1016/j.scienta.2020.109473

Abou-Zeid, H. M., & Moustafa, Y. (2014). *Physiological and Cytogenetic Responses of Wheat and Barley to Silver Nanopriming Treatment.* IMSEAR. https://imsear.searo.who.int/handle/123456789/168212

Adhikary, S., Biswas, B., Chakraborty, D., Timsina, J., Pal, S., Chandra Tarafdar, J., Banerjee, S., Hossain, A., & Roy, S. (2022). Seed priming with selenium and zinc NPs modifies germination, growth, and yield of direct-seeded rice (Oryza sativa L.). *Scientific Reports, 12*(1), 1–14. doi:10.1038/s41598-022-11307-4

Ahmed, B., Dwivedi, S., Abdin, M. Z., Azam, A., Al-Shaeri, M., Khan, M. S., Saquib, Q., Al-Khedhairy, A. A., & Musarrat, J. (2017). Mitochondrial and Chromosomal Damage Induced by Oxidative Stress in Zn2+ Ions, ZnO-Bulk and ZnO-NPs treated Allium cepa roots. *Scientific Reports*, *7*(1), 40685. Advance online publication. doi:10.1038rep40685 PMID:28120857

Ali, S., Mehmood, A., & Khan, N. (2021). Uptake, Translocation, and Consequences of Nanomaterials on Plant Growth and Stress Adaptation. *Journal of Nanomaterials*, *2021*, 1–17. doi:10.1155/2021/6677616

Ashraf, M., & Foolad, M. R. (2007). Roles of glycine betaine and proline in improving plant abiotic stress resistance. *Environmental and Experimental Botany*, *59*(2), 206–216. doi:10.1016/j.envexpbot.2005.12.006

Ashraf, S. A., Siddiqui, A. J., Elkhalifa, A. E. O., Khan, M. I., Patel, M., Alreshidi, M., Moin, A., Singh, R., Snoussi, M., & Adnan, M. (2021). Innovations in nanoscience for the sustainable development of food and agriculture with implications on health and environment. *The Science of the Total Environment*, *768*, 144990. doi:10.1016/j.scitotenv.2021.144990 PMID:33736303

Bruce, T. J. A., Matthes, M. C., Napier, J. A., & Pickett, J. A. (2007). Stressful "memories" of plants: Evidence and possible mechanisms. *Plant Science*, *173*(6), 603–608. doi:10.1016/j.plantsci.2007.09.002

Chen, Z., Newman, I., Zhou, M., Mendham, N., Zhang, G., & Shabala, S. (2005). Screening plants for salt tolerance by measuring K+ flux: A case study for barley. *Plant, Cell & Environment*, *28*(10), 1230–1246. doi:10.1111/j.1365-3040.2005.01364.x

Das, K., & Roychoudhury, A. (2014). Reactive oxygen species (ROS) and response of antioxidants as ROS-scavengers during environmental stress in plants. *Frontiers in Environmental Science*, *2*(DEC), 53. doi:10.3389/fenvs.2014.00053

De La Torre-Roche, R., Cantu, J., Tamez, C., Zuverza-Mena, N., Hamdi, H., Adisa, I. O., Elmer, W., Gardea-Torresdey, J., & White, J. C. (2020). Seed Biofortification by Engineered Nanomaterials: A Pathway to Alleviate Malnutrition? *Journal of Agricultural and Food Chemistry*, *68*(44), 12189–12202. doi:10.1021/acs.jafc.0c04881 PMID:33085897

De Villiers, A. J., Von Teichman, I., Van Rooyen, M. W., & Theron, G. K. (1996). Salinity-induced changes in anatomy, stomatal counts and photosynthetic rate of Atriplex semibaccata R. Br. *South African Journal of Botany*, *62*(5), 270–276. doi:10.1016/S0254-6299(15)30656-6

Di Girolamo, G., & Barbanti, L. (2012). Treatment conditions and biochemical processes influencing seed priming effectiveness. *Italian Journal of Agronomy*, *7*(2), e25–e25. doi:10.4081/ija.2012.e25

Dvořák, P., Krasylenko, Y., Zeiner, A., Šamaj, J., & Takáč, T. (2021). Signaling Toward Reactive Oxygen Species-Scavenging Enzymes in Plants. *Frontiers in Plant Science*, *11*, 2178. doi:10.3389/fpls.2020.618835 PMID:33597960

El-Badri, A. M., Batool, M., Mohamed, I. A. A., Wang, Z., Khatab, A., Sherif, A., Ahmad, H., Khan, M. N., Hassan, H. M., Elrewainy, I. M., Kuai, J., Zhou, G., & Wang, B. (2021). Antioxidative and metabolic contribution to salinity stress responses in two rapeseed cultivars during the early seedling stage. *Antioxidants*, *10*(8). doi:10.3390/antiox10081227 PMID:34439475

El-Zohri, M., Al-Wadaani, N. A., & Bafeel, S. O. (2021). Foliar sprayed green zinc oxide NPs mitigate drought-induced oxidative stress in tomato. *Plants*, *10*(11), 2400. doi:10.3390/plants10112400 PMID:34834763

Elmer, W., Ma, C., & White, J. (2018). NPs for plant disease management. *Current Opinion in Environmental Science & Health*, *6*, 66–70. doi:10.1016/j.coesh.2018.08.002

Faisal, S., Jan, H., Shah, S. A., Shah, S., Khan, A., Akbar, M. T., Rizwan, M., Jan, F., Wajidullah, Akhtar, N., Khattak, A., & Syed, S. (2021). Green Synthesis of Zinc Oxide (ZnO) NPs Using Aqueous Fruit Extracts of Myristica fragrans: Their Characterizations and Biological and Environmental Applications. *ACS Omega*, *6*(14), 9709–9722. doi:10.1021/acsomega.1c00310 PMID:33869951

FAO WFP, & IFAD. (2012). The State of Food Insecurity in the World 2012. *Economic Growth Is Necessary but Not Sufficient to Accelerate Reduction of Hunger and Malnutrition*. FAO.

Farooq, M., Hussain, M., Habib, M. M., Khan, M. S., Ahmad, I., Farooq, S., Siddique, K. H. M., Farooq, M., Hussain, M., Habib, M. M., Khan, M. S., Ahmad, I., Farooq, S., & Siddique, K. H. M. (2020). Influence of seed priming techniques on grain yield and economic returns of bread wheat planted at different spacings. *Crop & Pasture Science*, *71*(8), 725–738. doi:10.1071/CP20065

Filippou, P., Bouchagier, P., Skotti, E., & Fotopoulos, V. (2014). Proline and reactive oxygen/nitrogen species metabolism is involved in the tolerant response of the invasive plant species Ailanthus altissima to drought and salinity. *Environmental and Experimental Botany*, *97*, 1–10. doi:10.1016/j.envexpbot.2013.09.010

Flowers, T. J., & Colmer, T. D. (2015). Plant salt tolerance: Adaptations in halophytes. *Annals of Botany*, *115*(3), 327–331. doi:10.1093/aob/mcu267 PMID:25844430

Gaafar, R. M., Diab, R. H., Halawa, M. L., El-Shanshory, A. R., El-Shaer, A., & Hamouda, M. M. (2020). Role of Zinc Oxide NPs in Ameliorating Salt Tolerance in Soybean. *Egyptian Journal of Botany*, *60*(3), 733–747. doi:10.21608/ejbo.2020.26415.1475

Gapper, C., & Dolan, L. (2006). Control of Plant Development by Reactive Oxygen Species. *Plant Physiology*, *141*(2), 341–345. doi:10.1104/pp.106.079079 PMID:16760485

Hasanuzzaman, M., Borhannuddin Bhuyan, M. H. M., Anee, T. I., Parvin, K., Nahar, K., Al Mahmud, J., & Fujita, M. (2019). Regulation of Ascorbate-Glutathione Pathway in Mitigating Oxidative Damage in Plants under Abiotic Stress. *Antioxidants, 8*(9), 384. doi:10.3390/antiox8090384

He, X., Deng, H., & Hwang, H. (2019). The current application of nanotechnology in food and agriculture. *Yao Wu Shi Pin Fen Xi, 27*(1), 1–21. doi.org/10.1016/j.jfda.2018.12.002 PMID:30648562

Hosseinpour, A., Haliloglu, K., Cinisli, K. T., Ozkan, G., Ozturk, H. I., Pour-Aboughadareh, A., & Poczai, P. (2020). Application of Zinc Oxide NPs and Plant Growth Promoting Bacteria Reduces Genetic Impairment under Salt Stress in Tomato (Solanum lycopersicum L. 'Linda'). *Agriculture, 10*(11), 521. doi:10.3390/agriculture10110521

Hussain, A., Ali, S., Rizwan, M., Zia ur Rehman, M., Javed, M. R., Imran, M., Chatha, S. A. S., & Nazir, R. (2018). Zinc oxide NPs alter the wheat physiological response and reduce the cadmium uptake by plants. *Environmental Pollution*, *242*, 1518–1526. doi:10.1016/j.envpol.2018.08.036 PMID:30144725

Hussain, M., Ahmad, S., Hussain, S., Lal, R., Ul-Allah, S., & Nawaz, A. (2018). Rice in Saline Soils: Physiology, Biochemistry, Genetics, and Management. *Advances in Agronomy*, *148*, 231–287. doi:10.1016/bs.agron.2017.11.002

Ibrahim, E. A. (2016). Seed priming to alleviate salinity stress in germinating seeds. *Journal of Plant Physiology*, *192*, 38–46. doi:10.1016/j.jplph.2015.12.011 PMID:26812088

Jia, J., Bai, J., Wang, W., Yin, S., Zhang, G., Zhao, Q., Wang, X., Liu, X., & Cui, B. (2020). Salt stress alters the short-term responses of nitrous oxide emissions to the nitrogen addition in salt-affected coastal soils. *The Science of the Total Environment*, *742*, 140124. doi:10.1016/j.scitotenv.2020.140124 PMID:32623179

Jiménez, L., Martínez, C., Pérez, I., & López, F. (1997). Biobleaching procedures for pulp from agricultural residues using Phanerochaete chrysosporium and enzymes. *Process Biochemistry (Barking, London, England)*, *32*(4), 297–304. doi:10.1016/S0032-9592(96)00090-8

Jisha, K. C., Vijayakumari, K., & Puthur, J. T. (2012). Seed priming for abiotic stress tolerance: an overview. *Acta Physiologiae Plantarum*, *35*(5), 1381–1396. doi:10.1007/s11738-012-1186-5

Kah, M., Tufenkji, N., & White, J. C. (2019). Nano-enabled strategies to enhance crop nutrition and protection. *Nature Nanotechnology*, *14*(6), 532–540. doi:10.1038/s41565-019-0439-5

Kaivosoja, J., Pesonen, L., Kleemola, J., Pölönen, I., Salo, H., Honkavaara, E., Saari, H., Mäkynen, J., & Rajala, A. (2013). A case study of a precision fertilizer application task generation for wheat based on classified hyperspectral data from UAV combined with farm history data. *Remote Sensing for Agriculture, Ecosystems, and Hydrology XV*, *8887*, 88870H. doi:10.1117/12.2029165

Kalteh, M., Alipour, Z. T., Ashraf, S., Marashi Aliabadi, M., & Falah Nosratabadi, A. (2018). Effect of silica NPs on Basil (Ocimum basilicum) Under Salinity Stress. *Journal of Chemical Health Risks*, *4*(3), 49–55. doi:10.22034/JCHR.2018.544075

Kandil, A.A Sharief, A.E., & Ahmed, S. (2012, December 20). Sharief. (2012). Germination and Seedling Growth of Some Chickpea Cultivars (Cicer arietinum L.) under Salinity Stress. *Journal of Basic and Applied Sciences*, *8*(2), 561–571. doi:10.6000/1927-5129.2012.08.02.49

Kar, R. K. (2015). ROS signaling: Relevance with site of production and metabolism of ROS. *Reactive Oxygen Species and Oxidative Damage in Plants Under Stress*, 115–125. https://doi.org/ doi:10.1007/978-3-319-20421-5_5/COVER

Khan, M. N., Mobin, M., Abbas, Z. K., AlMutairi, K. A., & Siddiqui, Z. H. (2017). Role of nanomaterials in plants under challenging environments. *Plant Physiology and Biochemistry*, *110*, 194–209. doi:10.1016/j.plaphy.2016.05.038 PMID:27269705

Khan, M. R., Adam, V., Rizvi, T. F., Zhang, B., Ahamad, F., Jośko, I., Zhu, Y., Yang, M., & Mao, C. (2019). Nanoparticle–Plant Interactions: Two-Way Traffic. *Small*, *15*(37), 1901794. doi:10.1002mll.201901794 PMID:31318142

Khodakovskaya, M. V., Kim, B. S., Kim, J. N., Alimohammadi, M., Dervishi, E., Mustafa, T., & Cernigla, C. E. (2013). Carbon Nanotubes as Plant Growth Regulators: Effects on Tomato Growth, Reproductive System, and Soil Microbial Community. *Small*, *9*(1), 115–123. doi:10.1002mll.201201225 PMID:23019062

Kuriyama, R., & Sakai, H. (1974). Role of tubulin-SH groups in polymerization to microtubules. Functional-SH groups in tubulin for polymerization. *Journal of Biochemistry*, *76*(3), 651–654. doi:10.1093/oxfordjournals.jbchem.a130609 PMID:4474165

Lahiani, M. H., Dervishi, E., Chen, J., Nima, Z., Gaume, A., Biris, A. S., & Khodakovskaya, M. V. (2013). Impact of carbon nanotube exposure to seeds of valuable crops. *ACS Applied Materials & Interfaces*, *5*(16), 7965–7973. doi:10.1021/am402052x PMID:23834323

Läuchli, A., & Grattan, S. R. (2007). Plant growth and development under salinity stress. *Advances in Molecular Breeding Toward Drought and Salt Tolerant Crops*, 1–32. https://doi.org/ doi:10.1007/978-1-4020-5578-2_1/COVER

Lee, C. K., Ibrahim, D., & Che Omar, I. (2013). Enzymatic deinking of various types of waste paper: Efficiency and characteristics. *Process Biochemistry (Barking, London, England)*, *48*(2), 299–305. doi:10.1016/j.procbio.2012.12.015

Lee, C. W., Mahendra, S., Zodrow, K., Li, D., Tsai, Y. C., Braam, J., & Alvarez, P. J. J. (2010). Developmental phytotoxicity of metal oxide NPs to Arabidopsis thaliana. *Environmental Toxicology and Chemistry*, *29*(3), 669–675. doi:10.1002/etc.58 PMID:20821493

Liu, Y., Cao, X., Yue, L., Wang, C., Tao, M., Wang, Z., & Xing, B. (2022). Foliar-applied cerium oxide nanomaterials improve maize yield under salinity stress: Reactive oxygen species homeostasis and rhizobacteria regulation. *Environmental Pollution*, *299*, 118900. doi:10.1016/j.envpol.2022.118900 PMID:35085650

Lutts, S., Benincasa, P., Wojtyla, L., S, S. K., Pace, R., Lechowska, K., Quinet, M., & Malgorzata, G. (2016). Seed Priming: New Comprehensive Approaches for an Old Empirical Technique. *New Challenges in Seed Biology - Basic and Translational Research Driving Seed Technology*. doi:10.5772/64420

Madhumathi, K., Sudheesh Kumar, P. T., Abhilash, S., Sreeja, V., Tamura, H., Manzoor, K., Nair, S. V., & Jayakumar, R. (2010). Development of novel chitin/nanosilver composite scaffolds for wound dressing applications. *Journal of Materials Science. Materials in Medicine*, *21*(2), 807–813. doi:10.100710856-009-3877-z PMID:19802687

Magdolenova, Z., Collins, A., Kumar, A., Dhawan, A., Stone, V., & Dusinska, M. (2014). Mechanisms of genotoxicity. A review of in vitro and in vivo studies with engineered NPs. *Nanotoxicology*, *8*(3), 233–278. doi:10.3109/17435390.2013.773464 PMID:23379603

Mahakham, W., Sarmah, A. K., Maensiri, S., & Theerakulpisut, P. (2017). Nanopriming technology for enhancing germination and starch metabolism of aged rice seeds using phytosynthesized silver NPs. *Scientific Reports, 7*(1), 1–21. doi:10.1038/s41598-017-08669-5

Mahdy, A. M., Sherif, F. K., Elkhatib, E. A., Fathi, N. O., & Ahmed, M. H. (2020). Seed priming in NPs of water treatment residual can increase the germination and growth of cucumber seedling under salinity stress. *Journal of Plant Nutrition, 43*(12), 1862–1874. doi:10.1080/01904167.2020.1750647

Maity, S., Chatterjee, A., Guchhait, R., De, S., & Pramanick, K. (2020). Cytogenotoxic potential of a hazardous material, polystyrene microparticles on Allium cepa L. *Journal of Hazardous Materials, 385*, 121560. doi:10.1016/j.jhazmat.2019.121560 PMID:31732349

Manjunatha, S. B., Biradar, D. P., & Aladakatti, Y. R. (2016). Nanotechnology and its applications in agriculture: A review. *Journal of Farm Sciences, 29*(1), 1–13.

Mazhar, M. W., Ishtiaq, M., Hussain, I., Parveen, A., Bhatti, K. H., Azeem, M., Thind, S., Ajaib, M., Maqbool, M., Sardar, T., Muzammil, K., & Nasir, N. (2022). Seed nano-priming with Zinc Oxide NPs in rice mitigates drought and enhances agronomic profile. *PLoS One, 17*(3), e0264967. doi:10.1371/journal.pone.0264967 PMID:35324949

Mohamed, A. K. S. H., Qayyum, M. F., Abdel-Hadi, A. M., Rehman, R. A., Ali, S., & Rizwan, M. (2017a). *Interactive effect of salinity and silver NPs on photosynthetic and biochemical parameters of wheat.* Taylor & Francis. doi:10.1080/03650340.2017.1300256

Mohamed, A. K. S. H., Qayyum, M. F., Abdel-Hadi, A. M., Rehman, R. A., Ali, S., & Rizwan, M. (2017b). Interactive effect of salinity and silver NPs on photosynthetic and biochemical parameters of wheat. Taylor & Francis. doi:10.1080/03650340.2017.1300256

Mohamed, I. A. A., Shalby, N., El-Badri, A. M. A., Saleem, M. H., Khan, M. N., Nawaz, M. A., Qin, M., Agami, R. A., Kuai, J., Wang, B., & Zhou, G. (2020). Stomata and Xylem Vessels Traits Improved by Melatonin Application Contribute to Enhancing Salt Tolerance and Fatty Acid Composition of Brassica napus L. Plants. *Agronomy 2020, Vol. 10, Page 1186, 10*(8), 1186. doi:10.3390/agronomy10081186

Munns, R. (2005). Genes and salt tolerance: Bringing them together. *The New Phytologist, 167*(3), 645–663. doi:10.1111/j.1469-8137.2005.01487.x PMID:16101905

Munns, R., & Tester, M. (2008a). Mechanisms of salinity tolerance. *Annual Review of Plant Biology*, *59*(1), 651–681. doi:10.1146/annurev.arplant.59.032607.092911 PMID:18444910

Munns, R., & Tester, M. (2008b). Mechanisms of Salinity Tolerance. *Annual Reviews*, *59*, 651–681. doi:10.1146/annurev.arplant.59.032607.092911

Naz, H., Akram, N. A., & Ashraf, M. (2016). Impact of ascorbic acid on growth and some physiological attributes of cucumber (Cucumis sativus) plants under water-deficit conditions. *Pakistan Journal of Botany*, *48*(3), 877–883.

Ozgur, R., Uzilday, B., Sekmen, A. H., Turkan, I., Ozgur, R., Uzilday, B., Sekmen, A. H., & Turkan, I. (2013). Reactive oxygen species regulation and antioxidant defence in halophytes. *Functional Plant Biology*, *40*(9), 832–847. doi:10.1071/FP12389 PMID:32481154

Parihar, P., Singh, S., Singh, R., Singh, V. P., & Prasad, S. M. (2015). Effect of salinity stress on plants and its tolerance strategies: A review. *Environmental Science and Pollution Research International*, *22*(6), 4056–4075. doi:10.100711356-014-3739-1 PMID:25398215

Rahman, A., Mostofa, M. G., Alam, M. M., Nahar, K., Hasanuzzaman, M., & Fujita, M. (2015). Calcium Mitigates Arsenic Toxicity in Rice Seedlings by Reducing Arsenic Uptake and Modulating the Antioxidant Defense and Glyoxalase Systems and Stress Markers. *BioMed Research International*, *2015*, 1–12. doi:10.1155/2015/340812 PMID:26798635

Rai-Kalal, P., & Jajoo, A. (2021). Priming with zinc oxide NPs improve germination and photosynthetic performance in wheat. *Plant Physiology and Biochemistry*, *160*, 341–351. doi:10.1016/j.plaphy.2021.01.032 PMID:33548801

Rajput, V., Minkina, T., Mazarji, M., Shende, S., Sushkova, S., Mandzhieva, S., Burachevskaya, M., Chaplygin, V., Singh, A., & Jatav, H. (2020). Accumulation of NPs in the soil-plant systems and their effects on human health. *Annals of Agricultural Science*, *65*(2), 137–143. doi:10.1016/j.aoas.2020.08.001

Rajput, Vishnu D., Minkina, T., Kumari, A., Harish, Singh, V. K., Verma, K. K., Mandzhieva, S., Sushkova, S., Srivastava, S., & Keswani, C. (2021). Coping with the Challenges of Abiotic Stress in Plants: New Dimensions in the Field Application of NPs. *Plants 2021*, *10*(6), 1221. doi:10.3390/plants10061221

Rajput, V. D., Chen, Y., & Ayup, M. (2015). Effects of high salinity on physiological and anatomical indices in the early stages of Populus euphratica growth. *Russian Journal of Plant Physiology*, *62*(2), 229–236. doi:10.1134/S1021443715020168

Rastogi, A., Tripathi, D. K., Yadav, S., Chauhan, D. K., Živčák, M., Ghorbanpour, M., El-Sheery, N. I., & Brestic, M. (2019a). Application of silicon NPs in agriculture. *3 Biotech, 9*(3). doi:10.1007/s13205-019-1626-7

Rastogi, A., Tripathi, D. K., Yadav, S., Chauhan, D. K., Živčák, M., Ghorbanpour, M., El-Sheery, N. I., & Brestic, M. (2019b). Application of silicon NPs in agriculture. *3 Biotech, 9*(3). doi:10.1007/s13205-019-1626-7

S, T., SK, T., & R, K. (2018). Bio-Nanotechnology and its Role in Agriculture and Food Industry. *Molecular and Genetic Medicine, 12*(1), 1–5. doi:10.4172/1747-0862.1000324

Sadat-Hosseini, M., Naeimi, A., Boroomand, N., Aalifar, M., Farajpour, M., & Sari, I. (2022, October 28). (123 C.E.). Alleviating the adverse effects of salinity on Roselle plants by green synthesized NPs. *Scientific Reports, 12*(1), 18165. doi:10.103841598-022-22903-9 PMID:36307527

Saddiq, M. S., Iqbal, S., Afzal, I., Ibrahim, A. M. H., Bakhtavar, M. A., & Hafeez, M. B. Jahanzaib, & Maqbool, M. M. (2019). *Mitigation of salinity stress in wheat (Triticum aestivum L.) seedlings through physiological seed enhancements.* Taylor & Francis. doi:10.1080/01904167.2019.1609509

Scott, N. R., Chen, H., & Cui, H. (2018). Nanotechnology Applications and Implications of Agrochemicals toward Sustainable Agriculture and Food Systems. *Journal of Agricultural and Food Chemistry*, *66*(26), 6451–6456. doi:10.1021/acs.jafc.8b00964 PMID:29879842

Shafiq, F., Iqbal, M., Ali, M., & Ashraf, M. A. (2019a). Seed Pre-treatment with Polyhydroxy Fullerene NPs Confer Salt Tolerance in Wheat Through Upregulation of H2O2 Neutralizing Enzymes and Phosphorus Uptake. *Journal of Soil Science and Plant Nutrition*, *19*(4), 734–742. doi:10.100742729-019-00073-4

Shafiq, F., Iqbal, M., Ali, M., & Ashraf, M. A. (2019b). Seed Pre-treatment with Polyhydroxy Fullerene NPs Confer Salt Tolerance in Wheat Through Upregulation of H2O2 Neutralizing Enzymes and Phosphorus Uptake. *Journal of Soil Science and Plant Nutrition*, *19*(4), 734–742. doi:10.100742729-019-00073-4

Shah, T., Latif, S., Saeed, F., Ali, I., Ullah, S., Abdullah Alsahli, A., Jan, S., & Ahmad, P. (2021). Seed priming with titanium dioxide NPs enhances seed vigor, leaf water status, and antioxidant enzyme activities in maize (Zea mays L.) under salinity stress. *Journal of King Saud University. Science*, *33*(1), 101207. doi:10.1016/j.jksus.2020.10.004

Sharma, D., Afzal, S., & Singh, N. K. (2021). Nanopriming with phytosynthesized zinc oxide NPs for promoting germination and starch metabolism in rice seeds. *Journal of Biotechnology*, *336*, 64–75. doi:10.1016/j.jbiotec.2021.06.014 PMID:34116127

Sharma, P., Jha, A. B., Dubey, R. S., & Pessarakli, M. (2012). Reactive Oxygen Species, Oxidative Damage, and Antioxidative Defense Mechanism in Plants under Stressful Conditions. *Journal of Botany (Egypt)*, *2012*, 1–26. doi:10.1155/2012/217037

Shukla, P., Chaurasia, P., Younis, K., Qadri, O. S., Faridi, S. A., & Srivastava, G. (2019). Nanotechnology in sustainable agriculture: Studies from seed priming to post-harvest management. *Nanotechnology for Environmental Engineering*, *4*(1), 1–15. doi:10.100741204-019-0058-2

Siddaiah, C. N., Prasanth, K. V. H., Satyanarayana, N. R., Mudili, V., Gupta, V. K., Kalagatur, N. K., Satyavati, T., Dai, X. F., Chen, J. Y., Mocan, A., Singh, B. P., & Srivastava, R. K. (2018). Chitosan NPs having higher degree of acetylation induce resistance against pearl millet downy mildew through nitric oxide generation. *Scientific Reports, 8*(1), 1–14. doi:10.1038/s41598-017-19016-z

Singh, A., Rajput, V., Singh, A. K., Sengar, R. S., Singh, R. K., & Minkina, T. (2021). Transformation Techniques and Their Role in Crop Improvements: A Global Scenario of GM Crops. *Policy Issues in Genetically Modified Crops*, 515–542. doi:10.1016/B978-0-12-820780-2.00023-6

Singh, A., Singh Sengar, R., Singh, R., Shahi, U. P., & Yadav, M. K., Vaishali, Gangwar, L. K., & Rajput, V. D. (. (2022). Effects of zinc oxide NPs for promoting seed germination of rice (Oryza sativa L.) under salinity stress. *Ecology. Environmental Conservation*, *28*, 254–259. doi:10.53550/EEC.2022.v28i07s.042

Singh, H., Sharma, A., Bhardwaj, S. K., Arya, S. K., Bhardwaj, N., & Khatri, M. (2021). Recent advances in the applications of nano-agrochemicals for sustainable agricultural development. *Environmental Science. Processes & Impacts*, *23*(2), 213–239. doi:10.1039/D0EM00404A PMID:33447834

Singh, N., & Bhardwaj, R. D. (2016). Ascorbic acid alleviates water deficit induced growth inhibition in wheat seedlings by modulating levels of endogenous antioxidants. *Biologia, 71*(4), 402–413. doi:10.1515/biolog-2016-0050

Singh, V. K., Singh, R., Tripathi, S., Devi, R. S., Srivastava, P., Singh, P., Kumar, A., & Bhadouria, R. (2020). Seed priming: state of the art and new perspectives in the era of climate change. *Climate Change and Soil Interactions*, 143–170. doi:10.1016/B978-0-12-818032-7.00006-0

Singh, Y., Meher, J. G., Raval, K., Khan, F. A., Chaurasia, M., Jain, N. K., & Chourasia, M. K. (2017). Nanoemulsion: Concepts, development and applications in drug delivery. *Journal of Controlled Release*, *252*, 28–49. doi:10.1016/j.jconrel.2017.03.008 PMID:28279798

Slama, I., Abdelly, C., Bouchereau, A., Flowers, T., & Savouré, A. (2015). Diversity, distribution and roles of osmoprotective compounds accumulated in halophytes under abiotic stress. *Annals of Botany*, *115*(3), 433–447. doi:10.1093/aob/mcu239 PMID:25564467

Tanou, G., Molassiotis, A., & Diamantidis, G. (2009). Induction of reactive oxygen species and necrotic death-like destruction in strawberry leaves by salinity. *Environmental and Experimental Botany*, *65*(2–3), 270–281. doi:10.1016/j.envexpbot.2008.09.005

Usman, M., Farooq, M., Wakeel, A., Nawaz, A., Cheema, S. A., Rehman, H., Ashraf, I., & Sanaullah, M. (2020). Nanotechnology in agriculture: Current status, challenges and future opportunities. *The Science of the Total Environment*, *721*, 137778. doi:10.1016/j.scitotenv.2020.137778 PMID:32179352

Van Zelm, E., Zhang, Y., & Testerink, C. (2020). *Salt Tolerance Mechanisms of Plants*. Annual Reviews. doi:10.1146/annurev-arplant-050718-100005

Vannini, C., Domingo, G., Onelli, E., De Mattia, F., Bruni, I., Marsoni, M., & Bracale, M. (2014). Phytotoxic and genotoxic effects of silver NPs exposure on germinating wheat seedlings. *Journal of Plant Physiology*, *171*(13), 1142–1148. doi:10.1016/j.jplph.2014.05.002 PMID:24973586

Verma, K. K., Song, X.-P., Joshi, A., Rajput, V. D., Singh, M., Sharma, A., Singh, R. K., Li, D.-M., Arora, J., Minkina, T., & Li, Y.-R. (2022). Nanofertilizer Possibilities for Healthy Soil, Water, and Food in Future: An Overview. *Frontiers in Plant Science*, *13*, 865048. Advance online publication. doi:10.3389/fpls.2022.865048 PMID:35677230

Verma, K. K., Song, X. P., Joshi, A., Tian, D. D., Rajput, V. D., Singh, M., Arora, J., Minkina, T., & Li, Y. R. (2022). Recent Trends in Nano-Fertilizers for Sustainable Agriculture under Climate Change for Global Food Security. *Nanomaterials (Basel, Switzerland)*, *12*(1), 173. doi:10.3390/nano12010173 PMID:35010126

Vishwakarma, K., Upadhyay, N., Kumar, N., Tripathi, D. K., Chauhan, D. K., Sharma, S., & Sahi, S. (2018). Potential Applications and Avenues of Nanotechnology in Sustainable Agriculture. *Nanomaterials in Plants, Algae, and Microorganisms, 1*, 473–500. doi:10.1016/B978-0-12-811487-2.00021-9

Wang, P., Lombi, E., Zhao, F. J., & Kopittke, P. M. (2016). Nanotechnology: A New Opportunity in Plant Sciences. *Trends in Plant Science, 21*(8), 699–712. doi:10.1016/j.tplants.2016.04.005 PMID:27130471

Wang, W. N., Tarafdar, J. C., & Biswas, P. (2013). Nanoparticle synthesis and delivery by an aerosol route for watermelon plant foliar uptake. *Journal of Nanoparticle Research, 15*(1), 1–13. doi:10.100711051-013-1417-8

Wu, L., Huo, W., Yao, D., & Li, M. (2019). Effects of solid matrix priming (SMP) and salt stress on broccoli and cauliflower seed germination and early seedling growth. *Scientia Horticulturae, 255*, 161–168. doi:10.1016/j.scienta.2019.05.007

Zahra, N., Shaukat, K., Hafeez, M. B., Raza, A., Hussain, S., Chaudhary, M. T., Akram, M. Z., Kakavand, S. N., Saddiq, M. S., & Wahid, A. (2021). Physiological and Molecular Responses to High, Chilling, and Freezing Temperature in Plant Growth and Production: Consequences and Mitigation Possibilities. *Harsh Environment and Plant Resilience*, 235–290. https://doi.org/ doi:10.1007/978-3-030-65912-7_10/COVER

Zhao, L., Peralta-Videa, J. R., Varela-Ramirez, A., Castillo-Michel, H., Li, C., Zhang, J., Aguilera, R. J., Keller, A. A., & Gardea-Torresdey, J. L. (2012). Effect of surface coating and organic matter on the uptake of CeO2 NPs by corn plants grown in soil: Insight into the uptake mechanism. *Journal of Hazardous Materials, 225–226*, 131–138. doi:10.1016/j.jhazmat.2012.05.008 PMID:22633924

Zhu, J. K. (2016). Abiotic Stress Signaling and Responses in Plants. *Cell, 167*(2), 313–324. doi:10.1016/j.cell.2016.08.029 PMID:27716505

Zulfiqar, F., & Ashraf, M. (2021). NPs potentially mediate salt stress tolerance in plants. *Plant Physiology and Biochemistry, 160*, 257–268. doi:10.1016/j.plaphy.2021.01.028 PMID:33529801

Section 7

Future Role of Nanopriming-Based Approaches for Food Security and Sustainable Agriculture

Chapter 14
Nanotechnology for Sustainability and Food Security in Agriculture:
A Nanopriming Long Story in Short

Omkar Singh
https://orcid.org/0000-0003-1971-185X
Department of Soil Science, Sardar Vallabhbhai Patel University of Agriculture and Technology, India

ABSTRACT

Sustainable techniques are crucial to ensuring the availability of nourishing food while reducing environmental effects due to the growing global population and increasing demand for agricultural resources. By utilizing nanomaterials and nanodevices to improve crop yield, resource efficiency, and pest management, nanotechnology offers creative solutions. The chapter covers the potential advantages of nanotechnology in agriculture, including the precise pest control abilities of nanopesticides and the controlled-release qualities of nanofertilizers to enhance nutrient delivery and use. Additionally, it emphasizes the use of nanosensors for continuous assessment of crop health and environmental factors, as well as nanoscale imaging methods for early disease detection and treatment. Resource efficiency, precision agriculture, waste reduction, and improved food security are made possible by the integration of nanotechnology into agriculture. However, it also highlights how crucial it is to address the safety, moral, and legal issues raised by the application of nanotechnology in farming.

INTRODUCTION

Agriculture is just one of the many industries where nanotechnology, the science, and engineering of materials and devices at the nanoscale, has enormous potential.

DOI: 10.4018/978-1-6684-7232-3.ch014

The use of nanotechnology in agriculture provides creative answers to pressing problems like sustainability and food security(Singh et al. 2022). It is crucial to adopt sustainable practices that ensure the availability of nourishing food while minimizing environmental impact due to the growing global population and increasing demand for agricultural resources (Rajput et al. 2018 and Singh et al., 2022). This chapter examines how nanotechnology can be used in agriculture to promote sustainability and food security, highlighting its potential advantages and addressing any related worries.

Development and application of nanomaterials and nanodevices in agriculture are necessary to improve crop production, resource efficiency, and pest control. Agricultural applications benefit from the special characteristics that nanomaterials display at the nanoscale, such as increased surface area, high reactivity, and tailored functionality. By controlling the release of nutrients in a targeted manner, for example, nanofertilizers can enhance nutrient delivery and utilization while minimizing nutrient loss and maximizing crop uptake (Guo et al., 2018). The use of conventional chemical pesticides and the associated environmental risks can be reduced by using nanopesticides, which have the potential to provide precise pest control (Chaud et al., 2021).

Crop health, soil conditions, and environmental parameters can all be monitored in real-time thanks to the use of nanosensors in agriculture. For precision farming and the best resource management, nanosensors can identify changes in pH, moisture, temperature, and nutrient levels (Rai et al., 2012). Additionally, new insights into plant physiology and disease diagnostics are provided by nanoscale imaging techniques, which facilitate early disease intervention and management (Worrall et al., 2019).

Using nanotechnology to improve sustainability in agriculture covers a wide range of topics. Utilizing nanomaterials with controlled-release properties, which ensure precise delivery of nutrients and agrochemicals to plants, can increase resource efficiency (Singh et al., 2022). This lessens waste, has a smaller negative impact on the environment, and maximizes resource use (Verma et al., 2022). Nanomaterials can also be used for wastewater treatment and remediation, which helps with water resource recycling and conservation in agriculture (Kumar et al., 2017).

Nanotechnology has made precision agriculture possible, enabling farmers to make decisions based on current data. Information on soil moisture, nutrient levels, and pest infestation is provided by nanosensors embedded in the ground or built into unmanned aerial vehicles (UAVs), enabling targeted interventions and reducing inputs (Wang et al., 2020). Using nanorobots to deliver fertilizers, pesticides, or other agricultural inputs precisely and locally has the potential to minimize waste and maximize effectiveness (Wang et al., 2019).

Recycling and waste reduction are essential elements of sustainable agriculture. By removing contaminants and pollutants from irrigation water, nanomaterial-based filtration systems can improve water quality and lower the chance of soil and groundwater contamination (Prasad et al., 2017). Agricultural products' shelf lives can be extended by using nanocomposites and nano-enabled packaging materials, which also reduce waste and post-harvest losses (Singh et al., 2017).

Agriculture's top priority is to ensure food security. Innovative methods for managing disease and pests are made possible by nanotechnology. Pests can be effectively controlled while having a minimal negative impact on the environment thanks to nanopesticides with improved targeting capabilities and decreased toxicity (Asif et al., 2022). Antimicrobial agents can be controlled in their release through nanoencapsulation, which boosts their effectiveness and slows the emergence of resistance (Li et al., 2021). Additionally, by increasing the bioavailability of nutrients and creating nano-delivery systems for targeted nutrient delivery, nanotechnology can enhance crop quality and nutrition (Arora et al., 2022). Food safety can be maintained throughout the supply chain by using nanoscale sensors integrated into food packaging to monitor product quality, spot spoilage, and ensure freshness.

Nanotechnology has a lot of potential for solving the problems with sustainability and food security that the agricultural sector is currently facing. Resource management, pest control, and disease management are made possible by the integration of nanomaterials, nanodevices, and nanosensors. However, the application of nanotechnology in agriculture necessitates careful attention to regulatory, moral, and safety issues. Sustainable and safe agricultural systems can be established with continued research, interdisciplinary cooperation, and responsible nanotechnology use.

BRIEF OVERVIEW OF NANOTECHNOLOGY AND ITS APPLICATIONS IN AGRICULTURE

Agriculture has a wide range of uses for nanotechnology, which is the manipulation and engineering of materials at the nanoscale. Agricultural problems can now be addressed in novel ways thanks to the special qualities exhibited by nanomaterials at the nanoscale, such as increased surface area, high reactivity, and tailored functionality. Innovative approaches to crop production, resource management, pest control, and food safety are provided by nanotechnology.

Through the creation and use of nanomaterials, nanodevices, and nanosensors, nanotechnology has applications in agriculture (Kumar et al. 2022) . For instance, nanofertilizers have drawn interest due to their capacity to increase the effectiveness of nutrient uptake and delivery in plants. These nanoscale fertilizers have the ability to regulate nutrient release, ensuring crops receive targeted and ongoing nutrient supplies (Guo et al., 2018). With this method, nutrient loss is minimized, plant nutrient uptake is maximized, and the environmental damage caused by excessive fertilizer application is reduced.

Similar to how nanopesticides present a possible substitute for traditional chemical pesticides. Because of the targeted and controlled release of active pesticide ingredients made possible by nanotechnology, pests can be effectively controlled with fewer risks to the environment (Chaud et al., 2021). Crop health, soil conditions, and environmental factors can all be monitored in real-time when nanosensors are used in agriculture. For precision farming and resource management, nanosensors can identify changes in temperature, moisture, pH, and nutrient levels (Rai et al., 2012).

Additionally, nanotechnology has a big impact on improving agricultural sustainability. Reduce waste and maximize resource use with the precise delivery of nutrients and agrochemicals to plants thanks to the controlled-release properties of nanomaterials (Verma et al., 2022). Nanomaterials can be used in soil and water remediation to get rid of contaminants and pollutants, which helps with resource recycling and conservation in agriculture (Kumar et al., 2017 and Rajput et al. 2022). In addition, nanotechnology makes it possible to create intelligent packaging materials that can increase the shelf life of agricultural products, lowering food waste and post-harvest losses (Singh et al., 2017).

Agriculture must prioritize ensuring food security, and nanotechnology presents fresh solutions in this regard. Effective pest control is provided by nanopesticides while the environmental impact is reduced. Nanopesticides improve crop yield and quality by precisely targeting pests and reducing the need for excessive chemical application (Asif et al., 2022). Additionally, nanotechnology makes it easier to create nano-delivery systems that increase nutrient bioavailability and specifically target nutrient delivery to plants, improving crop nutrition (Arora et al., 2022). Food safety can be maintained throughout the supply chain by using nanoscale sensors integrated into food packaging to monitor product quality, spot spoilage, and ensure freshness.

IMPORTANCE OF SUSTAINABILITY AND FOOD SECURITY IN AGRICULTURE

Two important pillars that will influence the future of agriculture are sustainability and food security. The need for sustainable agricultural practices that can meet current needs without compromising the ability of future generations to meet their own needs grows as the world's population continues to rise and environmental challenges become more severe. Nevertheless, addressing hunger, malnutrition, and disparities in access to wholesome food is only possible by ensuring food security. Food security and sustainability in agriculture must work together for the sake of both people and the environment.

The limited nature of natural resources is one of the main reason's sustainability is important in agriculture. Agriculture depends on essential resources that are scarce and susceptible to degradation, such as land, water, and energy. Unsustainable practices, such as the excessive use of chemical fertilizers, overuse of water, and deforestation, can result in soil erosion, a shortage of fresh water, and the extinction of biodiversity. On the other hand, sustainable agricultural practices work to maintain and improve the quality of these resources for long-term productivity and ecosystem health.

Environmentally friendly methods are encouraged by sustainable agriculture. For instance, using natural fertilizers and pest control techniques when farming organically lessens the need for synthetic inputs, reduces soil erosion, and improves soil health. Crop rotation and minimal tillage are examples of conservation agriculture practices that help retain soil moisture, reduce erosion, and enhance the structure and fertility of the soil. By conserving resources, preserving ecosystems, and lowering greenhouse gas emissions, these practices aid in the long-term sustainability of agricultural systems.

The availability, affordability, and accessibility of healthy food for all people falls under the umbrella of food security, on the other hand. In order to achieve food security, one must address issues with distribution, access, and utilization in addition to raising food production. Because it increases productivity, diversifies food sources, and lowers post-harvest losses, sustainable agriculture is essential to ensuring global food security.

By maximizing resource use and reducing waste, sustainable agricultural practices increase productivity. By monitoring crop conditions, optimizing irrigation, and applying inputs more effectively, farmers are able to increase yields while using fewer resources, thanks to precision agriculture technologies like the use of sensors, drones, and data analytics. The cultivation of a variety of crops and varieties that are resistant to pests, diseases, and changing climates is another aspect of sustainable agriculture that is emphasized. Food source diversification improves food security by reducing reliance on a single crop and expanding dietary options.

A key component of ensuring food security is minimizing post-harvest losses. Losses from spoilage, pests, and improper handling are minimized by sustainable post-harvest management techniques like proper storage, processing, and transportation. By extending the shelf life of food products, new packaging technologies can improve availability and cut waste. Examples include modified atmosphere packaging and materials with nanotechnology built in.

Furthermore, agricultural sustainability benefits the economy and general well-being of rural communities. Farmers can increase resilience to the effects of climate change, increase production efficiency, and improve livelihoods by implementing sustainable practices. Additionally, sustainable agriculture fosters rural development and job opportunities while assisting local economies.

NANOTECHNOLOGY IN AGRICULTURE

With a wide range of uses in agriculture, nanotechnology has emerged as a promising field. It entails the manipulation and engineering of materials at the nanoscale in order to create novel answers to a range of agricultural problems.

DEFINITION OF NANOTECHNOLOGY

Nanotechnology is the branch of science and technology concerned with the control and manipulation of matter at the nanoscale, which typically falls between 1 and 100 nanometers (nm). It focuses on the creation, use, and application of materials, structures, tools, and systems with special capabilities resulting from their nanoscale dimensions. In order to create and manipulate materials and devices with improved or novel properties at the nanoscale, nanotechnology uses principles from physics, chemistry, biology, and engineering.

PRINCIPLES OF NANOTECHNOLOGY

1. Size-dependent Properties: Materials can display special physical, chemical, optical, and electronic properties at the nanoscale that are distinct from those of their bulk counterparts. These characteristics appear as a result of surface effects and quantum confinement, which become important at nanoscale dimensions. Materials' altered properties, such as improved reactivity, increased surface energy, and modified optical absorption, result from the ratio of surface area to volume increasing as the size of the material decreases.
2. Surface-to-Volume Ratio: Nanotechnology takes advantage of nanomaterials' high surface-to-volume ratio. The large surface area of nanoscale structures allows for more efficient interactions with molecules, ions, and other substances. With increased surface interactions enhancing efficiency and reactivity, this property is essential for applications like catalysis, sensing, and drug delivery.
3. Self-Assembly and Bottom-up Approaches: Self-assembly, a process whereby nanoscale components spontaneously organize themselves into desired structures, is frequently used in nanotechnology. At the nanoscale, this theory enables the formation of intricate and useful structures. Building structures and devices from the inside out, atom by atom or molecule by molecule, enables precise control over their characteristics and capabilities.
4. Multidisciplinary Approach: The fields of physics, chemistry, materials science, biology, and engineering are all used in nanotechnology. In order to develop

new functionalities and applications, entails comprehending and modifying materials at the atomic and molecular level. Nanotechnology's interdisciplinary nature makes it possible to combine expertise and methodologies from various disciplines to create novel solutions.

5. Nanoscale Fabrication Techniques: To manipulate and engineer materials at the nanoscale, nanotechnology relies on specialized fabrication techniques. These methods include top-down methods that scale down larger structures to the nanoscale, like lithography and etching. Molecular self-assembly and synthesis are examples of bottom-up methods that assemble nanoscale structures from their constituent parts.
6. Application-specific Design: Nanotechnology places a strong emphasis on creating materials, tools, and systems that are specifically suited to certain uses. Researchers can tailor the conductivity, strength, and reactivity of materials to meet the needs of a variety of applications, including electronics, energy storage, healthcare, environmental remediation, and agriculture.

These ideas work together to advance the field of nanotechnology and enable scientists and engineers to take advantage of the special characteristics of nanoscale materials for a range of scientific and engineering developments.

NANOMATERIALS AND THEIR APPLICATIONS IN AGRICULTURE

Nanomaterials, or materials with nanoscale structures or components, have drawn a lot of interest for use in agriculture. Their special qualities and capabilities offer potential remedies for a variety of agricultural problems, including crop production, pest control, and food safety. The followings are some essential nanomaterials and their agricultural uses for them:

NANOFERTILIZERS FOR IMPROVED NUTRIENT DELIVERY

The distribution of nutrients in agriculture could be greatly enhanced by the use of nanofertilizers, a form of fertilizer based on nanomaterials. The targeted and regulated release of nutrients made possible by these nanomaterials results in improved nutrient uptake efficiency by plants. Here are some important characteristics and advantages of nanofertilizers for better nutrition delivery:

1. Controlled Release: Nutrients are released gradually and under controlled conditions via nanofertilizers. They can be programmed to react to particular environmental cues such as soil moisture, pH, or plant nutrition requirement. By minimizing fertilizer loss through leaching and runoff and optimizing crop nutrient uptake, this controlled release guarantees that the plants receive a consistent and long-lasting supply of nutrients.
2. Enhanced Nutrient Uptake: Nanofertilizers can connect and contact plant roots more effectively due to their small size and high surface area to volume ratio. The expanded surface area makes it easier for roots to efficiently adsorb and absorb nutrients, increasing the effectiveness of nutrient intake. Plant growth, yield, and overall nutrient-use efficiency are all improved as a result of this increased nutrient uptake.
3. Targeted Delivery: Targeting particular plant tissues or organs is possible with the use of functionalized or coated nanofertilizers. Nutrients can be directed to certain root zones or plant parts where they are most needed by altering the surface characteristics of nanoparticles. By delivering nutrients to the right places, this tailored delivery prevents wasting and optimizes nutrient distribution throughout plants.
4. Nutrient Stabilization: Nanofertilizers may protect nutrients from outside forces that might cause them to deteriorate or be lost. Nutrients are protected from interactions with soil particles, moisture, or reactive compounds that could result in nutrient immobilization or leaching by being enclosed within nanomaterials. The nutrients are stabilized to ensure their continued availability and longer release for plant uptake.
5. Reduced Environmental Impact: Nanofertilizers help to reduce environmental pollution because of their regulated release and enhanced nutrient uptake efficiency. By delivering fertilizers precisely, excess nutrients are applied less frequently, which limits nutrient runoff into water bodies and the resultant harm to aquatic ecosystems. The less frequent fertilizer applications required also lessen the possibility of nutrient imbalances and related environmental issues.
6. Customizable Formulations: Based on the type of crop, the stage of growth, and the soil characteristics, nanofertilizers can be specially formulated to satisfy particular nutrient needs. They can be made to supply a mix of vital nutrients, trace nutrients, or the precise ratios of nutrients required for healthy plant development. Due to the exact nutrient management made possible by this personalization, farmers are able to address certain nutrient deficits and improve crop production.

NANOPESTICIDES FOR TARGETED PEST CONTROL

Nanopesticides, a class of formulations for pesticides based on nanomaterials, have demonstrated potential for offering precise and efficient pest control in agriculture. These nanoparticles have special qualities that allow for regulated release, better penetration, and increased pest-fighting effectiveness of active chemicals. The following are some essential features and gains of using nanopesticides for selective pest control:

1. Targeted Delivery: Specific pests or plant components can be the target of nanopesticides, resulting in a more targeted and effective management strategy. Nanopesticides can specifically interact with the physiology of the pest or cling to particular plant surfaces by altering the surface qualities or encapsulating the active chemicals into nanoparticles, boosting their affinity and efficiency against the target pests.
2. Controlled Release: The active chemicals in nanopesticides can be designed to release slowly and steadily. This controlled release makes sure that pests are exposed to the active chemicals for a long time, increasing their effectiveness and lowering the frequency of pesticide applications. Additionally, it aids in minimizing environmental pollution and lowers the possibility of pest populations gaining resistance.
3. Improved Penetration: Nanopesticides' nanoscale-sized particles allow for greater penetration of plant tissues or pest exoskeletons. The active components are able to reach the target sites with more efficiency thanks to the enhanced penetration, which also increases their bioavailability and increases their contact with pests. In order to minimize the usage of pesticides and lessen their negative effects on the environment, nanopesticides can display greater efficacy even at reduced application rates.
4. Reduced Environmental Impact: The environmental damage caused by conventional pesticide formulations may be lessened with the use of nanopesticides. A more confined and constrained dispersion of active substances results from their focused delivery and controlled release properties. This lessens the chance of water and soil pollution by minimizing off-target effects on beneficial creatures like pollinators and natural predators.
5. Enhanced Stability: The stability and shelf life of active substances can be enhanced using nanopesticides. Nanopesticides shield the active ingredients from deterioration brought on by environmental variables including light, moisture, and temperature changes by encapsulating them inside nanoparticles. By lowering the need for frequent reapplication and increasing their cost-efficiency, this improved stability ensures the pesticides' long-term effectiveness.

6. Reduced Chemical Load: A more effective application of active ingredients is made possible by the targeted delivery and controlled release characteristics of nanopesticides. Nanopesticides reduce the overall chemical burden in the environment by delivering pesticides specifically to pests or specific plant tissues. This decrease in pesticide use promotes sustainable pest management techniques, which may enhance human health, ecosystem health, and the long-term viability of agriculture.

NANOSENSORS FOR REAL-TIME MONITORING OF CROP HEALTH AND SOIL CONDITIONS

In agriculture, nanosensors have become potent tools for real-time monitoring of crop health and soil conditions. Numerous indicators, like as nutrient levels, soil moisture, pH, temperature, and disease markers, can be detected and quantified by these nanoscale devices. Nanosensors empower farmers to make knowledgeable decisions about crop protection, irrigation, and fertilization by delivering precise and fast information. The following are some significant features and gains of using nanosensors for on-the-spot monitoring in agriculture:

1. Accurate Monitoring: Target analytes can be detected with excellent sensitivity and selectivity using nanosensors. They can interact effectively with analytes found in soil or plant tissues due to their small size and high surface-to-volume ratio. Farmers may evaluate the current state of key indicators essential to crop growth and health thanks to this precision monitoring.
2. Real-time Data: Nanosensors deliver in-the-moment information on crop and soil conditions, enabling quick management decisions. Because nanosensors are capable of continuous monitoring, changes in crucial parameters are quickly identified, enabling immediate adjustments to irrigation schedules, fertilizer applications, or pest control strategies. This real-time feedback loop promotes precision farming techniques and aids in resource efficiency.
3. Non-destructive Monitoring: Non-destructive monitoring, made possible by nanosensors, permits repeated observations without endangering the plants or upsetting the soil ecosystems. This non-destructive nature reduces plant stress and disturbance and makes it possible to track crop health and growth metrics over an extended period of time. Non-destructive monitoring also makes it easier to monitor temporal changes and assess the effectiveness of a treatment over time.
4. Remote Sensing: In order to collect data from numerous locations at once, nanosensors can be connected with wireless communication systems and

remote sensing platforms. This capacity makes it possible to monitor and aggregate data on a broad scale, giving you a complete picture of the situation on the ground. In large agricultural areas, remote sensing eliminates labour-intensive activities and offers a cost-effective way to monitor crop health and soil conditions.

5. Early Disease Detection: Prior to the appearance of obvious symptoms, nanosensors can identify disease indicators or harmful organisms. In order to stop the spread of infections and reduce crop losses, early disease diagnosis enables farmers to use effective disease management options, such as targeted treatments or quarantine measures. Nanosensors aid in enhanced disease management and general crop health by enabling early intervention.
6. Improved Resource Management: Nanosensors' real-time data improves agricultural resource management. Farmers can optimize irrigation and fertilization operations and reduce water and nutrient waste by monitoring variables like soil moisture and nutrient levels. This careful resource management lessens the environmental effect associated with excessive resource consumption and promotes more environmentally friendly farming methods.

BENEFITS AND POTENTIAL RISKS OF NANOTECHNOLOGY IN AGRICULTURE

The field of agriculture can benefit from and take advantage of several aspects of nanotechnology. However, it's crucial to consider both the advantages and any possible risks related to its application. The following are some of the main advantages and potential risks of using nanotechnology in agriculture:

NANOTECHNOLOGY'S ADVANTAGES FOR AGRICULTURE

1. Increased Crop Productivity: Solutions based on nanotechnologies, such as nanofertilizers and nanopesticides, can boost nutrient delivery, insect management, and plant growth regulation, resulting in increased crop productivity and yield.
2. Precision Agriculture: Real-time monitoring of crop health, soil conditions, and environmental parameters is made possible by nanosensors and nanoscale imaging technologies, facilitating precision agriculture methods and improving resource management.
3. Lessened Environmental influence: Using nanotechnology, agricultural operations may have a smaller negative influence on the environment.

Nanopesticides and nanofertilizers with controlled release can limit the release of chemicals into the environment, lowering pollution and enhancing agricultural sustainability.

4. Nanotechnology plays a part in guaranteeing the quality and safety of food. Nanoscale sensors and biosensors are able to quickly and sensitively detect pollutants, pathogens, and spoilage indicators in food products.
5. Sustainable Resource Management: Through accurate monitoring and tailored distribution methods, nanotechnology enables more effective use of resources, such as water and fertilizers. This encourages agriculture to use sustainable resource management techniques.

POTENTIAL RISKS AND CONCERNS OF NANOTECHNOLOGY IN AGRICULTURE

1. Environmental Impact: Concerns about the potential ecological effects of nanomaterials are raised by their discharge into the environment. The fate, behavior, and possible buildup of nanomaterials in soil, water, and living things, as well as their long-term effects on ecosystems, must all be understood.
2. Human Health and Safety: It is important to carefully evaluate the possible risks of nanomaterial exposure for both consumers and agricultural workers. Toxicology, bioavailability, and long-term health impacts of nanomaterials utilized in agricultural applications must all be assessed.
3. Regulatory Challenges: Regulatory frameworks face difficulties due to the special traits and actions of nanomaterials. It is essential to establish suitable rules and guidelines to guarantee the responsible and safe application of nanotechnology in agriculture.
4. The Fate of Nanomaterials in the Food Chain: Analyzing the fate of nanomaterials in the food chain is crucial for determining their safety and potential effects on human health.
5. Ethical Considerations: Concerns about fair access to technology, intellectual property rights, and potential inequalities in its adoption that are related to the use of nanotechnology in agriculture must be addressed.
6. Public Perception and Acceptance: The successful application of nanotechnology in agriculture depends on its widespread acceptance and comprehension. It is essential that the general public is informed about the advantages, dangers, and ethical issues related to nanotechnology.

ENHANCING SUSTAINABILITY THROUGH NANOTECHNOLOGY

Nanotechnology has become a potent instrument with the potential to alter several industries, including agriculture, in recent years. Due to its special characteristics at the nanoscale level, there are chances to improve agricultural operations' sustainability and solve issues with food security, resource scarcity, and environmental impact. Innovative approaches to increase resource efficiency, reduce waste, and support sustainable agriculture practices can be developed by utilizing nanotechnology. The specifics of how nanotechnology can improve agricultural sustainability will be covered in this essay.

RESOURCE EFFICIENCY AND REDUCED ENVIRONMENTAL IMPACT

Controlled-Release Nanofertilizers For Optimized Nutrient Utilization

Controlled-release nanofertilizers have become a viable agricultural tool for better crop output and optimum nutrient use. These nanoscale formulations are made to release nutrients gradually over a long period of time, more effectively meeting the nutrient needs of plants. Controlled-release nanofertilizers provide a continuous supply of nutrients, which has various advantages:

1. Increased Nutrient Use Efficiency: By distributing nutrients in a regulated and targeted way, controlled-release nanofertilizers aid in increasing nutrient use efficiency. Nanofertilizers release nutrients slowly and in response to plant demand, unlike conventional fertilizers, which frequently cause nutrient losses owing to leaching or volatilization. As a result, nutrient losses are reduced and nutrient absorption is increased, ensuring that nutrients are available when plants need them. Crop productivity is enhanced as a result, and nutrient utilization efficiency as a whole is raised.
2. Lessened Environmental Impact: The environmental effects of excessive fertilizer use are one of the main issues in agriculture. When used excessively or improperly, conventional fertilizers can cause nutrient runoff, water pollution, and ecological deterioration. By lowering the possibility of nutrient loss and minimizing the risk of nutrient leakage into water bodies, controlled-release nanofertilizers provide a more environmentally acceptable alternative. The slow-release technique minimizes overapplication and the accompanying

environmental effects while enabling plants to utilize the nutrients more efficiently.

3. Limited Nutrient Availability in Difficult Conditions: Controlled-release nanofertilizers are especially useful in difficult conditions when nutrients are scarce. For instance, conventional fertilizers may be easily leached away from sandy soils with inadequate nutrient retention capacity, resulting in nutritional shortages. With their controlled-release qualities, nanofertilizers can aid in ensuring that plants in such conditions always have access to nutrients. As a result, crop yields are improved and healthy plant growth and stress tolerance are encouraged.
4. Decreased Fertilizer Application Frequency: Nanofertilizers' slow-release properties decelerate fertilizer application. Nanofertilizers are able to deliver continuous nutrient release for a longer period of time than conventional fertilizers, which frequently need numerous applications during the growing season. This lessens the risk of nutrient overload in the soil while also saving labor and cutting application expenses. The less frequent spraying helps promote sustainable agriculture practices and the efficient use of resources.
5. Customized Nutrient Formulations: Nanotechnology enables the creation of nutrient formulations that are specifically suited to the needs of particular crops and soil types. Nutrient release may be carefully regulated to correspond to the growth phases and nutrient needs of various crops by adjusting the composition and structure of nanofertilizers. The optimal balance of nutrients is delivered to plants at the appropriate moment thanks to this customization, which maximizes nutrient uptake and reduces nutrient waste.

Nanoencapsulation of Agrochemicals for Reduced Environmental Contamination

Agrochemical nanoencapsulation is a cutting-edge strategy that has come to light recently due to its potential to lessen environmental contamination brought on by the usage of conventional agrochemicals. For the control of pests and weeds in agriculture, agrochemicals such as pesticides and herbicides are crucial. However, due to their extensive use, they may kill non-target creatures and cause unexpected environmental effects such soil erosion and water body contamination. To lessen these problems, nanoencapsulation offers a viable option.

Active agrochemical components are enclosed in nanoscale carriers, such as nanoparticles or nanocapsules, by the process of nanoencapsulation. The encapsulation technique offers the following benefits:

1. Controlled Release: Agrochemicals can be released in a regulated manner over an extended period of time thanks to nano encapsulation. A prolonged and precise delivery to the intended location of action is made possible by the nanocarriers' slow release of the encapsulated active substances. With fewer agrochemicals needed and less frequently applied thanks to this controlled release system, there is less chance of environmental contamination.
2. Targeted Delivery: In order to minimize exposure to non-target organisms, the nanocarriers utilized in nanoencapsulation can be created to selectively target pests or weeds. Agrochemicals can be released selectively by functionalized nanocarriers in response to certain stimuli, such as pH or enzymatic activity present in the intended pests or weeds. This method of tailored distribution increases the effectiveness of agrochemicals, lowers environmental contamination, and lessens off-target impacts.
3. Enhanced Stability and Persistence: By shielding agrochemicals from degradation brought on by environmental variables like UV radiation and microbial activity, nanoencapsulation improves the stability and persistence of agrochemicals. The encapsulated active substances are protected from premature deterioration by the nanocarriers, which function as a physical barrier. This improved stability makes the agrochemicals more effective for a longer period of time, requiring fewer applications and exposing the environment to less of them.
4. Lower Dosage and Environmental Impact: Nanoencapsulation makes it possible for agrochemicals to be delivered effectively at lower dosages. In order to effectively manage pests or weeds, fewer active chemicals must be used overall thanks to the controlled release and targeted delivery mechanisms. This dosage reduction helps to significantly reduce the dangers of environmental contamination for ecosystems, including non-target creatures and water supplies.
5. Improved Safety and Compatibility: By making agrochemicals less harmful to organisms other than the ones they are intended to affect, nanoencapsulation can make them safer and more compatible. The encapsulation procedure can change the active components' physicochemical characteristics, reducing their bioavailability and minimizing their negative effects on beneficial organisms. The impact on pollinators, natural predators, and other beneficial organisms is reduced as a result of this targeted selective killing, which promotes biodiversity and ecological harmony in agricultural settings.

Precision Agriculture and Smart Farming

1. Nanosensors and nanoscale imaging for precise monitoring and decision-making

Agriculture has undergone a transformation thanks to the development of nanosensors and nanoscale imaging technologies that allow for precise monitoring and decision-making. Real-time monitoring of critical factors like soil moisture, nutrient levels, pH, temperature, and gas concentrations is made possible by nanosensors based on nanomaterials or nanostructures. With the help of this ongoing monitoring, farmers are better equipped to manage resources and ensure the healthiest possible crop yields.

Additionally, high-resolution imaging of crops, plant tissues, and soil particles is possible using nanoscale imaging techniques including scanning electron microscopy (SEM) and atomic force microscopy (AFM). These imaging techniques offer in-depth perceptions into soil characteristics, nutrient uptake patterns, insect or disease infestations, and cellular structures. The early diagnosis of plant stress, nutrient deficits, or pathogen infections is made possible by the great precision and accuracy of nanoscale imaging, allowing for prompt interventions and focused management techniques.

Farmers can maximize the use of resources like water and fertilizers based on real-time data by utilizing nanosensors and nanoscale imaging. As a result, resource management is enhanced, environmental impact is diminished, and agricultural cost-effectiveness is raised. Additionally, the identification of pathogens, pests, or disease markers in plants is made possible by nanosensors that have been combined with certain biomarkers or nanomaterials. Early identification enables quick response to stop the future spread and reduce crop losses, which contributes to more efficient disease and pest control.

Nanorobots for Targeted Delivery and Site-Specific Interventions

Nanorobots, which combine robotics and nanotechnology, has enormous potential for site-specific treatments and targeted distribution in a variety of industries, including agriculture. These tiny robotic devices, which are frequently in the nanoscale range, can be made to precisely carry out predetermined duties while navigating through challenging surroundings. Here is a succinct comment on the application of nanorobots for site-specific therapies and targeted delivery:

Nanorobots provide previously unheard-of capabilities for precise delivery of chemicals like fertilizers, insecticides, or even genetic material to certain areas within soil or plants. It is possible to guide nanorobots to deliver these materials exactly where they are needed by functionalizing them with the proper payloads. With less waste and less pesticide use, and less negative influence on non-target organisms, this tailored delivery method improves the effectiveness of therapies.

Additionally, nanorobots are capable of site-specific micro- or nanoscale interventions, enabling precise operations catered to the requirements of plants and agricultural systems. Nanorobots, for instance, can be used to carry out targeted therapies, such as eradicating pests or sick tissues, modifying plant cells, or even helping with pollination procedures. Nanorobots' capacity to execute these interventions at a very limited scale improves the efficacy of therapies, lowers the possibility of collateral damage, and encourages the use of sustainable agriculture techniques.

Nanorobots can be made to collect data and provide real-time monitoring of crop health, soil conditions, and environmental parameters in addition to targeted delivery and site-specific therapies. Nanorobots with sensors and imaging capabilities can gather important data and wirelessly send it for analysis. Precision monitoring, prompt decision-making, and effective resource management are all made possible in agriculture by this data-driven methodology.

It is crucial to remember that there are still many technological and moral issues that need to be resolved before nanorobots in agriculture can be used in a practical way. Thorough investigation and careful thought are required for problems such power supply, navigation control, biocompatibility, and potential environmental implications.

WASTE REDUCTION AND RECYCLING

Recycling and waste reduction are essential components of sustainable development, and nanotechnology presents exciting potential solutions in these areas. Here are some succinct observations on recycling and waste reduction using nanotechnology:

1. Improved Waste Treatment: Waste treatment procedures are made more effective and efficient by nanotechnology. Organic contaminants, heavy metals, and industrial wastes can all be degraded or transformed using nanomaterials like nanoparticles and nanocatalysts. The special characteristics of nanomaterials, such as their high surface area and reactivity, improve the effectiveness and lower the environmental impact of waste treatment processes like catalysis, adsorption, and photocatalysis.
2. Nanoscale Sensors for Waste Monitoring: Nanosensors can be used to detect and monitor waste parameters in real-time. These sensors, which are frequently included in monitoring systems, have the capacity to identify and measure particular waste constituents or contaminants, providing useful information for waste management and recycling procedures. More precise and sensitive monitoring is made possible by nanoscale sensors, allowing for quicker reactions and better waste management.

3. Nanomaterials for garbage Separation and Recycling: Nanotechnology provides creative ways to separate and recycle garbage. The effectiveness of recycling systems can be increased by using membranes and filters made of nanomaterials to selectively separate and recover valuable elements from waste streams. Nanomaterials can also be used as catalysts to change trash into useful goods, assisting in the conversion of waste into valuable resources.
4. Waste-to-Energy Conversion with Nanotechnology: Nanotechnology is used in the waste-to-energy conversion process. Through techniques like microbial fuel cells, thermoelectric conversion, or solar energy conversion, nanomaterials, such as nanocatalysts or nanostructured electrodes, can improve energy generation from waste materials. These nano-enabled methods have the potential to produce energy from the trash in an efficient and sustainable manner, lowering reliance on fossil fuels and minimizing environmental effects.
5. Improved Resource Recovery: The recovery of valuable resources from waste products is made easier by nanotechnology. Nanomaterials can increase the effectiveness of extraction procedures like leaching or solvent extraction, making it possible to recover rare earth elements, precious metals, and other important components from industrial waste, mining tailings, and electronic trash. By decreasing the need for virgin materials and the depletion of natural resources, this resource recovery helps to create a more circular economy.

By strengthening trash treatment procedures, providing real-time monitoring, improving waste sorting and recycling techniques, and aiding resource recovery, nanotechnology offers creative solutions for waste reduction and recycling. However, it is crucial to take into account the possible effects of nanomaterials on the environment and human health throughout their life cycles and to ensure the responsible and sustainable application of nanotechnology in recycling and waste management methods.

ENSURING FOOD SECURITY WITH NANOTECHNOLOGY

Food security is a problem that affects the entire world, and nanotechnology may be able to help. Here are some succinct remarks on using nanotechnology to ensure food security:

1. Improved Crop Production: Opportunities for improving crop output and raising agricultural productivity exist thanks to nanotechnology. The targeted delivery of nutrients and pesticides made possible by nanomaterials, such as nanofertilizers and nanopesticides, ensures their effective utilization and

reduces loss. This focused strategy improves crop yields, encourages healthy plant growth, and optimizes resource use.

2. Enhanced Nutrient Management: Particularly in nutrient-deficient soils, nanotechnology facilitates effective nutrient management. A controlled and prolonged delivery of nutrients to plants is made possible by the high surface area and controlled-release features of nanofertilizers, which also reduce nutrient loss through leaching and volatilization. This decreases the need for fertilizer, increases fertilizer uptake efficiency, and lessens environmental contamination.
3. Disease and Pest Management: Effective disease and pest management methods benefit from nanotechnology. Incorporating nanoparticles or nanocapsules, nanopesticides allow for the targeted delivery of pest control agents to particular pest or disease areas, hence minimizing the need for overuse of pesticides. This focused strategy reduces environmental deterioration and the emergence of chemical resistance, encouraging environmentally friendly pest control methods.
4. Food Quality and Safety: Nanotechnology helps maintain food safety and quality along the whole supply chain. Real-time monitoring of food quality indicators, such as freshness, spoilage, and contamination, is made possible by nanosensors embedded into packing materials or processing machinery. As a result, prompt actions are made possible, food waste is reduced, and food safety is guaranteed.
5. Post-Harvest Preservation: Solutions for post-harvest preservation and prolonging the shelf life of agricultural products is provided by nanotechnology. For protection against moisture loss, microbial development, and exterior spoiling factors, fruits, vegetables, and grains can be coated with nanomaterials or nanocomposites. This method of food preservation increases product quality, lowers post-harvest losses, and extends the time that food is available.
6. Sustainable Agriculture Practices: Nanotechnology encourages effective resource use, lowers waste, and lessens environmental consequences, which helps sustainable agricultural methods. Precision agriculture is made possible by nanomaterials, which allows for the more effective use of resources like water, fertilizer, and pesticides while reducing the harm they cause to ecosystems and the environment.

Nanotechnology has the potential to make a significant contribution to maintaining food security by enhancing crop yield, nutritional management, disease and insect control, food quality and safety, post-harvest preservation, and supporting sustainable agricultural practices. To create sustainable and fair food systems, it is crucial to take into account the ethical, social, and environmental effects of nanotechnology in agriculture and assure its responsible application.

POST-HARVEST PRESERVATION AND STORAGE

Agricultural produce's quality is maintained and its shelf life is extended through post-harvest preservation and storage. Innovative post-harvest preservation techniques made possible by nanotechnology provide sustainable agriculture and global food security. Here are some succinct notes on nanotechnology-based post-harvest preservation and storage:

1. Nanocoating: Agricultural items can be coated with nanocoating's to provide a barrier that stops moisture loss, gas exchange, and microbial contamination. While being stored and transported, grains, fruits, and vegetables retain more of their quality and freshness because of the improved barrier qualities of nanomaterials such as nanoparticles and nanocomposites.
2. Controlled-Release Systems: Nanotechnology makes it possible to create controlled-release applications for use after harvest. In order to prevent microbial development, lessen oxidative stress, and prolong the shelf life of perishable foods, bioactive chemicals, such as antioxidants or antimicrobial agents, can be enclosed in nanocarriers or nanocapsules and released gradually.
3. Smart Packaging: Nanotechnology aids in the creation of systems for smart packaging that track changes in food quality and react accordingly. Nanosensors built into packing materials can identify signs of spoilage like temperature, pH, or gas composition. These sensors give instantaneous feedback on food quality and allow for prompt action to stop deterioration and reduce food waste.
4. Ethylene Management: Fruits and vegetables ripen and age more quickly as a result of ethylene, a hormone produced naturally by plants. Systems for ethylene scavenging are available thanks to nanotechnology and nanomaterials like zeolites or metal-organic frameworks. By selectively adsorbing and expelling ethylene gas, these nanomaterials delay fruit and vegetable ripening and lengthen produce's shelf life.
5. Antibacterial Nanomaterials: Packaging materials or storage spaces can contain nanomaterials with antibacterial capabilities, such as silver nanoparticles or nanoparticles laden with essential oils. These nanoparticles prolong the shelf life of agricultural products by preventing the growth of diseases and bacteria that cause food deterioration.
6. Monitoring Food Quality Parameters: Non-destructive monitoring of food quality parameters is made possible by nanoscale imaging and sensing

technology. Produce can be evaluated for its internal properties, maturity, and quality using methods like hyperspectral imaging, atomic force microscopy, or nanosensors without sacrificing its integrity. As a result, decision-making and quality control are made easier during post-harvest processing and storage.

Sustainable agriculture and food security can be achieved by applying nanotechnology in post-harvest preservation and storage. In order to lower post-harvest losses, improve food quality, and ensure the availability of nutrient-dense produce for extended periods of time, technologies such as nanocoatings, controlled-release systems, smart packaging, ethylene management, antimicrobial nanomaterials, and quality monitoring systems are used. To ensure their responsible usage for sustainable and equitable food systems, it is crucial to address the possible safety and regulatory concerns related to nanoparticles.

CHALLENGES AND FUTURE DIRECTIONS

While nanotechnology has enormous potential for sustainable agriculture and food security, there are a number of obstacles that must be overcome in order to fully realize these advantages. Here are a few obstacles to sustainable agriculture and food security as well as potential paths for nanotechnology:

1. Safety and Regulation: The security of nanomaterials and their possible effects on the environment and human health continue to be major worries. To comprehend the long-term consequences of nanomaterials in agricultural systems, assure their safe application, and develop suitable rules and standards for their use, more research is required.
2. Cost-effectiveness: The expense of nanotechnology-based solutions may prevent them from being widely used in agriculture. Scaling up manufacturing procedures, improving synthesis techniques, and finding economical ways to incorporate nanomaterials into agricultural practices should be the main areas of future research and development.
3. Environmental Impact: Despite the possibility of resource-saving and ecologically benign solutions provided by nanotechnology, it is crucial to identify and reduce any potential negative environmental effects. Studies should look into how nanomaterials behave in the environment, whether they might bioaccumulate in living things, and how they affect ecosystems and unintended organisms.
4. Ethical and Social Considerations: The ethical and societal ramifications of each new technology must be carefully considered. Responsible and inclusive

use of nanotechnology in agriculture depends on ensuring equal access to nanotechnology-based solutions, addressing any knowledge gaps and disparities, and integrating stakeholders in the decision-making process.

5. Scaling up and Practical Implementation: Scalable and realistic implementation strategies are necessary for the successful transfer of nanotechnology from the lab to actual agricultural systems. Future studies should concentrate on streamlining application techniques, fusing nanotechnology with conventional agriculture processes, and solving the practical difficulties of widespread application.
6. Public Perception and Acceptance: It is crucial to increase public acceptance of and understanding of nanotechnology in agriculture. To effectively inform stakeholders, customers, and the general public about the potential advantages, safety precautions, and sustainable elements of nanotechnology.
7. Multidisciplinary Collaboration: Working together among scientists, engineers, legislators, business leaders, and farmers is necessary to address the issues and realize the potential of nanotechnology in agriculture. Knowledge exchange, creativity, and the creation of useable solutions will be facilitated through promoting transdisciplinary research and developing partnerships.

Future developments in nanotechnology for sustainable agriculture and food security ought to concentrate on overcoming these difficulties while looking into fresh ideas. The development of ethical and efficient applications of nanotechnology to boost agricultural output, reduce environmental impact, and guarantee global food security will be made possible by ongoing research, collaboration, and knowledge-sharing.

CONCLUSION

Finally, nanotechnology offers a potent and cutting-edge strategy to tackle the crucial issues of sustainability and food security in agriculture. Nanotechnology has the potential to significantly improve resource efficiency, increase crop yields, lessen environmental impact, and ensure the availability of healthy food. It can be used in a variety of agricultural practices, from crop production to post-harvest preservation.

Through the use of nanopesticides, accurate and targeted pest management, and nanosensors for real-time crop health and soil condition monitoring, nanotechnology offers precise and focused interventions in agriculture. These developments support better resource management, less waste, and increased production.

By providing waste reduction and recycling procedures, eco-friendly nanomaterial development, and filtration and purification technologies, the integration of nanotechnology in agriculture also addresses environmental concerns. These methods encourage the wise use of resources, reduce pollution, and support soil and water quality preservation.

Additionally, nanotechnology is essential for the post-harvest preservation and storage of agricultural products, prolonging their shelf lives, minimizing post-harvest losses, and maintaining food security. Innovative approaches to maintaining product quality, lowering spoilage, and decreasing food waste are provided by nanocomposites for packaging and nanoencapsulation of agrochemicals.

Although nanotechnology has a lot of potential, it is crucial to recognize and deal with the implementation issues. Careful thought must be given to safety issues, moral implications, legal frameworks, and the affordability of solutions based on nanotechnology. To ensure the ethical and sustainable use of nanomaterials in agriculture, extensive research and risk evaluations are required.

REFERENCES

Arora, S., Murmu, G., Mukherjee, K., Saha, S., & Maity, D. (2022). A comprehensive overview of nanotechnology in sustainable agriculture. *Journal of Biotechnology*, *355*, 21–41. doi:10.1016/j.jbiotec.2022.06.007 PMID:35752390

Asif, M., Islam, S., & Malik, A. (2022). *M., Mahdi Dar, Z., Masood, A., Shafi, S., Rasid, B., & Sidique, S.* Nano Pesticides Application in Agriculture and Their Impact on Environment. IntechOpen., doi:10.5772/intechopen.100690

Chaud, M., Souto, E. B., Zielinska, A., Severino, P., Batain, F., Oliveira-Junior, J., & Alves, T. (2021). Nanopesticides in Agriculture: Benefits and Challenge in Agricultural Productivity, Toxicological Risks to Human Health and Environment. *Toxics*, *9*(6), 131. doi:10.3390/toxics9060131 PMID:34199739

Guo, H., White, J. C., Wang, Z., & Xing, B. (2018). Nano-enabled fertilizers to control the release and use efficiency of nutrients. *Current Opinion in Environmental Science & Health*, *6*, 77–83. doi:10.1016/j.coesh.2018.07.009

Kumar, P., Singh, A., Rajput, V. D., Yadav, A. K. S., Kumar, P., Singh, A. K., & Minkina, T. (2022). Role of artificial intelligence, sensor technology, big data in agriculture: next-generation farming. *Bioinformatics in Agriculture*, 625–639. doi:10.1016/B978-0-323-89778-5.00035-0

Kumar, V., Guleria, P., & Mehta, S. K. (2017). Nanosensors for evaluating the quality and safety of food. *Environmental Chemistry Letters*, *15*, 165–177. doi:10.100710311-017-0616-4

Le, H., Karakasyan, C., Jouenne, T., Le Cerf, D., & Dé, E. (2021). Application of Polymeric Nanocarriers for Enhancing the Bioavailability of Antibiotics at the Target Site and Overcoming Antimicrobial Resistance. *Applied Sciences (Basel, Switzerland)*, *11*(22), 10695. doi:10.3390/app112210695

Prasad, R., Bhattacharyya, A., & Nguyen, Q. D. (2017). Nanotechnology in Sustainable Agriculture: Recent Developments, Challenges, and Perspectives. *Frontiers in Microbiology*, *8*, 264976. doi:10.3389/fmicb.2017.01014 PMID:28676790

Rai, M., & Ingle, A. (2012). Role of nanotechnology in agriculture with special reference to management of insect pests. *Applied Microbiology and Biotechnology*, *94*(2), 287–293. doi:10.100700253-012-3969-4 PMID:22388570

Rajput, V. D., Minkina, T., Upadhyay, S. K., Kumari, A., Ranjan, A., Mandzhieva, S., Sushkova, S., Singh, R. K., & Verma, K. K. (2022). Nanotechnology in the Restoration of Polluted Soil. *Nanomaterials (Basel, Switzerland)*, *12*(5), 769. doi:10.3390/nano12050769 PMID:35269257

Rajput, V. D., Minkina, T. M., Behal, A., Sushkova, S. N., Mandzhieva, S., Singh, R., Gorovtsov, A., Tsitsuashvili, V. S., Purvis, W. O., Ghazaryan, K. A., & Movsesyan, H. S. (2018). Effects of zinc-oxide nanoparticles on soil, plants, animals and soil organisms: A review. *Environmental Nanotechnology, Monitoring & Management*, *9*, 76–84. doi:10.1016/j.enmm.2017.12.006

Singh, A., Sengar, R. S., Rajput, V. D., Minkina, T., & Singh, R. K. (2022). Zinc Oxide Nanoparticles Improve Salt Tolerance in Rice Seedlings by Improving Physiological and Biochemical Indices. *Agriculture*, *12*(7), 1014. doi:10.3390/agriculture12071014

Singh, A., Sengar, R. S., Shahi, U. P., Rajput, V. D., Minkina, T., & Ghazaryan, K. A. (2022, December 26). Prominent Effects of Zinc Oxide Nanoparticles on Roots of Rice (Oryza sativa L.) Grown under Salinity Stress. *Stresses*, *3*(1), 33–46. doi:10.3390tresses3010004

Singh, A., Singh, A. K., Rawat, S., Pal, N., Rajput, V. D., Minkina, T., Sharma, R., Singh, N. P., & Tripathi, J. N. (2022). Satellite-Based Quantification of Methane Emissions from Wetlands and Rice Paddies Ecosystems in North and Northeast India. *Hydrobiology*, *1*(3), 317–330. doi:10.3390/hydrobiology1030023

Singh, T., Shukla, S., Kumar, P., Wahla, V., Bajpai, V. K., & Rather, I. A. (2017). Application of Nanotechnology in Food Science: Perception and Overview. *Frontiers in Microbiology*, *8*, 268461. doi:10.3389/fmicb.2017.01501 PMID:28824605

Verma, K. K., Song, P., Joshi, A., Tian, D., Rajput, V. D., Singh, M., Arora, J., Minkina, T., & Li, R. (2022). Recent Trends in Nano-Fertilizers for Sustainable Agriculture under Climate Change for Global Food Security. *Nanomaterials (Basel, Switzerland)*, *12*(1), 173. doi:10.3390/nano12010173 PMID:35010126

Worrall, E. A., Hamid, A., Mody, K. T., Mitter, N., & Pappu, H. R. (2018). Nanotechnology for Plant Disease Management. *Agronomy (Basel)*, *8*(12), 285. doi:10.3390/agronomy8120285

Chapter 15
Nanotechnology:
A New Innovation in Biofortification and Food Safety

Shivangi Singh
Sardar Vallabhbhai Patel University of Agriculture and Technology, India

Omkar Singh
Sardar Vallabhbhai Patel University of Agriculture and Technology, India

Uday Pratap Shahi
Sardar Vallabhbhai Patel University of Agriculture and Technology, India

ABSTRACT

Nanotechnology brings new prospects to increase food quality and safety through nano-enabled delivery systems, nanosensors, smart packaging and processing assistance. Nanomaterials display unique features at the 1-100 nanometer range that surpass constraints of classic breeding and detection methods. Iron, zinc, and vitamin A nano-encapsulates and nanoparticles can boost crops through tailored nutritional delivery. Nanosensors and smart packaging with antimicrobial nanoparticle coatings offer rapid, ultrasensitive pathogen and toxin detection while enhancing food stability. Nano-filters utilizing photocatalysts remove pollutants from agricultural water. However, the putative hazards from nanoparticle toxicity and accumulation require continuous investigation. Regulations and public acceptance factors are also crucial for sustainable progress. If developed properly, nanotechnologies can considerably boost food security through augmented micronutrition, antibacterial capabilities, and safety monitoring from production to consumption.

INTRODUCTION

Malnutrition affects over 2 billion people globally, particularly in developing countries where diets lack essential micronutrients like iron, zinc, vitamin A, and

DOI: 10.4018/978-1-6684-7232-3.ch015

iodine (WHO, 2018). Biofortification aims to increase micronutrients in staple crops through plant breeding and biotechnology (Bouis & Saltzman, 2017). However, traditional breeding approaches are limited in the amount of micronutrients they can increase in crops (De Steur et al., 2015). Meanwhile, food safety issues like microbial contamination continue to impact public health and food security worldwide. Each year, nearly 1 in 10 people fall ill due to contaminated food (WHO, 2020). Pathogens like Salmonella, Listeria, and E. coli affect food safety at multiple points from production to consumption (Newell et al., 2010). Nanotechnology, the engineering of materials at the nanoscale level, presents innovative solutions to enhance both the nutritional content and safety of food through biofortication and antimicrobial applications (Sozer & Kokini, 2020).

Nanotechnology involves manipulating materials at the nanoscale, which is between 1-100 nanometers (Mousavi and Rezaei, 2011). At this small scale, materials exhibit unique physical, chemical, and biological properties that can be leveraged in applications like medicine, energy, and agriculture (Liu et al., 2019). In agriculture, nanomaterials are being studied for their potential to enhance crop nutrition through precise engineering and delivery of fertilizers, pesticides, and micronutrients (Nair et al., 2010). Nanoparticles and nanocapsules can protect nutrients and chemicals, allowing for controlled release directly in plant tissues and cells (Servin et al., 2015). This precision helps improve the efficacy and reduce potential toxicity of agrochemicals compared to conventional application methods (Paris et al., 2020). Nanotechnology is also being applied to develop nanosensors that can detect plant pathogens and contaminants in soil and water (Mishra et al., 2021). Smart nanotechnology-enabled packaging systems can monitor freshness and food safety from farm to fork (Emamifar et al., 2010). Overall, nanotechnology aims to sustainably improve agricultural productivity and food quality. Malnutrition and foodborne illnesses remain persistent global challenges that impact billions of lives. While conventional breeding and food safety practices have seen incremental improvements, revolutionary approaches are needed to enhance the nutritional content and safety of our food supply. Nanotechnology, the engineering of materials at the atomic and molecular scale between 1-100 nanometers, presents numerous opportunities in this regard (Mousavi and Rezaei, 2011). At the nanoscale, materials exhibit unique optical, electrical, and biological properties that can overcome limitations of traditional food production methods. Nanoparticles, nanocapsules, and nano-enabled sensors are now demonstrating significant potential for biofortification and enhanced food safety through precise delivery of nutrients, antimicrobial capabilities, and ultrasensitive contaminant detection (DeRosa et al., 2010). Key innovations in using nanotechnology for biofortification include encapsulated micronutrient nano-delivery systems, nanoparticle plant fortificants, and nano-formulated fertilizers to boost health-promoting nutrients in foods (Kanmani and Rhim, 2014; Lahiani et al., 2013). For improving food safety, nanotechnology advancements include nanomaterial-based pathogen/toxin sensors, antimicrobial packaging, photocatalytic nano-filters, and smart nano-enabled tracking systems (Emamifar et al., 2010; Yam et al., 2005). This chapter provides an overview of these major developments and applications of nanotechnology that can enhance both the nutritional value and safety of our food supply in revolutionary new ways.

NANOTECHNOLOGY FOR BIOFORTIFICATION

Nanotechnology presents opportunities to enhance the nutritional content of foods through biofortification. Biofortification aims to increase essential micronutrients like iron, zinc, vitamin A, and iodine in staple crops (Bouis and Saltzman, 2017). Traditional breeding approaches have had limited success in significantly boosting levels of these micronutrients in plants. However, nano-enabled delivery systems and fortificants can increase nutrient content and bioavailability.

Using Nano-Encapsulates to Deliver Micronutrients

Polymeric nanoparticles and nanocapsules can encapsulate iron, zinc, vitamin A, and other micronutrients, protecting them until release inside plant tissues (Kanmani and Rhim, 2014). The nano-encapsulates enhance nutrient absorption compared to traditional soil fertilization. Nanocapsules also allow for controlled, sustained release directly within plant cells and organs over time (Nair et al., 2010).

Iron, Zinc, and Vitamin "A" Nanoparticle Delivery Systems

Specific delivery systems for key micronutrients include:

- Iron: Iron oxide nanoparticles applied to peanut plants increased iron levels in pods and haulms by over 20% compared to iron sulfate controls (Rui et al., 2016).
- Zinc: Zinc oxide nanoparticles as a foliar spray elevated zinc content in wheat grains by over 40% more than bulk zinc applications (Adhikari et al., 2016).
- Vitamin A: Nanodispersions with beta-carotene increased vitamin A content in sweet potato tubers by up to 60% through coatings on storage roots (Liu et al., 2014).

Benefits Over Traditional Breeding and Supplementation

Nano-biofortification provides benefits in speed, precision, and efficiency over traditional crop breeding and nutrient supplementation methods (DeRosa et al., 2010). Breeding biofortified crops takes over a decade, while nanotechnology solutions can boost nutrients in existing varieties immediately. Nanomaterials also allow for controlled, targeted delivery in plant tissues compared to imprecise fertilizer applications.

Case Studies on Nano-Biofortified Crops

- Rice: Zinc oxide nanoparticles foliar sprays increased grain zinc levels by 50% compared to bulk zinc applications (Pradhan et al., 2013).
- Wheat: Iron nanoparticles increased grain iron content by over 25% when applied to soil during growth (Linglan et al., 2011).
- Soybean: Foliar iron nano-chelates raised seed iron content by 94% over conventional iron fertilization (Rui et al., 2018).

Further field trials and safety studies are still needed, but these case studies demonstrate the potential for nanotechnology to rapidly biofortify staple food crops.

NANOTECHNOLOGY FOR IMPROVED FOOD SAFETY

Advancements in nanotechnology are providing innovative solutions to enhance food safety by detecting pathogens and toxins, inhibiting microbial growth, and tracking food from farm to fork. Nanomaterials possess unique optical, electrical, and antibacterial properties at the nanoscale that can improve food safety in ways not possible with conventional approaches (Duncan, 2011).

Nano-Sensors for Pathogen Detection

Rapid detection of foodborne pathogens like Salmonella, Listeria, and E. coli is critical for preventing outbreaks and recalls. Conventional detection methods require complex analytical equipment, intensive labor, and days to get results. Nanotechnology is enabling portable, affordable sensors that can detect bacteria and toxins in just minutes (Yáñez-Sedeno et al., 2018). Silicon nanowire biosensors can detect E. coli and Salmonella by binding antibodies to nanowire surfaces and measuring conductance changes when pathogens attach (Zhang et al., 2022). Carbon nanotube and nanoparticle-based sensors also offer ultrasensitive detection of bacteria by detecting gases or physicochemical changes from pathogen growth (Fu et al., 2019). Handheld nano-enabled devices allow on-site pathogen screening all along the food production chain.

Smart Packaging with Nanomaterial Coatings

Incorporating nanoparticles like silver, zinc oxide, and titanium dioxide into packaging materials gives them antimicrobial properties to inhibit bacterial growth and contamination of foods (Tankhiwale and Bajpai, 2009). Nanosilver has seen the most use for its broad-spectrum antibacterial activity against common foodborne pathogens. Studies found a 99% reduction in E. coli and Staphylococcus aureus counts on nanosilver coated polyethylene films over 5 days of cold storage (Emamifar et al., 2010). Active packaging with nanomaterials can extend shelf life and reduce safety risks from spoilage microbes or pathogens. Nanocomposite films also provide improved mechanical strength and gas barrier properties to prevent food degradation (Rhim et al., 2013).

Nano-Filters for Contaminant Removal

Nanoscale filters utilizing nanoparticles, nanofibers, and photocatalysts allow removal of contaminants like pesticides, parasites, and heavy metals from foods and agricultural rinse water (Mahdi et al., 2021). Titanium and zinc oxide nano-photocatalysts degrade chemical residues through photocatalytic reactions activated by light (Pelaez et al., 2012). Electrospun nanofiber membranes offer high-throughput filtration to remove protozoan parasites like Cryptosporidium from fresh produce wash water (Mhawi et al., 2022).

Tracking Food Products With Nanotechnology

RFID tags and nano-sensors incorporated into packaging can track food products through the supply chain and monitor time-temperature fluctuations to prevent spoilage (Yam et al., 2005). If limits are exceeded, smart packaging containing nanomaterials can visibly indicate spoiled or unsafe foods. Nanobarcodes and nanoparticles with biosensors also allow rapid on-site testing for pathogens or toxins in the field (Chen et al., 2020). These nanotechnology tracking and indicator systems enhance traceability and food safety. While showing great potential, questions remain on the toxicity, accumulation, and migration of nanomaterials in foods and the environment. Adequate risk assessment and safety testing is vital as nanotechnology provides revolutionary new ways to enhance food safety.

RISK ASSESSMENT OF NANOTECHNOLOGY IN FOOD SYSTEMS

The unique properties of engineered nanomaterials that allow for innovative applications in agriculture and food also raise concerns regarding potential toxicity risks. Adequate safety assessment and regulation is vital to ensure responsible development of nanotechnology in the food supply chain (Bumbudsanpharoke and Ko, 2015).

Toxicity of Nanoparticles

The ultra-small size and high surface area to volume ratio of nanoparticles allows them to interact closely with biological tissues and cells. This can make otherwise inert materials toxic (Sharifi et al., 2012). Common nanoparticles like metal oxides and silver nanoparticles have displayed toxicity in mammalian cells related to oxidative stress, DNA damage, inflammation, and mitochondrial disruption (Elsaesser and Howard, 2012). However, toxicity depends on physiochemical properties of the specific nanomaterial and experimental model. Nanoparticle coatings, size, shape, charge, and aggregation all affect biocompatibility and safety (Podila and Brown, 2013). Dose, route of administration, and duration of exposure also determine the toxicity profile. Comprehensive studies under realistic exposure conditions are still needed for many common engineered nanomaterials used in food applications.

Accumulation and Interactions in Food Systems

A major concern is the accumulation of nanoparticles in agricultural soils, crops, and the food chain after repeated applications for pesticides or fertilizers. Nanomaterials may interact with plants and substrates differently than bulk materials. Studies found greater zinc uptake in soil microbes with exposure to zinc oxide nanoparticles compared to bulk zinc compounds (Gupta et al., 2021). Nanosilver applied as a fungicide accumulated more in soil and had higher toxicity to beneficial soil microbiota than conventional silver fungicides (Schlich et al., 2013). The persistence of some types of nanoparticles in the environment remains unclear. More research is still needed on the environmental fate and ecological impacts of chronic low dose exposures.

Regulation and Safety Evaluation

The U.S. Food and Drug Administration (FDA) currently regulates nanotechnology applications using existing frameworks, depending on the category of products

(Chaudhry et al., 2008). Food additives containing nanomaterials require FDA approval and rigorously validated safety data. The FDA also recommends specific nano-enabled products like nano-coatings in packaging undergo consultations to identify data needs and safety issues. The European Food Safety Authority (EFSA) provides guidance on risk assessment of nanotechnology applications in the food and agriculture sectors (More et al., 2021). Required elements include detailed characterization of nanoparticle physiochemical properties, absorption and distribution in tissues, and toxicology profiles considering realistic exposure doses. Ongoing initiatives aim to develop targeted testing guidelines and validate appropriate models to evaluate nanomaterial hazards (Amenta et al., 2015).

Public Perceptions of Health Risks

Beyond established toxicology endpoints, public perceptions of potential nanotechnology risks in food also deserve consideration in responsible development and communication strategies (Gupta et al., 2015). Surveys indicate consumer views on nano-enabled food and agriculture products depend on the specific applications and benefits. But transparent communication, labeling, and oversight help build public trust in emerging technologies (Conti et al., 2012). Rigorous safety and risk research Side-by-side with innovative applications will support the sustainable advancement of nanotechnology in enhancing global food systems.

CONCLUSION AND FUTURE OUTLOOK

The major innovations and applications of nanotechnology show great promise to enhance both the nutritional content and safety of our food supply. Nano-enabled delivery systems and fortificants provide opportunities to boost levels of critical micronutrients like iron, zinc, and vitamin A in staple crops through biofortification. The precision, control, and efficiency of nanoscale nutrient delivery has advantages over conventional breeding and supplementation approaches. For improving food safety, nanomaterials and nanosensors allow ultrasensitive detection of pathogens, toxins, and contaminants throughout the production chain. Antimicrobial nano-coatings and nano-filters also enable mitigation of microbial and chemical hazards in foods. Nanotechnology tracking systems can monitor time-temperature and freshness indicators in real-time to prevent spoilage and waste. However, there remain challenges and open questions as nanotechnology aims to scale up in the agriculture and food sectors. Toxicity concerns over the long-term impacts of nanoparticle accumulation and environmental interactions need to be rigorously evaluated, especially for chronic exposures. Public acceptance and standardized regulations are also important considerations in responsible development and communication about nanotechnology uses in the food system.

In the future, research needs to continue advancing the selectivity, sensitivity, and cost-effectiveness of nano-enabled detection, filtration, and nutrient delivery systems. Field trials assessing efficacy and risks will be critical as promising technologies move from lab prototypes toward commercialization. If safety and sustainability are maintained as priorities, nanotechnology can usher in a new era of optimized nutrition and food safety to support food security worldwide. But collaboration between stakeholders from scientists to consumers will be vital in translating nano-innovations into positive real-world impacts.

REFERENCES

Adhikari, T., Kundu, S., & Biswas, A. K. (2016). Zinc complexed chitosan/TPP nanoparticles improves micronutrient content and growth of wheat (Triticum aestivum L.) plant. *Agronomy (Basel)*, *6*(3), 51.

Amenta, V., Aschberger, K., Arena, M., Bouwmeester, H., Moniz, F. B., Brandhoff, P., & Peters, R. J. (2015). Regulatory aspects of nanotechnology in the agri/feed/food sector in EU and non-EU countries. *Regulatory Toxicology and Pharmacology*, *73*(2), 463–476. doi:10.1016/j.yrtph.2015.06.016 PMID:26169479

Bouis, H. E., & Saltzman, A. (2017). Improving nutrition through biofortification: A review of evidence from HarvestPlus, 2003 through 2016. *Global Food Security*, *12*, 49–58. doi:10.1016/j.gfs.2017.01.009 PMID:28580239

Bumbudsanpharoke, N., & Ko, S. (2015). *Journal of Nanoparticle Research*, *17*(6), 1–19.

Chaudhry, Q., Scotter, M., Blackburn, J., Ross, B., Boxall, A., Castle, L., & Watkins, R. (2008). *Food Additives and Contaminants*, *25*(10), 1224–1243.

Chen, H., Chang, Y., Lin, Z., & Chu, Y. (2020). Nanoparticle-enabled paper moisture sensor for intelligent food packaging. *ACS Applied Materials & Interfaces*, *12*(46), 52001–52009.

Conti, J. A., Killpack, K. A., & Ogorzalek, T. L. (2012)... *Journal of Food Science*, *77*(7), R93–R104. PMID:22515252

De Steur, H., Blancquaert, D., Strobbe, S., Lambert, W., Gellynck, X., & Van Der Straeten, D. (2015). Status and market potential of transgenic biofortified crops. *Nature Biotechnology*, *33*(1), 25–29. doi:10.1038/nbt.3110 PMID:25574631

DeRosa, M. C., Monreal, C., Schnitzer, M., Walsh, R., & Sultan, Y. (2010). Nanotechnology in fertilizers. *Nature Nanotechnology*, *5*(2), 91–91. doi:10.1038/nnano.2010.2 PMID:20130583

Duncan, T. V. (2011). Applications of nanotechnology in food packaging and food safety: Barrier materials, antimicrobials and sensors. *Journal of Colloid and Interface Science*, *363*(1), 1–24. doi:10.1016/j.jcis.2011.07.017 PMID:21824625

Elsaesser, A., & Howard, C. V. (2012). Toxicology of nanoparticles. *Advanced Drug Delivery Reviews*, *64*(2), 129–137. doi:10.1016/j.addr.2011.09.001 PMID:21925220

Emamifar, A., Kadivar, M., Shahedi, M., & Soleimanian-Zad, S. (2010). Evaluation of nanotechnology application in the enhancement of antioxidant and antimicrobial activities of green tea. *Journal of Food Science*, *75*(4), N62–N67.

Fu, X., Chen, L., Chu, Z., Wang, J., Zhang, X., & Qin, Y. (2019). Nanotechnology in agriculture, livestock, and aquaculture in China. A review. *Agronomy for Sustainable Development*, *39*(3), 35.

Gupta, N., Fischer, A. R., & Frewer, L. J. (2015). *Comprehensive Reviews in Food Science and Food Safety*, *14*(5), 559–576.

Gupta, P., Dhankher, O. P., & Bahadur, J. (2021). *Journal of Hazardous Materials*, *403*, 123855. doi:10.1016/j.jhazmat.2020.123855 PMID:33264931

Kanmani, P., & Rhim, J. W. (2014). Antimicrobial and physical-mechanical properties of agar-based films incorporated with grapefruit seed extract. *Carbohydrate Polymers*, *102*, 708–716. doi:10.1016/j.carbpol.2013.10.099 PMID:24507339

Lahiani, M. H., Dervishi, E., Chen, J., Nima, Z., Gaume, A., Biris, A. S., & Khodakovskaya, M. V. (2013). Impact of carbon nanotube exposure to seeds of valuable crops. *ACS Applied Materials & Interfaces*, *5*(16), 7965–7973. doi:10.1021/am402052x PMID:23834323

Linglan, M., Chao, L., Jiangong, L., & Chunyang, L. (2011). *Journal of Agricultural and Food Chemistry*.

Liu, Q., Zhao, Y., Wan, Y., Xiao, L., & Fang, Y. (2014). *Food Chemistry*.

Liu, R., Lal, R., & Evangelou, V. P. (Eds.). (2019). Applications of nanotechnology in agri-environmental sciences. Vol. 2: Nanotechnology for sustainable agricultural production. CRC Press.

Mahdi, Z., Bahrami, A., Jonidi Jafari, A., Rezaei, M., Farzi, G., & Asadi, A. (2021). Application of nanotechnology in detection and removal of chemical contaminants in food. *Food Control*, *123*, 107858.

Mhawi, A., Doherty, W. O., Kamalvand, M., Williams, D. G., Clifford, M., & Zahid, M. (2022). Electrospun nanofibers for water treatment and food applications: A review. *The Science of the Total Environment*, *805*, 150213.

Mishra, A., Tripathi, S., Sharma, S. K., Srivastava, A., Dwivedi, P. D., Mishra, P. K., & Das, D. K. (2021). Applications of nanotechnology in agricultural system. *Frontiers in nanotechnology*, 3, 30.

More, S. J., Bampidis, V., Benford, D., Bennekou, S. H., Bragard, C., Halldorsson, T. I., & Silano, V. (2021). *EFSA Journal*, *19*(4), e06585. PMID:33976718

Mousavi, S. R., & Rezaei, M. (2011). Nanotechnology in agriculture and food production. *Journal of Applied Environmental and Biological Sciences*, *1*(10), 414–419.

Nair, R., Varghese, S. H., Nair, B. G., Maekawa, T., Yoshida, Y., & Kumar, D. S. (2010). *Plant Science*.

Newell, D. G., Koopmans, M., Verhoef, L., Duizer, E., Aidara-Kane, A., Sprong, H., Opsteegh, M., Langelaar, M., Threfall, J., Scheutz, F., der Giessen, J., & Kruse, H. (2010). Food-borne diseases—The challenges of 20 years ago still persist while new ones continue to emerge. *International Journal of Food Microbiology*, *139*, S3–S15. doi:10.1016/j.ijfoodmicro.2010.01.021 PMID:20153070

Paris, J. L., de la Luz Cádiz-Gurrea, M., León-González, M. E., Pérez-Rodríguez, F., & Segura-Carretero, A. (2020). State-of-the-art of nanotechnology applications in the food sector. *Food Chemistry*, *331*, 127431.

Pelaez, M., Nolan, N. T., Pillai, S. C., Seery, M. K., Falaras, P., Kontos, A. G., Dunlop, P. S. M., Hamilton, J. W. J., Byrne, J. A., O'Shea, K., Entezari, M. H., & Dionysiou, D. D. (2012). A review on the visible light active titanium dioxide photocatalysts for environmental applications. *Applied Catalysis B: Environmental*, *125*, 331–349. doi:10.1016/j.apcatb.2012.05.036

Podila, R., & Brown, J. M. (2013). *Toxicology and Applied Pharmacology*, *270*(3), 132–141.

Pradhan, S., Patra, P., Das, S., Chandra, S., Mitra, S., Dey, K. K., & Chattopadhyay, D. (2013). *Environmental Science & Technology*.

Rhim, J. W., Park, H. M., & Ha, C. S. (2013). Bio-nanocomposites for food packaging applications. *Progress in Polymer Science*, *38*(10-11), 1629–1652. doi:10.1016/j.progpolymsci.2013.05.008

Rui, M., Hao, Y., Zhang, Z., Xing, B., Zhang, J., Zou, F., & Tang, X. (2018). *Journal of Agricultural and Food Chemistry*.

Rui, M., Ma, C., Hao, Y., Guo, J., Rui, Y., Tang, X., & Fan, F. (2016). *Frontiers in Plant Science*.

Schlich, K., Klawonn, T., Terytze, K., & Hund-Rinke, K. (2013). *Journal of Environmental Quality*, *42*(1), 30–36. doi:10.2134/jeq2012.0196 PMID:23673736

Servin, A., Elmer, W., Mukherjee, A., De la Torre-Roche, R., Hamdi, H., White, J. C., Bindraban, P., & Dimkpa, C. (2015). A review of the use of engineered nanomaterials to suppress plant disease and enhance crop yield. *Journal of Nanoparticle Research*, *17*(2), 1–21. doi:10.100711051-015-2907-7

Sharifi, S., Behzadi, S., Laurent, S., Forrest, M. L., Stroeve, P., & Mahmoudi, M. (2012). Toxicity of nanomaterials. *Chemical Society Reviews*, *41*(6), 2323–2343. doi:10.1039/C1CS15188F PMID:22170510

Sozer, N., & Kokini, J. L. (2020). Nanotechnology and its applications in the food sector. *Trends in Biotechnology*, *38*(1), 82–89. PMID:19135747

Tankhiwale, R., & Bajpai, S. K. (2009). Preparation, characterization and antibacterial applications of ZnO-nanoparticles coated polyethylene films for food packaging. *Colloids and Surfaces. B, Biointerfaces*, *90*, 16–20. doi:10.1016/j.colsurfb.2011.09.031 PMID:22015180

World Health Organization. (2018). *Malnutrition*. WHO. https://www.who.int/news-room/fact-sheets/detail/malnutrition

World Health Organization. (2020). *Food safety*. WHO. https://www.who.int/news-room/fact-sheets/detail/food-safety

Yam, K. L., Takhistov, P. T., & Miltz, J. (2005). Intelligent packaging: Concepts and applications. *Journal of Food Science*, *70*(1), R1–R10. doi:10.1111/j.1365-2621.2005.tb09052.x

Yáñez-Sedeno, P., Pingarron, J. M., Villalonga, R., & Pingarrón, J. (2018). Electrochemical biosensors for pathogen detection directly in food samples. A review. *Analytical and Bioanalytical Chemistry*, *410*(12), 2953–2960.

Zhang, X., Niu, H., Zhang, Y., Cai, Z., Zhang, X., Zhang, N., & Liu, Y. (2022). Ultrasensitive SERS Nanosensor for On-Site Detection of Escherichia coli O157: H7 with a Limit of Quantitation Below 10 cfu/mL. *Analytical Chemistry*, *94*(5), 2543–2552. doi:10.1021/acs.analchem.6b00054 PMID:35080383

Chapter 16
Nanopriming Revolutionizing Agribiotechnology for Sustainable Food Security

Varsha Gupta
Bureau of Indian Standards, India

Santosh Kumar Gupta
National Institute of Plant Genome Research, India

ABSTRACT

The most promising branch of science and technology is nanotechnology. This area is constantly developing new tools and ideas for the development of sustainable agriculture. Nanoparticles (NPs) are used in agriculture due to their distinct physicochemical characteristics that revolutionize agriculture farming and global food security. One of the innovative seed-priming techniques is nano-priming, which helps to improve seed germination, seed growth, and yield by providing resistance to various stresses in crop plants. Nano-priming is a considerably more effective method compared to all other seed-priming methods. The interaction of nanomaterials with soil components is strongly determined in terms of soil quality and plant growth. NPs would be applied to improve the quality and reduce post-harvest loss of agricultural products by extending their shelf life, particularly for fruits and vegetables. This chapter assesses the available literature on nanotechnology and comprises a collection of information, opinions, and perspectives on a single platform.

INTRODUCTION

Agriculture is the crucial support of the developing economy and provides food for a better life. The Global population is projected to be 9 billion by 2050. It requires a

DOI: 10.4018/978-1-6684-7232-3.ch016

60% increase in food to meet demand (Alexandratos and Bruinsma, 2012; Godfray et al., 2010). Food and agriculture's biodiversity comprises livestock, crops, forestry, and aquaculture systems that sustain human beings (https://population.un.org/wpp/Publications/Files/WPP2019). The need for increased crop production and better food security is now a critical area of research for agricultural science. The advancements in science and technology provide new ideas and solutions to combat severe problems in the farming sector. In emerging nanotechnology, efficient nanoformulations are continuously produced for sustainable agriculture (Fraceto et al., 2016; Mittal et al., 2020). Plant health management and soil improvement would be possible using a variety of nanoparticle-based formulations, including nano-sized pesticides, herbicides, fungicides, and fertilizers. Agriculture is the backbone of the Indian national economy. It is the 'engine for growth in India. India has the second largest agricultural land in the world, with 157.35 million hectares after the USA. It has more agricultural land than China, producing 40 per cent more rice and wheat than India.

Regarding fruits and vegetables, India is the second largest producer in the World after China. India produces 106.19 million tons of rice yearly from 44 million hectares of land. It has been reported that around 60.3 per cent of India's land area is under use as agricultural land (FAO, 2013). India produces 93.51 million tons of wheat from 29.65 million hectares. Overall, India's yield rate is 3.15 tonnes per hectare. If we compare the yield rate, this rate is better than Brazil (2.73 tonnes per hectare) but lower than South Africa and China (3.40 & 4.9 t/ha), respectively.

In 1974 during World Food Conference, "food security" term came into actuality with a proper definition focused on supply. Food security is a condition to which we can relate the food supply and individuals' access. Food security is the availability of adequate world food supplies of basic foodstuffs to endure a steady expansion of food consumption and to counterbalance fluctuations in production and prices. Later, there was some modification in the explanation when in 1996 World Food Summit stated that food security only exists when all people constantly have physical and economic admittance to sufficient, safe. Nutritious food to complete their dietary needs and food preferences for a healthy life. WHO states that three pillars determine food security, availability, access, and use. First, there is a direct relationship between the food supply and its production, distribution, and exchange. Second, the determination of Food production is related to various factors, including land ownership, crop selection, breeding, livestock breeding, harvesting, and management. Third, food demand is influenced by several forces, like population growth, income levels, urbanization, lifestyles, and preferences. Achieve food security. There are various challenges and factors. These are the Global Water Crisis, land degradation, Climate change, Plant diseases, and other environmental factors. Many pests, conditions, and weeds attack crops.

The world must produce 40 per cent more food by 2030 despite limited land and water availability, less energy, fertilizers, and pesticides to overcome the Global Food Security challenges. With the help of the latest technology developed by modern scientific approaches, the world is ready to solve such a crucial dispute. There are significant areas where future research and innovation have the potential to pay bonuses. These are crop improvement, crop protection, fisheries and aquaculture, sustainable livestock farming, etc. Our biggest challenge with time is to improve agricultural production and decrease yield loss using modern tools and techniques of molecular biology and biotechnology to feed a growing population worldwide. The global population is estimated to increase by 7 to 9 billion by 2050. Biotic stresses resulting from extensive crop losses due to plant diseases hit the capacity of agricultural food production, with ~15% of global yield loss due to plant disease (Oerke, 2006). Research development in model plants like Arabidopsis, rice, Medicago, etc., has significantly enhanced our understanding of plant innate immune perception and signaling. Genome sequencing and advancements in the analysis are the basic knowledge gained in plant model species at a genome scale in crops.

This book chapter comprises a collection of information, opinions, and perspectives at a single platform from research articles based on the issues of global food security challenges based on Worldwide yield loss and its relation with biotic stresses, which includes interaction between crop plants and a variety of pathogens including fungi, bacteria, viruses, and aphids along with abiotic stresses like drought, salinity, and other environmental factors and (Stergiopoulos and Gordon, 2014). Although nano-priming might effectively induce seed germination and attribute tolerance to mild and intermittent stresses, however not been studied in detail. The studies on nano-priming in plants have been considered an emerging field, and the available reports induce us to provide a review of the potential of nano-priming in crop plants. This chapter briefly describes the nano-priming approaches, their mechanisms, and their potential crop benefits.

Nanotechnology and Agriculture

Nanotechnology has led to a new scientific revolution. The building block of nanotechnology are those particles synthesized at the nanoscale with at least a one-dimension size of less than 100 nm. This technology likely provides an innovative platform which helps to balance agricultural production and environmental sustainability. Based on its many potential benefits, researchers get attracted. These benefits are stress tolerance and increased crop production, absorption of enriched nutrients from the soil, and depletion in agricultural inputs leading to food safety and security. The critical role of nanotechnology in agriculture is to limit the use of chemicals. These fertilizers minimize the loss of nutrients during fertilization and provide tolerance against various biotic and abiotic stress, leading to deteriorated yield. This technology will help to improve the agriculture and food industries through recently developed nanotools. These are nano fertilizers, nano pesticides, and nanosensors. These tools can be utilized for monitoring products and nutrient levels for better productivity and to facilitate a safe atmosphere against various biotic and abiotic challenges. Nano fertilizers increase crop yield through the improved strategy of the uptake of nutrients. This helps to lower production overall costs and agricultural resolution. More about the underlying mechanisms of nano fertilizers are the topic of research and development globally.

Crops can better absorb nanomaterials (NMs) via seed priming in seeds, stomata as a foliar spray, and root as exogenously applied. It was reported that nano fertilizers were 18–29% more efficient than conventional chemical fertilizers (Kah, 2015). Glycine max productivity could be increased up to 20–33% by using calcium and phosphorus hydroxyapatite NPs, which is more than any traditional phosphorus fertilizer (Liu and Lal, 2015). A 10% increase was recorded in leaf chlorophyll content by applying iron, phosphorus, and nitrogen in the southern plant. Many other NPs, including zinc oxide (ZnO), cerium oxide (CeO2), silicon oxide (SiO2), CNTs (carbon nanotubes), and titanium oxide (TiO2), resulted from improved plant growth and stress tolerance. SiO2 and TiO2 have upgraded seed germination and nitrogen fixation with enhanced Glycine max growth. Similarly, CNTs are widely used for plant growth improvement in vegetables (i.e., cucumber, rape, tomato, etc.) and as fertilizer in other crops like corn and soybean. Nanoparticles are widely used as pesticides to overcome the stress of biotic factors in agriculture. These particles minimize environmental pollution since they are utilized less and more infrequently than conventional pesticides. Studies have suggested that silver (Ag) NPs can control pathogens like *Fusarium culmorum*, *Botrytis cinerea*, *Biploaris sorokinniana*, and *Megnaporthe grisea*. Other nano pesticides include copper (Cu) NPs, silica NMs, and Si–Ag NPs, reported to control pest diseases such as powdery mildew.

Furthermore, they said that gold (Au) NPs, nanodots, magnetic NPs, and carbon nanostructures are used as nanosensors. Nano biosensors are defined by their biological receptors with unique specialities towards correspondent analysts such as DNA or protein. Nano biosensors comprise nano-sized components that act as bio-receptors on a sensor and send signals to recognition elements to recognize single or multiplex solutes.

The nano priming approach for sustainable intensification is related to production in farming. This help to increase the yield without adverse environmental impact while cultivating the same agricultural area (The Royal Society, 2009).

Nano fertilizers could provide a better solution for modern agriculture. The plant uptake nutrients in these fertilizers by encapsulating the nanomaterials or coating them with a thin polymer film. This may be a promising advance to combat food scarcity and environmental degradation. Nanoparticles containing nutrients can enhance plant growth and improve the efficiency of conventional fertilizers. Using nano fertilizers led to an increase in plant potential than traditional fertilizers. It is established and reported that nano fertilizers showed better results compared to conventional and chemical fertilizers. Nano fertilizers can fulfil the requirements of plant roots, promoting disease resistance, enhancing plant growth, improving plant stability, and encouraging deeper rooting of the crops (Table 1).

Nano Fungicides and Pesticides

More than 70% of plant diseases are fungal diseases. They can reduce the yield by up to 100% in some crops. Conventional treatment by chemical fungicides harms crops' health and is a significant source of environmental pollution. Nano fungicides provide valuable solutions to reduce these problems depending on the size and shape of the NPs. Nano fungicides are supplied with phytocompounds and biocontrol agents. Some commercially available nanomaterial-based fungicides are Subdue MAXX and Cruiser MaXX (Prasad et al., 2017). Some other nano fungicides like Silver, TiO2, carbon, silica and alumino-silicates have been considered effective antifungal agents (Nikunj et al., 2014). It was reported that AgNPs and fluconazole have antifungal activity toward fungi such as *Fusarium semitectum*, *Phoma herbarum* and *Phoma glomerata* (Gajbhiye et al., 2009). Various other nanomaterials, Ag2S NPs, and Ag NPs have also been used as nano fungicides against *Aspergillus niger*, *Fusarium oxysporum*, *Magnaporthe grisea*, *Bipolaris sorokiniana*, *Fusarium culmorum* and *Colletotrichum* (Ahmad et al., 2022). Nanotechnology-based newly developed nanoparticles are used as nano-pesticides to overcome chemical pesticides in agriculture. Various commercially available nano pesticides are nano green, nano-pole insecticides, encapsulated plant protection agents (PRIMO Maxx and Karate® ZEON), insecticides such as cruiser MaXX, nano-pole insecticides, etc. In addition to these nanomaterials, copper oxide (CuO NPs), zinc oxide (ZnO NPs), magnesium hydroxide (MgOH NPs), magnesium oxide (MgO NPs) and silica dioxide (SiO2 NPs) (Table1). These nano pesticides are highly effective against various reported harmful pests in agriculture (Rankic et al., 2021; Thabet et al., 2021; Kah et al., 2019; Ahmad et al., 2022).

Nanoparticles for Soil Improvement

It was reported in various studies that hydrogels, nano clays, and nano zeolites were used to enhance the water-holding capacity of the soil (Sekhon, 2014). Various nanomaterials, including both the types organic (polymer and carbon nanotubes) and inorganic (nanometals and metal oxides), have also been used to absorb environmental contaminants (Khin et al., 2012). Using these nanomaterials can enhance the soil remediation capacity and reduce the time and costs of the treatments (Table 1).

Nanomaterials for Improved Plant Growth

Nanomaterials like carbon nanotubes and nanoparticles of Au, SiO2, ZnO, and TiO2 can significantly contribute to amending the development of plants. This development is started by enhancing elemental uptake and use of nutrients (Khota et al., 2012). These nanomaterials can impact plants based on their composition, concentration, size, surface charge, and physical-chemical properties (Ma et al., 2010; Lambreva et al., 2015). Developing new protocols and using different analytical techniques (such as microscopy, magnetic resonance imaging, and fluorescence spectroscopy) could considerably contribute to understanding the interactions between plants and nanomaterials. It was reported that if nanoparticles may not supply the nutrients necessary for plant growth, they can still enhance photosynthesis and support plants through other mechanisms (Zhao et al., 2020). Various NPs were crucial in plant growth and development (Table 1).

Nanotechnology in Plant Genetic Engineering

It is the era of plant biotechnology, where the most robust discipline is plant genetic engineering. This is an essential tool to improve crop plants' quality regarding their yield and enhancement in secondary metabolite contents. The well-established delivery methods for any desirable changes in the genome is either *Agrobacterium*-mediated and or particle bombardment through gene-gun transformation methods. These are widely employed techniques in plants. The limitations of these techniques open new horizons for plant transformation. A new approach based on nanoparticle-mediated NPs) genetic change recently opened the door to shaping modern agriculture systems by removing these restrictions. The transformation method boosts a new plant biotechnology strategy to promote a sustainable agriculture system. Recently, gene transfer and protein delivery using this technique have proven their crucial role in crop engineering. NPs can deliver DNA, RNA, and proteins, in addition to chloroplast-targeted transgene delivery recently reported. RNA interference (RNAi) is widely employed to protect and enhance crop quality. This technique delivers double-stranded RNA or small interfering (si) RNA in the nucleus affecting the normal gene coding process. NPs-mediated RNA delivery is most widely employed due to its high specificity and low-cost development. The applied NMs in the RNAi delivery system protect siRNA and help them to reach the target site with high efficacy (Zhang et al., 2017; Yan et al., 2021; Ahmad et al., 2022).

Table 1. Nano fertilizers used for improved plant growth and protect plants from stress

Nano fertilizers	**Role of nano fertilizers in plant growth**
Fe_2O_3, SiO_2 NPs	Increased germination rate and activity index
Fe_3O_4 NPs, Mg NPs, ZnO NPs,	Increased biomass, photosynthetic rate, chlorophyll content
Fe_2O_3, SiO_2 NPs, Mn NPs	Increased total protein, photophosphorylation, and oxygen evolution
Mg NPs, ZnO NPs, FeOx NPS, Fe_3O_4 NPs	Increased root length, number of tips and root biomass
ZnO NPs	Increased grain yield and Zn content
Fe_2O_3 NPs	Increased sugar and protein content
Nano-regulators	
C60	Alleviate oxidative stress
CeO_2 NP	Alleviate oxidative stress, increase photosynthesis, enhance carbon assimilation, and increase crop yields.
Ce-based NPs	Against *Fusarium wilt* suppression on tomato
Mg-based NPs	Against *Ralstonia solacearum*, it exerts a good effect on tobacco bacterial wilt and inhibits bacterial growth on *Xanthomonas alfalfa*, *Pseudomonas syringae*, and *E coli.*
Si-based NPs	Resistance to *Aspergillus* species, suppress Fusarium wilt.
Ag-based NPs	Reduced the severity of bacterial spot disease and reduced the number of nematodes, with antifungal solid activity against *Bipolaris sorokiniana*
Cu-based NPs	Have antifungal activity against plant pathogenic fungi *Fusarium oxysporum*, C. lunata, *A. alternata*, and *P. destructive*, suppressed fungal disease, reducing the severity of *C. michigenesis*
Carbon-based NPs	Have antifungal activity against two important plant pathogenic fungi
Ti-based NPs	The photochemically active properties and antibacterial characteristics

Nanotechnology in Plant-Pathogen Interaction, Detection, and Crop Protection

Plant Pathogens are responsible for severely reducing crop yields and continuous food supply (Lopez et al., 2003). It was estimated that insect pests, weeds and plant diseases caused severe losses of approximately USD 2000 billion annually. Based on the symptoms of leaves, stems, and fruits, the diagnosis of plant diseases could be initiated (Yvon et al., 2009). Plant diseases are initially symptomless; due to this reason, early diagnosis is not possible. Various molecular biology and immunological tools have been employed to detect the disease. These are polymerase chain reaction (PCR), ELISA-based direct dot-blot immunoassay (DTBIA), DNA-based methods and immunoassays due to limitations such as low sensitivity, time-consuming and high initial sample quantity requirement. Nanotechnology, NPs and QDs (quantum dots) have become significant tools for quickly detecting plant pathogens with high precision. Nanoparticles can detect pathogens directly or embedded in any chip or be used to detect the disease's specific compounds. More newly developed techniques like biosensor-based antibodies and DNA, nanoimaging and nanopore DNA sequencing are also utilized to recognize pathogens with a high specificity rate and rapid detection (Lopez et al., 2009). Diagnostic kits based on nanotechnology have the potential to detect smart plant pathogens. These kits can better help farmers to control and prevent epidemic plant diseases. Therefore, rapid detection of plant diseases through nanomaterials-based kits could provide solutions and benefit farmers.

CONCLUSION AND PERCEPTIONS

Nanomaterials can provide tolerance to crops in an eco-friendly manner. Various biotic and abiotic stresses account for most crop yield losses worldwide. Nanopriming should be given high research priority in plant science. This chapter examined the use of NPs to promote plant growth and disease resistance under stress and growth conditions and hopes for promising applications in agriculture. Developing various chemical platforms like surface chemistry could provide multifunctional properties on metal oxide NPs, leading to their broader utilization in modern agriculture. NPs can provide better solutions to agricultural needs, which will be crucial in using nanotechnology in sustainable agriculture. The safe use of NPs in contemporary agriculture must be critically evaluated due to the fate, transport, and environmental risk. The porosity of the plant cell wall is ~15 nm. This could provide good translocation efficiency of nanoparticles. The zeta potential values of nanoparticles are highly associated with the efficiency in penetrating chloroplast membranes. The concentration of nanoparticles and the cultivation mode should be well characterized before their use. These characterizations could allow the researchers to understand its effect and use it along with its crucial role in improved plant performance and stress tolerance.

REFERENCES

Ahmad, Z., Tahseen, S., Wasi, A., Ganie, I. B., Shahzad, A., Emamverdian, A., Ramakrishnan, M., & Ding, Y. (2022, August 3). Nanotechnological Interventions in Agriculture. *Nanomaterials (Basel, Switzerland)*, *12*(15), 2667. doi:10.3390/nano12152667 PMID:35957097

Alexandratos, N., & Bruinsma, J. (2012). *World Agriculture Towards 2030/2050: The 2012 Revision.* (ESA Working Paper No. 12-03). Food and Agriculture Organization of the United Nations, Rome.

Fraceto, L. F., Grillo, R., de Medeiros, G. A., Scognamiglio, V., Rea, G., & Bartolucci, C. (2016). Nanotechnology in agriculture: Which innovation potential does it have? *Frontiers in Environmental Science*, *4*, 20. doi:10.3389/fenvs.2016.00020

Gajbhiye, M., Kesharwani, J., Ingle, A., Gade, A., & Rai, M. (2009). Fungus-mediated synthesis of silver nanoparticles and their activity against pathogenic fungi in combination with fluconazole. *Nanomedicine; Nanotechnology, Biology, and Medicine*, *5*(4), 382–386. doi:10.1016/j.nano.2009.06.005 PMID:19616127

Godfray, H. C., Beddington, J. R., Crute, I. R., Haddad, L., Lawrence, D., Muir, J. F., Pretty, J., Robinson, S., Thomas, S. M., & Toulmin, C. (2010). Food security: The challenge of feeding 9 billion people. *Science*, *327*(5967), 812–818. doi:10.1126cience.1185383 PMID:20110467

Kah, M. (2015). Nanopesticides and Nanofertilizers: Emerging Contaminants or Opportunities for Risk Mitigation? *Frontiers in Chemistry*, *3*, 64. doi:10.3389/fchem.2015.00064 PMID:26636068

Kah, M., Tufenkji, N., & White, J. C. (2019). Nano–enabled strategies to enhance crop nutrition and protection. *Nature Nanotechnology*, *14*(6), 532–540. doi:10.103841565-019-0439-5 PMID:31168071

Khin, M. M., Nair, A. S., Babu, V. J., Murugan, R., & Ramakrishna, S. (2012). A review on nanomaterials for environmental remediation. *Energy & Environmental Science*, *5*(8), 8075–8109. doi:10.1039/c2ee21818f

Khota, L. R., Sankarana, S., Majaa, J. M., Ehsania, R., & Schuster, E. W. (2012). Applications of nanomaterials in agricultural production and crop protection: A review. *Crop Protection (Guildford, Surrey)*, *35*, 64–70. doi:10.1016/j.cropro.2012.01.007

Lambreva, M. D., Lavecchia, T., Tyystjärvi, E., Antal, T. K., Orlanducci, S., Margonelli, A., & Rea, G. (2015). Potential of carbon nanotubes in algal biotechnology. *Photosynthesis Research*, *125*(3), 451–471. doi:10.100711120-015-0168-z PMID:26113435

Liu, R., & Lal, R. (2015). Potentials of engineered nanoparticles as fertilizers for increasing agronomic productions. *The Science of the Total Environment*, *514*, 131–139. doi:10.1016/j.scitotenv.2015.01.104 PMID:25659311

López, M. M., Llop, P., Olmos, A., Marco-Noales, E., Cambra, M., & Bertolini, E. (2009). Are molecular tools solving the challenges posed by detection of plant pathogenic bacteria and viruses? *Current Issues in Molecular Biology*, *11*, 13–46. PMID:18577779

Ma, X., Geiser-Lee, J., Deng, Y., & Kolmakov, A. (2010). Interactions between engineered nanoparticles (ENPs) and plants: Phytotoxicity, uptake and accumulation. *The Science of the Total Environment*, *408*(16), 3053–3061. doi:10.1016/j.scitotenv.2010.03.031 PMID:20435342

Mittal, D., Kaur, G., Singh, P., Yadav, K., & Ali, S. A. (2020). Nanoparticle-Based Sustainable Agriculture and Food Science: Recent Advances and Future Outlook. *Front. Nanotechnol.*, *2*, 579954. doi:10.3389/fnano.2020.579954

Nikunj, P., Purvi, D., Niti, P., Anamika, J., & Gautam, H. K. (2014). Agronanotechnology for plant fungal disease management: A review. *International Journal of Current Microbiology and Applied Sciences*, *3*, 71–84.

Oerke, E.-C. (2006). Crop losses to pests. *Journal of Agricultural Science*, *144*(1), 31–43. doi:10.1017/S0021859605005708

Prasad, R., Bhattacharyya, A., & Nguyen, Q. D. (2017). Nanotechnology in Sustainable Agriculture: Recent Developments, Challenges, and Perspectives. *Frontiers in Microbiology*, *8*, 1014. doi:10.3389/fmicb.2017.01014 PMID:28676790

Rankic, I., Zelinka, R., Ridoskova, A., Gagic, M., Pelcova, P., & Huska, D. (2021). Nano/micro particle in conjunction with microalgae extract as novel insecticides against Mealworm beetles, Tenebrio molitor. *Scientific Reports*, *11*(1), 17125. doi:10.103841598-021-96426-0 PMID:34429471

Sekhon, B. S. (2014). Nanotechnology in agri-food production: An overview. *Nanotechnology, Science and Applications*, *7*, 31. doi:10.2147/NSA.S39406 PMID:24966671

Stergiopoulos, I., & Gordon, T. R. (2014). Cryptic fungal infections: The hidden agenda of plant pathogens. *Frontiers in Plant Science*, *5*, 506. doi:10.3389/fpls.2014.00506 PMID:25309571

Thabet, A. F., Boraei, H. A., Galal, O. A., El-Samahy, M. F. M., Mousa, K. M., Zhang, Y. Z., Tuda, M., Helmy, E. A., Wen, J., & Nozaki, T. (2021). Silica nanoparticles as pesticide against insects of different feeding types and their non-target attraction of predators. *Scientific Reports*, *11*(1), 14484. doi:10.103841598-021-93518-9 PMID:34262071

The Royal Society. (2009). *Reaping the Benefits: Science and the Sustainable Intensification of Global Agriculture*. The Royal Society.

World Population Prospects. (2022). *United Nations*. Department of Economic and Social Affairs.

Yan, S., Ren, B. Y., & Shen, J. (2021). Nanoparticle-mediated double-stranded RNA delivery system: A promising approach for sustainable pest management. *Insect Science*, *28*(1), 21–34. doi:10.1111/1744-7917.12822 PMID:32478473

Yvon, M., Thébaud, G., Alary, R., & Labonne, G. (2009). Specific detection and quantification of the phytopathogenic agent 'Candidatus Phytoplasma prunorum'. *Molecular and Cellular Probes*, *23*(5), 227–234. doi:10.1016/j.mcp.2009.04.005 PMID:19401232

Zhang, J., Khan, S. A., Heckel, D. G., & Bock, R. (2017). Next-generation insectresistant plants: RNAi-mediated crop protection. *Trends in Biotechnology*, *35*(9), 871–882. doi:10.1016/j.tibtech.2017.04.009 PMID:28822479

Zhao, L., Lu, L., Wang, A., Zhang, H., Huang, M., Wu, H., Xing, B., Wang, Z., & Ji, R. (2020). Nanobiotechnology in Agriculture: Use of Nanomaterials To Promote Plant Growth and Stress Tolerance. *Journal of Agricultural and Food Chemistry*, *68*(7), 1935–1947. doi:10.1021/acs.jafc.9b06615 PMID:32003987

Compilation of References

Abbasi Khalaki, M., Moameri, M., Asgari Lajayer, B., & Astatkie, T. (2020). Influence of nano-priming on seed germination and plant growth of forage and medicinal plants. *Plant Growth Regulation, 93*(93), 13–28.

Abbasi Khalaki, M., Moameri, M., Asgari Lajayer, B., & Astatkie, T. (2020). Influence of Nano-Priming on Seed Germination and Plant Growth of Forage and Medicinal Plants. *Plant Growth Regulation, 931*(93), 13–28. doi:10.100710725-020-00670-9

Abbas, Q., Liu, G., Yousaf, B., Ali, M. U., Ullah, H., Munir, M. A. M., Ahmed, R., & Rehman, A. (2020). Biochar-assisted transformation of engineered-cerium oxide nanoparticles: Effect on wheat growth, photosynthetic traits and cerium accumulation. *Ecotoxicology and Environmental Safety, 187*, 109845. doi:10.1016/j.ecoenv.2019.109845 PMID:31654865

Abdel Latef, A. A. H., Abu Alhmad, M. F., & Abdelfattah, K. E. (2017). The Possible Roles of Priming with ZnO Nanoparticles in Mitigation of Salinity Stress in Lupine (*Lupinus termis*) Plants. *Journal of Plant Growth Regulation, 36*(1), 60–70. doi:10.100700344-016-9618-x

AbdElgawad, H., Zinta, G., Hegab, M. M., Pandey, R., Asard, H., & Abuelsoud, W. (2016). High salinity induces different oxidative stress and antioxidant responses in maize seedlings organs. *Frontiers in Plant Science, 7*(MAR2016), 276. https://doi.org/ doi:10.3389/FPLS.2016.00276/BIBTEX

Abdelmigid, H. M., Alyamani, A. A., Hussien, N. A., Morsi, M. M., & Alhumaidi, A. (2022). Integrated approaches for adsorption and incorporation testing of green-synthesized TiO2NPs mediated by seed-priming technology in *Punica granatum* L. *Agronomy (Basel), 12*(7), 1601. doi:10.3390/agronomy12071601

Abedi, S., Iranbakhsh, A., Oraghi Ardebili, Z., & Ebadi, M. (2020). Seed priming with cold plasma improved early growth, flowering, and protection of Cichorium intybus against selenium nanoparticle. *J. Theor. Appl. Phys., 14*(2), 113–119. doi:10.100740094-020-00371-8

Abideen, Z., Waqif, H., Munir, N., El-Keblawy, A., Hasnain, M., Radicetti, E., Mancinelli, R., Nielsen, B. L., & Haider, G. (2022). Algal-mediated nanoparticles, phycochar, and biofertilizers for mitigating abiotic stresses in plants: A review. *Agronomy (Basel), 12*(8), 1788. doi:10.3390/agronomy12081788

Abid, M., Zhang, Y. J., Li, Z., Bai, D. F., Zhong, Y. P., & Fang, J. B. (2020). Effect of Salt stress on growth, physiological and biochemical characters of Four kiwifruit genotypes. *Scientia Horticulturae, 271*, 109473. doi:10.1016/j.scienta.2020.109473

Abobatta, W. F. (2018). Nanotechnology application in agriculture. *Acta Scientific Agriculture, 2*(6).

Abou-Zeid, H. M., & Moustafa, Y. (2014). *Physiological and Cytogenetic Responses of Wheat and Barley to Silver Nanopriming Treatment.* IMSEAR. https://imsear.searo.who.int/handle/123456789/168212

Acharya, P., Jayaprakasha, G.K., Crosby, K.M., Jifon, J.L., & Patil, B.S. (2019). Green-Synthesized Nanoparticles Enhanced Seedling Growth, Yield, and Quality of Onion (*Allium cepa* L.). *ACS Sustain. Chem. Eng.* 7, 14580-14590. https://doi.org/ SCHEMENG.9B02180/SUPPL_FILE/SC9B02180_SI_001.PDF. doi:10.1021/ACSSU

Acharya, P., Jayaprakasha, G. K., Crosby, K. M., Jifon, J. L., & Patil, B. S. (2019). Green-Synthesized Nanoparticles Enhanced Seedling Growth, Yield, and Quality of Onion (*Allium cepa* L.). *ACS Sustainable Chemistry & Engineering, 7*(17), 14580–14590. doi:10.1021/acssuschemeng.9b02180

Acharya, P., Jayaprakasha, G. K., Crosby, K. M., Jifon, J. L., & Patil, B. S. (2020). Nanoparticle-Mediated Seed Priming Improves Germination, Growth, Yield, and Quality of Watermelons (*Citrullus lanatus*) at multi-locations in Texas. *Scientific Reports, 101*(10), 1–16. doi:10.103841598-020-61696-7 PMID:32193449

Adhikari, T., Kundu, S., & Biswas, A. K. (2016). Zinc complexed chitosan/TPP nanoparticles improves micronutrient content and growth of wheat (Triticum aestivum L.) plant. *Agronomy (Basel), 6*(3), 51.

Adhikari, T., Kundu, S., & Rao, A. S. (2013). Impact of SiO2 and Mo nano particles on seed germination of rice (Oryza sativa L.). *Int J Agric Food Sci Technol, 4*(8), 809–816.

Adhikary, S., Biswas, B., Chakraborty, D., Timsina, J., Pal, S., Chandra Tarafdar, J., Banerjee, S., Hossain, A., & Roy, S. (2022). Seed priming with selenium and zinc NPs modifies germination, growth, and yield of direct-seeded rice (Oryza sativa L.). *Scientific Reports, 12*(1), 1–14. doi:10.1038/s41598-022-11307-4

Adio, S. O., Omar, M. H., Asif, M., & Saleh, T. A. (2017). Arsenic and selenium removal from water using biosynthesized nanoscale zero-valent iron: A factorial design analysis. *Process Safety and Environmental Protection, 107*, 518–527. doi:10.1016/j.psep.2017.03.004

Adisa, I. O., Pullagurala, V. L. R., Peralta-Videa, J. R., Dimkpa, C. O., Elmer, W. H., Gardea-Torresdey, J. L., & White, J. C. (2019). Recent advances in nano-enabled fertilizers and pesticides: A critical review of mechanisms of action. *Environmental Science. Nano, 6*(7), 2002–2030. doi:10.1039/C9EN00265K

Adugna, G. (2016). A review on impact of compost on soil properties, water use and crop productivity. *Academic Research Journal of Agricultural Science and Research, 4*(3), 93–104.

Afsah-Hejri, L., Jinap, S., Hajeb, P., Radu, S., & Shakibazadeh, S. (2013). A review on mycotoxins in food and feed: Malaysia case study. *Comprehensive Reviews in Food Science and Food Safety*, *12*(6), 629–651. doi:10.1111/1541-4337.12029 PMID:33412719

Afzal, S., Sharma, D., & Singh, N. K. (2021). Eco-friendly synthesis of phytochemical-capped iron oxide nanoparticles as nano-priming agent for boosting seed germination in rice (Oryza sativa L.). *Environmental Science and Pollution Research International*, *28*(30), 40275–40287. doi:10.100711356-020-12056-5 PMID:33447981

Ahmad, R., Mohsin, M., Ahmad, T., & Sardar, M. (2015). Alpha amylase assisted synthesis of TiO_2 nanoparticles: Structural characterization and application as antibacterial agents. *Journal of Hazardous Materials*, *283*, 171–177. doi:10.1016/j.jhazmat.2014.08.073 PMID:25270329

Ahmadvand, G., Soleimani, F., Saadatian, B., & Pouya, M. (2012). Effect of seed priming with potassium nitrate on germination and emergence traits of two soybean cultivars under salinity stress conditions. *American-Eurasian Journal of Agricultural & Environmental Sciences*, *12*, 769–777.

Ahmad, Z. A., Yeap, S. K., Ali, A. M., Ho, W. Y., Alitheen, N. B., & Hamid, M. (2012). scFv antibody: Principles and clinical application. *Clinical & Developmental Immunology*, *2012*, 980250. doi:10.1155/2012/980250 PMID:22474489

Ahmad, Z., Tahseen, S., Wasi, A., Ganie, I. B., Shahzad, A., Emamverdian, A., Ramakrishnan, M., & Ding, Y. (2022, August 3). Nanotechnological Interventions in Agriculture. *Nanomaterials (Basel, Switzerland)*, *12*(15), 2667. doi:10.3390/nano12152667 PMID:35957097

Ahmed, B., Dwivedi, S., Abdin, M. Z., Azam, A., Al-Shaeri, M., Khan, M. S., Saquib, Q., Al-Khedhairy, A. A., & Musarrat, J. (2017). Mitochondrial and Chromosomal Damage Induced by Oxidative Stress in Zn2+ Ions, ZnO-Bulk and ZnO-NPs treated Allium cepa roots. *Scientific Reports*, *7*(1), 40685. Advance online publication. doi:10.1038rep40685 PMID:28120857

Ahmed, S., & Kumar, S. (2020). Seed coating with fungicides and various treatments for protection of crops: A review. *Int J Agric Environ Sustain*, *2*(1), 6–13.

Ahuja, R., Sidhu, A., & Bala, A. (2019). Synthesis and Evaluation of Iron (Ii) Sulfide Aqua Nanoparticles (FeS-NPs) against Fusarium Verticillioides Causing Sheath Rot and Seed Discoloration of Rice. *European Journal of Plant Pathology*, *155*(1), 163–171. doi:10.100710658-019-01758-3

Ain, Q. T., Siddique, K., Bawazeer, S., Ali, I., Mazhar, M., Rasool, R., Mubeen, B., Ullah, F., Unar, A., & Jafar, T. H. (2023). Adaptive mechanisms in quinoa for coping in stressful environments: An update. *PeerJ*, *11*, e14832. doi:10.7717/peerj.14832 PMID:36883058

Akl, B. A., Nader, M. M., & El-Saadony, M. T. (2020). Biosynthesis of silver nanoparticles by Serratia marcescens ssp sakuensis and its antibacterial application against some pathogenic bacteria. *Journal of Agricultural Chemistry and Biotechnology*, *11*(11), 1–8. doi:10.21608/jacb.2020.76656

Akter, L., Fakir, O. A., Alam, M. K., Islam, M. U., Chakraborti, P., Alam, M. J., Rashid, M. H., Begum, M., & Kader, M. A. (2018). Amelioration of Salinity Stress in Maize Seed Germination and Seedling Growth Attributes through Seed Priming. *Open J. Soil Sci*, *08*(5), 137–146. doi:10.4236/ojss.2018.85011

Alabdallah, N. M., & Hasan, M. M. (2021). Plant-based green synthesis of silver nanoparticles and its effective role in abiotic stress tolerance in crop plants. *Saudi Journal of Biological Sciences*, *28*(10), 5631–5639. doi:10.1016/j.sjbs.2021.05.081 PMID:34588874

Al-Babili, S., Ye, X., Lucca, P., Potrykus, I., & Beyer, P. (2001). Biosynthesis of beta-carotene (provitamin A) in rice endosperm achieved by genetic engineering. *Novartis Foundation Symposium*, *236*, 219–228. PMID:11387982

Alexandratos, N., & Bruinsma, J. (2012). *World Agriculture Towards 2030/2050: The 2012 Revision.* (ESA Working Paper No. 12-03). Food and Agriculture Organization of the United Nations, Rome.

Alexandratos, N., & Bruinsma, J. (2012). *World agriculture towards 2030/2050: the 2012 revision*. FAO.

Alexoaei, A. P., Robu, R. G., Cojanu, V., Miron, D., & Holobiuc, A.-M. (2022). Good Practices in Reforming the Common Agricultural Policy to Support the European Green Deal–A Perspective on the Consumption of Pesticides and Fertilizers. *Amfiteatru Economic*, *24*(60), 525–545. doi:10.24818/EA/2022/60/525

Alharbi, N. S., Bhakyaraj, K., Gopinath, K., Govindarajan, M., Kumuraguru, S., Mohan, S., & Benelli, G. (2016). Gum-mediated fabrication of eco-friendly gold nanoparticles promoting cell division and pollen germination in plant cells. *Journal of Cluster Science*, *28*(1), 507–517. doi:10.100710876-016-1130-8

Aliferis, K. A., & Jabaji, S. (2011). Metabolomics – A robust bioanalytical approach for the discovery of the modes-of-action of pesticides: A review. *Pesticide Biochemistry and Physiology*, *100*(2), 105–117. doi:10.1016/j.pestbp.2011.03.004

Ali, M. H., Sobze, J. M., Pham, T. H., Nadeem, M., Liu, C., Galagedara, L., Cheema, M., & Thomas, R. (2020). Carbon Nanotubes Improved the Germination and Vigor of Plant Species from Peatland Ecosystem Via Remodeling the Membrane Lipidome. *Nanomaterials (Basel, Switzerland)*, *10*(9), 1852. doi:10.3390/nano10091852 PMID:32947854

Ali, S., Mehmood, A., & Khan, N. (2021). Uptake, translocation, and consequences of nanomaterials on plant growth and stress adaptation. *Journal of Nanomaterials*, *2021*, 1–17. doi:10.1155/2021/6677616

Al-Juthery, H. W., Lahmod, N. R., & Al-Taee, R. A. (2021). Intelligent, nano-fertilizers: a new technology for improvement nutrient use efficiency (article review). *IOP Conference Series: Earth and Environmental Science*. IOP Science.

Almutairi, Z. M. (2016). Influence of silver nano-particles on the salt resistance of tomato (*Solanum lycopersicum*) during germination. *International Journal of Agriculture and Biology, 18*, 125–129. doi:10.17957/IJAB/15.0114

Al-Samarrai, A. M. (2012). Nanoparticles as alternative to pesticides in management plant diseases-a review. *International Journal of Scientific and Research Publications*, *2*(4), 1–4.

Al-Selwey, W. A., Alsadon, A. A., Ibrahim, A. A., Labis, J. P., & Seleiman, M. F. (2023). Effects of zinc oxide and silicon dioxide nanoparticles on physiological, yield, and water use efficiency traits of potato grown under water deficit. *Plants*, *12*(1), 218. doi:10.3390/plants12010218 PMID:36616345

Altieri, M. A., Nicholls, C. I., Henao, A., & Lana, M. A. (2015). Agroecology and the design of climate change-resilient farming systems. *Agronomy for Sustainable Development*, *35*(3), 869–890. doi:10.100713593-015-0285-2

Altieri, M. A., Nicholls, C. I., & Montalba, R. (2017). Technological approaches to sustainable agriculture at a crossroads: An agroecological perspective. *Sustainability (Basel)*, *9*(3), 349. doi:10.3390u9030349

Ambhorkar, P., Wang, Z., Ko, H., Lee, S., Koo, K. I., Kim, K., & Cho, D. I. D. (2018). Nanowire-based biosensors: From growth to applications. *Micromachines*, *2018*(9), 679. doi:10.3390/mi9120679 PMID:30572645

Amenta, V., Aschberger, K., Arena, M., Bouwmeester, H., Moniz, F. B., Brandhoff, P., & Peters, R. J. (2015). Regulatory aspects of nanotechnology in the agri/feed/food sector in EU and non-EU countries. *Regulatory Toxicology and Pharmacology*, *73*(2), 463–476. doi:10.1016/j.yrtph.2015.06.016 PMID:26169479

Amna, A. B., Azeem, M. A., Qayyum, A., Mustafa, G., Ahmad, M. A., & Chaudhary, H. J. (2022). Bio-fabricated silver nanoparticles: A sustainable approach for augmentation of plant growth and pathogen control. In Sustainable agriculture reviews 53: nanoparticles: a new tool to enhance stress tolerance (pp. 345-371). Cham: Springer International Publishing.

Amooaghaie, R., Saeri, M. R., & Azizi, M. (2015). Synthesis, characterization and biocompatibility of silver nanoparticles synthesized from Nigella sativa leaf extract in comparison with chemical silver nanoparticles. *Ecotoxicology and Environmental Safety*, *120*, 400–408. doi:10.1016/j.ecoenv.2015.06.025 PMID:26122733

Anita, S., & Rao, D. P. (2014). Enhancement of seed germination and plant growth of wheat, maize, peanut and garlic using multiwalled carbon nanotubes. *European Chemical Bulletin*, *3*(5), 502–504.

An, J., Hu, P., Li, F., Wu, H., Shen, Y., White, J. C., Tian, X., Li, Z., & Giraldo, J. P. (2020). Emerging investigator series: Molecular mechanisms of plant salinity stress tolerance improvement by seed priming with cerium oxide nanoparticles. *Environmental Science. Nano*, *7*(8), 2214–2228. doi:10.1039/D0EN00387E

Annamalai, J., Ummalyma, S. B., Pandey, A., & Bhaskar, T. (2021). Recent trends in microbial nanoparticle synthesis and potential application in environmental technology: A comprehensive review. *Environmental Science and Pollution Research International, 28*(36), 49362–49382. doi:10.100711356-021-15680-x PMID:34331227

Anwar, T. A., Qureshi, H., Parveen, N., Mahmood, S., Haider, M. Z., Mumtaz, S., & Ahmed, A. E. (2022). Herbicidal effectiveness of wild poisonous plant Rhazya stricta using different media by the sandwich method. *Pakistan Journal of Botany, 55*(10.30848).

Araya-Castro, K., Chao, T. C., Durán-Vinet, B., Cisternas, C., Ciudad, G., & Rubilar, O. (2020). Green synthesis of copper oxide nanoparticles using protein fractions from an aqueous extract of Brown Algae Macrocystis pyrifera. *Processes (Basel, Switzerland), 9*(9), 78. doi:10.3390/pr9010078

Arif, R., & Uddin, R. (2021). A review on recent developments in the biosynthesis of silver nanoparticles and its biomedical applications. *Medical Devices & Sensors, 4*(4), 10158. doi:10.1002/mds3.10158

Arif, U., Bhatti, K. H., Ajaib, M., Wagay, N. A., Majeed, M., Zeb, J., Hameed, A., & Kiani, J. (2021). Ethnobotanical indigenous knowledge of Tehsil Charhoi, District Kotli, Azad Jammu and Kashmir, Pakistan. *Ethnobotany Research and Applications, 22*. doi:10.32859/era.22.50.1-24

Arnott, A., Galagedara, L., Thomas, R., Cheema, M., & Sobze, J. M. (2021). The potential of rock dust nanoparticles to improve seed germination and seedling vigor of native species: A review. *The Science of the Total Environment, 775*, 145139. doi:10.1016/j.scitotenv.2021.145139

Arora, N. K., Fatima, T., Mishra, I., Verma, M., Mishra, J., & Mishra, V. (2018). Environmental sustainability: Challenges and viable solutions. *Environmental Sustainability, 1*(4), 309–340. doi:10.100742398-018-00038-w

Arora, S., Murmu, G., Mukherjee, K., Saha, S., & Maity, D. (2022). A comprehensive overview of nanotechnology in sustainable agriculture. *Journal of Biotechnology, 355*, 21–41. doi:10.1016/j.jbiotec.2022.06.007 PMID:35752390

Arshad, F., Waheed, M., Harun, N., Fatima, K., Khan, B. A., Fatima, K., Abbas, Z., Jabeen, S., & Majeed, M. (2022). Indigenous farmer's perception about fodder and foraging species of Semi-arid lowlands of Pakistan: A case study of District Kasur, Pakistan. *Taiwania, 67*(4), 510–523. doi:10.6165/tai.2022.67.510

Ashraf, M., & Foolad, M. R. (2005). Pre-sowing seed treatment—A shotgun approach to improve germination, plant growth, and crop yield under saline and non-saline conditions. *Advances in Agronomy, 88*, 223–271. doi:10.1016/S0065-2113(05)88006-X

Ashraf, M., & Foolad, M. R. (2007). Roles of glycine betaine and proline in improving plant abiotic stress resistance. *Environmental and Experimental Botany, 59*(2), 206–216. doi:10.1016/j.envexpbot.2005.12.006

Ashraf, S. A., Siddiqui, A. J., Elkhalifa, A. E. O., Khan, M. I., Patel, M., Alreshidi, M., Moin, A., Singh, R., Snoussi, M., & Adnan, M. (2021). Innovations in nanoscience for the sustainable development of food and agriculture with implications on health and environment. *The Science of the Total Environment*, *768*, 144990. doi:10.1016/j.scitotenv.2021.144990 PMID:33736303

Asif, M., Islam, S., & Malik, A. (2022). *M., Mahdi Dar, Z., Masood, A., Shafi, S., Rasid, B., & Sidique, S.* Nano Pesticides Application in Agriculture and Their Impact on Environment. IntechOpen., doi:10.5772/intechopen.100690

Atkinson, P. M., & Tate, J. J. (2000). *Advances in Remote Sensing and GIS Analysis*. Wiley.

Avellan, A., Schwab, F., Masion, A., Chaurand, P., Borschneck, D., Vidal, V., Rose, J., Santaella, C., & Levard, C. (2017). Nanoparticle Uptake in Plants: Gold Nanomaterial Localized in Roots of Arabidopsis thaliana by X-ray Computed Nanotomography and Hyperspectral Imaging. *Environmental Science & Technology*, *51*(15), 8682–8691. doi:10.1021/acs.est.7b01133 PMID:28686423

Awasthi, A., Bansal, S., Jangir, L. K., Awasthi, G., Awasthi, K. K., & Awasthi, K. (2017). Effect of ZnO Nanoparticles on Germination of *Triticum aestivum* Seeds. *Macromolecular Symposia*, *376*(1), 1700043. doi:10.1002/masy.201700043

Awasthi, G., Nagar, V., Mandzhieva, S., Minkina, T., Sankhla, M. S., Pandit, P. P., & Srivastava, S. (2022). Sustainable amelioration of heavy metals in soil ecosystem: Existing developments to emerging trends. *Minerals (Basel)*, *12*(1), 85. doi:10.3390/min12010085

Azeem, A., Abbas, N., Azeem, S., Iqbal, Z., & Ul-Allah, S. (2023). Physiological and Molecular Mechanism of Nanoparticles Induced Tolerance in Plants. In *Emerging Contaminants and Plants: Interactions, Adaptations and Remediation Technologies* (pp. 233–248). Springer International Publishing. doi:10.1007/978-3-031-22269-6_9

Azimi, F., Oraei, M., Gohari, G., Panahirad, S., & Farmarzi, A. (2021). Chitosan-selenium nanoparticles (Cs–Se NPs) modulate the photosynthesis parameters, antioxidant enzymes activities and essential oils in Dracocephalum moldavica L. under cadmium toxicity stress. *Plant Physiology and Biochemistry*, *167*, 257–268. doi:10.1016/j.plaphy.2021.08.013 PMID:34391200

Babajani, A., Iranbakhsh, A., Oraghi Ardebili, Z., & Eslami, B. (2019). Seed Priming with Non-thermal Plasma Modified Plant Reactions to Selenium or Zinc Oxide Nanoparticles: Cold Plasma as a Novel Emerging Tool for Plant Science. *Plasma Chemistry and Plasma Processing*, *39*(1), 21–34. doi:10.100711090-018-9934-y

Babu, P. J. (2021). Nanotechnology mediated intelligent and improved food packaging. *International Nano Letters*, (12), 1–14.

Baghaienezhad, M., Boroghani, M., & Anabestani, R. (2020). Silver nanoparticles synthesis by coffee residues extract and their antibacterial activity. *Nanomedicine Research Journal*, (5), 29–34.

Bailly, C. (2019). The signalling role of ROS in the regulation of seed germination and dormancy. *The Biochemical Journal*, *476*(20), 3019–3032. doi:10.1042/BCJ20190159 PMID:31657442

Baird, V., Abbott, A., Ballard, R., Sosinski, B., & Rajapakse, S. (1997). Current Topics in Plant Molecular Biology: Technology Transfer of Plant Biotechnology. CRC Press, Boca Raton. P. Gresshoff (Ed.), DNA Diagnostics in Horticulture, 111–130.

Bamal, D., Singh, A., Chaudhary, G., Kumar, M., Singh, M., Rani, N., Mundlia, P., & Sehrawat, A. R. (2021). Silver nanoparticles biosynthesis, characterization, antimicrobial activities, applications, cytotoxicity and safety issues: An updated review. *Nanomaterials (Basel, Switzerland)*, *11*(8), 2086. doi:10.3390/nano11082086 PMID:34443916

Bano, A., Gupta, A., Rai, S., Fatima, T., Sharma, S., & Pathak, N. (2021). Mechanistic role of reactive oxygen species and its regulation via the antioxidant system under environmental stress. *Plant Stress Physiology—Perspectives in Agriculture*, 1-18.

Barbieri, M., Sappa, G., & Nigro, A. (2018). Soil Pollution: Anthropogenic versus Geogenic Contributions over Large Areas of the Lazio Region. *Journal of Geochemical Exploration*, *195*, 78–86. doi:10.1016/j.gexplo.2017.11.014

Barone, A. (2004). Molecular marker-assisted selection for potato breeding. *American Journal of Potato Research*, *81*(2), 111–117. doi:10.1007/BF02853608

Bartolucci, C., Antonacci, A., Arduini, F., Moscone, D., Fraceto, L., Campos, E., Attaallah, R., Amine, A., Zanardi, C., Cubillana-Aguilera, L. M., Palacios Santander, J. M., & Scognamiglio, V. (2020). Green nanomaterials fostering agrifood sustainability. *Trends in Analytical Chemistry*, *125*, 115840. doi:10.1016/j.trac.2020.115840

Basten, C. J., Weir, B. S., & Zeng, Z.-B. (1994). Zmap-a QTL cartographer. J.S.G.C. Smith.

Basten, C., Weir, B., & Zeng, Z.-B. (2001). *QTL Cartographer*. Department of Statistics, North Carolina State University.

Beattie, A., Larsen, J., Michaels, T., & Pauls, K. (2003). Mapping quantitative trait loci for a common bean (*Phaseolus vulgaris L.*) ideotype. *Genome*, *46*(3), 411–422. doi:10.1139/g03-015 PMID:12834057

Begum, P., & Fugetsu, B. (2012). Phytotoxicity of multi-walled carbon nanotubes on red spinach (Amaranthus tricolor L.) and the role of ascorbic acid as an antioxidant. *Journal of Hazardous Materials*, *243*, 212–222. doi:10.1016/j.jhazmat.2012.10.025 PMID:23146354

Begum, P., & Fugetsu, B. (2013). Induction of cell death by graphene in Arabidopsis thaliana (Columbia ecotype) T87 cell suspensions. *Journal of Hazardous Materials*, *260*, 1032–1041. doi:10.1016/j.jhazmat.2013.06.063 PMID:23892171

Begum, P., Ikhtiari, R., & Fugetsu, B. (2011). Graphene phytotoxicity in the seedling stage of cabbage, tomato, red spinach, and lettuce. *Carbon*, *49*(12), 3907–3919. doi:10.1016/j.carbon.2011.05.029

Behl, T., Kaur, I., Sehgal, A., Singh, S., Sharma, N., Bhatia, S., Al-Harrasi, A., & Bungau, S. (2022). The dichotomy of nanotechnology as the cutting edge of agriculture: Nano-farming as an asset versus nanotoxicity. *Chemosphere*, *288*, 132533. doi:10.1016/j.chemosphere.2021.132533 PMID:34655646

Bello-Bello, J. J., Chavez-Santoscoy, R. A., Lecona-Guzman, C. A., Bogdanchikova, N., Salinas-Ruíz, J., Gomez-Merino, F. C., & Pestryakov, A. (2017). Hormetic response by silver nanoparticles on in vitro multiplication of sugarcane (Saccharum spp. Cv. Mex 69-290) using a temporary immersion system. *Dose-Response*, *15*(4), 1559325817744945. doi:10.1177/1559325817744945 PMID:29238274

B. J. Benkel, W. F. Chesnais, J. P. Gibson, B. W. Kennedy, & E. B. Burnside (Eds.). *Proceedings of the 5th World Congress on Genetics Applied to Livestock Production: Computing Strategies and Software*. Published by the Organizing Committee.

Bentsink, L., & Koornneef, M. (2008). Seed dormancy and germination. *The Arabidopsis Book / American Society of Plant Biologists*, *6*, e0119. doi:10.1199/tab.0119 PMID:22303244

Bergmann, C. P., & Machado, F. (2015). *Carbon Nanomaterials as Adsorbents for Environmental and Biological Applications*. Springer. doi:10.1007/978-3-319-18875-1

Bhardwaj, D., Ansari, M. W., Sahoo, R. K., & Tuteja, N. (2014). Biofertilizers function as key player in sustainable agriculture by improving soil fertility, plant tolerance and crop productivity. *Microbial Cell Factories*, *13*(1), 1–10. doi:10.1186/1475-2859-13-66 PMID:24885352

Bhatt, D., Bhatt, M.D., Nath, M., Dudhat, R., Sharma, M., & Bisht, D.S. (2020). Application of nanoparticles in overcoming different environmental stresses. Protective chemical agents in the amelioration of plant abiotic stress. *Biochemistry and Molcular Perspect.* 635–654.

Bhattacharjee, R., Kumar, L., Mukerjee, N., Anand, U., Dhasmana, A., Preetam, S., Bhaumik, S., Sihi, S., Pal, S., Khare, T., Chattopadhyay, S., El-Zahaby, S. A., Alexiou, A., Koshy, E. P., Kumar, V., Malik, S., Dey, A., & Proćków, J. (2022). The emergence of metal oxide nanoparticles (NPs) as a phytomedicine: A two-facet role in plant growth, nano-toxicity and anti-phyto-microbial activity. *Biomedicine and Pharmacotherapy*, *155*, 113658. doi:10.1016/j.biopha.2022.113658 PMID:36162370

Bhavyasree, P. G., & Xavier, T. S. (2020). Green synthesis of Copper Oxide/Carbon nanocomposites using the leaf extract of Adhatoda vasica Nees, their characterization and antimicrobial activity. *Heliyon*, *6*(6), 03323. doi:10.1016/j.heliyon.2020.e03323 PMID:32072042

Bhowmick, M. K., Biswas, P. K., Sen, P., & Bhattacharyya, P. (2010). Studies on seed priming, row spacing and foliar nutrition in chickpea under rainfed conditions in West Bengal, India. In: *Proceedings of the international seminar on climate change and environmental challenges of 21st century, December* (pp 7–9). Research Gate.

Bhowmick, M. K., Biswas, P. K., Sen, P., & Bhattacharyya, P. (2010, December). Studies on seed priming, row spacing and foliar nutrition in chickpea under rainfed conditions in West Bengal, India. In *Proceedings of the international seminar on climate change and environmental challenges of 21st century, December* (pp. 7-9). Research Gate.

Bienert, G. P., & Chaumont, F. (2014). Aquaporin-facilitated transmembrane diffusion of hydrogen peroxide. *Biochimica et Biophysica Acta*, *1840*(5), 1596–1604. doi:10.1016/j.bbagen.2013.09.017 PMID:24060746

Bisht, N., & Chauhan, P. S. (2020). Excessive and disproportionate use of chemicals cause soil contamination and nutritional stress. *Soil Contamination-Threats and Sustainable Solutions*, 1-10.

Bolstad, P. (2002). GIS Fundaments. Eider Press.

Bombo, A.B., Pereira, A.E.S., Lusa, M.G., De Medeiros Oliveira, E., De Oliveira, J.L., Campos, E.V.R., De Jesus, M.B., Oliveira, H.C., Fraceto, L.F., & Mayer, J.L.S. (2019). A Mechanistic View of Interactions of a Nanoherbicide with Target Organism. *J. Agric. Food Chem., 67*, 4453-4462. DOI: https://doi.org/. 9B00806.SOCIAL.JPEG_V03 doi:10.1021/ACS.JAFC.9B00806/ASSET/IMAGES/ACS.JAFC

Bombo, A. B., Pereira, A. E. S., Lusa, M. G., De Medeiros Oliveira, E., De Oliveira, J. L., Campos, E. V. R., De Jesus, M. B., Oliveira, H. C., Fraceto, L. F., & Mayer, J. L. S. (2019). A Mechanistic View of Interactions of a Nanoherbicide with Target Organism. *Journal of Agricultural and Food Chemistry*, *67*(16), 4453–4462. doi:10.1021/acs.jafc.9b00806 PMID:30933503

Borovitz, J. O., & Chory, J. (2004). Genomics tools for QTL analysis and gene discovery. *Current Opinion in Plant Biology*, *7*(2), 132–136. doi:10.1016/j.pbi.2004.01.011 PMID:15003212

Bose, B., Kumar, M., Singhal, R. K., & Mondal, S. (2018). *Impact of seed priming on the modulation of physico-chemical and molecular processes during germination, growth, and development of crops. Advances in seed priming*. Springer. doi:10.1007/978-981-13-0032-5_2

Bouis, H. E., & Saltzman, A. (2017). Improving nutrition through biofortification: A review of evidence from HarvestPlus, 2003 through 2016. *Global Food Security*, *12*, 49–58. doi:10.1016/j.gfs.2017.01.009 PMID:28580239

Bouwmeester, H., Dekkers, S., Noordam, M. Y., Hagens, W. I., Bulder, A. S., De Heer, C., Ten Voorde, S. E., Wijnhoven, S. W., Marvin, H. J., & Sips, A. J. (2009). Review of health safety aspects of nanotechnologies in food production. *Regulatory Toxicology and Pharmacology*, *53*(1), 52–62. doi:10.1016/j.yrtph.2008.10.008 PMID:19027049

Brady, N. C., & Well, R. R. (2002). *The Nature and Properties of Soils (Thirteen Edition). Peerson Prentice Hall, Peerson Education, Inc. Published in India by Dorling Kindersley*. Pvt. Ltd.

Braslavsky, I., Hebert, B., Kartalov, E., & Quake, S. R. (2003). Sequence information can be obtained from single DNA molecules. *Proceedings of the National Academy of Sciences of the United States of America*, *100*(7), 3960–396. doi:10.1073/pnas.0230489100 PMID:12651960

Bravo Cadena, M., Preston, G. M., Van der Hoorn, R. A. L., Flanagan, N. A., Townley, H. E., & Thompson, I. P. (2018). Enhancing Cinnamon Essential Oil Activity by Nanoparticle Encapsulation to Control Seed Pathogens. *Industrial Crops and Products*, *124*, 755–764. doi:10.1016/j.indcrop.2018.08.043

Brenchley, R., Spannag, M., Pfeifer, M., Barker, G. L. A., D'Amore, R., Allen, A. M., & McKenzie, N. (2012). Analysis of the bread wheat genome using whole-genome shotgun sequencing. *Nature*, *491*(7426), 705–710. doi:10.1038/nature11650 PMID:23192148

Bruce, T. J. A., Matthes, M. C., Napier, J. A., & Pickett, J. A. (2007). Stressful "memories" of plants: Evidence and possible mechanisms. *Plant Science*, *173*(6), 603–608. doi:10.1016/j.plantsci.2007.09.002

Brummer, E. C., Barber, W. T., Collier, S. M., Cox, T. S., Johnson, R., Murray, S. C., Olsen, R. T., Pratt, R. C., & Thro, A. M. (2011). Plant breeding for harmony between agriculture and the environment. *Frontiers in Ecology and the Environment*, *9*(10), 561–568. doi:10.1890/100225

Bumbudsanpharoke, N., & Ko, S. (2015). *Journal of Nanoparticle Research*, *17*(6), 1–19.

Burrough, P. A. (1998). *Principles of Geographical Information Systems for Land Resources Assessment* (2nd ed.). Oxford University Press.

Camara, M. C., Campos, E. V. R., Monteiro, R. A., Do Espirito Santo Pereira, A., De Freitas Proença, P. L., & Fraceto, L. F. (2019). Development of stimuli-responsive nano-based pesticides: Emerging opportunities for agriculture. *Journal of Nanobiotechnology*, *17*(1), 1–19. doi:10.118612951-019-0533-8 PMID:31542052

Camerel, F., Gabriel, J. C. P., Batail, P., Davidson, P., Lemaire, B., Schmutz, M., Gulik-Krzywicki, T., & Bourgaux, C. (2002). Original Single Walled Nanotubules Based on Weakly Interacting Covalent Mineral Polymers, 1∞[Nb2PS10-] in N-Methylformamide. *Nano Letters*, *2*(4), 403–407. doi:10.1021/nl010090l

Canas, J. E., Long, M., Nations, S., Vadan, R., Dai, L., Luo, M., Ambikapathi, R., Lee, E. H., & Olszyk, D. (2008). Effects of functionalized and nonfunctionalized single-walled carbon nanotubes on root elongation of select crop species. *Environmental Toxicology and Chemistry*, *27*(9), 1922–1931. doi:10.1897/08-117.1 PMID:19086209

Cao, Z., Rossi, L., Stowers, C., Zhang, W., Lombardini, L., & Ma, X. (2018). The impact of cerium oxide nanoparticles on the physiology of soybean (Glycine max (L.) Merr.) under different soil moisture conditions. *Environmental Science and Pollution Research International*, *25*(1), 930–939. doi:10.100711356-017-0501-5 PMID:29076022

Carbone, M., Donia, D. T., Sabbatella, G., & Antiochia, R. (2016). Silver nanoparticles in polymeric matrices for fresh food packaging. *Journal of King Saud University. Science*, *28*(4), 273–279. doi:10.1016/j.jksus.2016.05.004

Carrillo-Reche, J., Vallejo-Marín, M., & Quilliam, R. S. (2018). Quantifying the potential of 'on-farm' seed priming to increase crop performance in developing countries. A meta-analysis. *Agronomy for Sustainable Development*, *38*(6), 1–14. doi:10.100713593-018-0536-0

Carrillo-Reche, J., Vallejo-Marín, M., & Quilliam, R. S. (2018). Quantifying the potential of "on-farm" seed priming to increase crop performance in developing countries. A meta-analysis. Agron. *Journal of Sustainable Development*, *38*, 1–14.

Cembrowska-Lech, D., & Rybak, K. (2023). Nanopriming of Barley Seeds—A Shotgun Approach to Improve Germination under Salt Stress Conditions by Regulating of Reactive Oxygen Species. *Plants*, *12*(2), 405. doi:10.3390/plants12020405 PMID:36679118

Chahal, S., Singh, S., Kumar, A., & Kumar, P. (2020). Oxygen-deficient lanthanum doped cerium oxide nanoparticles for potential applications in spintronics and photocatalysis. *Vacuum*, *177*, 109395. doi:10.1016/j.vacuum.2020.109395

Chand, K., Jiao, C., Lakhan, M. N., Shah, A. H., Kumar, V., Fouad, D. E., Chandio, M. B., Maitlo, A. A., Ahmed, M., & Cao, D. (2021). Green synthesis, characterization and photocatalytic activity of silver nanoparticles synthesized with Nigella sativa seed extract. *Chemical Physics Letters*, *763*, 138218. doi:10.1016/j.cplett.2020.138218

Chandrasekaran, U., Luo, X., Wang, Q., & Shu, K. (2020). Are there unidentified factors involved in the germination of nanoprimed seeds? *Frontiers in Plant Science*, *11*, 832. doi:10.3389/fpls.2020.00832 PMID:32587599

Chaturvedi, V., & Verma, P. (2015). Fabrication of silver nanoparticles from leaf extract of Butea monosperma (Flame of Forest) and their inhibitory effect on bloom-forming cyanobacteria. *Bioresources and Bioprocessing*, *2*(2), 1–8. doi:10.118640643-015-0048-6

Chaudhary, P., Fatima, F., & Kumar, A. (2020). Relevance of nanomaterials in food packaging and its advanced future prospects. *Journal of Inorganic and Organometallic Polymers and Materials*, *30*(12), 5180–5192. doi:10.100710904-020-01674-8 PMID:32837459

Chaudhry, Q., Scotter, M., Blackburn, J., Ross, B., Boxall, A., Castle, L., & Watkins, R. (2008). *Food Additives and Contaminants*, *25*(10), 1224–1243.

Chaud, M., Souto, E. B., Zielinska, A., Severino, P., Batain, F., Oliveira-Junior, J., & Alves, T. (2021). Nanopesticides in Agriculture: Benefits and Challenge in Agricultural Productivity, Toxicological Risks to Human Health and Environment. *Toxics*, *9*(6), 131. doi:10.3390/toxics9060131 PMID:34199739

Chauhan, P. S., Singh, J., & Kavita, A. (2012). Chironji: A promising tree fruits subtropics. *HortFlora Research Spectrum*, *1*(3), 375–379.

Chauhan, R., Reddy, A., & Abraham, J. (2015). Biosynthesis of silver and zinc oxide nanoparticles using Pichia fermentans JA2 and their antimicrobial property. *Applied Nanoscience*, *5*(5), 63–71. doi:10.100713204-014-0292-7

Chauhan, R., Yadav, H. O., & Sehrawat, N. (2020). Nanobioremediation: A new and a versatile tool for sustainable environmental cleanup-Overview. *Journal of Materials & Environmental Sciences*, *11*(4), 564–573.

Chau, N. H., Doan, Q. H., Chu, T. H., Nguyen, T. T., Dao Trong, H., & Ngo, Q. B. (2019). Effects of Different Nanoscale Microelement-Containing Formulations for Presowing Seed Treatment on Growth of Soybean Seedlings. *Journal of Chemistry*, *2019*, 1–8. doi:10.1155/2019/8060316

Chen, Y., Wang, D., Zhu, X., Zheng, X., & Feng, L. (2012). Long-term effects of copper nanoparticles on wastewater biological nutrient removal and N2O generation in the activated sludge process. *Environ. Sci. Technol.* 46, 12452-12458. DOI: https://doi.org/ 302646Q doi:10.1021/ES

Chen, G., Qiu, J., Liu, Y., Jiang, R., Cai, S., Liu, Y., Zhu, F., Zeng, F., Luan, T., & Ouyang, G. (2015). Carbon Nanotubes Act as Contaminant Carriers and Translocate within Plants. *Scientific Reports*, *5*(1), 1–9. doi:10.1038rep15682 PMID:26498499

Chen, H., Chang, Y., Lin, Z., & Chu, Y. (2020). Nanoparticle-enabled paper moisture sensor for intelligent food packaging. *ACS Applied Materials & Interfaces*, *12*(46), 52001–52009.

Chen, K., Fessehaie, A., & Arora, R. (2012). Dehydrin metabolism is altered during seed osmopriming and subsequent germination under chilling and desiccation in Spinacia oleracea L. cv. Bloomsdale: Possible role in stress tolerance. *Plant Science*, *183*, 27–36. doi:10.1016/j.plantsci.2011.11.002 PMID:22195574

Chen, N., Yao, P., Zhang, W., Zhang, Y., Xin, N., Wei, H., & Zhao, C. (2022). Selenium nanoparticles: Enhanced nutrition and beyond. *Critical Reviews in Food Science and Nutrition*, 1–12. PMID:35848122

Chen, Y., Wang, D., Zhu, X., Zheng, X., & Feng, L. (2012). Long-term effects of copper nanoparticles on wastewater biological nutrient removal and N_2O generation in the activated sludge process. *Environmental Science & Technology*, *46*(22), 12452–12458. doi:10.1021/es302646q PMID:23110389

Chen, Z., Newman, I., Zhou, M., Mendham, N., Zhang, G., & Shabala, S. (2005). Screening plants for salt tolerance by measuring K+ flux: A case study for barley. *Plant, Cell & Environment*, *28*(10), 1230–1246. doi:10.1111/j.1365-3040.2005.01364.x

Chokheli, V. A., Shmaraeva, A. N., Kornienko, I. V., Stepanenko, V. V., Rajput, V. D., Singh, A., Kumari, A., Dmitriev, P. A., Azarov, A. S., Minkina, T. M., & Varduni, T. V. (2022). Assessment of genetic diversity and DNA fingerprinting of rare species of the genus Crambe (Brassicaceae). *Crop & Pasture Science*. Advance online publication. doi:10.1071/CP22027

Choudhary, R. C., Kumaraswamy, R. V., Kumari, S., Sharma, S. S., Pal, A., Raliya, R., Biswas, P., & Saharan, V. (2019). Zinc encapsulated chitosan nanoparticle to promote maize crop yield. *International Journal of Biological Macromolecules*, *127*, 126–135. doi:10.1016/j.ijbiomac.2018.12.274 PMID:30610949

Churchill, G. A., & Doerge, R. W. (1994). Empirical threshold values for quantitative trait mapping. *Genetics*, *138*(3), 963–971. doi:10.1093/genetics/138.3.963 PMID:7851788

Collins, H. M., Panozzo, J. F., Logue, S. J., Jefferies, S. P., & Barr, A. R. (2003). Mapping and validation of chromosome regions associated with high malt extract in barley (*Hordeum vulgare L.*). *Australian Journal of Agricultural Research*, *54*(12), 1223–1240. doi:10.1071/AR02201

Colmer, J., O'Neill, C. M., Wells, R., Bostrom, A., Reynolds, D., Websdale, D., Shiralagi, G., Lu, W., Lou, Q., Le Cornu, T., Ball, J., Renema, J., Flores Andaluz, G., Benjamins, R., Penfield, S., & Zhou, J. (2020). SeedGerm: A cost-effective phenotyping platform for automated seed imaging and machine-learning based phenotypic analysis of crop seed germination. *The New Phytologist*, *228*(2), 778–793. doi:10.1111/nph.16736 PMID:32533857

Conti, J. A., Killpack, K. A., & Ogorzalek, T. L. (2012).. . *Journal of Food Science*, *77*(7), R93–R104. PMID:22515252

Córdoba-Cañero, D., Roldán-Arjona, T., & Ariza, R. R. (2014). Arabidopsis ZDP DNA 3′-phosphatase and ARP endonuclease function in 8-oxoG repair initiated by FPG and OGG 1 DNA glycosylases. *The Plant Journal*, *79*(5), 824–834. doi:10.1111/tpj.12588 PMID:24934622

Core, L. J., Waterfall, J. J., & Lis, J. T. (2008). Nascent RNA sequencing reveals widespread pausing and divergent initiation at human promoters. *Science*, *322*(5909), 1845–1848. doi:10.1126cience.1162228 PMID:19056941

Crane, H., Kinderman, E., & Malhotra, R. (2010). *A cubic mile of oil: realities and options for averting the looming global energy crisis*. Oxford University Press. doi:10.1093/oso/9780195325546.001.0001

Dahal, K., Li, X.-Q., Tai, H., Creelman, A., & Bizimungu, B. (2019). Improving potato stress tolerance and tuber yield under a climate change scenario–a current overview. *Frontiers in Plant Science*, *10*, 563. doi:10.3389/fpls.2019.00563 PMID:31139199

Dahle, J. T., & Arai, Y. (2015). Environmental geochemistry of cerium: Applications and toxicology of cerium oxide nanoparticles. *International Journal of Environmental Research and Public Health*, *12*(2), 1253–1278. doi:10.3390/ijerph120201253 PMID:25625406

Dai, Y., Chen, F., Yue, L., Li, T., Jiang, Z., Xu, Z., Wang, Z., & Xing, B. (2020). Uptake, Transport, and Transformation of CeO2 Nanoparticles by Strawberry and Their Impact on the Rhizosphere Bacterial Community. *ACS Sustainable Chemistry & Engineering*, *8*(12), 4792–4800. doi:10.1021/acssuschemeng.9b07422

Dantas de Medeiros, A., Silva, L., Ribeiro, J. P., Ferreira, K., Fim Rosas, J., Santos, A., & Barboza da Silva, C. (2020). Machine Learning for Seed Quality Classification: An Advanced Approach Using Merger Data from FT-NIR Spectroscopy and X-ray Imaging. *Sensors (Basel)*, *20*(15), 4319. doi:10.339020154319 PMID:32756355

Daphedar, A. B., Kakkalameli, S. B., Melappa, G., Taranath, T. C., Srinivasa, C., Shivamallu, C., Syed, A., Marraiki, N., Elgorban, A. M., & Veerapur, R. (2021). Genotoxic assay of silver and zinc oxide nanoparticles synthesized by leaf extract of Garcinia livingstonei T. Anderson: A comparative study. *Pharmacognosy Magazine*, (17), 114.

Das, A., & Das, B. (2019). Nanotechnology a potential tool to mitigate abiotic stress in crop plants. *Abiotic and biotic stress in plants*. InTech Open.

Das, C. K., Jangir, H., Kumar, J., Verma, S., Mahapatra, S. S., Philip, D., Srivastava, G., & Das, M. (2018). Nano-pyrite seed dressing: A sustainable design for NPK equivalent rice production. Nanotechnol. *Environ. Eng*, *31*(3), 1–14. doi:10.100741204-018-0043-1

Das, K., & Roychoudhury, A. (2014). Reactive oxygen species (ROS) and response of antioxidants as ROS-scavengers during environmental stress in plants. *Frontiers in Environmental Science*, *2*, 53. doi:10.3389/fenvs.2014.00053

Das, M., & Smita, S. S. (2018). Biosynthesis of silver nanoparticles using bark extracts of Butea monosperma (Lam.) Taub. and study of their antimicrobial activity. *Applied Nanoscience*, *8*(8), 1059–1067. doi:10.100713204-018-0721-0

De La Torre-Roche, R., Cantu, J., Tamez, C., Zuverza-Mena, N., Hamdi, H., Adisa, I. O., Elmer, W., Gardea-Torresdey, J., & White, J. C. (2020). Seed Biofortification by Engineered Nanomaterials: A Pathway to Alleviate Malnutrition? *Journal of Agricultural and Food Chemistry*, *68*(44), 12189–12202. doi:10.1021/acs.jafc.0c04881 PMID:33085897

De Oliveira, J. L., Campos, E. V. R., Germano-Costa, T., Lima, R., Vechia, J. F. D., Soares, S. T., de Andrade, D. J., Gonçalves, K. C., do Nascimento, J., Polanczyk, R. A., & Fraceto, L. F. (2019). Association of Zein Nanoparticles with Botanical Compounds for Effective Pest Control Systems. *Pest Management Science*, *75*(7), 1855–1865. doi:10.1002/ps.5338 PMID:30653814

De Rossi, A., Rigueto, C. V., Dettmer, A., Colla, L. M., & Piccin, J. S. (2020). Synthesis, characterization, and application of Saccharomyces cerevisiae/alginate composites beads for adsorption of heavy metals. *Journal of Environmental Chemical Engineering*, *8*(4), 104009. doi:10.1016/j.jece.2020.104009

De Steur, H., Blancquaert, D., Strobbe, S., Lambert, W., Gellynck, X., & Van Der Straeten, D. (2015). Status and market potential of transgenic biofortified crops. *Nature Biotechnology*, *33*(1), 25–29. doi:10.1038/nbt.3110 PMID:25574631

De Villiers, A. J., Von Teichman, I., Van Rooyen, M. W., & Theron, G. K. (1996). Salinity-induced changes in anatomy, stomatal counts and photosynthetic rate of Atriplex semibaccata R. Br. *South African Journal of Botany*, *62*(5), 270–276. doi:10.1016/S0254-6299(15)30656-6

Dehkourdi, E. H., Chehrazi, M., Hosseini, H., & Hosseini, M. (2014). The effect of anatase nanoparticles (TiO2) on pepper seed germination (Capsicum annum L.). [IJB]. *International Journal of Biosciences*, *4*(5), 141–145.

Dekkers, S., Krystek, P., Peters, R. J., Lankveld, D. P., Bokkers, B. G., van Hoeven-Arentzen, P. H., Bouwmeester, H., & Oomen, A. G. (2011). Presence and risks of nanosilica in food products. *Nanotoxicology*, *5*(5), 393–405. doi:10.3109/17435390.2010.519836 PMID:20868236

Delfani, M., Baradarn Firouzabadi, M., Farrokhi, N., & Makarian, H. (2014). Some physiological responses of black-eyed pea to iron and magnesium nanofertilizers. *Communications in Soil Science and Plant Analysis*, *45*(4), 530–540. doi:10.1080/00103624.2013.863911

Dell'Aquila, A. (2009). Digital Imaging Information Technology Applied to Seed Germination Testing: A Review. *Agronomy for Sustainable Development*, *29*(1), 213–221. doi:10.1051/agro:2008039

Demil, B., & Lecocq, X. (2010). Business model evolution: In search of dynamic consistency. *Long Range Planning*, *43*(2-3), 227–246. doi:10.1016/j.lrp.2010.02.004

Demirer, G. S., Silva, T. N., Jackson, C. T., Thomas, J. B., Ehrhardt, D. W., Rhee, S. Y., Mortimer, J. C., & Landry, M. P. (2021). Nanotechnology to advance CRISPR–Cas genetic engineering of plants. *Nanotechnology*, (16), 243–250. PMID:33712738

Dera, M. W., & Teseme, W. B. (2020). Review on the Application of Food Nanotechnology in Food Processing. *American Journal of Engineering and Technology Management*, *5*(5), 41–47. doi:10.11648/j.ajetm.20200502.12

DeRosa, M. C., Monreal, C., Schnitzer, M., Walsh, R., & Sultan, Y. (2010). Nanotechnology in fertilizers. *Nature Nanotechnology*, *5*(2), 91–91. doi:10.1038/nnano.2010.2 PMID:20130583

Dhal, P., Sahu, G., Dhal, A., Mohanty, S., & Dash, S. K. (2022). Priming of vegetable seeds: A review. *The Pharma Innovation Journal*, *11*(2), 519–525.

Dhand, V., Soumya, L., Bharadwaj, S., Chakra, S., Bhatt, D., & Sreedhar, B. (2016). Green synthesis of silver nanoparticles using Coffea arabica seed extract and its antibacterial activity. *Materials Science and Engineering C*, *58*, 36–43. doi:10.1016/j.msec.2015.08.018 PMID:26478284

Dhewa, T. (2015). Nanotechnology applications in agriculture: an update. *Octa Journal of Environmental Research, 3*(2).

Dhir, B. (2021). Nanofertilizers and their Applications. *New frontiers of nanomaterials in environmental science*, 229-241.

Dhivahar, J., Khusro, A., Elancheran, L., Agastian, P., Al-Dhabi, N. A., Esmail, G. A., Arasu, M. V., Kim, Y. O., Kim, H., & Kim, H. J. (2020). Photo-mediated biosynthesis and characterization of silver nanoparticles using bacterial xylanases as reductant: Role of synthesized product (Xyl-AgNPs) in fruits juice clarification. *Surfaces and Interfaces*, (21), 100747.

Di Girolamo, G., & Barbanti, L. (2012). Treatment conditions and biochemical processes influencing seed priming effectiveness. *Italian Journal of Agronomy*, *7*(2), e25–e25. doi:10.4081/ija.2012.e25

Dietz, K. J., & Herth, S. (2011). Plant nanotoxicology. *Trends in Plant Science*, *16*(11), 582–589. doi:10.1016/j.tplants.2011.08.003 PMID:21906987

Dikshit, P. K., Kumar, J., Das, A. K., Sadhu, S., Sharma, S., Singh, S., Gupta, P. K., & Kim, B. S. (2021). Green Synthesis of Metallic Nanoparticles: Applications and Limitations. *Catalysts, 11*(8), 902. doi:10.3390/catal11080902

Dileep Kumar, G., Raja, K., Natarajan, N., Govindaraju, K., & Subramanian, K. S. (2020). Invigouration Treatment of Metal and Metal Oxide Nanoparticles for Improving the Seed Quality of Aged Chilli Seeds (*Capsicum annum* L.). *Materials Chemistry and Physics, 242*, 122492. doi:10.1016/j.matchemphys.2019.122492

Dimkpa, C. O., Fugice, J., Singh, U., & Lewis, T. D. (2020). Development of fertilizers for enhanced nitrogen use efficiency–Trends and perspectives. *The Science of the Total Environment, 731*, 139113. doi:10.1016/j.scitotenv.2020.139113 PMID:32438083

Dimkpa, C. O., McLean, J. E., Britt, D. W., & Anderson, A. J. (2015). Nano-CuO and interaction with nano-ZnO or soil bacterium provide evidence for the interference of nanoparticles in metal nutrition of plants. *Ecotoxicology (London, England), 24*(1), 119–129. doi:10.100710646-014-1364-x PMID:25297564

do Espirito Santo Pereira, A., Caixeta Oliveira, H., Fernandes Fraceto, L., & Santaella, C. (2021). Nanotechnology potential in seed priming for sustainable agriculture. *Nanomaterials (Basel, Switzerland), 11*(2), 267. doi:10.3390/nano11020267 PMID:33498531

Dos Santos, C. A., Ingle, A. P., & Rai, M. (2020). The emerging role of metallic nanoparticles in food. *Applied Microbiology and Biotechnology, 104*(6), 2373–2383. doi:10.100700253-020-10372-x PMID:31989225

Drobek, M., Frąc, M., & Cybulska, J. (2019). Plant biostimulants: Importance of the quality and yield of horticultural crops and the improvement of plant tolerance to abiotic stress—A review. *Agronomy (Basel), 9*(6), 335. doi:10.3390/agronomy9060335

Dubey, J., & Thakur, M. (2020). Pollution: Pesticides in Natural Ecosystems. In Managing Global Resources and Universal Processes (pp. 111-118). CRC Press.

Duncan, T. V. (2011). Applications of nanotechnology in food packaging and food safety: Barrier materials, antimicrobials and sensors. *Journal of Colloid and Interface Science, 363*(1), 1–24. doi:10.1016/j.jcis.2011.07.017 PMID:21824625

Duran, N. M., Savassa, S. M., Lima, R. G., de Almeida, E., Linhares, F. S., van Gestel, C. A. M., & Pereira de Carvalho, H. W. (2017). X-ray Spectroscopy Uncovering the Effects of Cu Based Nanoparticle Concentration and Structure on Phaseolus vulgaris Germination and Seedling Development. *Journal of Agricultural and Food Chemistry, 65*(36), 7874–7884. doi:10.1021/acs.jafc.7b03014 PMID:28817280

Durán, N., Nakazato, G., & Seabra, A. B. (2016). Antimicrobial activity of biogenic silver nanoparticles, and silver chloride nanoparticles: An overview and comments. *Applied Microbiology and Biotechnology, 100*(15), 6555–6570. doi:10.100700253-016-7657-7 PMID:27289481

Dutta, P. (2018). Seed priming: New vistas and contemporary perspectives. In *Advances in seed priming* (pp. 3–22). Springer. doi:10.1007/978-981-13-0032-5_1

Dvořák, P., Krasylenko, Y., Zeiner, A., Šamaj, J., & Takáč, T. (2021). Signaling Toward Reactive Oxygen Species-Scavenging Enzymes in Plants. *Frontiers in Plant Science*, *11*, 2178. doi:10.3389/fpls.2020.618835 PMID:33597960

Eagles, H., Bariana, H., Ogbonnaya, F., Rebetzke, G., Hollamby, G., Henry, R., Henschke, P., & Carter, M. (2001). Implementation of markers in Australian wheat breeding. *Australian Journal of Agricultural Research*, *52*(12), 1349–1356. doi:10.1071/AR01067

Ealia, S. A. M., & Saravanakumar, M. P. (2017) A Review on the Classification, Characterisation, Synthesis of Nanoparticles and Their Application. In *IOP Conference Series: Materials Science and Engineering*. IOP Publishing: Bristol, UK.

Echeverría, I., López-Caballero, M. E., Gómez-Guillén, M. C., Mauri, A. N., & Montero, M. P. (2018). Active nanocomposite films based on soy proteins-montmorillonite-clove essential oil for the preservation of refrigerated bluefin tuna (Thunnus thynnus) fillets. *International Journal of Food Microbiology*, *266*, 142–149. doi:10.1016/j.ijfoodmicro.2017.10.003 PMID:29216554

El-Badri, A. M., Batool, M., Mohamed, I. A. A., Wang, Z., Khatab, A., Sherif, A., Ahmad, H., Khan, M. N., Hassan, H. M., Elrewainy, I. M., Kuai, J., Zhou, G., & Wang, B. (2021). Antioxidative and metabolic contribution to salinity stress responses in two rapeseed cultivars during the early seedling stage. *Antioxidants*, *10*(8). doi:10.3390/antiox10081227 PMID:34439475

El-Badri, A. M., Batool, M., Mohamed, I. A., Khatab, A., Sherif, A., Wang, Z., Salah, A., Nishawy, E., Ayaad, M., Kuai, J., Wang, B., & Zhou, G. (2021). Modulation of salinity impact on early seedling stage via nano-priming application of zinc oxide on rapeseed (Brassica napus L.). *Plant Physiology and Biochemistry*, *166*, 376–392. doi:10.1016/j.plaphy.2021.05.040 PMID:34153882

Elemike, E. E., Uzoh, I. M., Onwudiwe, D. C., & Babalola, O. O. (2019). The role of nanotechnology in the fortification of plant nutrients and improvement of crop production. *Applied Sciences (Basel, Switzerland)*, *9*(3), 499. doi:10.3390/app9030499

Elizabath, A., Babychan, M., Mathew, A. M., & Syriac, G. M. (2019). Application of nanotechnology in agriculture. *Int. J. Pure Appl. Biosci*, *7*(2), 131–139. doi:10.18782/2320-7051.6493

Elmer, W., Ma, C., & White, J. (2018). NPs for plant disease management. *Current Opinion in Environmental Science & Health*, *6*, 66–70. doi:10.1016/j.coesh.2018.08.002

Elsaesser, A., & Howard, C. V. (2012). Toxicology of nanoparticles. *Advanced Drug Delivery Reviews*, *64*(2), 129–137. doi:10.1016/j.addr.2011.09.001 PMID:21925220

Elsakhawy, T., Omara, A. E. D., Alshaal, T., El-Ramady, H., Ghazi, A., El-Nahrawy, S., & Elhawat, N. (2018). Nanomaterials and plant abiotic stress in agroecosystems. *Environmental Biodiversity and Soil Security*, *2*(2), 73–94. doi:10.21608/jenvbs.2018.3897.1030

Elsheery, N. I., Helaly, M. N., El-Hoseiny, H. M., & Alam-Eldein, S. M. (2020). Zinc oxide and silicone nanoparticles to improve the resistance mechanism and annual productivity of salt-stressed mango trees. *Agronomy (Basel), 10*(4), 558. doi:10.3390/agronomy10040558

El-Zohri, M., Al-Wadaani, N. A., & Bafeel, S. O. (2021). Foliar sprayed green zinc oxide NPs mitigate drought-induced oxidative stress in tomato. *Plants, 10*(11), 2400. doi:10.3390/plants10112400 PMID:34834763

Emamifar, A., Kadivar, M., Shahedi, M., & Soleimanian-Zad, S. (2010). Evaluation of nanotechnology application in the enhancement of antioxidant and antimicrobial activities of green tea. *Journal of Food Science, 75*(4), N62–N67.

Ernest, V., Shiny, P. J., Mukherjee, A., & Chandrasekaran, N. (2012). Silver nanoparticles: A potential nanocatalyst for the rapid degradation of starch hydrolysis by α-amylase. *Carbohydrate Research, 352*, 60–64. doi:10.1016/j.carres.2012.02.009 PMID:22405762

Faisal, M., Saquib, Q., Alatar, A. A., Al-Khedhairy, A. A., Hegazy, A. K., & Musarrat, J. (2013). Phytotoxic hazards of NiO-nanoparticles in tomato: A study on mechanism of cell death. *Journal of Hazardous Materials, 250*, 318–332. doi:10.1016/j.jhazmat.2013.01.063 PMID:23474406

Faisal, S., Jan, H., Shah, S. A., Shah, S., Khan, A., Akbar, M. T., Rizwan, M., Jan, F., Wajidullah, Akhtar, N., Khattak, A., & Syed, S. (2021). Green Synthesis of Zinc Oxide (ZnO) NPs Using Aqueous Fruit Extracts of Myristica fragrans: Their Characterizations and Biological and Environmental Applications. *ACS Omega, 6*(14), 9709–9722. doi:10.1021/acsomega.1c00310 PMID:33869951

Falk, K. G., Jubery, T. Z., Mirnezami, S. V., Parmley, K. A., Sarkar, S., Singh, A., Ganapathysubramanian, B., & Singh, A. K. (2020). Computer vision and machine learning enabled soybean root phenotyping pipeline. *Plant Methods, 16*(1), 5. doi:10.118613007-019-0550-5 PMID:31993072

Falsini, S., Clemente, I., Papini, A., Tani, C., Schiff, S., Salvatici, M. C., Petruccelli, R., Benelli, C., Giordano, C., Gonnelli, C., & Ristori, S. (2019). When Sustainable Nanochemistry Meets Agriculture: Lignin Nanocapsules for Bioactive Compound Delivery to Plantlets. *ACS Sustainable Chemistry & Engineering, 7*(24), 19935–19942. doi:10.1021/acssuschemeng.9b05462

Fan, Y., Clavel, G., & Pinna, N. (2018). Effect of passivating Al2O3 thin films on MnO2/carbon nanotube composite lithium-ion battery anodes. *Journal of Nanoparticle Research, 20*(8), 216. doi:10.100711051-018-4315-2

FAO WFP, & IFAD. (2012). The State of Food Insecurity in the World 2012. *Economic Growth Is Necessary but Not Sufficient to Accelerate Reduction of Hunger and Malnutrition.* FAO.

Farahani, H. A., Sani, B., & Maroufi, K. (2012). The germination variations in fleawort (Plantago psyllium L.) by nano-particle. Research Gate.

Farias, B. V., Pirzada, T., Mathew, R., Sit, T. L., Opperman, C., & Khan, S. A. (2019). Electrospun Polymer Nanofibers as Seed Coatings for Crop Protection. *ACS Sustainable Chemistry & Engineering*, *7*(24), 19848–19856. doi:10.1021/acssuschemeng.9b05200

Farooq, M., Hussain, M., Habib, M. M., Khan, M. S., Ahmad, I., Farooq, S., Siddique, K. H. M., Farooq, M., Hussain, M., Habib, M. M., Khan, M. S., Ahmad, I., Farooq, S., & Siddique, K. H. M. (2020). Influence of seed priming techniques on grain yield and economic returns of bread wheat planted at different spacings. *Crop & Pasture Science*, *71*(8), 725–738. doi:10.1071/CP20065

Farooq, T., Akram, M. N., Hameed, A., Ahmed, T., & Hameed, A. (2022). Nanopriming-mediated memory imprints reduce salt toxicity in wheat seedlings by modulating physiobiochemical attributes. *BMC Plant Biology*, *22*(1), 540. doi:10.118612870-022-03912-2 PMID:36414951

Fatemi, M., Mollania, N., Momeni-Moghaddam, M., & Sadeghifar, F. (2018). Extracellular biosynthesis of magnetic iron oxide nanoparticles by Bacillus cereus strain HMH1: Characterization and in vitro cytotoxicity analysis on MCF-7 and 3T3 cell lines. *Journal of Biotechnology*, *270*, 1–11. doi:10.1016/j.jbiotec.2018.01.021 PMID:29407416

Favaretto, N., Norton, L., Joern, B., & Brouder, S. (2006). Gypsum amendment and exchangeable calcium and magnesium affecting phosphorus and nitrogen in runoff. *Soil Science Society of America Journal*, *70*(5), 1788–1796. doi:10.2136ssaj2005.0228

Fazeli-Nasab, B., Sayyed, R., Piri, R., & Rahmani, A. F. (2021). Biopriming and Nanopriming: green revolution wings to increase plant yield, growth, and development under stress condition and forward dimensions. *Antioxidants in Plant-Microbe Interaction*, 623-655.

Feitshans, I. L., & Sabatier, P. (2022). Global health impacts of nanotechnology law: Advances in safernano regulation. *Materials Today: Proceedings*, *985*(67), 994. doi:10.1016/j.matpr.2022.08.377

Fernández-Luqueño, F., Medina-Pérez, G., López-Valdez, F., Gutiérrez-Ramírez, R., Campos-Montiel, R. G., Vázquez-Núñez, E., Loera-Serna, S., Almaraz-Buendía, I., Del Razo-Rodríguez, O. E., & Madariaga-Navarrete, A. (2018). Use of agronanobiotechnology in the agro-food industry to preserve environmental health and improve the welfare of farmers. *Agricultural Nanobiotechnology: Modern Agriculture for a Sustainable Future*, 3-16.

Fernie, A. R., & Yan, J. (2019). *De novo* domestication: An alternative route toward new crops for the future. *Molecular Plant*, *12*(5), 615–631. doi:10.1016/j.molp.2019.03.016 PMID:30999078

Filho, J. M. (2015). Seed vigor testing: An overview of the past, present and future perspectives. *Scientia Agrícola*, *72*(4), 363–374. doi:10.1590/0103-9016-2015-0007

Filipoiu, D. C., Bungau, S. G., Endres, L., Negru, P. A., Bungau, A. F., Pasca, B., Radu, A.-F., Tarce, A. G., Bogdan, M. A., Behl, T., Nechifor, A. C., Hassan, S. S., & Tit, D. M. (2022). Characterization of the Toxicological Impact of Heavy Metals on Human Health in Conjunction with Modern Analytical Methods. *Toxics*, *10*(12), 716. doi:10.3390/toxics10120716 PMID:36548549

Filippou, P., Bouchagier, P., Skotti, E., & Fotopoulos, V. (2014). Proline and reactive oxygen/nitrogen species metabolism is involved in the tolerant response of the invasive plant species Ailanthus altissima to drought and salinity. *Environmental and Experimental Botany*, *97*, 1–10. doi:10.1016/j.envexpbot.2013.09.010

Flowers, T. J., & Colmer, T. D. (2015). Plant salt tolerance: Adaptations in halophytes. *Annals of Botany*, *115*(3), 327–331. doi:10.1093/aob/mcu267 PMID:25844430

Fosso-Kankeu, E., Mulaba-Bafubiandi, A. F., & Mishra, A. K. (2014). Prospects for immobilization of microbial sorbents on carbon nanotubes for biosorption: bioremediation of heavy metals polluted water. *Application of nanotechnology in water research, 39*.

Fraceto, L. F., Grillo, R., de Medeiros, G. A., Scognamiglio, V., Rea, G., & Bartolucci, C. (2016). Nanotechnology in agriculture: Which innovation potential does it have? *Frontiers in Environmental Science*, *4*, 20. doi:10.3389/fenvs.2016.00020

Fraser, P. D., Pinto, M. E., Holloway, D. E., & Bramley, P. M. (2000). Application of high-performance liquid chromatography with photodiode array detection to the metabolic profiling of plant isoprenoids. *The Plant Journal*, *24*(4), 551–558. doi:10.1046/j.1365-313x.2000.00896.x PMID:11115136

Friedrichs, S., Takasu, Y., Kearns, P., Dagallier, B., Oshima, R., Schofield, J., & Moreddu, C. (2019). An overview of regulatory approaches to genome editing in agriculture. *Biotechnol. Res. Innov.* 3, 208-220. DOI: https://doi.org/. 2019.07.001 doi:10.1016/J.BIORI

Friedrichs, S., Takasu, Y., Kearns, P., Dagallier, B., Oshima, R., Schofield, J., & Moreddu, C. (2019). An overview of regulatory approaches to genome editing in agriculture. *Biotechnol. Res. Innov.*, *3*(2), 208–220. doi:10.1016/j.biori.2019.07.001

Gaafar, R. M., Diab, R. H., Halawa, M. L., El-Shanshory, A. R., El-Shaer, A., & Hamouda, M. M. (2020). Role of Zinc Oxide Nanoparticles in Ameliorating Salt Tolerance in Soybean. *Egyptian Journal of Botany*, *60*(0), 733–747. doi:10.21608/ejbo.2020.26415.1475

Gajbhiye, M., Kesharwani, J., Ingle, A., Gade, A., & Rai, M. (2009). Fungus-mediated synthesis of silver nanoparticles and their activity against pathogenic fungi in combination with fluconazole. *Nanomedicine; Nanotechnology, Biology, and Medicine*, *5*(4), 382–386. doi:10.1016/j.nano.2009.06.005 PMID:19616127

Gamalero, E., & Glick, B. R. (2020). The use of plant growth-promoting bacteria to prevent nematode damage to plants. *Biology (Basel)*, *9*(11), 381. doi:10.3390/biology9110381 PMID:33171782

Gapper, C., & Dolan, L. (2006). Control of Plant Development by Reactive Oxygen Species. *Plant Physiology*, *141*(2), 341–345. doi:10.1104/pp.106.079079 PMID:16760485

Garcia, D., Arif, S., Zhao, Y., Zhao, S., Ming, L. C., & Huang, D. (2021). *Seed priming technology as a key strategy to increase crop plant production under adverse environmental conditions*. Research Gate.

Garg, M., Sharma, N., Sharma, S., Kapoor, P., Kumar, A., Chunduri, V., & Arora, P. (2018). Biofortified Crops Generated by Breeding, Agronomy, and Transgenic Approaches Are Improving Lives of Millions of People around the World. *Frontiers in Nutrition*, *5*, 12. doi:10.3389/fnut.2018.00012 PMID:29492405

Garg, R., & Maheshwari, S. (2023). SEED PRIMING AND FORTIFICATION OF SEEDS USING NANOTECHNOLOGY: A REVIEW. *EPH-International Journal of Agriculture and Environmental Research*, *9*(1), 11–17. doi:10.53555/eijaer.v9i1.68

Gebhardt, C., & Valkonen, J. P. T. (2001). Organization of genes controlling disease resistance in the potato genome. *Annual Review of Phytopathology*, *39*(1), 79–102. doi:10.1146/annurev.phyto.39.1.79 PMID:11701860

Genze, N., Bharti, R., Grieb, M., Schultheiss, S. J., & Grimm, D. G. (2020). Accurate machine learning-based germination detection, prediction and quality assessment of three grain crops. *Plant Methods*, *16*(1), 157. doi:10.118613007-020-00699-x PMID:33353559

George, M. L. C., Prasanna, B. M., Rathore, R. S., Setty, T. A. S., Kasim, F., Azrai, M., Vasal, S., Balla, O., Hautea, D., Canama, A., Regalado, E., Vargas, M., Khairallah, M., Jeffers, D., & Hoisington, D. (2003). Identification of QTLs conferring resistance to downy mildews of maize in Asia. *Theoretical and Applied Genetics*, *107*(3), 544–551. doi:10.100700122-003-1280-6 PMID:12759731

Ghafariyan, M. H., Malakouti, M. J., Dadpour, M. R., Stroeve, P., & Mahmoudi, M. (2013). Effects of Magnetite Nanoparticles on Soybean Chlorophyll. *Environmental Science & Technology*, *47*, 10645–10652. doi:10.1021/es402249b PMID:23951999

Ghasempour M, Iranbakhsh A, Ebadi M, & Oraghi Ardebili Z. (2019). Multi-walled carbon nanotubes improved growth, anatomy, physiology, secondary metabolism, and callus performance in *Catharanthus roseus*: an in vitro study. *Biotech.*. doi:10.1007/s13205-019-1934-y

Ghidan, A. Y., Al Antary, T. M., & Awwad, A. M. (2017b). Aphidicidal potential of green synthesized magnesium hydroxide nanoparticles using *Olea europaea* leaves extract. *Journal of Agricultural and Biological Science*, *12*(10), 293–301.

Ghidan, A. Y., Al-Antary, T. M., Salem, N. M., & Awwad, A. M. (2017a). Facile green synthetic route to the zinc oxide (ZnONPs) nanoparticles: Effect on green peach aphid and antibacterial activity. *Journal of Agricultural Science (Toronto)*, *9*(2), 131–138. doi:10.5539/jas.v9n2p131

Gilbertson, L. M., Pourzahedi, L., Laughton, S., Gao, X., Zimmerman, J. B., Theis, T. L., Westerhoff, P., & Lowry, G. V. (2020). Guiding the design space for nanotechnology to advance sustainable crop production. *Nature Nanotechnology*, *15*(9), 801–810. doi:10.103841565-020-0706-5 PMID:32572231

Giraldo, J. P., Wu, H., Newkirk, G. M., & Kruss, S. (2019). Nanobiotechnology approaches for engineering smart plant sensors. *Nature Nanotechnology*, *14*(6), 541–553. doi:10.103841565-019-0470-6 PMID:31168083

Gnanasekaran, D. k., Raja, K., Natarajan, N., Govindaraju, K., & Subramanian, K. (2019). Invigouration treatment of metal and metal oxide nanoparticles for improving the seed quality of aged chilli seeds (*Capsicum annum* L.). *Materials Chemistry and Physics*, *242*, 122492.

Godfray, H. C., Beddington, J. R., Crute, I. R., Haddad, L., Lawrence, D., Muir, J. F., Pretty, J., Robinson, S., Thomas, S. M., & Toulmin, C. (2010). Food security: The challenge of feeding 9 billion people. *Science*, *327*(5967), 812–818. doi:10.1126cience.1185383 PMID:20110467

Gohari, G., Panahirad, S., Sadeghi, M., Akbari, A., Zareei, E., Zahedi, S. M., Bahrami, M. K., & Fotopoulos, V. (2021). Putrescine functionalized carbon quantum dot (put-CQD) nanoparticles effectively prime grapevine (*Vitis vinifera* cv. 'Sultana') against salt stress. *BMC Plant Biology*, *21*(1), 120. doi:10.118612870-021-02901-1 PMID:33639848

Gomes, M. P., & Garcia, Q. S. (2013). Reactive oxygen species and seed germination. *Biologia*, *68*(3), 351–357. doi:10.247811756-013-0161-y

Gomez, A., Narayan, M., Zhao, L., Jia, X., Bernal, R. A., Lopez-Moreno, M. L., & Peralta-Videa, J. R. (2021). Effects of nano-enabled agricultural strategies on food quality: Current knowledge and future research needs. *Journal of Hazardous Materials*, *401*, 123385. doi:10.1016/j.jhazmat.2020.123385 PMID:32763688

Gong, C. Y., & Wang, T. (2013). Proteomic evaluation of genetically modified crops: Current status and challenges. *Frontiers in Plant Science*, *4*, 41. doi:10.3389/fpls.2013.00041 PMID:23471542

Gonzalez Guzman, M., Cellini, F., Fotopoulos, V., Balestrini, R., & Arbona, V. (2022). New approaches to improve crop tolerance to biotic and abiotic stresses. *Physiologia Plantarum*, *174*(1), e13547. doi:10.1111/ppl.13547 PMID:34480798

Gross, M. S., Bean, T. G., Hladik, M. L., Rattner, B. A., & Kuivila, K. M. (2020). Uptake, Metabolism, and Elimination of Fungicides from Coated Wheat Seeds in Japanese Quail (*Coturnix japonica*). *Journal of Agricultural and Food Chemistry*, *68*(6), 1514–1524. doi:10.1021/acs.jafc.9b05668 PMID:31977218

Grover, S., & Khan, A. S. (2014). Effect of ionizing radiation on some characteristics of seeds of wheat. *International Journal of Science and Technology Resources*, *3*, 2277–8616.

Gs, G., Seo, Y., Park, D., & Lee, D. (2010). Phytotoxicity of Carbon Nanotubes Assessed by Brassica Juncea and Phaseolus Mungo. *Journal of Nanoelectronics and Optoelectronics*, *5*(2), 157–160. doi:10.1166/jno.2010.1084

Guha, T., Gopal, G., Das, H., Mukherjee, A., & Kundu, R. (2021). Nanopriming with zero-valent iron synthesized using pomegranate peel waste: A "green" approach for yield enhancement in Oryza sativa L. cv. Gonindobhog. *Plant Physiology and Biochemistry*, *163*, 261–275. doi:10.1016/j.plaphy.2021.04.006 PMID:33872831

Guha, T., Ravikumar, K. V. G., Mukherjee, A., Mukherjee, A., & Kundu, R. (2018). Nanopriming with Zero Valent Iron (NZVI) Enhances Germination and Growth in Aromatic Rice Cultivar (*Oryza sativa* Cv. Gobindabhog L.). *Plant Physiology and Biochemistry*, *127*, 403–413. doi:10.1016/j.plaphy.2018.04.014 PMID:29679934

Guo, H., White, J. C., Wang, Z., & Xing, B. (2018). Nano-enabled fertilizers to control the release and use efficiency of nutrients. *Current Opinion in Environmental Science & Health*, *6*, 77–83. doi:10.1016/j.coesh.2018.07.009

Gupta, N., Fischer, A. R., & Frewer, L. J. (2015). *Comprehensive Reviews in Food Science and Food Safety*, *14*(5), 559–576.

Gupta, N., Singh, P. M., Sagar, V., Pandya, A., Chinnappa, M., Kumar, R., & Bahadur, A. (2022). Seed priming with ZnOand Fe3O4Nanoparticles alleviate the lead toxicity in Basella Alba L. through reduced lead uptake and regulation of ROS. *Plants*, *11*(17), 2227. doi:10.3390/plants11172227 PMID:36079609

Gupta, P., Dhankher, O. P., & Bahadur, J. (2021). *Journal of Hazardous Materials*, *403*, 123855. doi:10.1016/j.jhazmat.2020.123855 PMID:33264931

Gupta, P., Varshney, R., Sharma, P., & Ramesh, B. (1999). Molecular markers and their applications in wheat breeding. *Plant Breeding*, *118*(5), 369–390. doi:10.1046/j.1439-0523.1999.00401.x

Gupta, R., & Xie, H. (2018). Nanoparticles in daily life: Applications, toxicity and regulations. *Journal of Environmental Pathology, Toxicology and Oncology*, *37*(3), 209–230. doi:10.1615/JEnvironPatholToxicolOncol.2018026009 PMID:30317972

Hackett, C. (2002). Statistical methods for QTL mapping in cereals. *Plant Molecular Biology*, *48*(5/6), 585–599. doi:10.1023/A:1014896712447 PMID:11999836

Haggag, W. M., Abouziena, H., Abd-El-Kreem, F., & El Habbasha, S. (2015). Agriculture biotechnology for management of multiple biotic and abiotic environmental stress in crops. *Journal of Chemical and Pharmaceutical Research*, *7*(10), 882–889.

Haghighi, M., & da Silva, T. (2014). The effect of N-TiO2 on tomato, onion, and radish seed germination. *Journal of Crop Science and Biotechnology*, *17*(4), 221–227. doi:10.100712892-014-0056-7

Haghighi, M., & Silva, J. A. T. (2014). The effect of carbon nanotubes on the seed germination and seedling growth of four vegetable species. *Journal of Crop Science and Biotechnology*, *17*(4), 201–208. doi:10.100712892-014-0057-6

Haley, C., & Andersson, L. (1997). Linkage mapping of quantitative trait loci in plants and animals, P. Dear (Ed.), Genome mapping–A practical approach. Oxford University Press.

Hall, R. D. (2006). Plant metabolomics: From holistic hope to hype to hot topic. *The New Phytologist*, *169*(3), 453–468. doi:10.1111/j.1469-8137.2005.01632.x PMID:16411949

Hamad, A. F., Han, J. H., Kim, B. C., & Rather, I. A. (2018). The intertwine of nanotechnology with the food industry. *Saudi Journal of Biological Sciences, 25*(1), 27–30. doi:10.1016/j.sjbs.2017.09.004 PMID:29379352

Hamilton, A. J., Burry, K., Mok, H.-F., Barker, S. F., Grove, J. R., & Williamson, V. G. (2014). Give peas a chance? Urban agriculture in developing countries. A review. *Agronomy for Sustainable Development, 34*(1), 45–73. doi:10.100713593-013-0155-8

Handford, C. E., Dean, M., Henchion, M., Spence, M., Elliott, C. T., & Campbell, K. (2014). Implications of nanotechnology for the agri-food industry: Opportunities, benefits and risks. *Trends in Food Science & Technology, 40*(2), 226–241. doi:10.1016/j.tifs.2014.09.007

Handy, R. D., van den Brink, N., Chappell, M., Mühling, M., Behra, R., Dušinská, M., Simpson, P., Ahtiainen, J., Jha, A. N., Seiter, J., Bednar, A., Kennedy, A., Fernandes, T. F., & Riediker, M. (2012). Practical considerations for conducting ecotoxicity test methods with manufactured nanomaterials: What have we learnt so far? *Ecotoxicology (London, England), 21*(4), 933–972. doi:10.100710646-012-0862-y PMID:22422174

Haq, S. M., Tariq, A., Li, Q., Yaqoob, U., Majeed, M., Hassan, M., Fatima, S., Kumar, M., Bussmann, R. W., Farhan, M., Moazzam, U., & Aslam, M. (2022). *Influence of Edaphic Properties in Determining Forest Community Patterns of the Zabarwan Mountain Range in the Kashmir Himalayas*. MDPI, 1–15.

Haq, S. M., Yaqoob, U., Majeed, M., Amjad, M. S., Hassan, M., Ahmad, R., Waheed, M., Bussmann, R. W., Calixto, E. S., & Pro, J. (n.d.). *Quantitative ethnoveterinary study on plant resource utilization by indigenous communities in high-altitude regions*. Frontiers.

Hartl, D., & Jones, E. (2001). *Genetics: Analysis of Genes and Genomes*. Jones and Bartlett Publishers.

Hasanuzzaman, M., Borhannuddin Bhuyan, M. H. M., Anee, T. I., Parvin, K., Nahar, K., Al Mahmud, J., & Fujita, M. (2019). Regulation of Ascorbate-Glutathione Pathway in Mitigating Oxidative Damage in Plants under Abiotic Stress. *Antioxidants, 8*(9), 384. doi:10.3390/antiox8090384

Hassanisaadi, M., Barani, M., Rahdar, A., Heidary, M., Thysiadou, A., & Kyzas, G. Z. (2022). Role of agrochemical-based nanomaterials in plants: Biotic and abiotic stress with germination improvement of seeds. *Plant Growth Regulation, 97*(2), 375–418. doi:10.100710725-021-00782-w

He, C., Liu, D., & Lin, W. (2015). Nanomedicine applications of hybrid nanomaterials built from metal–ligand coordination bonds: Nanoscale metal–organic frameworks and nanoscale coordination polymers. *Chemical Reviews, 115*(19), 11079–11108. doi:10.1021/acs.chemrev.5b00125 PMID:26312730

Hernández-Díaz, J. A., Garza-García, J. J., Zamudio-Ojeda, A., León-Morales, J. M., López-Velázquez, J. C., & García-Morales, S. (2021). Plant-mediated synthesis of nanoparticles and their antimicrobial activity against phytopathogens. *Journal of the Science of Food and Agriculture, 101*(4), 1270–1287. doi:10.1002/jsfa.10767 PMID:32869290

He, X., Deng, H., & Hwang, H. (2019). The current application of nanotechnology in food and agriculture. *Yao Wu Shi Pin Fen Xi, 27*(1), 1–21. doi.org/10.1016/j.jfda.2018.12.002 PMID:30648562

He, Y., Du, Z., Ma, S., Cheng, S., Jiang, S., Liu, Y., Li, D., Huang, H., Zhang, K., & Zheng, X. (2016). Biosynthesis, antibacterial activity and anticancer effects against prostate cancer (PC-3) cells of silver nanoparticles using Dimocarpus Longan Lour. Peel extract. *Nanoscale Research Letters, 11*(11), 1–10. doi:10.118611671-016-1511-9 PMID:27316741

Hittalmani, S., Shashidhar, H. E., Bagali, P. G., Huang, N., Sidhu, J. S., Singh, V. P., & Khush, G. S. (2002). Molecular mapping of quantitative trait loci for plant growth, yield and yield related traits across three diverse locations in a doubled haploid rice population. *Euphytica, 125*(2), 207–214. doi:10.1023/A:1015890125247

Hofmann, T., Lowry, G. V., Ghoshal, S., Tufenkji, N., Brambilla, D., Dutcher, J. R., Gilbertson, L. M., Giraldo, J. P., Kinsella, J. M., Landry, M. P., Lovell, W., Naccache, R., Paret, M., Pedersen, J. A., Unrine, J. M., White, J. C., & Wilkinson, K. J. (2020). Technology readiness and overcoming barriers to sustainably implement nanotechnology-enabled plant agriculture. *Nature Food, 1*(7), 416–425. doi:10.103843016-020-0110-1

Hojjat, S. S., & Kamyab, M. (2017). The effect of silver nanoparticle on Fenugreek seed germination under salinity levels. *Russian Agricultural Sciences, 43*(1), 61–65. doi:10.3103/S1068367417010189

Holland, J. H. (1992). Genetic algorithms. *Scientific American, 267*(1), 66–73. doi:10.1038scientificamerican0792-66

Hong, G., Diao, S., Antaris, A. L., & Dai, H. (2015). Carbon nanomaterials for biological imaging and nanomedicinal therapy. *Chemical Reviews, 115*(19), 10816–10906. doi:10.1021/acs.chemrev.5b00008 PMID:25997028

Hong, J., Wang, L., Sun, Y., Zhao, L., Niu, G., Tan, W., & Gardea-Torresdey, J. L. (2016). Foliar applied nanoscale and microscale CeO2 and CuO alter cucumber (Cucumis sativus) fruit quality. *The Science of the Total Environment, 563*, 904–911. doi:10.1016/j.scitotenv.2015.08.029 PMID:26351199

Horton, A. A., Walton, A., Spurgeon, D. J., Lahive, E., & Svendsen, C. (2017). Microplastics in freshwater and terrestrial environments: Evaluating the current understanding to identify the knowledge gaps and future research priorities. *The Science of the Total Environment, 586*, 127–141. doi:10.1016/j.scitotenv.2017.01.190 PMID:28169032

Hossain, A., Kerry, R. G., Farooq, M., Abdullah, N., & Tofazzal Islam, M. (2020). Application of nanotechnology for sustainable crop production systems. *Nanotechnology for food, agriculture, and environment*, 135-159.

Hossain, Z., & Komatsu, S. (2013). Contribution of proteomic studies towards understanding plant heavy metal stress responses. *Frontiers in Plant Science, 3*, 310. doi:10.3389/fpls.2012.00310 PMID:23355841

Hossain, Z., Mustafa, G., & Komatsu, S. (2015). Plant Responses to Nanoparticle Stress. *International Journal of Molecular Sciences*, *16*(11), 26644–26653. doi:10.3390/ijms161125980 PMID:26561803

Hosseinpour, A., Haliloglu, K., Cinisli, K. T., Ozkan, G., Ozturk, H. I., Pour-Aboughadareh, A., & Poczai, P. (2020). Application of Zinc Oxide NPs and Plant Growth Promoting Bacteria Reduces Genetic Impairment under Salt Stress in Tomato (Solanum lycopersicum L. 'Linda'). *Agriculture, 10*(11), 521. doi:10.3390/agriculture10110521

Hu, P., An, J., Faulkner, M.M., Wu, H., Li, Z., Tian, X., Giraldo, J.P. (2020). Nanoparticle Charge and Size Control Foliar Delivery Efficiency to Plant Cells and Organelles. *ACS Nano, 14*, 7970-7986. https://doi.org/PL_FILE/NN9B09178_SI_017.AVI doi:10.1021/ACSNANO.9B09178/SUP

Huang, D., Huo, J., & Liao, W. (2021). Hydrogen sulfide: Roles in plant abiotic stress response and crosstalk with other signals. *Plant Science*, *302*, 110733. doi:10.1016/j.plantsci.2020.110733 PMID:33288031

Huang, S., Wang, L., Liu, L., Hou, Y., & Li, L. (2015). Nanotechnology in agriculture, livestock, and aquaculture in China. A review. *Agronomy for Sustainable Development*, *35*(2), 369–400. doi:10.100713593-014-0274-x

Huang, X., Zhu, Y., & Kianfar, E. (2021). Nano biosensors: Properties, applications and electrochemical techniques. *Journal of Materials Research and Technology*, *12*, 1649–1672. doi:10.1016/j.jmrt.2021.03.048

Hu, P., An, J., Faulkner, M. M., Wu, H., Li, Z., Tian, X., & Giraldo, J. P. (2020). Nanoparticle Charge and Size Control Foliar Delivery Efficiency to Plant Cells and Organelles. *ACS Nano*, *14*(7), 7970–7986. doi:10.1021/acsnano.9b09178 PMID:32628442

Hussain, S., Shaukat, M., Ashraf, M., Zhu, C., Jin, Q., & Zhang, J. (2019). Salinity stress in arid and semi-arid climates: Effects and management in field crops. *Climate change and agriculture, 13*.

Hussain, A., Ali, S., Rizwan, M., Zia ur Rehman, M., Javed, M. R., Imran, M., Chatha, S. A. S., & Nazir, R. (2018). Zinc oxide NPs alter the wheat physiological response and reduce the cadmium uptake by plants. *Environmental Pollution*, *242*, 1518–1526. doi:10.1016/j.envpol.2018.08.036 PMID:30144725

Hussain, A., Rizwan, M., Ali, Q., & Ali, S. (2019). Seed priming with silicon nanoparticles improved the biomass and yield while reduced the oxidative stress and cadmium concentration in wheat grains. *Environmental Science and Pollution Research International*, *26*(26), 7579–7588. doi:10.100711356-019-04210-5 PMID:30661166

Hussain, M., Ahmad, S., Hussain, S., Lal, R., Ul-Allah, S., & Nawaz, A. (2018). Rice in Saline Soils: Physiology, Biochemistry, Genetics, and Management. *Advances in Agronomy*, *148*, 231–287. doi:10.1016/bs.agron.2017.11.002

Hyllested, J. Æ., Palanco, M. E., Hagen, N., Mogensen, K. B., & Kneipp, K. (2015). Green preparation and spectroscopic characterization of plasmonic silver nanoparticles using fruits as reducing agents. *Beilstein Journal of Nanotechnology*, *6*(6), 293–299. doi:10.3762/bjnano.6.27 PMID:25821667

Ibanez, I., Gornish, E. S., Buckley, L., Debinski, D. M., Hellmann, J., Helmuth, B., HilleRisLambers, J., Latimer, A. M., Miller-Rushing, A. J., & Uriarte, M. (2013). Moving forward in global-change ecology: Capitalizing on natural variability. *Ecology and Evolution*, *3*(1), 170–181. doi:10.1002/ece3.433 PMID:23404535

Ibrahim, E. A. (2016). Seed priming to alleviate salinity stress in germinating seeds. *Journal of Plant Physiology*, *192*, 38–46. doi:10.1016/j.jplph.2015.12.011 PMID:26812088

Ibrahim, M. L., Ijah, U. J. J., Manga, S. B., Bilbis, L. S., & Umar, S. (2013). Production and partial characterization of biosurfactant produced by crude oil degrading bacteria. *International Biodeterioration & Biodegradation*, *81*, 28–34. doi:10.1016/j.ibiod.2012.11.012

Iijima, S. (1991). Helical microtubules of graphitic carbon. *Nature*, *354*(6348), 56–58. doi:10.1038/354056a0

Imtiaz, H., Shiraz, M., Mir, A. R., Siddiqui, H., & Hayat, S. (2023). Nano-Priming Techniques for Plant Physio-Biochemistry and Stress Tolerance. *Journal of Plant Growth Regulation*, 1–21. doi:10.100700344-023-10981-6

Ioannou, A., Gohari, G., Papaphilippou, P., Panahirad, S., Akbari, A., Dadpour, M. R., Krasia-Christoforou, T., & Fotopoulos, V. (2020). Advanced nanomaterials in agriculture under a changing climate: The way to the future? *Environmental and Experimental Botany*, *176*, 104048. doi:10.1016/j.envexpbot.2020.104048

Iqbal, S., Riaz, U., Murtaza, G., Jamil, M., Ahmed, M., Hussain, A., & Abbas, Z. (2021). Chemical fertilizers, formulation, and their influence on soil health. *Microbiota and Biofertilizers: A Sustainable Continuum for Plant and Soil Health*, 1-15.

Iranbakhsh, A., Oraghi Ardebili, Z., Oraghi Ardebili, N., Ghoranneviss, M., & Safari, N. (2018). Cold plasma relieved toxicity signs of nano zinc oxide in *Capsicum annuum* cayenne via modifying growth, differentiation, and physiology. *Acta Physiologiae Plantarum*, *40*(8), 154. doi:10.100711738-018-2730-8

Islam, F., Shohag, S., Uddin, M. J., Islam, M. R., Nafady, M. H., Akter, A., Mitra, S., Roy, A., Emran, T. B., & Cavalu, S. (2022). Exploring the journey of zinc oxide nanoparticles (ZnO-NPs) toward biomedical applications. *Materials (Basel)*, *15*(6), 2160. doi:10.3390/ma15062160 PMID:35329610

Islam, M., & Yasmin, R. (2017). Impact of aquaculture and contemporary environmental issues in Bangladesh. *International Journal of Fisheries and Aquatic Studies*, *5*, 100–107.

Itelima, J., Bang, W., Onyimba, I., Sila, M., & Egbere, O. (2018). Bio-fertilizers as key player in enhancing soil fertility and crop productivity. *RE:view*.

Itroutwar, P. D., Govindaraju, K., Tamilselvan, S., Kannan, M., Raja, K., & Subramanian, K. S. (2019). Seaweed-Based Biogenic ZnO Nanoparticles for Improving Agro-morphological Characteristics of Rice (*Oryza sativa* L.). *Journal of Plant Growth Regulation*, *39*(2), 717–728. doi:10.100700344-019-10012-3

Itroutwar, P. D., Govindaraju, K., Tamilselvan, S., Kannan, M., Raja, K., & Subramanian, K. S. (2019). Seaweed-based biogenic ZnO nanoparticles for improving agromorphological characteristics of rice (Oryza sativa L.). *Plant Growth Regulation*, *392*(39), 717–728.

Itroutwar, P. D., Kasivelu, G., Raguraman, V., Malaichamy, K., & Sevathapandian, S. K. (2020). Effects of biogenic zinc oxide nanoparticles on seed germination and seedling vigor of maize (*Zea mays*). *Biocatalysis and Agricultural Biotechnology*, *29*, 101778. doi:10.1016/j.bcab.2020.101778

Jadhav, R., Pawar, P., Choudhari, V., Topare, N., Raut-Jadhav, S., Bokil, S., & Khan, A. (2023). An overview of antimicrobial nanoparticles for food preservation. *Materials Today: Proceedings*, *72*, 204–216. doi:10.1016/j.matpr.2022.07.045

Jadoun, S., Arif, R., Jangid, N. K., & Meena, R. K. (2021). Green synthesis of nanoparticles using plant extracts: A review. *Environmental Chemistry Letters*, *19*(1), 355–374. doi:10.100710311-020-01074-x

Jahufer, M., Barret, B., Griffiths, A., & Woodfield, D. (2003). DNA fingerprinting and genetic relationships among white clover cultivars. J. Morton (Ed.), *Proceedings of the New Zealand Grassland Association*. Taieri Print Limited, Dunedin.

Jakhar, A. M., Aziz, I., Kaleri, A. R., Hasnain, M., Haider, G., Ma, J., & Abideen, Z. (2022). Nano-fertilizers: A sustainable technology for improving crop nutrition and food security. *NanoImpact*, *27*, 100411. doi:10.1016/j.impact.2022.100411 PMID:35803478

Jamil, M. D., Waheed, M., Akhtar, S., Bangash, N., Chaudhari, S. K., Majeed, M., Hussain, M., Ali, K., & Jones, D. A. (2022). Invasive Plants Diversity, Ecological Status, and Distribution Pattern in Relation to Edaphic Factors in Different Habitat Types of District Mandi Bahauddin, Punjab, Pakistan. *Sustainability (Basel)*, *14*(20), 13312. doi:10.3390u142013312

Jamkhande, P. G., Ghule, N. W., Bamer, A. H., & Kalaskar, M. G. (2019). Metal nanoparticles synthesis: An overview on methods of preparation, advantages and disadvantages, and applications. *Journal of Drug Delivery Science and Technology*, *53*, 101174. doi:10.1016/j.jddst.2019.101174

Jansen, R., & Stam, P. (1994). High resolution of quantitative traits into multiple loci via interval mapping. *Genetics*, *136*(4), 1447–1455. doi:10.1093/genetics/136.4.1447 PMID:8013917

Jasrotia, P., Kashyap, P. L., Bhardwaj, A. K., Kumar, S., & Singh, G. (2018). Scope and applications of nanotechnology for wheat production: A review of recent advances. *Wheat Barley Res*, *10*(1), 1–14.

Jeger, M., Beresford, R., Bock, C., Brown, N., Fox, A., Newton, A., Vicent, A., Xu, X., & Yuen, J. (2021). Global challenges facing plant pathology: Multidisciplinary approaches to meet the food security and environmental challenges in the mid-twenty-first century. *CABI Agriculture and Bioscience*, *2*(1), 1–18. doi:10.118643170-021-00042-x

Jia, J., Bai, J., Wang, W., Yin, S., Zhang, G., Zhao, Q., Wang, X., Liu, X., & Cui, B. (2020). Salt stress alters the short-term responses of nitrous oxide emissions to the nitrogen addition in salt-affected coastal soils. *The Science of the Total Environment*, *742*, 140124. doi:10.1016/j.scitotenv.2020.140124 PMID:32623179

Jiang, H. S., Qiu, X. N., Li, G. B., Li, W., & Yin, L. Y. (2014). Silver nanoparticles induced accumulation of reactive oxygen species and alteration of antioxidant systems in the aquatic plant Spirodela polyrhiza. *Environmental Toxicology and Chemistry*, *33*(6), 1398–1405. doi:10.1002/etc.2577 PMID:24619507

Jiang, Y., Zhou, P., Zhang, P., Adeel, M., Shakoor, N., Li, Y., Li, M., Guo, M., Zhao, W., Lou, B., Wang, L., Lynch, I., & Rui, Y. (2022). Green synthesis of metal-based nanoparticles for sustainable agriculture. *Environmental Pollution*, *309*, 119755. doi:10.1016/j.envpol.2022.119755 PMID:35839973

Jiménez, L., Martínez, C., Pérez, I., & López, F. (1997). Biobleaching procedures for pulp from agricultural residues using Phanerochaete chrysosporium and enzymes. *Process Biochemistry (Barking, London, England)*, *32*(4), 297–304. doi:10.1016/S0032-9592(96)00090-8

Jisha, K. C., Vijayakumari, K., & Puthur, J. T. (2012). Seed priming for abiotic stress tolerance: an overview. *Acta Physiologiae Plantarum, 35*(5), 1381–1396. doi:10.1007/s11738-012-1186-5

Jo, J. H., Singh, P., Kim, Y. J., Wang, C., Mathiyalagan, R., Jin, C. G., & Yang, D. C. (2015). Pseudomonas deceptionensis DC5-mediated synthesis of extracellular silver nanoparticles. *ArtifitialCells Nanomedicines and Biotechnology*, 1–6.

Johnson, R., & Puthur, J. T. (2021). Seed priming as a cost effective technique for developing plants with cross tolerance to salinity stress. *Plant Physiology and Biochemistry*, *162*, 247–257. doi:10.1016/j.plaphy.2021.02.034 PMID:33711718

Jordan, M. C., Somers, D. J., & Banks, T. W. (2007). Identifying regions of the wheat genome controlling seed development by mapping expression quantitative trait loci. *Plant Biotechnology Journal*, *5*(3), 442–453. doi:10.1111/j.1467-7652.2007.00253.x PMID:17388792

Joshi, A., Kaur, S., Dharamvir, K., Nayyar, H., & Verma, G. (2018). Multi-walled carbon nanotubes applied through seed-priming influence early germination, root hair, growth and yield of bread wheat (*Triticum aestivum* L.). *Journal of the Science of Food and Agriculture*, *98*, 3148–3160. doi:10.1002/jsfa.8818 PMID:29220088

Joshi, A., Kaur, S., Dharamvir, K., Nayyar, H., & Verma, G. (2018). Multi-walled carbon nanotubes applied through seed-priming influence early germination, root hair, growth and yield of bread wheat (*Triticumaestivum* L.) [PubMed] [Google Scholar]. *Journal of the Science of Food and Agriculture*, *98*(8), 3148–3160. PMID:29220088

Joshi, S., Ranjekar, P., & Gupta, V. (1999). Molecular markers in plant genome analysis. *Current Science*, *77*, 230–240.

Kader, M. A., & Lindberg, S. (2010). Cytosolic calcium and pH signaling in plants under salinity stress. *Plant Signaling & Behavior*, *5*(3), 233–238. doi:10.4161/psb.5.3.10740 PMID:20037468

Kah, M. (2015). Nanopesticides and Nanofertilizers: Emerging Contaminants or Opportunities for Risk Mitigation? *Frontiers in Chemistry*, *3*, 64. doi:10.3389/fchem.2015.00064 PMID:26636068

Kah, M., & Hofmann, T. (2014). Nanopesticide: Current trends and future priorities. *Environment International*, *63*, 224–235. doi:10.1016/j.envint.2013.11.015 PMID:24333990

Kah, M., Kookana, R. S., Gogos, A., & Bucheli, T. D. (2018). A Critical Evaluation of Nanopesticides and Nanofertilizers Against Their Conventional Analogues. *Nature Nanotechnology*, *13*(8), 677–684. doi:10.103841565-018-0131-1 PMID:29736032

Kah, M., Tufenkji, N., & White, J. C. (2019). Nano-Enabled Strategies to Enhance Crop Nutrition and Protection. *Nature Nanotechnology*, *14*(6), 532–540. doi:10.103841565-019-0439-5 PMID:31168071

Kaivosoja, J., Pesonen, L., Kleemola, J., Pölönen, I., Salo, H., Honkavaara, E., Saari, H., Mäkynen, J., & Rajala, A. (2013). A case study of a precision fertilizer application task generation for wheat based on classified hyperspectral data from UAV combined with farm history data. *Remote Sensing for Agriculture, Ecosystems, and Hydrology XV*, *8887*, 88870H. doi:10.1117/12.2029165

Kalteh, M., Alipour, Z. T., Ashraf, S., Marashi Aliabadi, M., & Falah Nosratabadi, A. (2018). Effect of silica NPs on Basil (Ocimum basilicum) Under Salinity Stress. *Journal of Chemical Health Risks*, *4*(3), 49–55. doi:10.22034/JCHR.2018.544075

Kamran, K., Kemmerling, B., Shutaywi, M., & Mashwani, Z. U. R. (2020). Nano zinc elicited biochemical characterization, nutritional assessment, antioxidant enzymes and fatty acid profiling of rapeseed. *PLoS One*, (15), 0241568. PMID:33170873

Kandhol, N., Singh, V. P., Ramawat, N., Prasad, R., Chauhan, D. K., Sharma, S., Grillo, R., Sahi, S., Peralta-Videa, J., & Tripathi, D. K. (2022). Nano-priming: Impression on the beginner of plant life. *Plant Stress*, *5*, 100091. doi:10.1016/j.stress.2022.100091

Kandil, A.A Sharief, A.E., & Ahmed, S. (2012, December 20). Sharief. (2012). Germination and Seedling Growth of Some Chickpea Cultivars (Cicer arietinum L.) under Salinity Stress. *Journal of Basic and Applied Sciences*, *8*(2), 561–571. doi:10.6000/1927-5129.2012.08.02.49

Kang, Y. J., Kim, S. K., Kim, M. Y., Lestari, P., Kim, K. H., Ha, B. K., Jun, T. H., Hwang, W. J., Lee, T., Lee, J., Shim, S., Yoon, M. Y., Jang, Y. E., Han, K. S., Taeprayoon, P., Yoon, N., Somta, P., Tanya, P., Kim, K. S., & Lee, S.-H. (2014). Genome sequence of mung bean and insights into evolution within Vigna species. *Nature Communications*, *5*(1), 5443. doi:10.1038/ncomms6443 PMID:25384727

Kanmani, P., & Rhim, J. W. (2014). Antimicrobial and physical-mechanical properties of agar-based films incorporated with grapefruit seed extract. *Carbohydrate Polymers*, *102*, 708–716. doi:10.1016/j.carbpol.2013.10.099 PMID:24507339

Kannaujia, R., Srivastava, C. M., Prasad, V., Singh, B. N., & Pandey, V. (2019). Phyllanthus emblica fruit extract stabilized biogenic silver nanoparticles as a growth promoter of wheat varieties by reducing ROS toxicity. *Plant Physiology and Biochemistry*, *142*, 460–471. doi:10.1016/j.plaphy.2019.08.008 PMID:31425972

Kar, R. K. (2015). ROS signaling: Relevance with site of production and metabolism of ROS. *Reactive Oxygen Species and Oxidative Damage in Plants Under Stress*, 115–125. https://doi.org/ doi:10.1007/978-3-319-20421-5_5/COVER

Kasote, D. M., Lee, J. H. J., Jayaprakasha, G. K., & Patil, B. S. (2019). Seed Priming with Iron Oxide Nanoparticles Modulate Antioxidant Potential and Defense-Linked Hormones in Watermelon Seedlings. *ACS Sustainable Chemistry & Engineering*, *7*(5), 5142–5151. doi:10.1021/acssuschemeng.8b06013

Kasote, D. M., Lee, J. H., Jayaprakasha, G. K., & Patil, B. S. (2021). Manganese oxide nanoparticles as safer seed priming agent to improve chlorophyll and antioxidant profiles in watermelon seedlings. *Nanomaterials (Basel, Switzerland)*, *11*(11), 1016. doi:10.3390/nano11041016 PMID:33921180

Kaushal, S., Nanda, S. S., Samal, S., & Yi, D. K. (2020). Strategies for the development of metallic-nanoparticle-based label-free biosensors and their biomedical applications. *ChemBioChem*, *21*(5), 576–600. doi:10.1002/cbic.201900566 PMID:31634410

Kearsey, M., & Pooni, H. (1996). *The genetical analysis of quantitative traits*. Chapman & Hall. doi:10.1007/978-1-4899-4441-2

Kelly, J. D., Gepts, P., Miklas, P. N., & Coyne, D. P. (2003). Tagging and mapping of genes and QTL and molecular marker-assisted selection for traits of economic importance in bean and cowpea. *Field Crops Research*, *82*(2-3), 135–154. doi:10.1016/S0378-4290(03)00034-0

Khairy, A. M., Tohamy, M. R. A., Zayed, M. A., Mahmoud, S. F., El-Tahan, A. M., El-Saadony, M. T., & Mesiha, P. K. (2022). Eco-friendly application of Nano-chitosan for controlling potato and tomato bacterial wilt. *Saudi Journal of Biological Sciences*, *29*(4), 2199–2209. doi:10.1016/j.sjbs.2021.11.041 PMID:35531227

Khan, A. M., Li, Q., Saqib, Z., Khan, N., Habib, T., Khalid, N., Majeed, M., & Tariq, A. (2022). MaxEnt Modelling and Impact of Climate Change on Habitat Suitability Variations of Economically Important Chilgoza Pine (Pinus gerardiana Wall.) in South Asia. *Forests*, *13*(5), 715. doi:10.3390/f13050715

Khan, I., Saeed, K., & Khan, I. (2019). Nanoparticles: Properties, applications and toxicities. *Arabian Journal of Chemistry*, *12*(7), 908–931. doi:10.1016/j.arabjc.2017.05.011

Khan, M. N., Li, Y., Fu, C., Hu, J., Chen, L., Yan, J., Khan, Z., Wu, H., & Li, Z. (2022). CeO2 nanoparticles seed priming increases salicylic acid level and ROS scavenging ability to improve rapeseed salt tolerance. *Global Challenges (Hoboken, NJ)*, *6*(7), 2200025. doi:10.1002/gch2.202200025 PMID:35860396

Khan, M. N., Mobin, M., Abbas, Z. K., AlMutairi, K. A., & Siddiqui, Z. H. (2017). Role of Nanomaterials in Plants under Challenging Environments. *Plant Physiology and Biochemistry*, *110*, 194–209. doi:10.1016/j.plaphy.2016.05.038 PMID:27269705

Khan, M. R., Adam, V., Rizvi, T. F., Zhang, B., Ahamad, F., Jośko, I., Zhu, Y., Yang, M., & Mao, C. (2019). Nanoparticle–Plant Interactions: Two-Way Traffic. *Small*, *15*(37), 1901794. doi:10.1002mll.201901794 PMID:31318142

Khin, M. M., Nair, A. S., Babu, V. J., Murugan, R., & Ramakrishna, S. (2012). A review on nanomaterials for environmental remediation. *Energy & Environmental Science*, *5*(8), 8075–8109. doi:10.1039/c2ee21818f

Khodakovskaya, M. V., De Silva, K., Biris, A. S., Dervishi, E., & Villagarcia, H. (2012). Carbon nanotubes induce growth enhancement of tobacco cells. *ACS Nano*, *6*(3), 2128–2135. doi:10.1021/nn204643g PMID:22360840

Khodakovskaya, M. V., De Silva, K., Nedosekin, D. A., Dervishi, E., Biris, A. S., Shashkov, E. V., Galanzha, E. I., & Zharov, V. P. (2011). Complex genetic, photothermal, and photoacoustic analysis of nanoparticle-plant interactions. *Proceedings of the National Academy of Sciences of the United States of America*, *108*(3), 1028–1033. doi:10.1073/pnas.1008856108 PMID:21189303

Khodakovskaya, M. V., Kim, B.-S., Kim, J. N., Alimohammadi, M., Dervishi, E., Mustafa, T., & Cernigla, C. E. (2013). Carbon Nanotubes as Plant Growth Regulators: Effects on Tomato Growth, Reproductive System, and Soil Microbial Community. *Small*, *9*(1), 115–123. doi:10.1002mll.201201225 PMID:23019062

Khodakovskaya, M., Dervishi, E., Mahmood, M., Xu, Y., Li, Z., Watanabe, F., & Biris, A. S. (2009). Carbon Nanotubes are Able to Penetrate Plant Seed Coat and Dramatically Affect Seed Germination and Plant Growth. *ACS Nano*, *3*(10), 3221–3227. doi:10.1021/nn900887m PMID:19772305

Khoja, A. A., Haq, S. M., Majeed, M., Hassan, M., Waheed, M., Yaqoob, U., Bussmann, R. W., Alataway, A., & Dewidar, A. Z. (2022). *Diversity*. Ecological and Traditional Knowledge of Pteridophytes in the Western Himalayas.

Khota, L. R., Sankarana, S., Majaa, J. M., Ehsania, R., & Schuster, E. W. (2012). Applications of nanomaterials in agricultural production and crop protection: A review. *Crop Protection (Guildford, Surrey)*, *35*, 64–70. doi:10.1016/j.cropro.2012.01.007

Kim, D. Y., Kadam, A., Shinde, S., Saratale, R. G., Patra, J., & Ghodake, G. (2018). Recent developments in nanotechnology transforming the agricultural sector: A transition replete with opportunities. *Journal of the Science of Food and Agriculture*, *98*(3), 849–864. doi:10.1002/jsfa.8749 PMID:29065236

Kirkby, E. A. (2023). Introduction, definition, and classification of nutrients. In *Marschner's Mineral Nutrition of Plants* (pp. 3–9). Elsevier. doi:10.1016/B978-0-12-819773-8.00016-2

Klaine, S. J., Koelmans, A. A., Horne, N., Carley, S., Handy, R. D., Kapustka, L., Nowack, B., & von der Kammer, F. (2012). *Paradigms to assess the environmental impact of manufactured nanomaterials* (Vol. 31). Citeseer.

Kobayashi, H., Yanaka, M., & Ikeda, T. M. (2015). Interaction of carbon nanohorns with plants: Uptake and biological effects. *Carbon, 81*, 607–619. doi:10.1016/j.carbon.2014.09.095

Koevoets, I. T., Venema, J. H., Elzenga, J. T. M., & Testerink, C. (2016). Roots withstanding their environment: Exploiting root system architecture responses to abiotic stress to improve crop tolerance. *Frontiers in Plant Science, 7*, 1335. doi:10.3389/fpls.2016.01335 PMID:27630659

Komyshev, E., Genaev, M., & Afonnikov, D. (2017). Evaluation of the SeedCounter, A Mobile Application for Grain Phenotyping. *Frontiers in Plant Science, 7*, 1990. doi:10.3389/fpls.2016.01990 PMID:28101093

Koohkan, H., & Maftoun, M. (2016). Effect of nitrogen–boron interaction on plant growth and tissue nutrient concentration of canola (Brassica napus L.). *Journal of Plant Nutrition, 39*(7), 922–931. doi:10.1080/01904167.2016.1143492

Korbekandi, H., Mohseni, S., Mardani Jouneghani, R., Pourhossein, M., & Iravani, S. (2016). Biosynthesis of silver nanoparticles using Saccharomyces cerevisiae. *Artificial Cells, Nanomedicine, and Biotechnology, 44*(1), 235–239. doi:10.3109/21691401.2014.937870 PMID:25101816

Kotresh, H., Fakrudin, B., Punnuri, S., Rajkumar, B., Thudi, M., Paramesh, H., Lohithswa, H., & Kuruvinashei, M. S. (2006). Identification of two RAPD markers genetically linked to a recessive allele of a Fusarium wilt resistance gene in pigeon pea (Cajanus cajan L.) Millsp.). *Euphytica, 149*(1-2), 113–120. doi:10.100710681-005-9059-2

Kubala, S., Garnczarska, M., Wojtyla, Ł., Clippe, A., Kosmala, A., Żmieńko, A., & Quinet, M. (2015). Deciphering priming-induced improvement of rapeseed (Brassica napus L.) germination through an integrated transcriptomic and proteomic approach. *Plant Science, 231*, 94–113. doi:10.1016/j.plantsci.2014.11.008 PMID:25575995

Kumar, B., Singhal, R. K., Chand, S., Chauhan, J., Kumar, V., Mishra, U. N., Hidangmayum, A., Singh, A., & Bose, B. (2022). Nanopriming in sustainable agriculture: recent advances, emerging challenges and future prospective. *New and Future Developments in Microbial Biotechnology and Bioengineering*, 339-365.

Kumar, P., Singh, A., Rajput, V. D., Yadav, A. K. S., Kumar, P., Singh, A. K., & Minkina, T. (2022) Role of artificial intelligence, sensor technology, big data in agriculture: next-generation farming, Bioinformatics in Agriculture. Academic Press.

Kumar, A., Gupta, K., Dixit, S., Mishra, K., & Srivastava, S. (2019). A review on positive and negative impacts of nanotechnology in agriculture. *International Journal of Environmental Science and Technology, 16*(4), 2175–2184. doi:10.100713762-018-2119-7

Kumar, P., Singh, A., Rajput, V. D., Yadav, A. K. S., Kumar, P., Singh, A. K., & Minkina, T. (2022). Role of artificial intelligence, sensor technology, big data in agriculture: next-generation farming. In *Bioinformatics in Agriculture* (pp. 625–639). Academic Press. doi:10.1016/B978-0-323-89778-5.00035-0

Kumar, S., Nehra, M., Dilbaghi, N., Marrazza, G., Tuteja, S. K., & Kim, K.-H. (2020). Nanovehicles for Plant Modifications towards Pest- and Disease-Resistance Traits. *Trends in Plant Science*, *25*(2), 198–212. doi:10.1016/j.tplants.2019.10.007 PMID:31780333

Kumar, V. K., Muthukrishnan, S., & Rajalakshmi, R. (2020). Phytostimulatory effect of phytochemical fabricated nanosilver (AgNPs) on Psophocarpus tetragonolobus (L.) DC. seed germination: An insight from antioxidative enzyme activities and genetic similarity studies. *Current Plant Biology*, *23*, 100158. doi:10.1016/j.cpb.2020.100158

Kumar, V., Guleria, P., & Mehta, S. K. (2017). Nanosensors for evaluating the quality and safety of food. *Environmental Chemistry Letters*, *15*, 165–177. doi:10.100710311-017-0616-4

Kumar, V., Singhal, R. K., Kumar, N., & Bose, B. (2020). *Micronutrient seed priming: A pragmatic approach towards abiotic stress management. New frontiers in stress management for durable agriculture*. Springer.

Kumar, Y., Singh, K. T. T., & Raliya, R. (2021). Nanofertilizers and their role in sustainable agriculture. *Annals of Plant and Soil Reserach*, *23*(3), 238–255. doi:10.47815/apsr.2021.10067

Kumawat, N., Kumar, R., Kumar, S., & Meena, V. S. (2017). *Nutrient solubilizing microbes (NSMs): its role in sustainable crop production. Agriculturally Important Microbes for Sustainable Agriculture* (Vol. 2). Applications in Crop Production and Protection.

Kuriyama, R., & Sakai, H. (1974). Role of tubulin-SH groups in polymerization to microtubules. Functional-SH groups in tubulin for polymerization. *Journal of Biochemistry*, *76*(3), 651–654. doi:10.1093/oxfordjournals.jbchem.a130609 PMID:4474165

Kusano, M., Fukushima, A., Kobayashi, M., Hayashi, N., Jonson, P., & Mortiz, T. (2007). Application of a metabolomics method combining one-dimensional and two-dimensional gas chromatography time of light/ mass spectrometry to metabolic phenotyping of natural variants in rice. *Journal of Chromatography. B, Analytical Technologies in the Biomedical and Life Sciences*, *855*(1), 71–79. doi:10.1016/j.jchromb.2007.05.002 PMID:17556050

Lahiani, M. H., Dervishi, E., Chen, J., Nima, Z., Gaume, A., Biris, A. S., & Khodakovskaya, M. V. (2013). Impact of Carbon Nanotube Exposure to Seeds of Valuable Crops. *ACS Applied Materials & Interfaces*, *5*(16), 7965–7973. doi:10.1021/am402052x PMID:23834323

Lal, R., Bouma, J., Brevik, E., Dawson, L., Field, D. J., Glaser, B., Hatano, R., Hartemink, A. E., Kosaki, T., Lascelles, B., Monger, C., Muggler, C., Ndzana, G. M., Norra, S., Pan, X., Paradelo, R., Reyes-Sánchez, L. B., Sandén, T., Singh, B. R., & Zhang, J. (2021). Soils and sustainable development goals of the United Nations: An International Union of Soil Sciences perspective. *Geoderma Regional*, *25*, e00398. doi:10.1016/j.geodrs.2021.e00398

Lambreva, M. D., Lavecchia, T., Tyystjärvi, E., Antal, T. K., Orlanducci, S., Margonelli, A., & Rea, G. (2015). Potential of carbon nanotubes in algal biotechnology. *Photosynthesis Research, 125*(3), 451–471. doi:10.100711120-015-0168-z PMID:26113435

Lander, E. S., Green, P., Abrahamson, J., Barlow, A., Daly, M. J., Lincoln, S. E., & Newburg, L. (1987). Mapmaker an interactive computer package for constructing primary genetic linkage maps of experimental and natural populations. *Genomics, 1*(2), 174–181. doi:10.1016/0888-7543(87)90010-3 PMID:3692487

Lander, E. S., & Kruglyak, L. (1995). Genetic dissection of complex traits: Guidelines for interpreting and reporting linkage results. *Nature Genetics, 11*(3), 241–247. doi:10.1038/ng1195-241 PMID:7581446

Lander, E., & Botstein, D. (1989). Mapping Mendelian factors underlying quantitative traits using RFLP linkage maps. *Genetics, 121*(1), 185–199. doi:10.1093/genetics/121.1.185 PMID:2563713

Larue, C., Pinault, M., Czarny, B., Georgin, D., Jaillard, D., Bendiab, N., Mayne-L'Hermite, M., Taran, F., Dive, V., & Carrière, M. (2012). 'Quantitative evaluation of multi-walled carbon nanotube uptake in wheat and rapeseed', Journal of Hazardous Materials. *Journal of Hazardous Materials, 227–228*, 155–163. doi:10.1016/j.jhazmat.2012.05.033 PMID:22652322

Lastochkina, O., Aliniaeifard, S., SeifiKalhor, M., Bosacchi, M., Maslennikova, D., & Lubyanova, A. (2022). Novel Approaches for Sustainable Horticultural Crop Production: Advances and Prospects. *Horticulturae, 8*(10), 910. doi:10.3390/horticulturae8100910

Läuchli, A., & Grattan, S. R. (2007). Plant growth and development under salinity stress. *Advances in Molecular Breeding Toward Drought and Salt Tolerant Crops*, 1–32. https://doi.org/ doi:10.1007/978-1-4020-5578-2_1/COVER

Lee, C. K., Ibrahim, D., & Che Omar, I. (2013). Enzymatic deinking of various types of waste paper: Efficiency and characteristics. *Process Biochemistry (Barking, London, England), 48*(2), 299–305. doi:10.1016/j.procbio.2012.12.015

Lee, C. W., Mahendra, S., Zodrow, K., Li, D., Tsai, Y. C., Braam, J., & Alvarez, P. J. J. (2010). Developmental phytotoxicity of metal oxide NPs to Arabidopsis thaliana. *Environmental Toxicology and Chemistry, 29*(3), 669–675. doi:10.1002/etc.58 PMID:20821493

Le, H., Karakasyan, C., Jouenne, T., Le Cerf, D., & Dé, E. (2021). Application of Polymeric Nanocarriers for Enhancing the Bioavailability of Antibiotics at the Target Site and Overcoming Antimicrobial Resistance. *Applied Sciences (Basel, Switzerland), 11*(22), 10695. doi:10.3390/app112210695

Lemmens, E., Deleu, L.J., De Brier, N., De Man, W.L., De Proft, M., Prinsen, E., & Delcour, J.A. (2019). The Impact of Hydro-Priming and Osmo-Priming on Seedling Characteristics, Plant Hormone Concentrations, Activity of Selected Hydrolytic Enzymes, and Cell Wall and Phytate Hydrolysis in Sprouted Wheat (*Triticum aestivum* L.). *ACS Omega, 4,* 22089-22100. https://doi.org/ _SI_001.PDF doi:10.1021/ACSOMEGA.9B03210/SUPPL_FILE/AO9B03210

Lemmens, E., Deleu, L. J., De Brier, N., De Man, W. L., De Proft, M., Prinsen, E., & Delcour, J. A. (2019). The Impact of Hydro-Priming and Osmo-Priming on Seedling Characteristics, Plant Hormone Concentrations, Activity of Selected Hydrolytic Enzymes, and Cell Wall and Phytate Hydrolysis in Sprouted Wheat (*Triticum aestivum* L.). *ACS Omega*, *4*(26), 22089–22100. doi:10.1021/acsomega.9b03210 PMID:31891089

Leng, X., Jia, H., Sun, X., Shangguan, L., Mu, Q., Wang, B., & Fang, J. (2015). Comparative Transcriptome Analysis of Grapevine in Response to Copper Stress. *Scientific Reports*, *5*(1), 17749. doi:10.1038rep17749 PMID:26673527

Lew, T. T. S., Sarojam, R., Jang, I.-C., Park, B. S., Naqvi, N. I., Wong, M. H., Singh, G. P., Ram, R. J., Shoseyov, O., Saito, K., Chua, N.-H., & Strano, M. S. (2020). Species-independent analytical tools for next-generation agriculture. *Nature Plants*, *6*(12), 1408–1417. doi:10.103841477-020-00808-7 PMID:33257857

Li, L., Fan, M., Brown, R. C., Van Leeuwen, J., Wang, J., Wang, W., Song, Y., & Zhang, P. (2006). Synthesis, Properties, and Environmental Applications of Nanoscale Iron-Based Materials: A Review. *Critical Reviews in Environmental Science and Technology*, *36*(5), 405–431. doi:10.1080/10643380600620387

Li, L., Zhang, Q., & Huang, D. (2014). A review of imaging techniques for plant phenotyping. *Sensors (Basel)*, *14*(11), 20078–20111. doi:10.3390141120078 PMID:25347588

Li, M., Liu, H., Dang, F., Hintelmann, H., Yin, B., & Zhou, D. (2020). Alteration of Crop Yield and Quality of Three Vegetables upon Exposure to Silver Nanoparticles in Sludge-Amended Soil. *ACS Sustainable Chemistry & Engineering*, *8*(6), 2472–2480. doi:10.1021/acssuschemeng.9b06721

Lincoln, S., Daly, M., & Lander, E. (1993a). *Constructing genetic linkage maps with MAPMAKER/EXP. Version 3.0* (3rd ed.). Whitehead Institute for Biomedical Research Technical Report.

Lincoln, S., Daly, M., & Lander, E. (1993b). *Mapping genes controlling quantitative traits using MAPMAKER/QTL. Version 1.1* (2nd ed.). Whitehead Institute for Biomedical Research Technical Report.

Lin, D., & Xing, B. (2007). Phytotoxicity of nanoparticles: Inhibition of seed germination and root growth. *Environmental Pollution*, *150*(2), 243–250. doi:10.1016/j.envpol.2007.01.016 PMID:17374428

Linglan, M., Chao, L., Jiangong, L., & Chunyang, L. (2011). *Journal of Agricultural and Food Chemistry.*

Li, R., He, J., Xie, H., Wang, W., Bose, S. K., Sun, Y., Hu, J., & Yin, H. (2019). Effects of chitosan nanoparticles on seed germination and seedling growth of wheat (*Triticum aestivum* L.*)*. *International Journal of Biological Macromolecules*, *126*, 91–100. doi:10.1016/j.ijbiomac.2018.12.118 PMID:30557637

Liu, R., Lal, R., & Evangelou, V. P. (Eds.). (2019). Applications of nanotechnology in agri-environmental sciences. Vol. 2: Nanotechnology for sustainable agricultural production. CRC Press.

Liu, S., Liu, Y., Yang, X., Tong, C., Edwards, D., Parkin, I.A., & Zhao, M. (2014). The Brassica oleracea genome reveals the asymmetrical evolution of polyploidy genomes. *Nature Communications*.

Liu, B. (1998). *Statistical Genomics: Linkage*. Mapping and QTL Analysis CRC Press.

Liu, C., Jian, Z., & Meng, X. (2013). Combining Float Car Data and Multispectral Satellite Images to Extract Road Features and Networks. In *Lecture Notes in Geoinformation and Cartography* (pp. 29–43). Springer Berlin Heidelberg. doi:10.1007/978-3-642-34203-5_

Liu, J., Li, G., Chen, L., Gu, J., Wu, H., & Li, Z. (2021). Cerium oxide nanoparticles improve cotton salt tolerance by enabling better ability to maintain cytosolic K+/Na+ ratio. *Journal of Nanobiotechnology*, *19*(1), 153. doi:10.118612951-021-00892-7 PMID:34034767

Liu, Q., Chen, B., Wang, Q., Shi, X., Xiao, Z., Lin, J., & Fang, X. (2009). Carbon nanotubes as molecular transporters for walled plant cells. *Nano Letters*, *9*(3), 1007–1010. doi:10.1021/nl803083u PMID:19191500

Liu, Q., Zhao, Y., Wan, Y., Xiao, L., & Fang, Y. (2014). *Food Chemistry*.

Liu, R., & Lal, R. (2015). Potentials of engineered nanoparticles as fertilizers for increasing agronomic productions. *The Science of the Total Environment*, *514*, 131–139. doi:10.1016/j.scitotenv.2015.01.104 PMID:25659311

Liu, Y., Cao, X., Yue, L., Wang, C., Tao, M., Wang, Z., & Xing, B. (2022). Foliar-applied cerium oxide nanomaterials improve maize yield under salinity stress: Reactive oxygen species homeostasis and rhizobacteria regulation. *Environmental Pollution*, *299*, 118900. doi:10.1016/j.envpol.2022.118900 PMID:35085650

Li, X., Baskin, J. M., & Baskin, C. C. (1999). Comparative morphology and physiology of fruit and seed development in the two shrubs Rhus aromatic and *R. glabra* (Anacardiaceae). *American Journal of Botany*, *86*(9), 1217–1225. doi:10.2307/2656769 PMID:10487809

Li, Y., Xu, X., Wu, Y., Zhuang, J., Zhang, X., Zhang, H., Lei, B., Hu, C., & Liu, Y. (2020). A review on the effects of carbon dots in plant systems. *Materials Chemistry Frontiers*, *4*(2), 437–448. doi:10.1039/C9QM00614A

Lockie, S., Fairley-Grenot, K., Ankeny, R., Botterill, L., Howlett, B., Mcbratney, A., Probyn, E., Sorrell, T., Sukkarieh, S., & Woodhead, I. (2020). *The future of agricultural technologies*. Australian Council of Learned Academies (ACOLA).

Lodeyro, A. F., Ceccoli, R. D., Karlusich, J. J. P., & Carrillo, N. (2012). The importance of flavodoxin for environmental stress tolerance in photosynthetic microorganisms and transgenic plants. Mechanism, evolution and biotechnological potential. *FEBS Letters*, *586*(18), 2917–2924. doi:10.1016/j.febslet.2012.07.026 PMID:22819831

López, M. M., Llop, P., Olmos, A., Marco-Noales, E., Cambra, M., & Bertolini, E. (2009). Are molecular tools solving the challenges posed by detection of plant pathogenic bacteria and viruses? *Current Issues in Molecular Biology*, *11*, 13–46. PMID:18577779

Lowe, R., Shirley, N., Bleackley, M., Dolan, S., & Shafee, T. (2017). Transcriptomics technologies. *PLoS Computational Biology, 13*(5), 1005457. doi:10.1371/journal.pcbi.1005457 PMID:28545146

Lowry, G. V., Avellan, A., & Gilbertson, L. M. (2019). Opportunities and Challenges For Nanotechnology in the Agri-Tech Revolution. *Nature Nanotechnology, 14*(6), 517–522. doi:10.103841565-019-0461-7 PMID:31168073

Lu, L., Huang, M., Huang, Y., Corvini, P. F.-X., Ji, R., & Zhao, L. (2020). Mn 3 O 4 nanozymes boost endogenous antioxidant metabolites in cucumber (Cucumis sativus) plant and enhance resistance to salinity stress. *Environmental Science. Nano, 7*(6), 1692–1703. doi:10.1039/D0EN00214C

Lutts, S., Benincasa, P., Wojtyla, L., S, S. K., Pace, R., Lechowska, K., Quinet, M., & Malgorzata, G. (2016). Seed Priming: New Comprehensive Approaches for an Old Empirical Technique. *New Challenges in Seed Biology - Basic and Translational Research Driving Seed Technology*. doi:10.5772/64420

Lu, Y., Song, S., Wang, R., Liu, Z., Meng, J., Sweetman, A. J., Jenkins, A., Ferrier, R. C., Li, H., Luo, W., & Wang, T. (2015). Impacts of soil and water pollution on food safety and health risks in China. *Environment International, 77*, 5–15. doi:10.1016/j.envint.2014.12.010 PMID:25603422

Machado, T. O., Grabow, J., Sayer, C., de Araújo, P. H., Ehrenhard, M. L., & Wurm, F. R. (2022). Biopolymer-based nanocarriers for sustained release of agrochemicals: A review on materials and social science perspectives for a sustainable future of agri-and horticulture. *Advances in Colloid and Interface Science, 303*, 102645. doi:10.1016/j.cis.2022.102645 PMID:35358807

Mackill, D. J., Nguyen, H. T., & Zhan, J. (1999). Use of molecular markers in plant improvement programs for rain fed lowland rice. *Field Crops Research, 64*(1-2), 177–185. doi:10.1016/S0378-4290(99)00058-1

Macovei, A., Garg, B., Raikwar, S., Balestrazzi, A., Carbonera, D., Buttafava, A., & Tuteja, N. (2014). Synergistic exposure of rice seeds to different doses of-ray and salinity stress resulted in increased antioxidant enzyme activities and gene-specific modulation of tc-ner pathway. *BioMed Research International, 2014*, 2014. doi:10.1155/2014/676934 PMID:24551849

Macovei, A., & Tuteja, N. (2013). Different expression of miRNAs targeting helicases in rice in response to low and high dose rate γ-ray treatments. *Plant Signaling & Behavior, 8*(8), 25128. doi:10.4161/psb.25128 PMID:23733055

Madhumathi, K., Sudheesh Kumar, P. T., Abhilash, S., Sreeja, V., Tamura, H., Manzoor, K., Nair, S. V., & Jayakumar, R. (2010). Development of novel chitin/nanosilver composite scaffolds for wound dressing applications. *Journal of Materials Science. Materials in Medicine, 21*(2), 807–813. doi:10.100710856-009-3877-z PMID:19802687

Magdolenova, Z., Collins, A., Kumar, A., Dhawan, A., Stone, V., & Dusinska, M. (2014). Mechanisms of genotoxicity. A review of in vitro and in vivo studies with engineered NPs. *Nanotoxicology, 8*(3), 233–278. doi:10.3109/17435390.2013.773464 PMID:23379603

Maggio, A., Bressan, R. A., Zhao, Y., Park, J., & Yun, D.-J. (2018). It's hard to avoid avoidance: Uncoupling the evolutionary connection between plant growth, productivity and stress "tolerance". *International Journal of Molecular Sciences*, *19*(11), 3671. doi:10.3390/ijms19113671 PMID:30463352

Mahakham, W., Sarmah, A. K., Maensiri, S., & Theerakulpisut, P. (2017). Nanopriming Technology For Enhancing Germination And Starch Metabolism Of Aged Rice Seeds Using Phytosynthesized Silver Nanoparticles. *Scientific Reports*, *7*(1), 1–21. doi:10.103841598-017-08669-5 PMID:28811584

Mahakham, W., Theerakulpisut, P., Maensiri, S., Phumying, S., & Sarmah, A. K. (2016). Environmentally benign synthesis of phytochemicals-capped gold nanoparticles as nanopriming agent for promoting maize seed germination. *The Science of the Total Environment*, *573*, 1089–1102. doi:10.1016/j.scitotenv.2016.08.120 PMID:27639594

Maharramov, A. M., Hasanova, U. A., Suleymanova, I. A., Osmanova, G. E., & Hajiyeva, N. E. (2019). The engineered nanoparticles in food chain: Potential toxicity and effects. *SN Applied Sciences*, *1*(11), 1–25. doi:10.100742452-019-1412-5

Mahdi, Z., Bahrami, A., Jonidi Jafari, A., Rezaei, M., Farzi, G., & Asadi, A. (2021). Application of nanotechnology in detection and removal of chemical contaminants in food. *Food Control*, *123*, 107858.

Mahdy, A. M., Sherif, F. K., Elkhatib, E. A., Fathi, N. O., & Ahmed, M. H. (2020). Seed priming in NPs of water treatment residual can increase the germination and growth of cucumber seedling under salinity stress. *Journal of Plant Nutrition*, *43*(12), 1862–1874. doi:10.1080/01904167.2020.1750647

Maity, A., Natarajan, N., Pastor, M., Vijay, D., Gupta, C. K., & Wasnik, V. K. (2018). Nanoparticles influence seed germination traits and seed pathogen infection rate in forage sorghum (*Sorghum bicolour*) and cowpea (*Vigna unguiculata*). *Indian Journal of Experimental Biology*, *56*, 363–372.

Maity, S., Chatterjee, A., Guchhait, R., De, S., & Pramanick, K. (2020). Cytogenotoxic potential of a hazardous material, polystyrene microparticles on Allium cepa L. *Journal of Hazardous Materials*, *385*, 121560. doi:10.1016/j.jhazmat.2019.121560 PMID:31732349

Majeed, M., Bhatti, K. H., Pieroni, A., Renata, S., Bussmann, R. W., Khan, A. M., Chaudhari, S. K., Aziz, M. A., & Amjad, M. S. (2021). *Gathered Wild Food Plants among Diverse Religious Groups in*. MDPI. http://scholar.google.com/scholar_url?url=https://www.mdpi.com/2304-8158/10/3/594/pdf&hl=sl&sa=X&d=16557763567181478619&ei=yrdPYOSADsbYmQHIqoSICA&scisig=AAGBfm3h_EDHFITvMVt55xuUBsvq3j_PUw&nossl=1&oi=scholaralrt&hist=RTqo9GEAAAAJ:3473121909733627244:AAGBfm

Majeed, M., Bhatti, K. H., & Amjad, M. S. (2021). Impact of climatic variations on the flowering phenology of plant species in jhelum district, punjab, pakistan. *Applied Ecology and Environmental Research*, *19*(5), 3343–3376. doi:10.15666/aeer/1905_33433376

Majeed, M., Bhatti, K. H., Pieroni, A., Sõukand, R., Bussmann, R. W., Khan, A. M., Chaudhari, S. K., Aziz, M. A., & Amjad, M. S. (2021). Gathered wild food plants among diverse religious groups in Jhelum District, Punjab, Pakistan. *Foods*, *10*(3), 1–28. doi:10.3390/foods10030594 PMID:33799901

Majeed, M., Lu, L., Anwar, M. M., Tariq, A., Qin, S., El-Hefnawy, M. E., El-Sharnouby, M., Li, Q., & Alasmari, A. (2023). Prediction of flash flood susceptibility using integrating analytic hierarchy process (AHP) and frequency ratio (FR) algorithms. *Frontiers in Environmental Science*, *10*, 1037547. doi:10.3389/fenvs.2022.1037547

Majeed, M., Lu, L., Haq, S. M., Waheed, M., Sahito, H. A., Fatima, S., Aziz, R., Bussmann, R. W., Tariq, A., & Ullah, I. (2022). *Spatiotemporal Distribution Patterns of Climbers along an Abiotic Gradient in Jhelum District*. doi:10.3390/f13081244

Majeed, M., Mahmood, A., Habib, T., Mushahid, M., Ali, H., Khan, N., & Ali, K. (2022). Vegetation analysis and environmental indicators of an arid tropical forest ecosystem of Pakistan. *Ecological Indicators*, *142*(May), 109291. doi:10.1016/j.ecolind.2022.109291

Majeed, M., Tariq, A., Haq, S. M., Waheed, M., Anwar, M. M., Li, Q., Aslam, M., Abbasi, S., Mousa, B. G., & Jamil, A. (2022). A Detailed Ecological Exploration of the Distribution Patterns of Wild Poaceae from the Jhelum District (Punjab), Pakistan. *Sustainability (Basel)*, *14*(7), 3786. doi:10.3390u14073786

Makarov, V. V., Love, A. J., Sinitsyna, O. V., Makarova, S. S., Yaminsky, I. V., Taliansky, M. E., & Kalinina, N. O. (2014). "Green" Nanotechnologies: Synthesis of Metal Nanoparticles Using Plants. *Acta Naturae (English Ed.)*, *6*(1), 35–44. doi:10.32607/20758251-2014-6-1-35-44 PMID:24772325

Malik, A., Mor, V. S., Tokas, J., Punia, H., Malik, S., Malik, K., Sangwan, S., Tomar, S., Singh, P., Singh, N., Himangini, Vikram, Nidhi, Singh, G., Vikram, Kumar, V., Sandhya, & Karwasra, A. (2020). Biostimulant-Treated Seedlings under Sustainable Agriculture: A Global Perspective Facing Climate Change. *Agronomy (Basel)*, *11*(1), 14. doi:10.3390/agronomy11010014

Malik, S. K., Chaudhary, R., Panwar, N. S., & Dhariwal, O. P. (2012). Genetic resources of Chironji (*Buchanania lanzan* Spreng.): A socio-economically important tree species of central India tribal population. *Genetic Resources and Crop Evolution*, *59*(4), 615–623. doi:10.100710722-012-9801-2

Mallakpour, S., & Azadi, E. (2021). Nanofiltration membranes for food and pharmaceutical industries. *Emergent Materials*, 1–15.

Mangala, T. & Bhirud, S. (n.d.). *Road Network Extraction from High Resolution Satellite Images based on LSE LBF Model*. Academia.

Manjaiah, K.M., Mukhopadhyay, R., Paul, R., Datta, S.C., Kumararaja, P., Sarkar, B. (2018). Clay Minerals And Zeolites For Environmentally Sustainable Agriculture. Modif. Clay Zeolite Nanocomposite Mater. *Environ. Pharm. Appl.*, 309–329. doi:10.1016/B978-0-12-814617-0.00008-6

Manjunatha, S., Biradar, D., & Aladakatti, Y. R. (2016). Nanotechnology and its applications in agriculture: A review. *J farm Sci, 29*(1), 1-13.

Manjunatha, R., Naik, D., & Usharani, K. (2019). Nanotechnology application in agriculture: A review. *Journal of Pharmacognosy and Phytochemistry*, *8*(3), 1073–1083.

Manjunatha, S. B., Biradar, D. P., & Aladakatti, Y. R. (2016). Nanotechnology and its applications in agriculture: A review. *Journal of Farm Sciences*, *29*(1), 1–13.

Manly, K. F., Cudmore Robert, H. Jr, & Meer, J. M. (2001). Map Manager QTX, cross-platform software for genetic mapping. *Mammalian Genome*, *12*(12), 930–932. doi:10.100700335-001-1016-3 PMID:11707780

Marmiroli, N., & White, J. C. (2016). Editorial: Nanotoxicology and Environmental Risk Assessment of Engineered Nanomaterials (ENMs) in Plants. *Frontiers in Plant Science*, *7*, 1370. doi:10.3389/fpls.2016.01370 PMID:27672390

Marra, M. A., Hillier, L., & Waterston, R. H. (1998). Expressed sequence tag-EST abolishing bridges between genomes. *Trends in Genetics*, *14*(1), 4–7. doi:10.1016/S0168-9525(97)01355-3 PMID:9448457

Marthandan, V., Geetha, R., Kumutha, K., Renganathan, V. G., Karthikeyan, A., & Ramalingam, J. (2020). Seed Priming: A Feasible Strategy to Enhance Drought Tolerance in Crop Plants. *International Journal of Molecular Sciences*, *21*(21), 8258. doi:10.3390/ijms21218258 PMID:33158156

Martínez-Ballesta, M. C., Zapata, L., Chalbi, N., & Carvajal, M. (2016). Multiwalled carbon nanotubes enter broccoli cells enhancing growth and water uptake of plants exposed to salinity. *Journal of Nanobiotechnology*, *14*(1), 42. doi:10.118612951-016-0199-4 PMID:27278384

Maswada, H. F., Djanaguiraman, M., & Prasad, P. V. V. (2018). Seed Treatment With Nano-Iron (III) Oxide Enhances Germination, Seeding Growth And Salinity Tolerance Of Sorghum. *Journal Agronomy & Crop Science*, *204*(6), 577–587. doi:10.1111/jac.12280

Mathur, S., Singh, D., & Ranjan, R. (2022). Remediation of heavy metal (loid) contaminated soil through green nanotechnology. *Frontiers in Sustainable Food Systems*, *6*, 932424. doi:10.3389/fsufs.2022.932424

Matyjaszczyk, E. (2017). Comparison between Seed and Foliar Treatment as a Tool in Integrated Pest Management. *Journal of Agricultural and Food Chemistry*, *65*(30), 6081–6086. doi:10.1021/acs.jafc.7b01095 PMID:28686448

Mauch-Mani, B., Baccelli, I., Luna, E., & Flors, V. (2017). Defense priming: An adaptive part of induced resistance. *Annual Review of Plant Biology*, *68*(1), 485–512. doi:10.1146/annurev-arplant-042916-041132 PMID:28226238

Ma, X., Geiser-Lee, J., Deng, Y., & Kolmakov, A. (2010). Interactions between engineered nanoparticles (ENPs) and plants: Phytotoxicity, uptake and accumulation. *The Science of the Total Environment*, *408*(16), 3053–3061. doi:10.1016/j.scitotenv.2010.03.031 PMID:20435342

Mazhar, M. W., Ishtiaq, M., Hussain, I., Parveen, A., Bhatti, K. H., Azeem, M., Thind, S., Ajaib, M., Maqbool, M., Sardar, T., Muzammil, K., & Nasir, N. (2022). Seed nano-priming with Zinc Oxide NPs in rice mitigates drought and enhances agronomic profile. *PLoS One*, *17*(3), e0264967. doi:10.1371/journal.pone.0264967 PMID:35324949

McCouch, S. R., & Doerge, R. W. (1995). QTL mapping in rice. *Trends in Genetics*, *11*(12), 482–487. doi:10.1016/S0168-9525(00)89157-X PMID:8533164

McCouch, S. R., & Doerge, R. W. (1995). QTL mapping in rice. *Trends*, *11*, 482–487. PMID:8533164

McDonald, M. B. (2000). Seed Priming. In M. Black & J. D. Bewley (Eds.), *Seed Technology and its Biological Basis* (pp. 287–325). Shefield Academic Press.

Mehlenbacher, S. A. (1995). Classical and molecular approaches to breeding fruit and nut crops for disease resistance. *HortScience*, *30*(3), 466–477. doi:10.21273/HORTSCI.30.3.466

Melchinger, A. E., Utz, H. F., & Schon, C. C. (1998). Quantitative trait locus (QTL) mapping using different testers and independent population samples in maize reveals low power of QTL detection and large bias in estimates of QTL effects. *Genetics*, *149*(1), 383–403. doi:10.1093/genetics/149.1.383 PMID:9584111

Memelink, J. (2005). Tailoring the plant metabolome without a loose stitch. *Trends in Plant Science*, *10*(7), 305–307. doi:10.1016/j.tplants.2005.05.006 PMID:15950519

Mhawi, A., Doherty, W. O., Kamalvand, M., Williams, D. G., Clifford, M., & Zahid, M. (2022). Electrospun nanofibers for water treatment and food applications: A review. *The Science of the Total Environment*, *805*, 150213.

Mhlongo, M. I., Piater, L. A., Steenkamp, P. A., Labuschagne, N., & Dubery, I. A. (2021). Metabolomic evaluation of tissue-specific defense responses in tomato plants modulated by PGPR-priming against Phytophthora capsici infection. *Plants*, *10*(8), 1530. doi:10.3390/plants10081530 PMID:34451575

Michelmore, R. (1995). Molecular approaches to manipulation of disease resistance genes. *Annual Review of Phytopathology*, *33*(1), 393–427. doi:10.1146/annurev.py.33.090195.002141 PMID:18999967

Mie, A., Andersen, H. R., Gunnarsson, S., Kahl, J., Kesse-Guyot, E., Rembiałkowska, E., Quaglio, G., & Grandjean, P. (2017). Human health implications of organic food and organic agriculture: A comprehensive review. *Environmental Health*, *16*(1), 1–22. doi:10.118612940-017-0315-4 PMID:29073935

Mildaziene, V., Ivankov, A., Sera, B., & Baniulis, D. (2022). Biochemical and physiological plant processes affected by seed treatment with non-thermal plasma. *Plants*, *11*(7), 856. doi:10.3390/plants11070856 PMID:35406836

Mishra, A., Tripathi, S., Sharma, S. K., Srivastava, A., Dwivedi, P. D., Mishra, P. K., & Das, D. K. (2021). Applications of nanotechnology in agricultural system. *Frontiers in nanotechnology*, 3, 30.

Mishra, S. N., Kumar, D., Kumar, B., & Tiwari, S. (2021). Assessing impact of varying climatic conditions on distribution of *Buchanania Cochinchinensis* in Jharkhand using species distribution modeling approach. *Current Research in Environmental Sustainability*, *V*, 3. doi:10.1016/j.crsust.2021.100025

Misson, M., Zhang, H., & Jin, B. (2015). Nanobiocatalyst advancements and bioprocessing applications. *Journal of the Royal Society, Interface*, *12*(102), 20140891. doi:10.1098/rsif.2014.0891 PMID:25392397

Mittal, A. K., Chisti, Y., & Banerjee, U. C. (2013). Synthesis Of Metallic Nanoparticles Using Plant Extracts. *Biotechnology Advances*, *31*(2), 346–356. doi:10.1016/j.biotechadv.2013.01.003 PMID:23318667

Mittal, D., Kaur, G., Singh, P., Yadav, K., & Ali, S. A. (2020). Nanoparticle-based sustainable agriculture and food science: Recent advances and future outlook. *Frontiers in Nanotechnology*, *2*, 579954. doi:10.3389/fnano.2020.579954

Mittler, R., & Blumwald, E. (2010). Genetic engineering for modern agriculture: Challenges and perspectives. *Annual Review of Plant Biology*, *61*(1), 443–462. doi:10.1146/annurev-arplant-042809-112116 PMID:20192746

Mochida, K., & Shinozaki, K. (2010). Genomics and Bioinformatics Resources for Crop Improvement. *Plant & Cell Physiology*, *51*(4), 497–523. doi:10.1093/pcp/pcq027 PMID:20208064

Moghanloo, M., Iranbakhsh, A., Ebadi, M., & Oraghi Ardebili, Z. (2019). Differential physiology and expression of phenylalanine ammonia lyase (PAL) and universal stress protein (USP) in the endangered species Astragalus fridae following seed priming with cold plasma and manipulation of culture medium with silica nanoparticles. *3 Biotech, 9*, 1-13.

Mohamad, N. A. N., Arham, N. A., Jai, J., & Hadi, A. (2014). Plant Extract as Reducing Agent in Synthesis of Metallic Nanoparticles: A Review. *Advanced Materials Research*, *832*, 350–355. . doi:10.4028/www.scientific.net/AMR.832.350

Mohamed, A. K. S. H., Qayyum, M. F., Abdel-Hadi, A. M., Rehman, R. A., Ali, S., & Rizwan, M. (2017a). *Interactive effect of salinity and silver NPs on photosynthetic and biochemical parameters of wheat.* Taylor & Francis. doi:10.1080/03650340.2017.1300256

Mohamed, I. A. A., Shalby, N., El-Badri, A. M. A., Saleem, M. H., Khan, M. N., Nawaz, M. A., Qin, M., Agami, R. A., Kuai, J., Wang, B., & Zhou, G. (2020). Stomata and Xylem Vessels Traits Improved by Melatonin Application Contribute to Enhancing Salt Tolerance and Fatty Acid Composition of Brassica napus L. Plants. *Agronomy 2020, Vol. 10, Page 1186, 10*(8), 1186. doi:10.3390/agronomy10081186

Mohammadinejad, R., Karimi, S., Iravani, S., & Varma, R. S. (2016). Plant-derived nanostructures: Types and applications. *Green Chemistry*, *18*(1), 20–52. doi:10.1039/C5GC01403D

Mohsenzadeh, F., & Rad, A. C. (2011). Application of nano-particles of Euphorbia Macroclada for bioremediation of heavy metal polluted environments. In *International Conference on Nanotechnology and Biosensors IPCBEE* (Vol. 25). Academia.

Mondal, A., Basu, R., Das, S., & Nandy, P. (2011). Beneficial role of carbon nanotubes on mustard plant growth: An agricultural prospect. *Journal of Nanoparticle Research*, *13*(10), 4519–4528. doi:10.100711051-011-0406-z

Montanha, G. S., Rodrigues, E. S., Marques, J. P. R., de Almeida, E., Colzato, M., & Pereira de Carvalho, H. W. (2020). Zinc Nanocoated Seeds: An Alternative To Boost Soybean Seed Germination And Seedling Development. *SN Applied Sciences*, *2*(5), 1–11. doi:10.100742452-020-2630-6

Moore, P. A. Jr, & Patrick, W. H. Jr. (1989). Manganese availability and uptake by rice in acid sulfate soils. *Soil Science Society of America Journal*, *53*(1), 104–109. doi:10.2136ssaj1989.03615995005300010019x

More, S. J., Bampidis, V., Benford, D., Bennekou, S. H., Bragard, C., Halldorsson, T. I., & Silano, V. (2021). *EFSA Journal*, *19*(4), e06585. PMID:33976718

Mousavi, S. R., & Rezaei, M. (2011). Nanotechnology in agriculture and food production. *Journal of Applied Environmental and Biological Sciences*, *1*(10), 414–419.

Mozgova, I., Mikulski, P., Pecinka, A., & Farrona, S. (2019). Epigenetic mechanisms of abiotic stress response and memory in plants. *Epigenetics in plants of agronomic importance: Fundamentals and applications: Transcriptional regulation and chromatin remodelling in plants*, 1-64.

Mukherjee, A., Majumdar, S., Servin, A. D., Pagano, L., Dhankher, O. P., & White, J. C. (2016). Carbon nanomaterials in agriculture: A critical review. *Frontiers in Plant Science*, *7*, 172. doi:10.3389/fpls.2016.00172 PMID:26941751

Mukherjee, A., Peralta-Videa, J. R., Bandyopadhyay, S., Rico, C. M., Zhao, L., & Gardea-Torresdey, J. L. (2014). Physiological effects of nanoparticulate ZnO in green peas (Pisum sativum L.) cultivated in soil. *Metallomics*, *6*(1), 132–138. doi:10.1039/C3MT00064H PMID:24190632

Mukherjee, D. (2019). Seed invigoration: An effective tool for conservation of endangered and valuable medicinal plant. *Advances in Plant Physiology*, *28*, 243–272.

Mukhopadhyay, S. S. (2014). Nanotechnology in agriculture: Prospects and constraints. *Nanotechnology, Science and Applications*, 63–71. doi:10.2147/NSA.S39409 PMID:25187699

Müller, K., Linkies, A., Vreeburg, R. A. M., Fry, S. C., Krieger-Liszkay, A., & Leubner-Metzger, G. (2009). In Vivo Cell Wall Loosening by Hydroxyl Radicals during Cress Seed Germination and Elongation Growth. *Plant Physiology*, *150*(4), 1855–1865. doi:10.1104/pp.109.139204 PMID:19493972

Munns, R. (2005). Genes and salt tolerance: Bringing them together. *The New Phytologist*, *167*(3), 645–663. doi:10.1111/j.1469-8137.2005.01487.x PMID:16101905

Munns, R., & Tester, M. (2008a). Mechanisms of salinity tolerance. *Annual Review of Plant Biology*, *59*(1), 651–681. doi:10.1146/annurev.arplant.59.032607.092911 PMID:18444910

Mu, Q., Jiang, G., Chen, L., Zhou, H., Fourches, D., Tropsha, A., & Yan, B. (2014). Chemical basis of interactions between engineered nanoparticles and biological systems. *Chemical Reviews*, *114*(15), 7740–7781. doi:10.1021/cr400295a PMID:24927254

Musial, J., Krakowiak, R., Mlynarczyk, D. T., Goslinski, T., & Stanisz, B. J. (2020). Titanium dioxide nanoparticles in food and personal care products—What do we know about their safety? *Nanomaterials (Basel, Switzerland)*, *10*(10), 1110. doi:10.3390/nano10061110 PMID:32512703

Nagargade, M., Tyagi, V., Kumar, D., Shukla, S., & Pathak, A. (2022). Nanofertilizers: Importance in Nutrient Management. In Nanotechnology in Agriculture and Environmental Science (pp. 69-80). CRC Press.

Naguib, D. M., & Abdalla, H. (2019). Metabolic Status during Germination of Nano Silica Primed Zea mays Seeds under Salinity Stress. *Journal of Crop Science and Biotechnology*, *22*(5), 415–423. doi:10.100712892-019-0168-0

Nair, R., Mohamed, M. S., Gao, W., Maekawa, T., Yoshida, Y., Ajayan, P. M., & Kumar, D. S. (2012). Effect of carbon nanomaterials on the germination and growth of rice plants. *Journal of Nanoscience and Nanotechnology*, *12*(3), 2212–2220. doi:10.1166/jnn.2012.5775 PMID:22755040

Nair, R., Varghese, S. H., Nair, B. G., Maekawa, T., Yoshida, Y., & Kumar, D. S. (2010). Nanoparticulate material delivery to plants. *Plant Science*, *179*(3), 154–163. doi:10.1016/j.plantsci.2010.04.012

Nair, R., Varghese, S. H., Nair, B. G., Maekawa, T., Yoshida, Y., & Kumar, D. S. (2010). *Plant Science*.

Najafi Disfani, M., Mikhak, A., Kassaee, M. Z., & Maghari, A. (2017). Effects Of Nano Fe/SiO_2 Fertilizers On Germination And Growth Of Barley And Maize. *Archives of Agronomy and Soil Science*, *63*(6), 817–826. doi:10.1080/03650340.2016.1239016

Nandhini, M., Rajini, S. B., Udayashankar, A. C., Niranjana, S. R., Lund, O. S., Shetty, H. S., & Prakash, H. S. (2019). Biofabricated Zinc Oxide Nanoparticles as an Eco-Friendly Alternative for Growth Promotion and Management of Downy Mildew of Pearl Millet. *Crop Protection (Guildford, Surrey)*, *121*, 103–112. doi:10.1016/j.cropro.2019.03.015

Nasrollahzadeh, M., Sajadi, S. M., Sajjadi, M., & Issaabadi, Z. (2019). An introduction to nanotechnology. In *Interface science and technology* (Vol. 28, pp. 1–27). Elsevier.

Nayban, G., Mandal, A. K., & De, B. K. (2017). Seed priming: A low-cost climate-resilient tool for improving germination, growth and productivity of mungbean. *SATSA Mukhaptra Annual Technical*, (21), 162–172.

Naz, H., Akram, N. A., & Ashraf, M. (2016). Impact of ascorbic acid on growth and some physiological attributes of cucumber (Cucumis sativus) plants under water-deficit conditions. *Pakistan Journal of Botany*, *48*(3), 877–883.

Nelson, J. C. (1997). Qgene—Software for marker-based genomic analysis and breeding. *Molecular Breeding, 3*(3), 239–245. doi:10.1023/A:1009604312050

Neme, K., Nafady, A., Uddin, S., & Tola, Y. B. (2021). Application of nanotechnology in agriculture, postharvest loss reduction and food processing: Food security implication and challenges. *Heliyon, 7*(7), 08539. doi:10.1016/j.heliyon.2021.e08539 PMID:34934845

Newell, D. G., Koopmans, M., Verhoef, L., Duizer, E., Aidara-Kane, A., Sprong, H., Opsteegh, M., Langelaar, M., Threfall, J., Scheutz, F., der Giessen, J., & Kruse, H. (2010). Food-borne diseases—The challenges of 20 years ago still persist while new ones continue to emerge. *International Journal of Food Microbiology, 139*, S3–S15. doi:10.1016/j.ijfoodmicro.2010.01.021 PMID:20153070

Nguyen, T., Hoang, V.-N., Le, T., Tran, T.-H., & Hai, V. (2018). A Vision Based Method for Automatic Evaluation of Germination Rate of Rice Seeds. In *1st International Conference on Multimedia Analysis and Pattern Recognition (MAPR)*, Ho Chi Minh, Vietnam. 10.1109/MAPR.2018.8337511

Niggli, U., Fließbach, A., Hepperly, P., & Scialabba, N. (2009). *Low greenhouse gas agriculture: mitigation and adaptation potential of sustainable farming systems* (Vol. 141). Food and Agriculture Organization of the United Nations.

Nikunj, P., Purvi, D., Niti, P., Anamika, J., & Gautam, H. K. (2014). Agronanotechnology for plant fungal disease management: A review. *International Journal of Current Microbiology and Applied Sciences, 3*, 71–84.

Nile, S. H., Thiruvengadam, M., Wang, Y., Samynathan, R., Shariati, M. A., Rebezov, M., Nile, A., Sun, M., Venkidasamy, B., Xiao, J., & Kai, G. (2022). Nano-priming as emerging seed priming technology for sustainable agriculture—Recent developments and future perspectives. *Journal of Nanobiotechnology, 20*(1), 1–31. doi:10.118612951-022-01423-8 PMID:35659295

Niratker, C., & Sailaja, D. (2018). Evaluation of genetic diversity of Chironji (*Buchnania Lanzan*) in Chhattisgarh. *International Journal of Research and Analytical Reviews, 5*, 4.

Nonogaki, H. (2014). Seed Dormancy And Germination-Emerging Mechanisms And New Hypotheses. *Frontiers in Plant Science, 5*, 233. doi:10.3389/fpls.2014.00233 PMID:24904627

Nonogaki, H., Bassel, G. W., & Bewley, J. D. (2010). Germination—Still a mystery. *Plant Science, 179*(6), 574–581. doi:10.1016/j.plantsci.2010.02.010

Noorhosseini, S. A., Jokar, N. K., & Damalas, C. A. (2017). Improving Seed Germination and Early Growth of Garden Cress (*Lepidium sativum*) and Basil (*Ocimum basilicum*) with Hydro-priming. *Journal of Plant Growth Regulation, 37*(1), 323–334. doi:10.100700344-017-9728-0

Noorhosseini, S. A., Jokar, N. K., & Damalas, C. A. (2017). Improving seed germination and early growth of garden cress (Lepidium sativum) and basil (Ocimum basilicum) with hydro-priming. *Journal of Plant Growth Regulation, 37*(37), 323–334.

Noor, R., Yasmin, H., Ilyas, N., Nosheen, A., Hassan, M. N., Mumtaz, S., Khan, N., Ahmad, A., & Ahmad, P. (2022). Comparative analysis of iron oxide nanoparticles synthesized from ginger *(Zingiber officinale)* and cumin seeds (*Cuminum cyminum*) to induce resistance in wheat against drought stress. *Chemosphere, 292*, 133201. doi:10.1016/j.chemosphere.2021.133201 PMID:34921860

Nuruzzaman, M., Rahman, M. M., Liu, Y., & Naidu, R. (2016). Nanoencapsulation, nano-guard for pesticides: A new window for safe application. *Journal of Agricultural and Food Chemistry, 64*(7), 1447–1483. doi:10.1021/acs.jafc.5b05214 PMID:26730488

Oerke, E.-C. (2006). Crop losses to pests. *Journal of Agricultural Science, 144*(1), 31–43. doi:10.1017/S0021859605005708

Okey-Onyesolu, C. F., Hassanisaadi, M., Bilal, M., Barani, M., Rahdar, A., Iqbal, J., & Kyzas, G. Z. (2021). Nanomaterials as nanofertilizers and nanopesticides: An overview. *ChemistrySelect, 6*(6), 8645–8663. doi:10.1002lct.202102379

Olalekan, R., Omidiji, A., Williams, E., Christianah, M., & Modupe, O. (2019). The roles of all tiers of government and development partners in environmental conservation of natural resource: A case study in Nigeria. *MOJ Ecology & Environmental Sciences, 4*(3), 114–121. doi:10.15406/mojes.2019.04.00142

Oliveira, H. C., Seabra, A. B., Kondak, S., Adedokun, O. P., & Kolbert, Z. (2023). Multilevel approach of plant-nanomaterial relationship: From cells to living ecosystems. *Journal of Experimental Botany, 74*(12), erad107. doi:10.1093/jxb/erad107

Oliver, M. A., & Gregory, P. (2015). Soil, food security and human health: A review. *European Journal of Soil Science, 66*(2), 257–276. doi:10.1111/ejss.12216

Omran, B. A., & Baek, K.-H. (2022). Valorization of agro-industrial biowaste to green nanomaterials for wastewater treatment: Approaching green chemistry and circular economy principles. *Journal of Environmental Management, 311*, 114806. doi:10.1016/j.jenvman.2022.114806 PMID:35240500

Onyekachi, O. G., Boniface, O. O., Gemlack, N. F., & Nicholas, N. (2019). The effect of climate change on abiotic plant stress: a review. *Abiotic and biotic stress in plants, 17.*

Oweis, T. (2018). Use of brackish and marginal water for irrigation in water-scarce areas Z. Gao, China Institute of Water Resources and Hydropower Research, China. In Water management for sustainable agriculture (pp. 253-276). Burleigh Dodds Science Publishing.

Ozgur, R., Uzilday, B., Sekmen, A. H., Turkan, I., Ozgur, R., Uzilday, B., Sekmen, A. H., & Turkan, I. (2013). Reactive oxygen species regulation and antioxidant defence in halophytes. *Functional Plant Biology, 40*(9), 832–847. doi:10.1071/FP12389 PMID:32481154

Pahalvi, H. N., Rafiya, L., Rashid, S., Nisar, B., & Kamili, A. N. (2021). *Chemical fertilizers and their impact on soil health. Microbiota and Biofertilizers* (Vol. 2). Ecofriendly Tools for Reclamation of Degraded Soil Environs.

Palocci, C., Valletta, A., Chronopoulou, L., Donati, L., Bramosanti, M., Brasili, E., Baldan, B., & Pasqua, G. (2017). Endocytic Pathways Involved In PLGA Nanoparticle Uptake By Grapevine Cells And Role Of Cell Wall And Membrane In Size Selection. *Plant Cell Reports*, *36*(12), 1917–1928. doi:10.100700299-017-2206-0 PMID:28913707

Pal, R., Mondal, T., Dev, R., Parihar, M., Kumar, A., Sharma, D., & Kant, L. (2023). 5 Role of Enzyme-Mimicking Nanoparticles in Crop Plants. *Biotic Stress Management of Crop Plants using. Nanomaterials (Basel, Switzerland)*, 71–85.

Pandey, G. (2018). Challenges and future prospects of agri-nanotechnology for sustainable agriculture in India. *Environmental Technology & Innovation*, *11*(11), 299–307. doi:10.1016/j.eti.2018.06.012

Pandey, S., Giri, K., Kumar, R., Mishra, G., & Raja Rishi, R. (2018). Nanopesticides: Opportunities in crop protection and associated environmental risks. *Proceedings of the National Academy of Sciences. India. Section B, Biological Sciences*, *88*(4), 1287–1308. doi:10.100740011-016-0791-2

Panpatte, D. G., Jhala, Y. G., Shelat, H. N., & Vyas, R. V. (2016). *Nanoparticles: The Next Generation Technology for Sustainable Agriculture. Microb. Inoculants Sustain. Agric. Product* (Vol. 2). Funct. Appl. doi:10.1007/978-81-322-2644-4_18

Panyuta, O., Belava, V., Fomaidi, S., Kalinichenko, O., Volkogon, M., & Taran, N. (2016). The Effect of Pre-Sowing Seed Treatment with Metal Nanoparticles on the Formation of the Defensive Reaction of Wheat Seedlings Infected with the Eyespot Causal Agent. *Nanoscale Research Letters*, *11*(1), 92. doi:10.118611671-016-1305-0 PMID:26880729

Paparella, S., Araújo, S. S., Rossi, G., Wijayasinghe, M., Carbonera, D., & Balestrazzi, A. (2015). Seed priming: State of the art and new perspectives. *Plant Cell Reports*, *34*(8), 1281–1293. doi:10.100700299-015-1784-y PMID:25812837

Parihar, P., Singh, S., Singh, R., Singh, V. P., & Prasad, S. M. (2015). Effect of salinity stress on plants and its tolerance strategies: A review. *Environmental Science and Pollution Research International*, *22*(6), 4056–4075. doi:10.100711356-014-3739-1 PMID:25398215

Paris, J. L., de la Luz Cádiz-Gurrea, M., León-González, M. E., Pérez-Rodríguez, F., & Segura-Carretero, A. (2020). State-of-the-art of nanotechnology applications in the food sector. *Food Chemistry*, *331*, 127431.

Parra-Arroyo, L., González-González, R. B., Castillo-Zacarías, C., Martínez, E. M. M., Sosa-Hernández, J. E., Bilal, M., Iqbal, H. M., Barceló, D., & Parra-Saldívar, R. (2022). Highly hazardous pesticides and related pollutants: Toxicological, regulatory, and analytical aspects. *The Science of the Total Environment*, *807*, 151879. doi:10.1016/j.scitotenv.2021.151879 PMID:34826476

Parveen, A., & Rao, S. (2015). Effect of Nanosilver on Seed Germination and Seedling Growth in *Pennisetum glaucum. Journal of Cluster Science*, *26*(3), 693–701. doi:10.100710876-014-0728-y

Paterson, A. H. (1996a). Making genetic maps. In A. H. Paterson (Ed.), *Genome Mapping in Plants, R. G. Landes Company, San Diego, California* (pp. 23–39). Academic Press.

Paterson, A., Tanksley, S., & Sorrels, M. E. (1991a). DNA markers in plant improvement. *Advances in Agronomy*, *44*, 39–90. doi:10.1016/S0065-2113(08)60578-7

Pattanayak, S., Mollick, M. M. R., Maity, D., Chakraborty, S., Dash, S. K., Chattopadhyay, S., Roy, S., Chattopadhyay, D., & Chakraborty, M. (2017). Butea monosperma bark extract mediated green synthesis of silver nanoparticles: Characterization and biomedical applications. *Journal of Saudi Chemical Society*, *21*(6), 673–684. doi:10.1016/j.jscs.2015.11.004

Pavoni, L., Perinelli, D. R., Bonacucina, G., Cespi, M., & Palmieri, G. F. (2020). An overview of micro-and nanoemulsions as vehicles for essential oils: Formulation, preparation and stability. *Nanomaterials (Basel, Switzerland)*, *10*(1), 135. doi:10.3390/nano10010135 PMID:31940900

Pelaez, M., Nolan, N. T., Pillai, S. C., Seery, M. K., Falaras, P., Kontos, A. G., Dunlop, P. S. M., Hamilton, J. W. J., Byrne, J. A., O'Shea, K., Entezari, M. H., & Dionysiou, D. D. (2012). A review on the visible light active titanium dioxide photocatalysts for environmental applications. *Applied Catalysis B: Environmental*, *125*, 331–349. doi:10.1016/j.apcatb.2012.05.036

Pereira, A. E. S., Oliveira, H. C., & Fraceto, L. F. (2019). Polymeric nanoparticles as an alternative for application of gibberellic acid in sustainable agriculture: A field study. *Scientific Reports*, *9*(1), 7135. doi:10.103841598-019-43494-y PMID:31073210

Pérez-de-Luque, A. (2017). Interaction Of Nanomaterials With Plants: What Do We Need For Real Applications In Agriculture? *Frontiers in Environmental Science*, *5*, 12. doi:10.3389/fenvs.2017.00012

Phelps, J., Carrasco, L. R., Webb, E. L., Koh, L. P., & Pascual, U. (2013). Agricultural intensification escalates future conservation costs. *Proceedings of the National Academy of Sciences of the United States of America*, *110*(19), 7601–7606. doi:10.1073/pnas.1220070110 PMID:23589860

Pilarska, A., Wysokowski, M., Markiewicz, E., & Jesionowski, T. (2013). Synthesis of magnesium hydroxide and its calcinates by a precipitation method with the use of magnesium sulfate and poly (ethylene glycols). *Powder Technology*, *235*, 148–157. doi:10.1016/j.powtec.2012.10.008

Pirzada, T., de Farias, B. V., Mathew, R., Guenther, R. H., Byrd, M. V., Sit, T. L., Pal, L., Opperman, C. H., & Khan, S. A. (2020). Recent Advances In Biodegradable Matrices For Active Ingredient Release In Crop Protection: Towards Attaining Sustainability In Agriculture. *Current Opinion in Colloid & Interface Science*, *48*, 121–136. doi:10.1016/j.cocis.2020.05.002 PMID:33013179

Plaksenkova, I., Kokina, I., Petrova, A., Jermalonoka, M., Gerbreders, V., & Krasovska, M. (2016). The Impact of Zinc Oxide Nanoparticles on Cytotoxicity, Genotoxicity, and MiRNA Expression in Barley (*Hordeum vulgare* L.) Seedlings. *Hindawi*. https://www. hindawi.com/journals/tswj/2020/6649746/

Podila, R., & Brown, J. M. (2013). *Toxicology and Applied Pharmacology*, *270*(3), 132–141.

Poudel, S., Poudel, B., Acharya, B., & Poudel, P. (2020). Pesticide use and its impacts on human health and environment. *Environ Ecosyst Sci*, *4*(1), 47–51. doi:10.26480/ees.01.2020.47.51

Pourkhaloee, A., Haghighi, M., Saharkhiz, M., Jouzi, H., & Mahdi, M. (2011). Carbon Nanotubes Can Promote Seed Germination via Seed Coat Penetration. *Seed Technology*, *33*, 155–169.

Pradhan, S., Patra, P., Das, S., Chandra, S., Mitra, S., Dey, K. K., & Chattopadhyay, D. (2013). *Environmental Science & Technology*.

Pramanik, P., Krishnan, P., Maity, A., Mridha, N., Mukherjee, A., & Rai, V. (2020). Application of nanotechnology in agriculture. *Environmental Nanotechnology*, *4*, 317–348.

Prasad, R., Bhattacharyya, A., & Nguyen, Q. D. (2017). Nanotechnology in Sustainable Agriculture: Recent Developments, Challenges, and Perspectives. *Frontiers in Microbiology*, *8*, 1014. doi:10.3389/fmicb.2017.01014 PMID:28676790

Prasad, R., Kumar, V., & Prasad, K. S. (2014). Nanotechnology in sustainable agriculture: Present concerns and future aspects. *African Journal of Biotechnology*, *13*(6), 705–713. doi:10.5897/AJBX2013.13554

Preisler, A. C., Carvalho, L. B., Saraiva-Santos, T., Verri, W. A. Jr, Mayer, J. L. S., Fraceto, L. F., Dalazen, G., & Oliveira, H. C. (2022). Interaction of Nanoatrazine and Target Organism: Evaluation of Fate and Photosystem II Inhibition in Hydroponically Grown Mustard (Brassica juncea) Plants. *Journal of Agricultural and Food Chemistry*, *70*(25), 7644–7652. doi:10.1021/acs.jafc.2c01601 PMID:35675570

Pulighe, G., & Lupia, F. (2020). Food first: COVID-19 outbreak and cities lockdown a booster for a wider vision on urban agriculture. *Sustainability (Basel)*, *12*(12), 5012. doi:10.3390u12125012

Qayyum, M. F., Rehman, M. Z., Ali, S., Rizwan, M., Naeem, A., Maqsood, M. A., Khalid, H., Rinklebe, J., & Ok, Y. S. (2017). Residual Effects Of Monoammonium Phosphate, Gypsum And Elemental Sulfur On Cadmium Phytoavailability And Translocation From Soil To Wheat In An Effluent Irrigated Field. *Chemosphere*, *174*, 515–523. doi:10.1016/j.chemosphere.2017.02.006 PMID:28189896

Quisumbing, A. R., & Pandolfelli, L. (2010). Promising approaches to address the needs of poor female farmers: Resources, constraints, and interventions. *World Development*, *38*(4), 581–592. doi:10.1016/j.worlddev.2009.10.006

Qureshi, A., Singh, D., & Dwivedi, S. (2018). Nano-fertilizers: A novel way for enhancing nutrient use efficiency and crop productivity. *International Journal of Current Microbiology and Applied Sciences*, *7*(2), 3325–3335. doi:10.20546/ijcmas.2018.702.398

Qu, Y., You, S., Zhang, X., Pei, X., Shen, W., Li, Z., Li, S., & Zhang, Z. (2018). Biosynthesis of gold nanoparticles using cell-free extracts of Magnusiomyces ingens LH-F1 for nitrophenols reduction. *Bioprocess and Biosystems Engineering*, *41*(3), 359–367. doi:10.100700449-017-1869-9 PMID:29188359

Ragaei, M., & Sabry, A.-H. (2014). Nanotechnology for insect pest control. *International Journal of Science, Environment and Technology*, *3*(2), 528–545.

Rahimi, D., Kartoolinejad, D., Nourmohammadi, K., & Naghdi, R. (2016). Increasing drought resistance of Alnus subcordata CA Mey. seeds using a Nano priming technique with multi-walled carbon nanotubes. *Journal of Forensic Sciences*, *62*, 269–278.

Rahimi, D., Kartoolinejad, D., Nourmohammadi, K., & Naghdi, R. (2016). Increasing drought resistance of Alnus subcordata CA Mey. seeds using a nano priming technique with multi-walled carbon nanotubes. *Journal of Forest Science*, *62*(6), 269–278. doi:10.17221/15/2016-JFS

Rahimi, S., Hatami, M., & Ghorbanpour, M. (2021). Silicon-nanoparticle mediated changes in seed germination and vigor index ofMarigold (Calendula officinalis L.) compared to silicate under PEG-induced drought stress. *Gesunde Pflanzen*, *73*(4), 575–589. doi:10.100710343-021-00579-x

Rahman, A., Mostofa, M. G., Alam, M. M., Nahar, K., Hasanuzzaman, M., & Fujita, M. (2015). Calcium Mitigates Arsenic Toxicity in Rice Seedlings by Reducing Arsenic Uptake and Modulating the Antioxidant Defense and Glyoxalase Systems and Stress Markers. *BioMed Research International*, *2015*, 1–12. doi:10.1155/2015/340812 PMID:26798635

Rahmani, N., Radjabian, T., & Soltani, B. M. (2020). Impacts of foliar exposure to multi-walled carbon nanotubes on physiological and molecular traits of Salvia verticillata L., as a medicinal plant. *Plant Physiology and Biochemistry*, *150*, 27–38. doi:10.1016/j.plaphy.2020.02.022 PMID:32109787

Rahman, M. H., Haque, K. S., & Khan, M. Z. H. (2021). A review on application of controlled released fertilizers influencing the sustainable agricultural production: A Cleaner production process. *Environmental Technology & Innovation*, *23*, 101697. doi:10.1016/j.eti.2021.101697

Rahman, M. S., Chakraborty, A., Mazumdar, S., Nandi, N. C., Bhuiyan, M. N. I., Alauddin, S. M., Khan, I. A., & Hossain, M. J. (2020). Effects Of Poly(Vinylpyrrolidone) Protected Platinum Nanoparticles On Seed Germination And Growth Performance Of *Pisum sativum*. *Nano-Structures & Nano-Objects*, *21*, 100408. doi:10.1016/j.nanoso.2019.100408

Rai-Kalal, P., & Jajoo, A. (2021). Priming with zinc oxide nanoparticles improve germination and photosynthetic performance in wheat. *Plant Physiology and Biochemistry*, *160*, 341–351. doi:10.1016/j.plaphy.2021.01.032 PMID:33548801

Rai-Kalal, P., Tomar, R. S., & Jajoo, A. (2021). H2O2 signaling regulates seed germination in ZnO nanoprimed wheat (Triticum aestivum L.) seeds for improving plant performance under drought stress. *Environmental and Experimental Botany*, *189*, 104561. doi:10.1016/j.envexpbot.2021.104561

Rai-Kalal, P., Tomar, R. S., & Jajoo, A. (2021). Seed nanopriming by silicon oxide improves drought stress alleviation potential in wheat plants. *Functional Plant Biology*, *48*(9), 905–915. doi:10.1071/FP21079 PMID:34366002

Rai, M., & Ingle, A. (2012). Role of nanotechnology in agriculture with special reference to management of insect pests. *Applied Microbiology and Biotechnology*, *94*(2), 287–293. doi:10.100700253-012-3969-4 PMID:22388570

Rai, R. S., Bajpai, V., Khan, M. I., Elboughdiri, N., Shanableh, A., & Luque, R. (2022). An eco-friendly approach on green synthesis, bio-engineering applications, and future outlook of ZnO nanomaterial: A critical review. *Environmental Research*, 114807. PMID:36455633

Raja, K., Sowmya, R., Sudhagar, R., Moorthy, P. S., Govindaraju, K., & Subramanian, K. S. (2019). Biogenic ZnO and Cu Nanoparticles to Improve Seed Germination Quality in Blackgram (*Vigna mungo*). *Materials Letters*, *235*, 164–167. doi:10.1016/j.matlet.2018.10.038

Rajam, K. S., Rani, M. E., Gunaseeli, R., & Munavar, M. H. (2017). Extracellular synthesis of silver nanoparticles by the fungus Emericella nidulans EV4 and its application. *Indian Journal of Experimental Biology*, (55), 262–265.

Rajput, V., Minkina, T., Sushkova, S., Behal, A., Maksimov, A., Blicharska, E., Ghazaryan, K., Movsesyan, H., Barsova, N. (2020). ZnO and CuO Nanoparticles: A Threat To Soil Organisms, Plants, And Human Health. *Environmental Geochemistry and Health*, 42(1), 147–158. doi:10.1007/s10653-019-00317-3

Rajput, V., Singh, A., Rawat, S., Sharma, R., Tatiana, M., Singh, A., Shuvaeva, V., & Nazarenko, O. (2022). Nanotechnology as the cutting edge for sustainable agriculture. In: Nanotechnology for Sustainable Agriculture An Innovative and Eco-Friendly Approach.

Rajput, V.D., Chen, Y., Ayup, M. (2015). Effects Of High Salinity On Physiological And Anatomical Indices In The Early Stages Of *Populus euphratica* Growth. *Russian Journal of Plant Physiology, 62*(2), 229–236. (https://doi.org/) doi:10.1134/S1021443715020168

Rajput, V.D., Minkina, T., Fedorenko, A., Chernikova, N., Hassan, T., Mandzhieva, S., Sushkova, S., Lysenko, V., Soldatov, M.A., Burachevskaya, M. (2021). Effects Of Zinc Oxide Nanoparticles On Physiological And Anatomical Indices In Spring Barley Tissues. *Nanomaterials, 11*(7), 1722. (https://doi.org/) doi:10.3390/nano11071722

Rajput, V.D., Singh, A., Minkina, T., Rawat, S., Mandzhieva, S., Sushkova, S., Shuvaeva, V., Nazarenko, O., Rajput, P., Komariah, Verma, K.K., Singh, A.K., Rao, M., & Upadhyay, S.K. (2021). Nano-Enabled Products: Challenges and Opportunities For Sustainable Agriculture. *Plants*, *10*. (https://doi.org/) doi:10.3390/plants10122727

Rajput, V.D., Singh, A., Minkina, T.M., Shende, S.S., Kumar, P., Verma, K.K., Bauer, T., Gorobtsova, O., Deneva, S., Sindireva, A. (2021). Potential Applications of Nanobiotechnology in Plant Nutrition and Protection for Sustainable Agriculture. *Nanotechnology in Plant Growth Promotion and Protection,* 79–92. (https://doi.org/) doi:10.1002/9781119745884.ch5

Rajput, Vishnu D., Minkina, T., Kumari, A., Harish, Singh, V. K., Verma, K. K., Mandzhieva, S., Sushkova, S., Srivastava, S., & Keswani, C. (2021). Coping with the Challenges of Abiotic Stress in Plants: New Dimensions in the Field Application of NPs. *Plants 2021, 10*(6), 1221. doi:10.3390/plants10061221

Rajput, V. D., Minkina, T. M., Behal, A., Sushkova, S. N., Mandzhieva, S., Singh, R., Gorovtsov, A., Tsitsuashvili, V. S., Purvis, W. O., Ghazaryan, K. A., & Movsesyan, H. S. (2018). Effects of zinc-oxide nanoparticles on soil, plants, animals and soil organisms: A review. *Environmental Nanotechnology, Monitoring & Management*, *9*, 76–84. doi:10.1016/j.enmm.2017.12.006

Rajput, V. D., Minkina, T., Feizi, M., Kumari, A., Khan, M., Mandzhieva, S., Sushkova, S., El-Ramady, H., Verma, K. K., Singh, A., Hullebusch, E. D., Singh, R. K., Jatav, H. S., & Choudhary, R. (2021). Effects of Silicon and Silicon-Based Nanoparticles on Rhizosphere Microbiome, Plant Stress and Growth. *Biology (Basel)*, *10*(10), 791. doi:10.3390/biology10080791 PMID:34440021

Rajput, V. D., Minkina, T., Upadhyay, S. K., Kumari, A., Ranjan, A., Mandzhieva, S., Sushkova, S., Singh, R. K., & Verma, K. K. (2022). Nanotechnology in the Restoration of Polluted Soil. *Nanomaterials (Basel, Switzerland)*, *12*(5), 769. doi:10.3390/nano12050769 PMID:35269257

Rajput, V. D., Tatiana, M., Kumari, A., Shende, S. S., Ranjan, A., Faizan, M., & Barakvov, A. (2022). A Review On Nanobioremediation Approaches For Restoration Of Contaminated Soil. *Eurasian Journal of Soil Science*, *11*(1), 43–60. doi:10.18393/ejss.990605

Rajput, V., Minkina, T., Fedorenko, A., Sushkova, S., Mandzhieva, S., Lysenko, V., Duplii, N., Fedorenko, G., Dvadnenko, K., & Ghazaryan, K. (2018). Toxicity Of Copper Oxide Nanoparticles On Spring Barley (*Hordeum sativum distichum*). *The Science of the Total Environment*, *645*, 1103–1113. doi:10.1016/j.scitotenv.2018.07.211 PMID:30248835

Rajput, V., Minkina, T., Mazarji, M., Shende, S., Sushkova, S., Mandzhieva, S., Burachevskaya, M., Chaplygin, V., Singh, A., & Jatav, H. (2020). Accumulation Of Nanoparticles In The Soil-Plant Systems And Their Effects On Human Health. *Annals of Agricultural Science*, *65*(2), 137–143. doi:10.1016/j.aoas.2020.08.001

Raliya, R., Biswas, P., & Tarafdar, J. C. (2015). TiO_2 nanoparticle biosynthesis and its physiological effect on mung bean (Vigna radiata L.). *Biotechnology Reports (Amsterdam, Netherlands)*, *5*(5), 22–26. doi:10.1016/j.btre.2014.10.009 PMID:28626678

Raliya, R., Saharan, V., Dimkpa, C., & Biswas, P. (2017). Nanofertilizer for precision and sustainable agriculture: Current state and future perspectives. *Journal of Agricultural and Food Chemistry*, (66), 6487–6503. PMID:28835103

Raliya, R., Tarafdar, J. C., & Biswas, P. (2016). Enhancing the Mobilization of Native Phosphorus in the Mung Bean Rhizosphere Using ZnO Nanoparticles Synthesized by Soil Fungi. *Journal of Agricultural and Food Chemistry*, *64*(16), 3111–3118. doi:10.1021/acs.jafc.5b05224 PMID:27054413

Ramachandraiah, K., Han, S. G., & Chin, K. B. (2015). Nanotechnology in meat processing and packaging: Potential applications—A review. *Asian-Australasian Journal of Animal Sciences*, *28*(2), 290–302. doi:10.5713/ajas.14.0607 PMID:25557827

Rameshraddy, P., Pavithra, G. J., Rajashekar Reddy, B. H., Salimath, M., Geetha, K. N., & Shankar, A. G. (2017). Zinc oxide nanoparticles increases Zn uptake, translocation in rice with positive effect on growth, yield and moisture stress tolerance. *Indian Journal of Plant Physiology / Official Publication of the Indian Society for Plant Physiology*, *22*(3), 287–294. doi:10.100740502-017-0303-2

Rankic, I., Zelinka, R., Ridoskova, A., Gagic, M., Pelcova, P., & Huska, D. (2021). Nano/micro particle in conjunction with microalgae extract as novel insecticides against Mealworm beetles, Tenebrio molitor. *Scientific Reports*, *11*(1), 17125. doi:10.103841598-021-96426-0 PMID:34429471

Rastogi, A., Tripathi, D.K., Yadav, S., Chauhan, D.K., Živčák, M., Ghorbanpour, M., El-Sheery, N.I., Brestic, M. (2019). Application of Silicon Nanoparticles in Agriculture. *3 Biotech*, *9*. (https://doi.org/) doi:10.1007/s13205-019-1626-7

Rastogi, A., Zivcak, M., Sytar, O., Kalaji, H. M., He, X., Mbarki, S., & Brestic, M. (2017). Impact of metal and metal oxide nanoparticles on plant: A critical review. *Frontiers in Chemistry*, *5*, 78. doi:10.3389/fchem.2017.00078 PMID:29075626

Ratan, Z. A., Haidere, M. F., Nurunnabi, M. D., Shahriar, S. M., Ahammad, A. J., Shim, Y. Y., Reaney, M. J., & Cho, J. Y. (2020). Green chemistry synthesis of silver nanoparticles and their potential anticancer effects. *Cancers (Basel)*, *12*(12), 855. doi:10.3390/cancers12040855 PMID:32244822

Ravichandran, V., Vasanthi, S., Shalini, S., Shah, S. A. A., & Harish, R. (2016). Green synthesis of silver nanoparticles using Atrocarpus altilis leaf extract and the study of their antimicrobial and antioxidant activity. *Materials Letters*, *180*, 264–267. doi:10.1016/j.matlet.2016.05.172

Rawat, L., Bisht, T., & Naithani, D. C. (2021). Plant disease management in organic farming system: strategies and challenges. *Emerging trends in plant pathology*, 611-642. NIH.

Raymond, A. W., & Felix, E. O. (2011). Heavy metals in contaminated soils: A review of sources, chemistry, risks and best available strategies for remediation. *International Scholarly Research Network ISRN Ecology. Article ID*, *402647*, 1–20.

Raza, A., Razzaq, A., Mehmood, S. S., Zou, X., Zhang, X., Lv, Y., & Xu, J. (2019). Impact of climate change on crops adaptation and strategies to tackle its outcome: A review. *Plants*, *8*(2), 34. doi:10.3390/plants8020034 PMID:30704089

Rehman, A., Saba, T., Kashif, M., Fati, S. M., Bahaj, S. A., & Chaudhry, H. (2022). A revisit of internet of things technologies for monitoring and control strategies in smart agriculture. *Agronomy (Basel)*, *12*(1), 127. doi:10.3390/agronomy12010127

Remédios, C., Rosário, F., & Bastos, V. (2012). Environmental nanoparticles interactions with plants: Morphological, physiological, and genotoxic aspects. *Le Journal de Botanique*.

Renaut, J., Hausman, J. F., & Wisniewski, M. E. (2006). Proteomics and low-temperature studies: Bridging the gap between gene expression and metabolism. *Physiologia Plantarum*, *126*(1), 97–109. doi:10.1111/j.1399-3054.2006.00617.x

Rexlin, J., Vijayakumar, S., Nilavukkarasi, M., Vidhya, E., Alharthi, N. S., Sajjad, M., Punitha, V., & Praseetha, P. (2022). Bioengineered ZnO nanoparticles as a nano priming agent in Cyamopsis tetragonoloba (L). Taub. to improve yield and disease resistance. *Applied Nanoscience*, 1–9. doi:10.100713204-022-02526-2

Rhaman, M. S., Rauf, F., Tania, S. S., & Khatun, M. (2020). Seed priming methods: Application in field crops and future perspectives. *Asian Journal of Research in Crop Science*, *5*(2), 8–19. doi:10.9734/ajrcs/2020/v5i230091

Rhaman, M. S., Tania, S. S., Imran, S., Rauf, F., Kibria, M. G., Ye, W., Hasanuzzaman, M., & Murata, Y. (2022). Seed priming with nanoparticles: An emerging technique for improving plant growth, development, and abiotic stress tolerance. *Journal of Soil Science and Plant Nutrition*, *22*(4), 4047–4062. doi:10.100742729-022-01007-3

Rhee, S. Y., Dickerson, J., & Xu, D. (2006). Bioinformatics and its applications in plant biology. *Annual Review of Plant Biology*, *57*(1), 335–360. doi:10.1146/annurev.arplant.56.032604.144103 PMID:16669765

Rhim, J. W., Park, H. M., & Ha, C. S. (2013). Bio-nanocomposites for food packaging applications. *Progress in Polymer Science*, *38*(10-11), 1629–1652. doi:10.1016/j.progpolymsci.2013.05.008

Ribaut, J.-M., & Hoisington, D. (1998). Marker-assisted selection: New tools and strategies. *Trends in Plant Science*, *3*(6), 236–239. doi:10.1016/S1360-1385(98)01240-0

Rico, C. M., Majumdar, S., Duarte-Gardea, M., Peralta-Videa, J. R., & Gardea-Torresdey, J. L. (2011). Interaction of nanoparticles with edible plants and their possible implications in the food chain. *Journal of Agricultural and Food Chemistry*, *59*(8), 3485–3498. doi:10.1021/jf104517j PMID:21405020

Rico, C. M., Morales, M. I., Barrios, A. C., McCreary, R., Hong, J., Lee, W. Y., & Gardea-Torresdey, J. L. (2013). Effect of cerium oxide nanoparticles on the quality of rice (Oryza sativa L.) grains. *Journal of Agricultural and Food Chemistry*, *61*(47), 11278–11285. doi:10.1021/jf404046v PMID:24188281

Rivero, R. M., Mittler, R., Blumwald, E., & Zandalinas, S. I. (2022). Developing climate-resilient crops: Improving plant tolerance to stress combination. *The Plant Journal*, *109*(2), 373–389. doi:10.1111/tpj.15483 PMID:34482588

Rizwan, M., Ali, S., Ali, B., Adrees, M., Arshad, M., Hussain, A., Zia ur Rehman, M., & Waris, A.A. (2019). Zinc And Iron Oxide Nanoparticles Improved The Plant Growth And Reduced The Oxidative Stress And Cadmium Concentration In Wheat. *Chemosphere*, *214*, 269–277. (https://doi.org/) doi:10.1016/j.chemosphere.2018.09.120

Roberts, C. M., Mcclean, C. J., Veron, J. E. N., Hawkins, J. P., Allen, G. R., McAllister, D. E., Mittermeier, C. G., Schueler, F. W., Spalding, M., Wells, F., Vynne, C., & Werner, T. B. (2002). Marine Biodiversity Hotspots and Conservation Priorities for Tropical Reefs. *Science*, *295*(5558), 1280–1284. doi:10.1126cience.1067728 PMID:11847338

Roberts, D. P., & Mattoo, A. K. (2018). Sustainable agriculture—Enhancing environmental benefits, food nutritional quality and building crop resilience to abiotic and biotic stresses. *Agriculture*, *8*(1), 8. doi:10.3390/agriculture8010008

Rommens, C. M. (2007). Intragenic crop improvement: Combining the benefits of traditional breeding and genetic engineering. *Journal of Agricultural and Food Chemistry*, *55*(11), 4281–4288. doi:10.1021/jf0706631 PMID:17488120

Rosegrant, M. W., Ringler, C., & Zhu, T. (2009). Water for agriculture: Maintaining food security under growing scarcity. *Annual Review of Environment and Resources*, *34*(1), 205–222. doi:10.1146/annurev.environ.030308.090351

Rui, M., Ma, C., Hao, Y., Guo, J., Rui, Y., Tang, X., & Fan, F. (2016). *Frontiers in Plant Science*.

S, T., SK, T., & R, K. (2018). Bio-Nanotechnology and its Role in Agriculture and Food Industry. *Molecular and Genetic Medicine, 12*(1), 1–5. doi:10.4172/1747-0862.1000324

Sadat-Hosseini, M., Naeimi, A., Boroomand, N., Aalifar, M., Farajpour, M., & Sari, I. (2022, October 28). (123 C.E.). Alleviating the adverse effects of salinity on Roselle plants by green synthesized NPs. *Scientific Reports*, *12*(1), 18165. doi:10.103841598-022-22903-9 PMID:36307527

Saddiq, M.S., Iqbal, S., Afzal, I., Ibrahim, A.M.H., Bakhtavar, M.A., Hafeez, M.B., & Jahanzaib, M.M. (2019). *Mitigation Of Salinity Stress In Wheat (Triticum aestivum L.) Seedlings through Physiological Seed Enhancements.* Taylor & Francis. (https://doi.org/) doi:10.1080/01904167.2019.1609509

Safdar, M., Kim, W., Park, S., Gwon, Y., Kim, Y. O., & Kim, J. (2022). Engineering plants with carbon nanotubes: A sustainable agriculture approach. *Journal of Nanobiotechnology*, *20*(1), 1–30. doi:10.118612951-022-01483-w PMID:35701848

Saha, G.C., Sarker, A., Chen, W., Vandemark, G.J., & Muehlbauer, F.J. (2013). Inheritance and linkage map positions of genes conferring agromorphological traits in *Lens culinaris* Medic. *International Journal of Agronomy*. doi:10.1155/2013/618926

Sahani, S., & Sharma, Y. C. (2021). Advancements in applications of nanotechnology in global food industry. *Food Chemistry*, *342*, 128318. doi:10.1016/j.foodchem.2020.128318 PMID:33189478

Saharan, V., Kumaraswamy, R. V., Choudhary, R. C., Kumari, S., Pal, A., Raliya, R., & Biswas, P. (2016). Cu-Chitosan Nanoparticle Mediated Sustainable Approach To Enhance Seedling Growth in Maize by Mobilizing Reserved Food. *Journal of Agricultural and Food Chemistry*, *64*(31), 6148–6155. doi:10.1021/acs.jafc.6b02239 PMID:27460439

Sahle, A., & Potting, J. (2013). Environmental life cycle assessment of Ethiopian rose cultivation. *The Science of the Total Environment*, *443*, 163–172. doi:10.1016/j.scitotenv.2012.10.048 PMID:23183227

Sajid, M., Basit, A., Ullah, I., Shah, S., Haleema, B., Nawab, K., Ahmad, I., Ahmad, N., Khan, A., & Muhammad, M. (2021). Screening of citrus rootstocks for salinity tolerance based on initial growth attributes and leaves proline content. *Element. Educ. Online*, *19*, 3128–3140.

Sajid, M., & Płotka-Wasylka, J. (2020). Nanoparticles: Synthesis, characteristics, and applications in analytical and other sciences. *Microchemical Journal*, *154*, 104623. doi:10.1016/j.microc.2020.104623

Sajyan, T. K., Alturki, S. M., & Sassine, Y. N. (2020). Nano-fertilizers and their impact on vegetables: Contribution of Nano-chelate Super Plus ZFM and Lithovit®-standard to improve salt-tolerance of pepper. *Annals of Agricultural Science*, *65*(2), 200–208. doi:10.1016/j.aoas.2020.11.001

Salam, A., Afridi, M. S., Javed, M. A., Saleem, A., Hafeez, A., Khan, A. R., Zeeshan, M., Ali, B., Azhar, W., Sumaira, Ulhassan, Z., & Gan, Y. (2022). Nano-priming against abiotic stress: A way forward towards sustainable agriculture. *Sustainability (Basel)*, *14*(22), 14880. doi:10.3390u142214880

Salama, D. M., Osman, S. A., Abd El-Aziz, M. E., Abd Elwahed, M. S. A., & Shaaban, E. A. (2019). Effect of Zinc Oxide Nanoparticles on the Growth, Genomic DNA, Production and the Quality of Common Dry Bean (*Phaseolus vulgaris*). *Biocatalysis and Agricultural Biotechnology*, *11*, 101083. doi:10.1016/j.bcab.2019.101083

Saleh, T. A. (2020). Nanomaterials: Classification, properties, and environmental toxicities. *Environmental Technology & Innovation*, *20*, 101067. doi:10.1016/j.eti.2020.101067

Salgotra, R. K., Gupta, B. B., & Stewart, C. N. Jr. (2014). From genomics to functional markers in the era of next-generation sequencing. *Biotechnology Letters*, *36*(3), 417–426. doi:10.100710529-013-1377-1 PMID:24129954

Samaj, J., Baluska, F., Voigt, B., Schlicht, M., Volkmann, D., & Menzel, D. (2004). Endocytosis, actin cytoskeleton, and signaling. *Plant Physiology*, *135*(3), 1150–1161. doi:10.1104/pp.104.040683 PMID:15266049

Saminathan, K. (2015). Biosynthesis of silver nanoparticles using soil Actinomycetes Streptomyces sp. *International Journal of Current Microbiology and Applied Sciences*, (4), 1073–1083.

Samir, K., & Lutz, W. (2017). The human core of the shared socioeconomic pathways: Population scenarios by age, sex and level of education for all countries to 2100. *Global Environmental Change*, *42*, 181–192. doi:10.1016/j.gloenvcha.2014.06.004 PMID:28239237

Sanger, F., Nicklen, S., & Coulson, A. R. (1977). DNA sequencing with chain-terminating inhibitors. *Proceedings of the National Academy of Sciences of the United States of America*, *74*(12), 5463–5467. doi:10.1073/pnas.74.12.5463 PMID:271968

Sanna, V., Pala, N., & Sechi, M. (2014). Targeted therapy using nanotechnology: Focus on cancer. *International Journal of Nanomedicine*, *9*, 467. PMID:24531078

Santaella, C., & Plancot, B. (2020). Interactions of Nanoenabled Agrochemicals with Soil Microbiome. In V. L. de Castro, R. Grillo, D. Ávila, H. Caixeta Oliveira, & R. Lima (Eds.), *Nanopesticides: From Research and Development to Mechanisms of Action and Sustainable Use in Agriculture; Fraceto, L.F., S.S* (pp. 137–163). Springer International Publishing. doi:10.1007/978-3-030-44873-8_6

Sanzari, I., Leone, A., & Ambrosone, A. (2019). Nanotechnology in Plant Science: To Make a Long Story Short. *Frontiers in Bioengineering and Biotechnology*, *7*, 120. doi:10.3389/fbioe.2019.00120 PMID:31192203

Saravanakumar, A., Ganesh, M., Jayaprakash, J., & Jang, H. T. (2015). Biosynthesis of silver nanoparticles using Cassia tora leaf extract and its antioxidant and antibacterial activities. *Journal of Industrial and Engineering Chemistry*, *28*, 277–281. doi:10.1016/j.jiec.2015.03.003

Saravi, S. S. S., & Shokrzadeh, M. (2011). Role of pesticides in human life in the modern age: a review. *Pesticides in the modern world-risks and benefits, 5*, 3-12.

Satalkar, P., Elger, B. S., & Shaw, D. M. (2016). Defining nano, nanotechnology and nanomedicine: Why should it matter? *Science and Engineering Ethics*, *22*(5), 1255–1276. doi:10.100711948-015-9705-6 PMID:26373718

Sathiyabama, M., & Muthukumar, S. (2020). Chitosan guar nanoparticle preparation and its in vitro antimicrobial activity towards phytopathogens of rice. *International Journal of Biological Macromolecules*, *153*, 297–304. doi:10.1016/j.ijbiomac.2020.03.001 PMID:32135260

Savassa, S. M., Duran, N. M., Rodrigues, E. S., de Almeida, E., van Gestel, C. A. M., Bompadre, T. F. V., & de Carvalho, H. W. P. (2018). Effects of ZnO Nanoparticles on *Phaseolus vulgaris* Germination and Seedling Development Determined by X-ray Spectroscopy. *ACS Applied Nano Materials*, *1*, 6414–6426. doi:10.1021/acsanm.8b01619

Savci, S. (2012). An agricultural pollutant: Chemical fertilizer. *International Journal of Environmental Sciences and Development*, *3*(1), 73–80. doi:10.7763/IJESD.2012.V3.191

Saxena, R., Tomar, R. S., & Kumar, M. (2016). Exploring nanobiotechnology to mitigate abiotic stress in crop plants. *Journal of Pharmaceutical Sciences and Research*, (8), 974.

Sayed, H., Kayyal, H., Ramsey, L., Ceccarelli, S., & Baum, M. (2002). Segregation distortion in doubled haploid lines of barley (*Hordeum vulgare L.*) detected by simple sequence repeat markers. *Euphytica*, *225*(2), 265–272. doi:10.1023/A:1015861610226

Schlich, K., Klawonn, T., Terytze, K., & Hund-Rinke, K. (2013). *Journal of Environmental Quality*, *42*(1), 30–36. doi:10.2134/jeq2012.0196 PMID:23673736

Schmitt, C. C., Moreira, R., Neves, R. C., Richter, D., Funke, A., Raffelt, K., Grunwaldt, J.-D., & Dahmen, N. (2020). From Agriculture Residue to Upgraded Product: The Thermochemical Conversion of Sugarcane Bagasse for Fuel and Chemical Products. *Fuel Processing Technology, 197*, 106199. doi:10.1016/j.fuproc.2019.106199

Scott, N.R., Chen, H., Cui, H. (2018). Nanotechnology Applications And Implications Of Agrochemicals Toward Sustainable Agriculture And Food Systems. *Journal of Agricultural and Food Chemistry, 66*, 6451–6456. (https://doi.org/) doi:10.1021/acs.jafc.8b00964

Seddighinia, F. S., Iranbakhsh, A., Oraghi Ardebili, Z., & Soleimanpour, S. (2021). Seed-priming with cold plasma and supplementation of nutrient solution with carbon nanotube enhanced carotenoid contents and the expression of psy and pds in Bitter melon (Momordica charantia). *Journal of Applied Botany and Food Quality, 94*, 7–14. doi:10.5073/JABFQ.2021.094.002

Seddighinia, F., Iranbakhsh, A., & Ardebili, O. (2020). Seed Priming with Cold Plasma and Multi-walled Carbon Nanotubes Modified Growth, Tissue Differentiation, Anatomy, and Yield in Bitter Melon (Momordica charantia). *Journal of Plant Growth Regulation, 39*(1), 87–98. Advance online publication. doi:10.100700344-019-09965-2

Sehgal, J. L., Saxena, R. K., & Vadivelu, S. (1987). Field. Manual- Soil Resource Mapping of Different States of India, *Technical Bulletin No. 13, NBSS&LUP, Nagpur.*

Sekhon, B. S. (2014). Nanotechnology in agri-food production: An overview. *Nanotechnology, Science and Applications*, 31–53. doi:10.2147/NSA.S39406 PMID:24966671

Seleiman, M. F., Almutairi, K. F., Alotaibi, M., Shami, A., Alhammad, B. A., & Battaglia, M. L. (2021). Nano-Fertilization as an Emerging Fertilization Technique: Why Can Modern Agriculture Benefit from Its Use? *Plants*, (10), 2. PMID:33375026

Serag, M. F., Kaji, N., Tokeshi, M., & Baba, Y. (2012). Introducing carbon nanotubes into living walled plant cells through cellulase-induced nanoholes. *RSC Advances, 2*(2), 398–400. doi:10.1039/C1RA00760B

Serag, M. F., Kaji, N., Tokeshi, M., & Baba, Y. (2015). In H. Manzer, H. Mohamed, & M. Firoz (Eds.), *Carbon nanotubes and modern nanoagriculture," in Nanotechnology and Plant Sciences* (pp. 183–201). Springer International Publishing.

Servin, A., Elmer, W., Mukherjee, A., De la Torre-Roche, R., Hamdi, H., White, J. C., Bindraban, P., & Dimkpa, C. (2015). A review of the use of engineered nanomaterials to suppress plant disease and enhance crop yield. *Journal of Nanoparticle Research, 17*(2), 1–21. doi:10.100711051-015-2907-7

Shafiq, F., Iqbal, M., Ali, M., & Ashraf, M. A. (2019a). Seed Pre-treatment with Polyhydroxy Fullerene NPs Confer Salt Tolerance in Wheat Through Upregulation of H2O2 Neutralizing Enzymes and Phosphorus Uptake. *Journal of Soil Science and Plant Nutrition, 19*(4), 734–742. doi:10.100742729-019-00073-4

Shah, T., Latif, S., Saeed, F., Ali, I., Ullah, S., Abdullah Alsahli, A., Jan, S., & Ahmad, P. (2021). Seed priming with titanium dioxide NPs enhances seed vigor, leaf water status, and antioxidant enzyme activities in maize (Zea mays L.) under salinity stress. *Journal of King Saud University. Science, 33*(1), 101207. doi:10.1016/j.jksus.2020.10.004

Shang, Y., Hasan, M. K., Ahammed, G. J., Li, M., Yin, H., & Zhou, J. (2019). Applications of nanotechnology in plant growth and crop protection: A review. *Molecules (Basel, Switzerland), 24*(14), 2558. doi:10.3390/molecules24142558 PMID:31337070

Sharifi, S., Behzadi, S., Laurent, S., Forrest, M. L., Stroeve, P., & Mahmoudi, M. (2012). Toxicity of nanomaterials. *Chemical Society Reviews, 41*(6), 2323–2343. doi:10.1039/C1CS15188F PMID:22170510

Sharma, S., Singh, V. K., Kumar, A., & Mallubhotla, S. (2019). Effect of nanoparticles on oxidative damage and antioxidant defense system in plants. *Molecular Plant Abiotic Stress: Biology and Biotechnology*, 315-333.

Sharma, B., Tiwari, S., Kumawat, K. C., & Cardinale, M. (2023). Nano-biofertilizers as bio-emerging strategies for sustainable agriculture development: Potentiality and their limitations. *The Science of the Total Environment, 860*, 160476. doi:10.1016/j.scitotenv.2022.160476 PMID:36436627

Sharma, D., Afzal, S., & Singh, N. K. (2021). Nanopriming with phytosynthesized zinc oxide nanoparticles for promoting germination and starch metabolism in rice seeds. *Journal of Biotechnology, 336*, 64–75. doi:10.1016/j.jbiotec.2021.06.014 PMID:34116127

Sharma, K., Singh, U. S., Sharma, P., Kumar, A., & Sharma, L. (2015). Seed treatments for sustainable agriculture-A review. *Journal of Applied and Natural Science, 7*(1), 521–539. doi:10.31018/jans.v7i1.641

Sharma, P., Jha, A. B., Dubey, R. S., & Pessarakli, M. (2012). Reactive Oxygen Species, Oxidative Damage, and Antioxidative Defense Mechanism in Plants under Stressful Conditions. *Journal of Botany (Egypt), 2012*, 1–26. doi:10.1155/2012/217037

Sharp, P. J., Johnston, S., Brown, G., McIntosh, R. A., Pallotta, M., Carter, M., Bariana, H. S., Khartkar, S., Lagudah, E. S., Singh, R. P., Khairallah, M., Potter, R., & Jones, M. G. K. (2001). Validation of molecular markers for wheat breeding. *Australian Journal of Agricultural Research, 52*(12), 1357–1366. doi:10.1071/AR01052

Shcherbakova, E. N., Shcherbakov, A. V., Andronov, E. E., Gonchar, L. N., Kalenskaya, S. M., & Chebotar, V. K. (2017). Combined Pre-Seed Treatment with Microbial Inoculants and Mo Nanoparticles Changes Composition of Root Exudates and Rhizosphere Microbiome Structure of Chickpea (*Cicer arietinum* L.) Plants. *Symbiosis, 73*(1), 57–69. doi:10.100713199-016-0472-1

Sheikhalipour, M., Mohammadi, S. A., Esmaielpour, B., Spanos, A., Mahmoudi, R., Mahdavinia, G. R., Milani, M. H., Kahnamoei, A., Nouraein, M., Antoniou, C., Kulak, M., Gohari, G., & Fotopoulos, V. (2023). Seedling nanopriming with selenium-chitosan nanoparticles mitigates the adverse effects of salt stress by inducing multiple defence pathways in bitter melon plants. *International Journal of Biological Macromolecules, 242*, 124923. doi:10.1016/j.ijbiomac.2023.124923 PMID:37211072

Shelar, A., Nile, S. H., Singh, A. V., Rothenstein, D., Bill, J., Xiao, J., Chaskar, M., Kai, G., & Patil, R. (2023). Recent advances in nano-enabled seed treatment strategies for sustainable agriculture: Challenges, risk assessment, and future perspectives. *Nano-Micro Letters, 15*(1), 54. doi:10.100740820-023-01025-5 PMID:36795339

Shelar, A., Singh, A. V., Maharjan, R. S., Laux, P., Luch, A., Gemmati, D., Tisato, V., Singh, S. P., Santilli, M. F., Shelar, A., Chaskar, M., & Patil, R. (2021). Sustainable agriculture through multidisciplinary seed nanopriming: Prospects of opportunities and challenges. *Cells, 10*(9), 2428. doi:10.3390/cells10092428 PMID:34572078

Shen, C. X., Zhang, Q. F., Li, J., Bi, F. C., & Yao, N. (2010). Induction of programmed cell death in Arabidopsis and rice by single-wall carbon nanotubes. *American Journal of Botany, 97*(10), 1602–1609. doi:10.3732/ajb.1000073 PMID:21616795

Shiferaw, B., Prasanna, B. M., Hellin, J., & Bänziger, M. (2011). Crops that feed the world 6. Past successes and future challenges to the role played by maize in global food security. *Food Security, 3*(3), 307–327. doi:10.100712571-011-0140-5

Shimada, D. (2015). Multi-level natural resources governance based on local community: A case study on semi-natural grassland in Tarōji, Nara, Japan. *The International Journal of the Commons, 9*(2), 486. doi:10.18352/ijc.510

Shukla, S. K., Solanki, K. R., Kumar, R. V., & Dwivedi, R. P. (1999). Stud ies on air layer ing in Buchanania lanzan (Chironji) and Basia latifolia (Mahua). Annual Report, NRCAF, Jhansi, p. 47.

Shukla, A. K., Behera, S. K., Pakhre, A., & Chaudhari, S. (2018). Micronutrients in soils, plants, animals and humans. *Indian Journal of Fertilisers, 14*(3), 30–54.

Shukla, P., Chaurasia, P., Younis, K., Qadri, O. S., Faridi, S. A., & Srivastava, G. (2019). Nanotechnology in sustainable agriculture: Studies from seed priming to post-harvest management. *Nanotechnology for Environmental Engineering, 4*(1), 1–15. doi:10.100741204-019-0058-2

Shukla, S. K., & Solanki, K. R. (2000). Studies on seed germination, plant survival and growth of Chironjee (Buchanania lanzan Spreng.). *Journal of Tropical Forestry, 16*, 44–49.

Si, S., Mandal, T.K. (2007). Tryptophan-Based Peptides To Synthesize Gold And Silver Nanoparticles: A Mechanistic And Kinetic Study. *Chemistry - A European Journal, 13*, 3160–3168. (https://doi.org/) doi:10.1002/chem.200601492

Siddaiah, C. N., Prasanth, K. V. H., Satyanarayana, N. R., Mudili, V., Gupta, V. K., Kalagatur, N. K., Satyavati, T., Dai, X. F., Chen, J. Y., Mocan, A., Singh, B. P., & Srivastava, R. K. (2018). Chitosan Nanoparticles Having Higher Degree Of Acetylation Induce Resistance Against Pearl Millet Downy Mildew Through Nitric Oxide Generation. *Scientific Reports*, *81*(8), 1–14. doi:10.103841598-017-19016-z PMID:29410438

Simonin, M., Colman, B. P., Tang, W., Judy, J. D., Anderson, S. M., Bergemann, C. M., Rocca, J. D., Unrine, J. M., Cassar, N., & Bernhardt, E. S. (2018). Plant and Microbial Responses to Repeated Cu (OH)2 Nanopesticide Exposures under Different Fertilization Levels in an Agro-Ecosystem. *Frontiers in Microbiology*, *9*, 9. doi:10.3389/fmicb.2018.01769 PMID:30108580

Singh, A., Rajput, V. D., Rawat, S., Sharma, R., Singh, A. K., Kumar, P., & Singh, S. (2022). Geoinformatics and Nanotechnological Approaches for Coping Up Abiotic and Biotic Stress in Crop Plants. *Sustainable Agriculture Systems and Technologies*, 337-359.

Singh, A., Rajput, V., Singh, A.K., Sengar, R.S., Singh, R.K., & Minkina, T. (2021). Transformation Techniques And Their Role In Crop Improvements: A Global Scenario Of Gm Crops. *Policy Issues in Genetically Modified Crops*, 515–542. (https://doi.org/) doi:10.1016/B978-0-12-820780-2.00023-6

Singh, A., Rajput, V.D., Rawat, S., Sharma, R., Singh, Anil Kumar, Kumar, P., Singh, Awani Kumar, Minkina, T., Singh, R.P., & Singh, S. (2022). Geoinformatics and Nanotechnological Approaches for Coping Up Abiotic and Biotic Stress in Crop Plants. *Sustainable Agriculture Systems and Technology,* 337–359. (https://doi.org/) doi:10.1002/9781119808565.ch17

Singh, A., Sharma, R., Rawat, S., Singh, A. K., Rajput, V. D., Fedorov, Y., Minkina, T., & Chaplygin, V. (2022). Nanomaterial-plant interaction: Views on the pros and cons. *Toxicity of Nanoparticles in Plants: An Evaluation of Cyto/Morpho-Physiological, Biochemical and Molecular Responses*, 47–68. doi:10.1016/B978-0-323-90774-3.00015-5

Singh, A., Singh, N.B., Hussain, I., Singh, H., Yadav, V. (2017). Synthesis And Characterization Of Copper Oxide Nanoparticles And Its Impact On Germination Of *Vigna radiata* (L.) R. Wilczek. *Trends in Phytochemical Research, 4*(2), 66–73. https://doi.org/ doi:10.22271/tpr.2017.v4.i2.034

Singh, N., & Bhardwaj, R. D. (2016). Ascorbic acid alleviates water deficit induced growth inhibition in wheat seedlings by modulating levels of endogenous antioxidants. *Biologia, 71*(4), 402–413. doi:10.1515/biolog-2016-0050

Singh, V. K., Singh, R., Tripathi, S., Devi, R. S., Srivastava, P., Singh, P., Kumar, A., & Bhadouria, R. (2020). Seed priming: state of the art and new perspectives in the era of climate change. *Climate Change and Soil Interactions*, 143–170. doi:10.1016/B978-0-12-818032-7.00006-0

Singh, A. V., Chandrasekar, V., Janapareddy, P., Mathews, D. E., Laux, P., Luch, A., Yang, Y., Garcia-Canibano, B., Balakrishnan, S., Abinahed, J., Al Ansari, A., & Dakua, S. P. (2021). Emerging Application of Nanorobotics and Artificial Intelligence To Cross the BBB: Advances in Design, Controlled Maneuvering, and Targeting of the Barriers. *ACS Chemical Neuroscience*, *12*(11), 1835–1853. doi:10.1021/acschemneuro.1c00087 PMID:34008957

Singh, A., Rajput, V. D., Rawat, S., Sharma, R., Singh, A. K., Kumar, P., Singh, A. K., Minkina, T., Singh, R. P., & Singh, S. (2022). *Geoinformatics and Nanotechnological Approaches for Coping Up Abiotic and Biotic Stress in Crop Plants. Sustainable Agriculture Systems and Technologies*. Wiley.

Singh, A., Rajput, V., Rawat, S., Singh, A. K., Bind, A., & Singh, A. (2020). Monitoring Soil Salinity and Recent Advances in Mechanism of Salinity Tolerance in Plants. *Biogeosystem Technique*, *7*(2), 66–87. doi:10.13187/bgt.2020.2.66

Singh, A., Rakesh Singh Sengar, R. S., Sharma, R., Rajput, P., & Singh, A. K. (2021a). Nano-Priming Technology for Sustainable Agriculture. *Biogeosystem Technique.*, *8*(2). doi:10.13187/bgt.2021.2.79

Singh, A., Sengar, R. S., Rajput, V. D., Minkina, T., & Singh, R. K. (2022). Zinc oxide nanoparticles improve salt tolerance in rice seedlings by improving physiological and biochemical indices. *Agriculture*, *12*(7), 1014. doi:10.3390/agriculture12071014

Singh, A., Sengar, R. S., Shahi, U. P., Rajput, V. D., Minkina, T., & Ghazaryan, K. A. (2023). Prominent Effects of Zinc Oxide Nanoparticles on Roots of Rice (Oryza sativa L.) Grown under Salinity Stress. *Stresses*, *3*(1), 33–46. doi:10.3390tresses3010004

Singh, A., Sengar, R. S., Sharma, R., Rajput, P., & Singh, A. K. (2021). Nano-priming technology for sustainable agriculture. *Biogeosystem Technique*, *8*(2), 79–92.

Singh, A., Shukla, N., Kabadwal, B., Tewari, A., & Kumar, J. (2018). Review on plant-Trichoderma-pathogen interaction. *International Journal of Current Microbiology and Applied Sciences*, *7*(2), 2382–2397. doi:10.20546/ijcmas.2018.702.291

Singh, A., Singh Sengar, R., Singh, R., Shahi, U. P., Yadav, M. K., Vaishali, Gangwar, L. K., & Rajput, V. D. (2022). Effects of zinc oxide nanoparticles for promoting seed germination of rice (*Oryza sativa* L.) under salinity stress. Ecology. *Environmental Conservation*, *28*, 254–259. doi:10.53550/EEC.2022.v28i07s.042

Singh, A., Singh, A. K., Rawat, S., Pal, N., Rajput, V. D., Minkina, T., Sharma, R., Singh, N. P., & Tripathi, J. N. (2022). Satellite-Based Quantification of Methane Emissions from Wetlands and Rice Paddies Ecosystems in North and Northeast India. *Hydrobiology*, *1*(3), 317–330. doi:10.3390/hydrobiology1030023

Singh, A., Singh, N., Hussain, I., Singh, H., & Singh, S. (2015). Plant-nanoparticle interaction: An approach to improve agricultural practices and plant productivity. *International Journal of Pharmaceutical Science Invention*, *4*(8), 25–40.

Singh, H., Jassal, R. K., Kang, J. S., Sandhu, S. S., Kang, H., & Grewal, K. (2015). Seed priming techniques in field crops-A review. *Agricultural Reviews (Karnal)*, *36*(4), 251–264. doi:10.18805/ag.v36i4.6662

Singh, H., Sharma, A., Bhardwaj, S. K., Arya, S. K., Bhardwaj, N., & Khatri, M. (2021). Recent advances in the applications of nano-agrochemicals for sustainable agricultural development. *Environmental Science. Processes & Impacts*, *23*(2), 213–239. doi:10.1039/D0EM00404A PMID:33447834

Singh, K. U., Kumar, A., Raja, L., Kumar, V., Vashney, N., & Chhetri, M. (2022). An artificial neural network-based pest identification and control in smart agriculture using wireless sensor networks. *Journal of Food Quality*, *2022*, 2022. doi:10.1155/2022/5801206

Singh, P., Kim, Y. J., Singh, H., Mathiyalagan, R., Wang, C., & Yang, D. C. (2015). Biosynthesis of anisotropic silver nanoparticles by Bhargavaea indica and their synergistic effect with antibiotics against pathogenic microorganisms. *Journal of Nanomaterials*, 1–10. doi:10.1155/2015/234741

Singh, P., Kim, Y. J., Wang, C., Mathiyalagan, R., El-Agamy Farh, M., & Yang, D. C. (2016). Biogenic silver and gold nanoparticles synthesized using red ginseng root extract, and their applications. *Artificial Cells, Nanomedicine, and Biotechnology*, (44), 811–816. PMID:25706249

Singh, S., Kumar, R., Panchal, R., & Tiwari, M. K. (2021). Impact of COVID-19 on logistics systems and disruptions in food supply chain. *International Journal of Production Research*, *59*(7), 1993–2008. doi:10.1080/00207543.2020.1792000

Singh, S., Sangwan, S., Sharma, P., Devi, P., & Moond, M. (2021). Nanotechnology for sustainable agriculture: An emerging perspective. *Journal of Nanoscience and Nanotechnology*, *21*(6), 3453–3465. doi:10.1166/jnn.2021.19012 PMID:34739798

Singh, T., Shukla, S., Kumar, P., Wahla, V., Bajpai, V. K., & Rather, I. A. (2017). Application of Nanotechnology in Food Science: Perception and Overview. *Frontiers in Microbiology*, *8*, 268461. doi:10.3389/fmicb.2017.01501 PMID:28824605

Singh, Y., Kaushal, S., & Sodhi, R. S. (2020). Biogenic synthesis of silver nanoparticles using *cyanobacterium Leptolyngbya* sp. WUC 59 cell-free extract and their effects on bacterial growth and seed germination. *Nanoscale Advances*, *2*(9), 3972–3982. doi:10.1039/D0NA00357C PMID:36132754

Singh, Y., Meher, J. G., Raval, K., Khan, F. A., Chaurasia, M., Jain, N. K., & Chourasia, M. K. (2017). Nanoemulsion: Concepts, development and applications in drug delivery. *Journal of Controlled Release*, *252*, 28–49. doi:10.1016/j.jconrel.2017.03.008 PMID:28279798

Sinha, B. B., & Dhanalakshmi, R. (2022). Recent advancements and challenges of Internet of Things in smart agriculture: A survey. *Future Generation Computer Systems*, *126*, 169–184. doi:10.1016/j.future.2021.08.006

Slama, I., Abdelly, C., Bouchereau, A., Flowers, T., & Savouré, A. (2015). Diversity, distribution and roles of osmoprotective compounds accumulated in halophytes under abiotic stress. *Annals of Botany*, *115*(3), 433–447. doi:10.1093/aob/mcu239 PMID:25564467

Smith, M. T., & Berjak, P. (2017). Deteriorative changes associated with the loss of viability of stored desiccation-tolerant and desiccation-sensitive seeds. In *Seed development and germination* (pp. 701–746). Routledge. doi:10.1201/9780203740071-26

Smith, P., Haberl, H., Popp, A., Erb, K., Lauk, C., Harper, R., Tubiello, F. N., de Siqueira Pinto, A., Jafari, M., Sohi, S., Masera, O., Böttcher, H., Berndes, G., Bustamante, M., Ahammad, H., Clark, H., Dong, H., Elsiddig, E. A., Mbow, C., & Rose, S. (2013). How much land-based greenhouse gas mitigation can be achieved without compromising food security and environmental goals? *Global Change Biology*, *19*(8), 2285–2302. doi:10.1111/gcb.12160 PMID:23505220

Snowdon, R., & Friedt, W. (2004). Molecular markers in Brassica oilseeds breeding: Current status and future possibilities. *Plant Breeding*, *123*(1), 1–8. doi:10.1111/j.1439-0523.2003.00968.x

Soetan, K., Olaiya, C., & Oyewole, O. (2010). The importance of mineral elements for humans, domestic animals and plants: A review. *African Journal of Food Science*, *4*(5), 200–222.

Soil Survey Staff. (1951). *Soil Survey Manual. U.S. Department of Agriculture. Handbook No. 18*. Oxford & IBH Publishing Co.

Sonawane, H., Arya, S., Math, S., & Shelke, D. (2021). Myco-synthesized silver and titanium oxide nanoparticles as seed priming agents to promote seed germination and seedling growth of *Solanum lycopersicum*: A comparative study. *International Nano Letters*, *11*(4), 371–379. doi:10.100740089-021-00346-w

Song, K., & He, X. (2021). How to improve seed germination with green nanopriming. *Seed Science and Technology*, *49*(2), 81–92. doi:10.15258st.2021.49.2.01

Sozer, N., & Kokini, J. L. (2020). Nanotechnology and its applications in the food sector. *Trends in Biotechnology*, *38*(1), 82–89. PMID:19135747

Spagnoletti, F. N., Spedalieri, C., Kronberg, F., & Giacometti, R. (2019). Extracellular Biosynthesis of Bactericidal Ag/AgCl Nanoparticles for Crop Protection Using the Fungus *Macrophomina Phaseolina*. *Journal of Environmental Management*, *231*, 457–466. doi:10.1016/j.jenvman.2018.10.081 PMID:30388644

Spielman-Sun, E., Avellan, A., Bland, G.D., Tappero, R.V., Acerbo, A.S., Unrine, J.M., Giraldo, J.P., Lowry, G.V. (2019). Nanoparticle Surface Charge Influences Translocation And Leaf Distribution In Vascular Plants With Contrasting Anatomy. *Environmental Science: Nano,* 6, 2508–2519. (https://doi.org/) doi:10.1039/C9EN00626E

Srinivasan, C., & Saraswathi, R. (2010). Nano-agriculture—Carbon nanotubes enhance tomato seed germination and plant growth. *Current Science*, *99*, 274–275.

Srivastava, A. K., & Bose, B. (2012). Effect of nitrate seed priming on phenology, growth rate and yield attributes in rice (*Oryza sativa L.*). *Vegetos Int J Plant Res*, *25*(2), 174–181.

Srivastava, A., & Rao, D. P. (2014). Enhancement of seed germination and plant growth of wheat, maize, peanut and garlic using multiwalled carbon nanotubes. *Chem. Bull*, *3*, 502–504.

Stam, P. (1993). Construction of integrated genetic linkage maps by means of a new computer package: JoinMap. *The Plant Journal*, *3*(5), 739–744. doi:10.1111/j.1365-313X.1993.00739.x

Stampoulis, D., Sinha, S. K., & White, J. C. (2009). Assay-dependent phytotoxicity of nanoparticles to plants. *Environmental Science & Technology*, *43*(24), 9473–9479. doi:10.1021/es901695c PMID:19924897

Stergiopoulos, I., & Gordon, T. R. (2014). Cryptic fungal infections: The hidden agenda of plant pathogens. *Frontiers in Plant Science*, *5*, 506. doi:10.3389/fpls.2014.00506 PMID:25309571

Stuber, C. W., Polacco, M., & Senior, M. L. (1999). Synergy of empirical breeding, marker-assisted selection, and genomics to increase crop yield potential. *Crop Science*, *39*(6), 1571–1583. doi:10.2135/cropsci1999.3961571x

Sundaria, N., Singh, M., Upreti, P., Chauhan, R., Jaiswal, J. P., & Kumar, A. (2019). Seed Priming with Iron Oxide Nanoparticles Triggers Iron Acquisition and Biofortification in Wheat (Triticum aestivum L.) Grains. *Journal of Plant Growth Regulation*, *38*(1), 122–131. doi:10.100700344-018-9818-7

Sun, L., Song, F., Zhu, X., Liu, S., Liu, F., & Wang, Y. (2020). Nano-ZnO alleviates drought stress via modulating the plant water use and carbohydrate metabolism in maize. *Archives of Agronomy and Soil Science*, 115.

Sun, Q., Li, J., & Le, T. (2018). Zinc oxide nanoparticle as a novel class of antifungal agents: Current advances and future perspectives. *Journal of Agricultural and Food Chemistry*, *66*(43), 11209–11220. doi:10.1021/acs.jafc.8b03210 PMID:30299956

Svetleva, D., Velcheva, M., & Bhowmik, G. (2003). Biotechnology as a useful tool in common bean (*Phaseolus vulgaris L.*) improvement: A review. *Euphytica*, *131*(2), 189–200. doi:10.1023/A:1023983831582

Sytar, O., Kumari, P., Yadav, S., Brestic, M., Rastogi, A. (2018). Phytohormone Priming: Regulator For Heavy Metal Stress In Plants. *Journal of Plant Growth Regulation*, *38*(2), 739–752. (https://doi.org/) doi:10.1007/s00344-018-9886-8

Sytar, O., Kumari, P., Yadav, S., Brestic, M., & Rastogi, A. (2018). Phytohormone priming: Regulator for heavy metal stress in plants. *Journal of Plant Growth Regulation*, *38*(38), 739–752.

Szőllősi, R., Molnár, Á., Kondak, S., & Kolbert, Z. (2020). Dual effect of nanomaterials on germination and seedling growth: Stimulation vs. phytotoxicity. *Plants*, *9*(12), 1745. doi:10.3390/plants9121745 PMID:33321844

Talukdar, S., Singha, P., Mahato, S., Shahfahad, Pal, S., Liou, Y.-A., & Rahman, A. (2020). Land-Use Land-Cover Classification by Machine Learning Classifiers for Satellite Observations—A Review. *Remote Sensing (Basel)*, *12*(7), 1135. doi:10.3390/rs12071135

Tanabata, T., Shibaya, T., Hori, K., Ebana, K., & Yano, M. (2012). SmartGrain: High-throughput phenotyping software for measuring seed shape through image analysis. *Plant Physiology*, *160*(4), 1871–1880. doi:10.1104/pp.112.205120 PMID:23054566

Tang, X., Mu, X., Shao, H., Wang, H., & Brestic, M. (2015). Global plant-responding mechanisms to salt stress: Physiological and molecular levels and implications in biotechnology. *Critical Reviews in Biotechnology*, *35*(4), 425–437. doi:10.3109/07388551.2014.889080 PMID:24738851

Tankhiwale, R., & Bajpai, S. K. (2009). Preparation, characterization and antibacterial applications of ZnO-nanoparticles coated polyethylene films for food packaging. *Colloids and Surfaces. B, Biointerfaces*, *90*, 16–20. doi:10.1016/j.colsurfb.2011.09.031 PMID:22015180

Tanksley, S. D. (1993). Mapping polygenes. *Annual Review of Genetics*, *27*(1), 205–233. doi:10.1146/annurev.ge.27.120193.001225 PMID:8122902

Tanou, G., Molassiotis, A., & Diamantidis, G. (2009). Induction of reactive oxygen species and necrotic death-like destruction in strawberry leaves by salinity. *Environmental and Experimental Botany*, *65*(2–3), 270–281. doi:10.1016/j.envexpbot.2008.09.005

Tan, X., Lin, C., & Fugetsu, B. (2009). Bunshi Fugetsu, Studies on toxicity of multi-walled carbon nanotubes on suspension rice cells. *Carbon*, *47*(15), 3479–3487. doi:10.1016/j.carbon.2009.08.018

Tao, R., You, C., Qu, Q., Zhang, X., Deng, Y., Ma, W., & Huang, C. (2023). Recent advances in the design of controlled-and sustained-release micro/nanocarriers of pesticide. *Environmental Science. Nano*, *10*(2), 351–371. doi:10.1039/D2EN00446A

Taran, M., Fateh, R., Rezaei, S., & Gholi, M. K. (2019). Isolation of arsenic accumulating bacteria from garbage leachates for possible application in bioremediation. *Iranian Journal of Microbiology*, *11*(1), 60. doi:10.18502/ijm.v11i1.707 PMID:30996833

Taran, N. Y., Gonchar, O. M., Lopatko, K. G., Batsmanova, L. M., Patyka, M. V., & Volkogon, M. V. (2014). The effect of colloidal solution of molybdenum nanoparticles on the microbial composition in rhizosphere of *Cicer arietinum* L. *Nanoscale Research Letters*, *9*(1), 289. doi:10.1186/1556-276X-9-289 PMID:25024677

Tassadduq, S. S., Akhtar, S., Waheed, M., Bangash, N., Nayab, D. E., Majeed, M., Abbasi, S., Muhammad, M., Alataway, A., Dewidar, A. Z., Elansary, H. O., & Yessoufou, K. (2022). Ecological Distribution Patterns of Wild Grasses and Abiotic Factors. *Sustainability (Basel)*, *14*(18), 11117. Advance online publication. doi:10.3390u141811117

Teh, C. Y., Budiman, P. M., Shak, K. P. Y., & Wu, T. Y. (2016). Recent advancement of coagulation–flocculation and its application in wastewater treatment. *Industrial & Engineering Chemistry Research*, *55*(16), 4363–4389. doi:10.1021/acs.iecr.5b04703

Thabet, A. F., Boraei, H. A., Galal, O. A., El-Samahy, M. F. M., Mousa, K. M., Zhang, Y. Z., Tuda, M., Helmy, E. A., Wen, J., & Nozaki, T. (2021). Silica nanoparticles as pesticide against insects of different feeding types and their non-target attraction of predators. *Scientific Reports*, *11*(1), 14484. doi:10.103841598-021-93518-9 PMID:34262071

Thakur, M., Sharma, P., & Anand, A. (2019). *Seed priming-induced early vigor in crops: An alternate strategy for abiotic stress tolerance. Priming and pretreatment of seeds and seedlings*. Springer.

The Royal Society. (2009). *Reaping the Benefits: Science and the Sustainable Intensification of Global Agriculture*. The Royal Society.

Timmusk, S., Seisenbaeva, G., & Behers, L. (2018). Titania (TiO2) nanoparticles enhance the performance of growth-promoting rhizobacteria. *Scientific Reports*, *8*(1), 617. doi:10.103841598-017-18939-x PMID:29330479

Tiwari, D. K., Dasgupta-Schubert, N., Villaseñor Cendejas, L. M., Villegas, J., Carreto Montoya, L., & Borjas García, S. E. (2013). Interfacing carbon nanotubes (CNT) with plants: Enhancement of growth, water and ionic nutrient uptake in maize (Zea mays) and implications for nanoagriculture. *Applied Nanoscience*, *4*(5), 577–591. doi:10.100713204-013-0236-7

Toksha, B., Sonawale, V. A. M., Vanarase, A., Bornare, D., Tonde, S., Hazra, C., Kundu, D., Satdive, A., Tayde, S., & Chatterjee, A. (2021). Nanofertilizers: A review on synthesis and impact of their use on crop yield and environment. *Environmental Technology & Innovation*, *24*, 101986. doi:10.1016/j.eti.2021.101986

Tripathi, S., Sonkar, S. K., & Sarkar, S. (2011). Growth stimulation of gram (Cicer arietinum) plant by water soluble carbon nanotubes. *Nanoscale*, *3*(3), 1176–1181. doi:10.1039/c0nr00722f PMID:21253651

Trivedi, M., Kedari, S., & Nikalje, G. C. (2022). Role of nanoparticles in remediation of contaminated soil. In *The Role of Nanoparticles in Plant Nutrition under Soil Pollution: Nanoscience in Nutrient Use Efficiency* (pp. 353–370). Springer International Publishing. doi:10.1007/978-3-030-97389-6_15

Tsuji, K. (2001). Microencapsulation of pesticides and their improved handling safety. *Journal of Microencapsulation*, *18*(2), 137–147. doi:10.1080/026520401750063856 PMID:11253931

Tuberosa, R., Salvi, S., Sanguineti, M. C., Maccaferri, M., Giuliani, S., & Landi, P. (2003). Searching for quantitative trait loci controlling root traits in maize: A critical appraisal. *Plant and Soil*, *255*(1), 35–54. doi:10.1023/A:1026146615248

Tudi, M., Daniel Ruan, H., Wang, L., Lyu, J., Sadler, R., Connell, D., Chu, C., & Phung, D. T. (2021). Agriculture development, pesticide application and its impact on the environment. *International Journal of Environmental Research and Public Health*, *18*(3), 1112. doi:10.3390/ijerph18031112 PMID:33513796

Ukwattage, N. L., Ranjith, P., & Bouazza, M. (2013). The use of coal combustion fly ash as a soil amendment in agricultural lands (with comments on its potential to improve food security and sequester carbon). *Fuel*, *109*, 400–408. doi:10.1016/j.fuel.2013.02.016

UN. (2017). 2017 HLPF Thematic review of SDG2. *High-level Political Forum on Sustainable Development*. UN.

UN. (n.d.). *Goal 2: End hunger, achieve food security and improved nutrition and promote sustainable agriculture — SDG Indicators*. UN. unstats.un.org.

United Nations. (2015) *Resolution adopted by the General Assembly on 25 September 2015, Transforming our world: the 2030 Agenda for Sustainable Development (A/RES/70/1)*. UN.

Upreti, D., & Patro, T. U. (2022). Polymer Nanocomposite Membranes for Water Remediation. In *Advances in Nanostructured Materials* (pp. 227–279). Springer. doi:10.1007/978-981-16-8391-6_13

Usman, M., Farooq, M., Wakeel, A., Nawaz, A., Cheema, S. A., Rehman, H., Ashraf, I., & Sanaullah, M. (2020). Nanotechnology in agriculture: Current status, challenges and future opportunities. *The Science of the Total Environment*, *721*, 137778. doi:10.1016/j.scitotenv.2020.137778 PMID:32179352

Utz, H., & Melchinger, A. (1996). PLABQTL: A program for composite interval mapping of QTL. *J Quant Trait Loci*, *2*(1).

Vaisman, L. & Wagner, H.D. (2006). The role of surfactants in dispersion of carbon nanotubes. *Advances in Colloid and Interface Science, 128,* 37-46.

Valletta, A., Chronopoulou, L., Palocci, C., Baldan, B., Donati, L., Pasqua, G. (2014). Poly(Lactic-Co-Glycolic) Acid Nanoparticles Uptake By *Vitis Vinifera* And Grapevine-Pathogenic Fungi. *Journal of Nanoparticle Research*, *16*(12), 2744. (https://doi.org/) doi:10.1007/s11051-014-2744-0

Van der Elst, K., & Williams, A. (2019). Environment and natural resources. In Agriculture & Food Systems to 2050: Global Trends, Challenges and Opportunities (pp. 193-214). World Scientific.

van Dijk, M., Morley, T., Rau, M. L., & Saghai, Y. (2021). A meta-analysis of projected global food demand and population at risk of hunger for the period 2010–2050. *Nature Food*, *2*(7), 494–501. doi:10.103843016-021-00322-9 PMID:37117684

Van Nguyen, D., Nguyen, H. M., Le, N. T., Nguyen, K. H., Le, H. M., Nguyen, A. T., Dinh, N. T. T., Hoang, S. A., & Van Ha, C. (2020). Copper Nanoparticle Application Enhances Plant Growth And Grain Yield In Maize Under Drought Stress Conditions. bioRxiv 2020.02.24.963132. doi:10.1101/2020.02.24.963132

Van Sanford, D., Anderson, J., Campbell, K., Costa, J., Cregan, P., Griffey, C., Hayes, P., & Ward, R. (2001). Discovery and deployment of molecular markers linked to *Fusarium* head blight resistance: An integrated system for wheat and barley. *Crop Science*, *41*(3), 638–644. doi:10.2135/cropsci2001.413638x

Van Wesenbeeck, C., Keyzer, M., Van Veen, W., & Qiu, H. (2021). Can China's overuse of fertilizer be reduced without threatening food security and farm incomes? *Agricultural Systems*, *190*, 103093. doi:10.1016/j.agsy.2021.103093

Van Zelm, E., Zhang, Y., & Testerink, C. (2020). *Salt Tolerance Mechanisms of Plants*. Annual Reviews. doi:10.1146/annurev-arplant-050718-100005

Vannini, C., Domingo, G., Onelli, E., De Mattia, F., Bruni, I., Marsoni, M., & Bracale, M. (2014). Phytotoxic and genotoxic effects of silver NPs exposure on germinating wheat seedlings. *Journal of Plant Physiology*, *171*(13), 1142–1148. doi:10.1016/j.jplph.2014.05.002 PMID:24973586

Varshney, R. K., Shi, C., Thudi, M., Mariac, C., Wallace, J., Qi, P., Zhang, H., Zhao, Y., Wang, X., Rathore, A., Srivastava, R. K., Chitikineni, A., Fan, G., Bajaj, P., Punnuri, S., Gupta, S. K., Wang, H., Jiang, Y., Couderc, M., & Xu, X. (2017). Pearl millet genome sequence provides a resource to improve agronomic traits in arid environments. *Nature Biotechnology*, *35*(10), 969–976. Advance online publication. doi:10.1038/nbt.3943 PMID:28922347

Varshney, R. K., Song, C., Saxena, R. K., Azam, S., Yu, S., Sharpe, A. G., Cannon, S., Baek, J., Rosen, B. D., Tar'an, B., Millan, T., Zhang, X., Ramsay, L. D., Iwata, A., Wang, Y., Nelson, W., Farmer, A. D., Gaur, P. M., Soderlund, C., & Cook, D. R. (2013). The draft genome sequence of chickpea (Cicer arietinum) provides a resource for trait improvement. *Nature Biotechnology*, *31*(3), 240–248. doi:10.1038/nbt.2491 PMID:23354103

Vejan, P., Abdullah, R., Khadiran, T., Ismail, S., & Nasrulhaq Boyce, A. (2016). Role of Plant Growth Promoting Rhizobacteria in Agricultural Sustainability-A Review. *Molecules (Basel, Switzerland)*, *21*(5), 573. doi:10.3390/molecules21050573 PMID:27136521

Vejan, P., Khadiran, T., Abdullah, R., & Ahmad, N. (2021). Controlled release fertilizer: A review on developments, applications and potential in agriculture. *Journal of Controlled Release*, *339*, 321–334. doi:10.1016/j.jconrel.2021.10.003 PMID:34626724

Velculescu, V. E., Vogelstein, B., & Kinzler, K. W. (2000). Analysing uncharted transcriptomes with SAGE. *Trends in Genetics*, *16*(10), 423–425. doi:10.1016/S0168-9525(00)02114-4 PMID:11050322

Veneklaas, E. J., Lambers, H., Bragg, J., Finnegan, P. M., Lovelock, C. E., Plaxton, W. C., Price, C. A., Scheible, W. R., Shane, M. W., White, P. J., & Raven, J. A. (2012). Opportunities for improving phosphorus-use efficiency in crop plants. *The New Phytologist*, *195*(2), 306–320. doi:10.1111/j.1469-8137.2012.04190.x PMID:22691045

Ventura, L., Donà, M., Macovei, A., Carbonera, D., Buttafava, A., Mondoni, A., & Balestrazzi, A. (2012). Understanding the molecular pathways associated with seed vigor. *Plant Physiology and Biochemistry*, *60*, 196–206. doi:10.1016/j.plaphy.2012.07.031 PMID:22995217

Verma, K. K., Song, X.-P., Joshi, A., Rajput, V. D., Singh, M., Sharma, A., Singh, R. K., Li, D.-M., Arora, J., Minkina, T., & Li, Y.-R. (2022). Nanofertilizer Possibilities for Healthy Soil, Water, and Food in Future: An Overview. *Frontiers in Plant Science*, *13*, 865048. Advance online publication. doi:10.3389/fpls.2022.865048 PMID:35677230

Verma, K. K., Song, X.-P., Joshi, A., Tian, D.-D., Rajput, V. D., Singh, M., Arora, J., Minkina, T., & Li, Y.-R. (2022). Recent trends in nano-fertilizers for sustainable agriculture under climate change for global food security. *Nanomaterials (Basel, Switzerland)*, *12*(1), 173. doi:10.3390/nano12010173 PMID:35010126

Verma, S. K., Das, A. K., Patel, M. K., Shah, A., Kumar, V., & Gantait, S. (2018). Engineered nanomaterials for plant growth and development: A perspective analysis. *The Science of the Total Environment*, *630*, 1413–1435. doi:10.1016/j.scitotenv.2018.02.313 PMID:29554761

Vickers, C. E., Gershenzon, J., Lerdau, M. T., & Loreto, F. (2009). A unified mechanism of action for volatile isoprenoids in plant abiotic stress. *Nature Chemical Biology*, *5*(5), 283–291. doi:10.1038/nchembio.158 PMID:19377454

Villa, F., Silva, D. F. D., Rotili, M. C. C., Herzog, N. F. M., & Malavasi, M. D. M. (2019). Seed physiological quality and harvest point of dovyalis fruits. *Pesquisa Agropecuária Tropical*, *49*, 49. doi:10.1590/1983-40632019v4954520

Villagarcia, H., Dervishi, E., De Silva, K., Biris, A. S., & Khodakovskaya, M. V. (2012). Surface Chemistry Of Carbon Nanotubes Impacts The Growth And Expression Of Water Channel Protein In Tomato Plants. *Small*, *8*(15), 2328–2334. doi:10.1002mll.201102661 PMID:22514121

Vishwakarma, K., Upadhyay, N., Kumar, N., Tripathi, D. K., Chauhan, D. K., Sharma, S., & Sahi, S. (2018). Potential applications and avenues of nanotechnology in sustainable agriculture. In *Nanomaterials in Plants, Algae, and Microorganisms*. Academic Press.

Vishwakarma, K., Upadhyay, N., Kumar, N., Tripathi, D. K., Chauhan, D. K., Sharma, S., & Sahi, S. (2018). Potential Applications and Avenues of Nanotechnology in Sustainable Agriculture. *Nanomaterials in Plants, Algae, and Microorganisms*, *1*, 473–500. doi:10.1016/B978-0-12-811487-2.00021-9

Visscher, P., Thompson, R., & Haley, C. (1996). Confidence intervals in QTL mapping by bootstrapping. *Genetics*, *143*(2), 1013–1020. doi:10.1093/genetics/143.2.1013 PMID:8725246

Vithanage, M., Herath, I., Almaroai, Y. A., Rajapaksha, A. U., Huang, L., Sung, J. K., Lee, S. S., & Ok, Y. S. (2017). Effects of carbon nanotube and biochar on bioavailability of Pb, Cu and Sb in multi-metal contaminated soil. *Environmental Geochemistry and Health*, *39*(6), 1409–1420. doi:10.100710653-017-9941-6 PMID:28332174

Voloshina, M. (2020). Monitoring Soil Salinity and Recent Advances in Mechanism of Salinity Tolerance in Plants. *Biogeosystem Tech.*, *7*, 66–87. https://doi.org/http://dxdoi.org/10.13187/bgt.2020.2.66

Vos, P., Hogers, R., Bleeker, M., Reijans, M., van de Lee, T., Hoernes, M., Frijters, A., Pot, J., Peleman, J., Kuiper, M., & Zabeau, M. (1995). AFLP: A new technique for DNA fingerprinting. *Nucleic Acids Research*, *23*(21), 4407–4414. doi:10.1093/nar/23.21.4407 PMID:7501463

Wang, C., Kim, Y. J., Singh, P., Mathiyalagan, R., Jin, Y., & Yang, D. C. (2016). Green synthesis of silver nanoparticles by *Bacillus methylotrophicus*, and their antimicrobial activity. *Artificial Cells, Nanomedicine, and Biotechnology*, *44*(4), 1127–1132. doi:10.3109/21691401.2015.1011805 PMID:25749281

Wang, D., Saleh, N. B., Byro, A., Zepp, R., Sahle-Demessie, E., Luxton, T. P., Ho, K. T., Burgess, R. M., Flury, M., White, J. C., & Su, C. (2022). Nano-enabled pesticides for sustainable agriculture and global food security. *Nature Nanotechnology*, *17*(4), 347–360. doi:10.103841565-022-01082-8 PMID:35332293

Wang, H., Kou, X., Pei, Z., Xiao, J. Q., Shan, X., & Xing, B. (2011). Physiological Effects Of Magnetite (Fe_3O_4) Nanoparticles On Perennial Ryegrass (*Lolium perenne* L.) And Pumpkin (*Cucurbita mixta*) Plants. *Nanotoxicology*, *5*(1), 30–42. doi:10.3109/17435390.2010.489206 PMID:21417686

Wang, P., Lombi, E., Zhao, F. J., & Kopittke, P. M. (2016). Nanotechnology: A New Opportunity in Plant Sciences. *Trends in Plant Science*, *21*(8), 699–712. doi:10.1016/j.tplants.2016.04.005 PMID:27130471

Wang, W. N., Tarafdar, J. C., & Biswas, P. (2013). Nanoparticle synthesis and delivery by an aerosol route for watermelon plant foliar uptake. *Journal of Nanoparticle Research*, *15*(1), 1–13. doi:10.100711051-013-1417-8

Wang, X., Han, H., Liu, X., Gu, X., Chen, K., & Lu, D. (2012). Multi-walled carbon nanotubes can enhance root elongation of wheat (Triticum aestivum) plants. *Journal of Nanoparticle Research*, *14*(6), 841. doi:10.100711051-012-0841-5

Wang, X., Jiang, J., Dou, F., Sun, W., & Ma, X. (2021). Simultaneous mitigation of arsenic and cadmium accumulation in rice (Oryza sativa L.) seedlings by silicon oxide nanoparticles under different water management schemes. *Paddy and Water Environment*, *19*(4), 569–584. doi:10.100710333-021-00855-6

Wang, Y., Deng, C., Rawat, S., Cota-Ruiz, K., Medina-Velo, I., & Gardea-Torresdey, J. L. (2021). Evaluation of the effects of nanomaterials on rice (Oryza sativa L.) responses: Underlining the benefits of nanotechnology for agricultural applications. *ACS Agricultural Science & Technology*, *1*(1), 44–54. doi:10.1021/acsagscitech.1c00030

Waqas Mazhar, M., Ishtiaq, M., Maqbool, M., Akram, R., Shahid, A., Shokralla, S., Al-Ghobari, H., Alataway, A., Dewidar, A. Z., El-Sabrout, A. M., & Elansary, H. O. (2022). Seed priming with iron oxide nanoparticles raises biomass production and agronomic profile of water-stressed flax plants. *Agronomy (Basel)*, *12*(5), 982. doi:10.3390/agronomy12050982

Weeks, J. J. Jr, & Hettiarachchi, G. M. (2019). A review of the latest in phosphorus fertilizer technology: Possibilities and pragmatism. *Journal of Environmental Quality*, *48*(5), 1300–1313. doi:10.2134/jeq2019.02.0067 PMID:31589737

Weidner, T., Yang, A., & Hamm, M. W. (2021). Energy optimisation of plant factories and greenhouses for different climatic conditions. *Energy Conversion and Management*, *243*, 114336. doi:10.1016/j.enconman.2021.114336

WHO. (2019). *Microplastics in drinking-water*. WHO.

WHO. (2020). *The state of food security and nutrition in the world 2020: transforming food systems for affordable healthy diets* (Vol. 2020). Food & Agriculture Org.

Wild, E., & Jones, K. C. (2009). Novel method for the direct visualization of in vivo nanomaterials and chemical interactions in plants. *Environmental Science & Technology*, *43*(14), 5290–5294. doi:10.1021/es900065h PMID:19708355

Willett, W., Rockström, J., Loken, B., Springmann, M., Lang, T., Vermeulen, S., Garnett, T., Tilman, D., DeClerck, F., Wood, A., Jonell, M., Clark, M., Gordon, L. J., Fanzo, J., Hawkes, C., Zurayk, R., Rivera, J. A., De Vries, W., Majele Sibanda, L., ... Murray, C. J. L. (2019). Food in the Anthropocene: The EAT–Lancet Commission on healthy diets from sustainable food systems. *Lancet*, *393*(10170), 447–492. doi:10.1016/S0140-6736(18)31788-4 PMID:30660336

Williams, K. J. (2003). The molecular genetics of disease resistance in barley. *Australian Journal of Agricultural Research*, *54*(12), 1065–1079. doi:10.1071/AR02219

Winter, P., & Kahl, G. (1995). Molecular marker technologies for plant improvement. *World Journal of Microbiology & Biotechnology*, *11*(4), 438–448. doi:10.1007/BF00364619 PMID:24414752

Wojtyla, Ł., Lechowska, K., Kubala, S., & Garnczarska, M. (2016). Different modes of hydrogen peroxide action during seed germination. *Frontiers in Plant Science*, *7*, 66. doi:10.3389/fpls.2016.00066 PMID:26870076

World Health Organization. (2018). *Malnutrition.* WHO. https://www.who.int/news-room/fact-sheets/detail/malnutrition

World Health Organization. (2020). *Food safety.* WHO. https://www.who.int/news-room/fact-sheets/detail/food-safety

World Population Clock. (n.d.). *World Population.* Worldometers.

World Population Prospects. (2022). *United Nations.* Department of Economic and Social Affairs.

Worrall, E. A., Hamid, A., Mody, K. T., Mitter, N., & Pappu, H. R. (2018). Nanotechnology for Plant Disease Management. *Agronomy (Basel)*, *8*(12), 285. doi:10.3390/agronomy8120285

Woyessa, D. (2022). Weed control methods used in agriculture. *American Journal of Life Science and Innovation*, *1*(1), 19–26. doi:10.54536/ajlsi.v1i1.413

Wright, B. (2014). Global biofuels: Key to the puzzle of grain market behavior. *The Journal of Economic Perspectives*, *28*(1), 73–98. doi:10.1257/jep.28.1.73

Wu, H., & Li, Z. (2022). Recent advances in nano-enabled agriculture for improving plant performance. *The Crop Journal*, *10*(1), 1–12. doi:10.1016/j.cj.2021.06.002

Wu, H., Tito, N., & Giraldo, J. P. (2017). Anionic cerium oxide nanoparticles protect plant photosynthesis from abiotic stress by scavenging reactive oxygen species. *ACS Nano*, *11*(11), 11283–11297. doi:10.1021/acsnano.7b05723 PMID:29099581

Wu, L., Huo, W., Yao, D., & Li, M. (2019). Effects of solid matrix priming (SMP) and salt stress on broccoli and cauliflower seed germination and early seedling growth. *Scientia Horticulturae*, *255*, 161–168. doi:10.1016/j.scienta.2019.05.007

Wu, M., Wu, J., & Gan, Y. (2020). The New Insight Of Auxin Functions: Transition From Seed Dormancy To Germination And Floral Opening In Plants. *Plant Growth Regulation*, *912*(91), 169–174. doi:10.100710725-020-00608-1

Wu, X., Li, N., Li, H., & Tang, H. (2014). An optimized method for NMR-based plant seed metabolomic analysis with maximized polar metabolite extraction efficiency, signal-to-noise ratio, and chemical shift consistency. *Analyst (London), 139*(7), 1769–1778. doi:10.1039/C3AN02100A PMID:24519438

Xi, L., Zhang, M., Zhang, L., Lew, T. T., & Lam, Y. M. (2022). Novel materials for urban farming. *Advanced Materials, 34*(25), 2105009. doi:10.1002/adma.202105009 PMID:34668260

Xin, X., Zhao, F., Judy, J. D., & He, Z. (2022). Copper stress alleviation in corn (Zea mays L.): Comparative efficiency of carbon nanotubes and carbon nanoparticles. *NanoImpact, 25*, 100381. doi:10.1016/j.impact.2022.100381 PMID:35559887

Xu, Y., Zhu, L., Xiao, J., Huang, N., & McCouch, S. R. (1997). Chromosomal regions associated with segregation distortion of molecular markers in F2, backcross, doubled haploid, and recombinant inbred populations in rice (*Oryza sativa L.*). *Molecular & General Genetics, 253*(5), 535–545. doi:10.1007004380050355 PMID:9065686

Yadav, S., Modi, P., Dave, A., Vijapura, A., Patel, D., & Patel, M. (2020). Effect of abiotic stress on crops. *Sustainable crop production, 3.*

Yadav, S. K., Lal, S., Yadav, S., Laxman, J., Verma, B., Sushma, M., Choudhary, R., Singh, P., Singh, S., & Sharma, V. (2019). Use of nanotechnology in agri-food sectors and apprehensions: An overview. *Seed Res, 47*(2), 99–149.

Yadu, B., Xalxo, R., Chandra, J., Kumar, M., Chandrakar, V., & Keshavkant, S. (2021). Applications of Nanomaterials to Enhance Plant Health and Agricultural Production. In *Plant Responses to Nanomaterials*. Springer. doi:10.1007/978-3-030-36740-4_1

Yam, K. L., Takhistov, P. T., & Miltz, J. (2005). Intelligent packaging: Concepts and applications. *Journal of Food Science, 70*(1), R1–R10. doi:10.1111/j.1365-2621.2005.tb09052.x

Yáñez-Sedeno, P., Pingarron, J. M., Villalonga, R., & Pingarrón, J. (2018). Electrochemical biosensors for pathogen detection directly in food samples. A review. *Analytical and Bioanalytical Chemistry, 410*(12), 2953–2960.

Yang, D., Tu, Y., Wang, X., Cao, C., Hu, Y., Shao, J., Weng, L., Mou, X., & Dong, X. (2021). A photo-triggered antifungal nanoplatform with efflux pump and heat shock protein reversal activity for enhanced chemo-photothermal synergistic therapy. *Biomaterials Science, 9*(9), 3293–3299. doi:10.1039/D1BM00457C PMID:33861264

Yan, S., Ren, B. Y., & Shen, J. (2021). Nanoparticle-mediated double-stranded RNA delivery system: A promising approach for sustainable pest management. *Insect Science, 28*(1), 21–34. doi:10.1111/1744-7917.12822 PMID:32478473

Yan, S., Zhao, L., Li, H., Zhang, Q., Tan, J., Huang, M., He, S., & Li, L. (2013, February 15). Single-walled carbon nanotubes selectively influence maize root tissue development accompanied by the change in the related gene expression. *Journal of Hazardous Materials, 246-247*, 110–118. doi:10.1016/j.jhazmat.2012.12.013 PMID:23291336

Yasmeen, F., Raja, N. I., Razzaq, A., & Komatsu, S. (2017). Proteomic and Physiological Analyses of Wheat Seeds Exposed to Copper and Iron Nanoparticles. *Biochimica et Biophysica Acta. Proteins and Proteomics*, *1865*(1), 28–42. doi:10.1016/j.bbapap.2016.10.001 PMID:27717896

Ye, Y., Cota-Ruiz, K., Hernandez-Viezcas, J. A., Valdes, C., Medina-Velo, I. A., Turley, R. S., Peralta-Videa, J. R., & Gardea-Torresdey, J. L. (2020). Manganese nanoparticles control salinity-modulated molecular responses in Capsicum annuum L. through priming: A sustainable approach for agriculture. *ACS Sustainable Chemistry & Engineering*, *8*(3), 1427–1436. doi:10.1021/acssuschemeng.9b05615

Young, G. J., & Blevins, R. D. (1981). Heavy metal concentrations in the Holston river basin (Tennessee). *Archives of Environmental Contamination and Toxicology*, *10*(5), 541–560. doi:10.1007/BF01054878 PMID:7305447

Young, N. D. (1996). QTL mapping and quantitative disease resistance in plants. *Annual Review of Phytopathology*, *34*(1), 479–501. doi:10.1146/annurev.phyto.34.1.479 PMID:15012553

Younis, S. A., Kim, K.-H., Shaheen, S. M., Antoniadis, V., Tsang, Y. F., Rinklebe, J., Deep, A., & Brown, R. J. (2021). Advancements of nanotechnologies in crop promotion and soil fertility: Benefits, life cycle assessment, and legislation policies. *Renewable & Sustainable Energy Reviews*, *152*, 111686. doi:10.1016/j.rser.2021.111686

Yousefi, E., Sheidaei, A., Mahdavi, M., Baniassadi, M., Baghani, M., & Faraji, G. (2017). Azadeh Sheidaei, Mostafa Mahdavi, Majid Baniassadi, Mostafa Baghani, Ghader Faraji, Effect of nanofiller geometry on the energy absorption capability of coiled carbon nanotube composite material. *Composites Science and Technology*, *153*, 222–231. doi:10.1016/j.compscitech.2017.10.025

Yousefi, S., Kartoolinejad, D., & Naghdi, R. (2017). Effects of priming with multi-walled carbon nanotubes on seed physiological characteristics of hopbush (Dodonaeaviscosa L.) under drought stress. *The International Journal of Environmental Studies*, *74*(4), 528–539. doi:10.1080/00207233.2017.1325627

Yuan, P., Yang, T., & Poovaiah, B. (2018). Calcium signaling-mediated plant response to cold stress. *International Journal of Molecular Sciences*, *19*(12), 3896. doi:10.3390/ijms19123896 PMID:30563125

Yücel, E. (2000a). Effects of different salt (NaCl), nitrate (KNO3) and acid (H2S04) concentrations on the germination of some Salvia species seeds. *Seed Science and Technology*, *28*, 853–860.

Yu, H., Park, J. Y., Kwon, C. W., Hong, S. C., Park, K. M., & Chang, P. S. (2018). An overview of nanotechnology in food science: Preparative methods, practical applications, and safety. *Journal of Chemistry*, *2018*, 1–10. doi:10.1155/2018/5427978

Yu, J., Hu, S., Wang, J., Wong, K. S., Li, S., Liu, B., Deng, Y., Dai, L., Zhou, Y., Zhang, X., Cao, M., Liu, J., Sun, J., Tang, J., Chen, Y., Huang, X., Lin, W., Ye, C., Tong, W., & Yang, H. (2002). A draft sequence of the rice genome (Oryza sativa L. ssp. indica). *Science*, *296*(5565), 79–92. doi:10.1126cience.1068037 PMID:11935017

Yu, K., Park, S., & Poysa, V. (2000). Marker-assisted selection of common beans for resistance to common bacterial blight: Efficacy and economics. *Plant Breeding*, *119*(5), 411–415. doi:10.1046/j.1439-0523.2000.00514.x

Yvon, M., Thébaud, G., Alary, R., & Labonne, G. (2009). Specific detection and quantification of the phytopathogenic agent 'Candidatus Phytoplasma prunorum'. *Molecular and Cellular Probes*, *23*(5), 227–234. doi:10.1016/j.mcp.2009.04.005 PMID:19401232

Zafar, S., Perveen, S., Kamran Khan, M., Shaheen, M. R., Hussain, R., Sarwar, N., Rashid, S., Nafees, M., Farid, G., Alamri, S., Shah, A. A., Javed, T., Irfan, M., & Siddiqui, M. H. (2022). Effect of zinc nanoparticles seed priming and foliar application on the growth and physio-biochemical indices of spinach (Spinacia oleracea L.) under salt stress. *PLoS One*, *17*(2), e0263194. doi:10.1371/journal.pone.0263194 PMID:35192615

Zahedifar, M., & Zohrabi, S. (2016). Germination And Seedling Characteristics Of Drought-Stressed Corn Seed As Influenced By Seed Priming With Potassium Nano-Chelate And Sulfate Fertilizers. *Acta Agriculturae Slovenica*, *107*(1), 113–128. doi:10.14720/aas.2016.107.1.12

Zahra, N., Shaukat, K., Hafeez, M. B., Raza, A., Hussain, S., Chaudhary, M. T., Akram, M. Z., Kakavand, S. N., Saddiq, M. S., & Wahid, A. (2021). Physiological and Molecular Responses to High, Chilling, and Freezing Temperature in Plant Growth and Production: Consequences and Mitigation Possibilities. *Harsh Environment and Plant Resilience*, 235–290. https://doi.org/doi:10.1007/978-3-030-65912-7_10/COVER

Zaman, M., Shahid, S. A., Heng, L., Shahid, S. A., Zaman, M., & Heng, L. (2018). Soil salinity: Historical perspectives and a world overview of the problem. *Guideline for salinity assessment, mitigation and adaptation using nuclear and related techniques*, 43-53.

Zaytseva, O., & Neumann, G. (2016). Carbon nanomaterials: Production, impact on plant development, agricultural and environmental applications. *Chemical and Biological Technologies in Agriculture*, *3*(1), 17. doi:10.118640538-016-0070-8

Zeng, Z.-B. (1993). Theoretical basis for separation of multiple linked gene effects in mapping quantitative trait loci. *Proceedings of the National Academy of Sciences of the United States of America*, *90*(23), 10972–10976. doi:10.1073/pnas.90.23.10972 PMID:8248199

Zhai, G., Gutowski, S. M., Walters, K. S., Yan, B., & Schnoor, J. L. (2015a). Charge, Size, and Cellular Selectivity for Multiwall Carbon Nanotubes by Maize and Soybean. *Environmental Science & Technology*, *49*(12), 7380–7390. doi:10.1021/acs.est.5b01145 PMID:26010305

Zhai, Y., Hunting, E. R., Liu, G., Baas, E., Peijnenburg, W. J. G. M., & Vijver, M. G. (2019). Compositional Alterations in Soil Bacterial Communities Exposed to TiO2 Nanoparticles Are Not Reflected in Functional Impacts. *Environmental Research*, *178*, 108713. doi:10.1016/j.envres.2019.108713 PMID:31518961

Zhang, D., Ma, X. L., Gu, Y., Huang, H., & Zhang, G. W. (2020). Green synthesis of metallic nanoparticles and their potential applications to treat cancer. *Frontiers in Chemistry*, *8*, 799. doi:10.3389/fchem.2020.00799 PMID:33195027

Zhang, F., Cui, Z., Fan, M., Zhang, W., Chen, X., & Jiang, R. (2011). Integrated soil–crop system management: Reducing environmental risk while increasing crop productivity and improving nutrient use efficiency in China. *Journal of Environmental Quality*, *40*(4), 1051–1057. doi:10.2134/jeq2010.0292 PMID:21712573

Zhang, H., Huang, M., Zhang, W., Gardea-Torresdey, J. L., White, J. C., Ji, R., & Zhao, L. (2020). Silver Nanoparticles Alter Soil Microbial Community Compositions and Metabolite Profiles in Unplanted and Cucumber-Planted Soils. *Environmental Science & Technology*, *54*(6), 3334–3342. doi:10.1021/acs.est.9b07562 PMID:32088952

Zhang, J., Khan, S. A., Heckel, D. G., & Bock, R. (2017). Next-generation insectresistant plants: RNAi-mediated crop protection. *Trends in Biotechnology*, *35*(9), 871–882. doi:10.1016/j.tibtech.2017.04.009 PMID:28822479

Zhang, W., Musante, C., White, J. C., Schwab, P., Wang, Q., Ebbs, S. D., & Ma, X. (2017). Bioavailability of cerium oxide nanoparticles to Raphanus sativus L. in two soils. *Plant Physiology and Biochemistry*, *110*, 185–193. doi:10.1016/j.plaphy.2015.12.013 PMID:26754029

Zhang, X., Niu, H., Zhang, Y., Cai, Z., Zhang, X., Zhang, N., & Liu, Y. (2022). Ultrasensitive SERS Nanosensor for On-Site Detection of Escherichia coli O157: H7 with a Limit of Quantitation Below 10 cfu/mL. *Analytical Chemistry*, *94*(5), 2543–2552. doi:10.1021/acs.analchem.6b00054 PMID:35080383

Zhao, L., Lu, L., Wang, A., Zhang, H., Huang, M., Wu, H., Xing, B., Wang, Z., & Ji, R. (2020). Nano-biotechnology in agriculture: Use of nanomaterials to promote plant growth and stress tolerance. *Journal of Agricultural and Food Chemistry*, *68*(7), 1935–1947. doi:10.1021/acs.jafc.9b06615 PMID:32003987

Zhao, L., Peralta-Videa, J. R., Varela-Ramirez, A., Castillo-Michel, H., Li, C., Zhang, J., Aguilera, R. J., Keller, A. A., & Gardea-Torresdey, J. L. (2012). Effect of surface coating and organic matter on the uptake of CeO2 NPs by corn plants grown in soil: Insight into the uptake mechanism. *Journal of Hazardous Materials*, *225–226*, 131–138. doi:10.1016/j.jhazmat.2012.05.008 PMID:22633924

Zhao, Q., Chen, S., & Dai, S. (2013). C4 photosynthetic machinery: Insights from maize chloroplast proteomics. *Frontiers in Plant Science*, *4*, 85. doi:10.3389/fpls.2013.00085 PMID:23596450

Zhou, J., Applegate, C., Alonso, A., Reynolds, D., Orford, S., Mackiewicz, M., Griffiths, S., Penfield, S., & Pullen, N. (2017). Leaf-GP: An open and automated software application for measuring growth phenotypes for arabidopsis and wheat. *Plant Methods*, *13*(1), 117. doi:10.118613007-017-0266-3 PMID:29299051

Zhu, J.-K. (2016). Abiotic stress signaling and responses in plants. *Cell*, *167*(2), 313–324. doi:10.1016/j.cell.2016.08.029 PMID:27716505

Zulfiqar, F. (2021). Effect of seed priming on horticultural crops. *Scientia Horticulturae*, *286*, 110197. doi:10.1016/j.scienta.2021.110197

Zulfiqar, F., & Ashraf, M. (2021). NPs potentially mediate salt stress tolerance in plants. *Plant Physiology and Biochemistry*, *160*, 257–268. doi:10.1016/j.plaphy.2021.01.028 PMID:33529801

Zulfiqar, F., Navarro, M., Ashraf, M., Akram, N. A., & Munné-Bosch, S. (2019). Nanofertilizer use for sustainable agriculture: Advantages and limitations. *Plant Science*, *289*, 110270. doi:10.1016/j.plantsci.2019.110270 PMID:31623775

Zuverza-Mena, N., Martínez-Fernández, D., Du, W., Hernandez-Viezcas, J. A., Bonilla-Bird, N., López-Moreno, M. L., Komárek, M., Peralta-Videa, J. R., & Gardea-Torresdey, J. L. (2017). Exposure of engineered nanomaterials to plants: Insights into the physiological and biochemical responses—A review. *Plant Physiology and Biochemistry*, *110*, 236–264. doi:10.1016/j.plaphy.2016.05.037 PMID:27289187

About the Contributors

Abhishek Singh is a postdoctoral scientist at Yerevan State University in the Republic of Armenia. He possesses a strong educational background in agriculture and biotechnology, having obtained a Ph.D. from the esteemed Sardar Vallabh Bhai Patel University of Agriculture and Technology in Meerut, UP, India, as well as an Integrated MSc in Biotechnology with a specialization in Plant studies from the prestigious Sam Higginbottom University of Agriculture, Technology, and Sciences in Allahabad, India.His current research focuses on studying the effects of salinity stress on various aspects of plants, including physio-biochemical characteristics, morphology, anatomy, ultrastructure of cellular and subcellular organelles, and DNA, RNA, and protein damage. Additionally, he investigates the bioaccumulation, uptake, translocation, and toxic effects of metallic nanoparticles on plants. Understanding "how nanoparticles interact with plants, soil, and the physio-biochemical, genomics, NanoOmics responses for sustainable agriculture" is a crucial area of agricultural science research, and Dr. Singh has provided significant insights in this domain.Dr. Singh has an impressive publication record, with 25 full-length articles, 10 books, 30 chapters (all available in the Scopus database), and 6 articles presented at scientific conferences. Moreover, he has contributed over 30 agriculture-related articles to various magazines, resulting in a total of 74 scientific publications. He is actively involved in peer review for reputable publications globally, earning a certificate as a peer reviewer from Springer, among others. Furthermore, he serves on the editorial boards of numerous journals and societies.In recognition of his accomplishments, the state government of Uttar Pradesh, India, honored Dr. Abhishek Singh with the "Appreciation Award 2021."

Vishnu D. Rajput holds excellent academic records in Agriculture and Biological Sciences, having earned degrees with distinction in B.Sc. Ag and M.Sc. Ag from CS Azad University of Agriculture & Technology, India. He further pursued his Ph.D. from the esteemed Chinese Academy of Sciences, China, and completed his Post-doc from Southern Federal University, Russia. Currently, he serves as an Associate Professor (Leading Researcher) at Southern Federal University, Russia. His ongoing research focuses on soil contaminations, particularly the study of potentially toxic elements and metallic nanoparticles. Dr. Rajput's investigations encompass the bioaccumulation, bio/geo-transformations, uptake, translocation, and toxic effects of metallic nanoparticles on various aspects of plant physiology, morphology, anatomy, the ultrastructure of cellular and subcellular organelles, cytomorphometric modifications, and DNA damage. He has extensively detailed the current state of research in environmental science, shedding light on how nanoparticles and heavy metals interact with plants, soil, microbial communities, and the broader environment. With an impressive record of 300 scientific publications, Dr. Rajput has authored 300 peer-reviewed highly-rated full-length articles, 20 books, 60 chapters (indexed in Scopus), and 55 conference articles. He is renowned internationally for his expertise and contribution to the field and has reviewed 200 manuscripts for reputable journals, earning him an outstanding reviewing certificate from Elsevier and Springer. Dr. Rajput also serves on the editorial boards of various high-impact journals. For his outstanding contributions to academic, creative research, and publication activities, Dr. Rajput has received the "Certificate for Appreciation 2019" and the "Certificate of Honor 2020" from Southern Federal University, Russia. He has also been recognized as a "Highly Qualified Specialist" by SFedU and the Russian Ministry of Internal Affairs, a prestigious status that acknowledges his expertise and accomplishments.

Tatiana Minkina is the esteemed Head of the Soil Science and Land Evaluation Department at Southern Federal University. She also holds the position of Head of the International Master's Degree Educational Program "Management and Estimation of Land Resources" from 2015 to 2022, which has received accreditation by ACQUIN. Her primary areas of scientific interest include Soil Science, Biogeochemistry of Trace Elements, and Environmental Soil Chemistry. Additionally, she specializes in soil monitoring, assessment, modeling, and remediation, utilizing physicochemical treatment methods. In recognition of her exceptional contributions to the educational process and the training of highly qualified specialists, she was awarded the Diploma of the Ministry of Education and Science of the Russian Federation in 2015. Currently, Professor Dr. Tatiana Minkina is actively involved in various projects funded by prestigious organizations like the Russian Scientific Foundation, Ministry of Education and Science of the Russian Federation, and the Russian Foundation of Basic Research.She holds memberships in several esteemed organizations, including the Expert Group of the Russian Academy of Science, the International Committee on Contamination Land, Eurasian Soil Science Societies, the International Committee on Protection of the Environment, and the International Scientific Committee of the International Conferences on Biogeochemistry of Trace Elements.Professor Dr. Tatiana Minkina's impressive research portfolio includes a total of 757 scientific publications, out of which 389 are in English. She serves as an invited editor for the Open Access Journal by MDPI, "Water," with ISSN 2073-4441 and an Impact Factor of 2.524. Additionally, she holds editorial board membership in esteemed journals such as "Geochemistry: Environment, Exploration, Analysis" with ISSN 1467-7873 and an Impact Factor of 1.109; 5yr IF: 1.769; SJR: 0.334; SNIP: 0.598, as well as "Eurasian Journal of Soil Science" with e-ISSN 2147 – 4249.Professor Dr. Tatiana Minkina's contributions to the field of Soil Science and Environmental Chemistry have earned her a well-deserved reputation as a respected and accomplished researcher in her domain. Scopus Author ID: 15063165400 Web of Science ResearcherID: AAC-3561-2020

Santosh Kumar Gupta is a highly skilled and accomplished plant biologist, recognized for his leadership in co-leading a research group and overseeing research management practices and activities at the esteemed DBT-National Institute of Plant Genome Research in New Delhi. His expertise lies in translational research activities, focusing on the development of genetically modified and genome-edited crops such as chickpea, tomato, and other important crops. With a proven track record of successfully completing projects and establishing fruitful collaborations, Santosh Kumar Gupta has demonstrated exceptional communication skills and a strong commitment to publishing high-impact research findings. His contributions have been widely recognized and acknowledged in numerous peer-reviewed international journals, where he has served as a Guest Associate Editor, Editor, Review Editor, and Reviewer. Some of these prestigious journals include Frontiers in Genetics, Genome Editing, Plant Science, BMC Plant Biology, MDPI Genes, PLOS One, PCTOC, Plant Gene, and Phytomedicine. In his academic career, Santosh Kumar Gupta has authored 25 articles in reputable and SCI-listed journals, including the Journal of Experimental Botany, Plant Biotechnology Journal, Plant Cell & Environment, Journal of Plant Growth Regulation, Plant Physiology and Biochemistry, Protoplasma, and Functional & Integrative Genomics. His significant contributions to scientific literature have not only advanced the understanding of plant biology but have also contributed to the development of novel and promising agricultural practices. One of his notable achievements is the recent publication of the pioneering report on in-planta genome editing in the phytoene desaturase gene (PDS) of chickpea, a groundbreaking advancement in the field of plant genetic engineering. Santosh Kumar Gupta's exceptional expertise, dedication, and exceptional research skills have firmly positioned him as a valuable asset to any scientific organization, where his contributions can continue to drive innovative research and development in plant biology and crop improvement.

Karen Ghazaryan is a distinguished academic figure, currently holding the esteemed position of Deputy Dean at the prestigious Faculty of Biology, Yerevan State University. With a remarkable 16-year journey in the realm of scientific research, he has carved an illustrious path in multiple domains, making significant contributions to the fields of Soil Ecology, Water Resources Monitoring and Management, Climate Change and its Environmental Consequences, Ecological Management of the Environment, Ecological Evaluation of the Environment, and Biodiversity Protection. His profound expertise is reflected in an impressive portfolio of more than 96 scientific research articles, a testament to his dedication and relentless pursuit of advancing knowledge and understanding in his areas of interest. Dr. Ghazaryan scholarly accomplishments extend to the realm of authoring influential books, further cementing his position as a thought leader and authority in his field. Apart from his scholarly achievements, Dr. Ghazaryan actively contributes to the progress of scientific publishing as an esteemed member of various editorial boards of peer-reviewed journals. His commitment to advancing research and fostering academic excellence has made a lasting impact on the scientific community, and his invaluable insights continue to shape the landscape of environmental sciences. Dr. Ghazaryan exemplary career, unwavering dedication, and remarkable contributions to academia make him an outstanding and respected figure in the realm of Soil Ecology and Environmental Sciences. His profound commitment to advancing knowledge and preserving the planet's ecological balance serves as an inspiration to aspiring researchers and scientists worldwide.

***

Ghulam Abbas is working Assistant Professor at Department of Biotechnology University of Agriculture, Dera Ismail Khan-29111, Khyber Pakhtunkhwa, Pakistan

Athanasios Alexiou has a BSc in Applied Mathematics from the National and Kapodistrian University of Athens, a Ph.D. in Bioinformatics Algorithms and Mathematical Modeling in Biology, and a two-year Postdoc Research Fellowship in Neuroinformatics Applications from the Ionian University. He currently collaborates with research teams from King Abdulaziz University (Saudi Arabia) and the University of Sharjah (United Arab Emirates). He also serves as a Faculty Member at the Department of Science and Engineering at the Novel Global Community Education Foundation (Educational platform registered in Australia) and a senior researcher at the AFNP Med Company in Austria. Dr. Alexiou's current research interest falls into Computational Biomedicine. It focuses on developing novel mathematical models and algorithms in neurodegenerative diseases, prognostic Bayesian tools for Alzheimer's disease, identifying molecular mechanisms of infectious diseases, especially the SARS-CoV-2 virus, and the ethical implications of scientific research. He authorizes approximately 200 peer-reviewed scientific papers/chapters/books in high-impact publications and 4 patents.

Muqarrab Ali has an affiliation with institution postal code, province, country: Department of Agronomy, Muhammad Nawaz Shareef University of Agriculture Multan, 66000 Punjab-Pakistan.

Akshita Barthwal is currently working as testing and operations manager in R&D Bayer crop science Ltd. U.P India pronounced active in digitalisation and mechanisation surroundings, completed her Ph.D in soil science in 2022 with teaching and testing equipment handling experience of 1.5 years from SHUATS U.P India. Having effective writing skills with 30 national and international research papers, 10 popular articles and near about 10 book chapters published also member of editorial board of esteemed journals covering the vast agricultural recent and sustainable topics. With good command in communication and hold in management she has participated in many national and international conferences and industrial trainings, organised conferences and trainings being a committee member, given lectures and talks in national platforms such All India Radio, International conferences and Institutions like SHUATS, BHU,LPU etc. addressing students and farmers about the sustainable researches in the favour of precision farming.

Richa Das pursued graduation and post graduation in biotechnology from Parul University. Currently, Das is involved in a research project at Parul University.

Allah Bakhsh Gulshan is Associate Professor at Department of Botany, Ghazi University-32200, Dera Ghazi Khan, Punjab, Pakistan

Mumtaz Hussain is working as deputy director at PARC, Arid Zone Research Institute, Bahawalpur-63100, Pakistan

Tanveer Hussain is working as Assistant Professor in Department of Botany, Mirpur University of Science and Technology (MUST) Mirpur-10250 (AJK), Pakistan. He has published 62 research papers with impact factor 55. He has also published 7 book chapters and 2 books. His 124 abstracts have been published in conference proceedings.

Muhammad Mazhar Iqbal Soil and Water Testing Laboratory CHINIOT 35400, Ayub Agricultural Research Institute, Department of Agriculture, Government of Punjab-Pakistan.

Richa Jaswal has completed a Ph.D. from CSKHPKV, Palampur and is currently working as Junior Research Fellow in the department of Soil Science.

Atin Kumar works as Assistant Professor at the School of Agriculture, Uttaranchal University Dehradun (Uttarakhand), India. Dr. Atin Kumar did his Graduation i.e. B.Sc. (Agriculture) and M.Sc. Agriculture (Agricultural Chemistry & Soil Science) from Chaudhary Charan Singh University, Meerut, in 2012 and Ph.D. (Soil Science) from, Sardar Vallabhbhai Patel University of Agriculture & Technology (SVPUA&T), Modipuram, Meerut (UP) in 2019. Dr. Atin Kumar served as assistant professor at Bhagwant University, Ajmer (Rajasthan), India respectively. He also participated in more than 7 International level and 33 National level workshops/conferences/trainings. He has published more than 37 research papers, articles, abstract, and book chapters. He has also received Young Researcher Award", Confers Fellow Award, and Global Scientist Award. Dr. Atin Kumar's expertise is in soil fertility, nutrient management and water quality. He has nearly 4 years of experience in teaching, research, consultancy and the provision of agricultural extension services. He is also supervised 11 M.Sc. (Agriculture) students and 2 Ph.D. (Soil Science) students.

Chiranjeev Kumar is currently pursuing a Ph.D. in the department of Soil Science in CSKHPKV, Palampur.

Sachin Kumar has completed my Ph.D. from CSKHPKV, Palampur and currently is working as Senior Research Fellow in the department of Centre for Geo-Informatics Research and Training.

Satendra Kumar works as a Professor & Head of the Department at the Department of Soil Science & Agricultural Chemistry, College of Agriculture, Sardar Vallabhbhai Patel University of Agriculture & Technology, Meerut, Uttar Pradesh, India. Dr. Satendra Kumar did his Graduation i.e. B.Sc. (Agriculture) and M.Sc. in Agriculture (Soil Science) from N.D. University of Agriculture & Technology, Kumarganj, Faizabad in 1989 and 1992, respectively, and Ph.D. (Soil Science and Agricultural Chemistry from C.S. Azad University of Agriculture & Technology, Kanpur, Uttar Pradesh, India. Dr. Satendra Kumar has more than 30-year experience in teaching and research. He also participated in more than 50 national/international seminar symposia some of them as invited lead lecture, and chaired as President, Vice Chairman, and Co-convener of conferences and technical season. He has published more than 70 research papers in National/International journals, One Book, 10 Book chapters, one practical manual, more than 88 abstracts, and 18 popular articles. He has a life membership of the Indian Society of Soil Science and 10 other scientific and professional societies. He has the Outstanding Achievement

Award-2018, ATDS-Fellow Award-2018, BRICPL Excellence Award in Teaching and Research-2017, Distinguished Scientist of the Year-2016, SAARS Excellence in Teaching Award-2015, 2nd position in Paper Presentation at the International Conference EIAEASSD-2018, Second prize in the All India Farmers fair and Agri Expo-2004, First prize in the "All India Farmers fair and Agri Expo-2007 at CSAUA&T, Kanpur. He is also supervised 17 M.Sc. (Agriculture) students and 6 Ph.D. students.

Shilpa Manhas completed a Ph.D. in Agronomy from CSKHPKV, Palampur and currently is working as project associate in the department of Centre for Geo-Informatics Research and Training, CSKHPKV, Palampur.

Maria Mehboob as visiting Lecturer, Department of Botany, University of Poonch, Rawalakot-10250, Azad Jummu and Kashmir, Pakistan.

Hasmik Movsesyan is an Associate Professor at the Department of Ecology and Nature Protection, Faculty of Biology, Yerevan State University, PhD in Biology (1990), Associate Professor (2008). Scale of professional interests: Ecology, Nature Protection, Environmental and Occupational Health, Environmental Chemistry and Toxicology, Biochemistry, Food Safety. Publications: 90 papers and abstracts.

Murad Muhammad is researcher State key laboratory of Desert and Oasis Ecology, Xinjiang Institute of Ecology and Geography, Chinese Academy of Sciences, Urumqi 830011, PR China

Murad Muhammad is a Research Scholar State key laboratory of Desert and Oasis Ecology, Xinjiang Institute of Ecology and Geography, Chinese Academy of Sciences.

Allah Nawaz Khan is working researcher in State Key Laboratory of Vegetation and Environmental Change, Institute of Botany, Chinese Academy of Sciences.

Sehar Nawaz, Centre of Agricultural Biochemistry and Biotechnology (CABB), University of Agriculture Faisalabad.

Tayyaba Naz is affiliated with the Saline Agriculture Research Centre, Institute of Soil and Environmental Sciences, University of Agriculture Faisalabad.

Divya Pandey is a PhD Research Scholar.

Shipra Parmar is a PhD Research Scholar.

Himani Sharma works as a research scholar at the School of Agriculture, Uttaranchal University, Dehradun (Uttarakhand), India. Himani Sharma did his graduation with B.Sc. (Hons.) Agriculture from Sher E Kashmir University of Agricultural Sciences and Technology, Jammu, and currently pursuing her master's degree of M.Sc. Ag. (Agronomy) from School of Agriculture, Uttaranchal University Dehradun (Uttarakhand), India. She also published 12 articles, 3 book chapters (Scopus Indexed), 2 trainings, and 3 national/international seminars/conferences. She has a life member of the Indian Society of Agriculture & Horticulture Research Development (ISAHRD), Chandigarh, Punjab, India. She has also received the Young Achiever Award for outstanding contribution to the field of Agronomy on the occasion of the international conference "Innovative Approaches in Agriculture, Horticulture & Allied Sciences" jointly organized by SGT University Gurugram and Just Agriculture, Magazine. She has expertise in soil fertility, nutrient management, and irrigation water quality.

Rajat Singh is currently associated with the School of Agriculture, Uttaranchal University Dehradun NAAC A+ as Assistant professor. He completed his B.Sc. (Agriculture) in 2016, his M.Sc. (Horticulture) in 2018, and completed his Doctorate in Horticulture with a Specialization in Vegetable Science. Dr. Rajat Singh has authored 1 Practical Manual, 21 Hindi and English's popular, 16 Research Paper and 3 review papers in SCI, Scopus, IEEE, and NAAS-rated journals, and 8 book chapters, 1 book, National and International Conferences, 3 Workshop and Editorial Board Membership awarding Science Letters (International Journal). He has 4 awards for his excellence in research.

Nafeesa Zahid working as Head of Department, Department of Botany, University of Kotli-11100, Azad Jammu and Kashmir, Pakistan.

Index

A

Abiotic stress 1, 13, 18, 29, 50, 55, 58, 60-62, 111, 115, 117, 129, 141, 143, 145, 153, 155-157, 175, 184, 186, 211, 213, 217, 222-223, 239, 241-242, 245, 247, 251, 257, 261, 268-269, 273, 280-283, 285, 288-289, 291, 294-295, 297-298, 303, 305-306, 309, 312-313, 355
Agricultural 1-3, 14, 16, 21-22, 25-26, 28, 30-31, 33-35, 39-40, 44-45, 48-49, 54, 59, 64, 67, 69-70, 74-78, 80, 83-86, 101, 105-106, 109-128, 130, 132-133, 135-140, 142-143, 146-150, 152-153, 157, 160, 169-170, 173-174, 176-178, 180-181, 183, 185-186, 193, 195-197, 202, 204, 209-210, 216, 218-219, 221, 223-225, 235, 239, 242-243, 245, 247, 252-253, 255, 263-264, 266-272, 274, 277-278, 283, 287, 289, 291-292, 294, 301, 303, 306, 309-311, 315-321, 325-327, 329, 331-337, 340-341, 344-345, 348-350, 352-355, 359-362
Agricultural yield 272, 294
Agriculture 1, 10, 13, 15, 17-19, 21-24, 26-28, 30-31, 33-35, 37-45, 47, 56, 65-66, 69-70, 75-78, 80-85, 87, 90, 92, 103, 105, 109-132, 134-146, 148, 151-157, 166-169, 175-177, 179, 181-182, 184-185, 188, 190, 195-198, 207, 214-219, 221-222, 235, 238-243, 245, 247-248, 250, 253-257, 259-260, 263-265, 268-272, 275, 277-283, 285-287, 289, 291-292, 301-303, 305-306, 308, 310-312, 315-321, 323-331, 333-341, 345-346, 348-349, 352-357, 359-362
antioxidant 5, 7, 10, 14, 21, 24, 26, 28, 35, 43, 49, 51, 56, 59-62, 64, 67, 72, 78, 99-100, 102, 118, 144, 146, 155-156, 167-168, 177, 214-217, 231, 238, 244-247, 249-252, 256, 264, 266, 270-273, 276-277, 279, 284, 293-294, 296-299, 302, 309-310, 348

B

Biotechnology 18, 35, 37, 40, 43, 78, 81, 84-85, 103, 130, 136, 142, 154, 156, 169, 173-174, 178, 180, 188, 190-191, 210, 213-216, 218, 256-257, 259, 261, 270, 283, 286, 288, 290, 311, 337-338, 341, 347, 350, 354, 357, 360, 362
biotic and abiotic stress 1, 55, 62, 115, 145, 155, 175, 186, 222, 241-242, 281, 355
Biotic stress 1, 6, 18, 41, 46, 59, 62, 116-117, 129, 190, 217, 223, 233, 241, 250, 273, 286, 292

C

carbon nanotubes 4, 13, 17, 25-26, 35, 42, 51, 54, 62, 65, 78, 86, 90, 93-109, 118, 125-126, 144, 155-157, 166, 209, 229, 237, 246-247, 250, 252, 256, 258, 261, 264, 293, 306, 355-357, 360
Climate change 21, 30, 37, 43, 80, 112, 115, 120, 123, 126, 131, 140, 142, 176, 218, 222, 236, 264, 279, 283, 286-287, 311-312, 319, 339, 353
Crop resistance 22, 116, 292

D

Developments 15, 17, 82, 110, 113, 127, 132, 135, 140, 156, 196, 211, 236, 238, 283, 285, 321, 336, 338, 342, 361

E

Environmental 1-3, 7, 11, 14-17, 22-23, 26, 28, 30-36, 38-39, 41, 43-44, 55, 57, 59, 61, 67, 69, 72, 75-77, 80, 82, 86-87, 91, 95, 102, 104-105, 107-109, 111-115, 117-118, 120, 122-124, 127-140, 143-144, 146-152, 154-155, 157, 161, 168-169, 176, 179-180, 184, 186-187, 190, 196, 202, 207-211, 213-215, 218, 222, 230, 233-237, 240-241, 243, 251, 255-257, 261, 263, 265-266, 268-271, 273-274, 276, 278-280, 282, 284, 286-288, 291-292, 300-301, 303-305, 307, 309, 311-312, 315-316, 318, 321-323, 325-326, 328-338, 345-346, 349-350, 353-356, 359-360

F

fertilizers 2, 10, 21, 27, 29, 37, 42, 44, 46, 53, 55, 57, 81, 87, 112-115, 118-129, 131, 135-136, 143, 148-149, 179, 196, 199, 201-202, 205, 208, 224, 235, 258, 267, 272, 275, 291-292, 316-317, 319, 322, 326-328, 330, 337, 341, 345, 348, 353-356, 358, 360
food security 13, 15, 21, 44, 76, 111-112, 115, 119, 121, 123, 131, 135, 137-141, 144, 152, 157, 159, 215, 218, 221, 235, 242, 282, 300-301, 312, 315-319, 327, 332-337, 339-341, 347, 352-354, 360
food stability 340

G

genes 2-3, 5-9, 46, 51, 62, 64, 96, 98-99, 116, 144, 151, 160, 162, 165-168, 170-172, 176-183, 186, 189, 231, 272, 275-276, 293, 297-298, 301-302, 308
genome sequences 175, 179, 181
Geoinformatics 18, 41, 190, 193-196, 217
Geoinformation Technology 193
germination 1-2, 4-5, 7-8, 10-11, 13-19, 21-22, 24-30, 32-33, 35-38, 40-42, 44-46, 48-51, 55-59, 61-62, 65-69, 71-85, 87, 91-92, 94-95, 97-107, 109, 112-113, 116-117, 127, 139, 143-145, 152-153, 155-156, 160, 166-169, 182, 190, 210, 218, 221-222, 230-239, 241-245, 247-261, 265-273, 275-279, 281, 283, 286, 288, 292-302, 306, 308-309, 311, 313, 352, 354-355

H

heavy metals 2, 28-29, 56, 123, 130, 149, 221-229, 233, 235-238, 248, 252, 273, 331, 344
Horticultural crops 160, 241, 280, 289

M

Marker-assisted breeding 159, 181
molecular breeding 159, 172, 185, 307
MWCNTs 90, 93-98, 100, 299

N

Nano priming 17, 51, 67, 101, 117, 159-160, 166-169, 221, 223, 245-246, 248, 250-251, 258, 276, 278, 287, 355
Nano seed Priming 249, 253
Nanomaterials 1, 3-4, 13, 15-16, 19, 22-23, 25-26, 30-31, 34-35, 38-39, 44-47, 51, 53, 63, 66, 69-71, 74, 80, 82, 90, 95, 99, 101-103, 105, 108-119, 125, 127-129, 132, 135, 140, 146-147, 155, 157, 207, 211-215, 218-219, 229-230, 235, 241, 245-246, 250, 253, 256, 259-260, 262, 264-265, 267-268, 270, 272, 277-281, 286-287, 289, 293, 303, 306-307, 312, 315-318, 320-323, 326, 330-335, 337-341, 343-346, 350, 352, 355-357, 359-360, 362
Nano-nutrients 113
nanoparticles 2-4, 6-7, 10-11, 13-14, 16-19, 21-42, 44-46, 51-66, 69-79, 81-88,

92-94, 96, 98, 101, 104-107, 111-112, 114-117, 119, 121-122, 124-129, 131-132, 134, 138, 141, 143-157, 160, 166-169, 190, 205, 207-218, 222-223, 229-230, 233, 235-240, 243, 245-261, 263-272, 275, 277-278, 280, 282, 284-288, 322-323, 328, 331, 333-335, 338, 340-345, 347-348, 352, 355-357, 359-361
nanopriming 1, 16-18, 21-22, 36, 40, 46, 50-51, 63, 77, 79, 85, 90, 104-105, 110, 142, 144-147, 150-152, 159, 175, 193, 209, 221, 245, 249, 251-253, 257, 259-260, 265-267, 270-273, 275, 277, 279-280, 283, 286-287, 290, 293-297, 301-302, 308, 311, 315, 352, 359
nano-seed priming 270-271
Nanosensors 45, 206, 264, 266, 274-275, 315-318, 324-325, 329-331, 333-337, 341, 346, 355
Nanotechnology 3, 21-23, 31, 34-36, 39-40, 44-45, 66, 70-71, 75-79, 82, 84-85, 88, 90, 101, 105-106, 111-118, 120-122, 126-127, 129-132, 134, 136-138, 142-143, 154-157, 204-205, 207-209, 211, 213-219, 222, 237-238, 241, 245, 253, 255, 259, 263-267, 269-271, 278, 280, 282-283, 285-287, 289, 292, 305-306, 308, 310-313, 315-321, 325-328, 330-335, 337-350, 352-354, 357-361
nanotubes 4, 13, 17, 26, 35, 42, 51, 54, 62, 65, 78, 86, 90, 93-109, 112, 118-119, 125-126, 144, 155-157, 166, 209, 229, 237, 246-247, 250, 252, 256, 258, 261, 264, 293, 306, 355-357, 360
NPs 3-4, 6-10, 12, 23-24, 27-28, 54, 56, 72, 88, 92, 114-115, 117-118, 121, 124, 126, 154, 209, 223, 229, 233, 235, 247-248, 254, 265-267, 269-273, 275-276, 278, 293, 299-302, 304-313, 352, 355-359

O

organism 33, 73, 136, 175-176, 178-179, 222

P

pesticides 2, 7, 10, 21, 33, 44-46, 53, 59, 108, 111-112, 114-115, 117, 119-120, 126-129, 135, 137, 144, 147, 149-150, 153, 157, 177, 179, 187, 196, 209, 222, 229, 272, 291-292, 316, 318, 323-324, 328, 332-333, 337, 341, 344-345, 353-356
photosynthesis 9-10, 27-28, 72, 93, 122-123, 143, 157, 184, 221, 233, 246, 252, 264, 266-267, 270, 273, 277-278, 292, 357, 360
Plant Pathogen 352

Q

Quantitative Trait Loci 162, 164-165, 169-171, 173-174, 188

S

Salt stress 18, 22, 27, 44, 55, 61-62, 64, 77, 233, 247, 252, 261, 266, 274, 276, 279, 288, 290, 292, 294, 300-302, 305, 313
Seed and Nanopriming 241
Seed germination 2, 4, 10-11, 13-14, 17, 22, 24-25, 28-29, 32-33, 35-38, 41, 48-51, 56-58, 61-62, 66-68, 71-72, 74, 77-83, 91, 94-107, 112, 117, 143-145, 152, 160, 166-169, 182, 190, 210, 218, 232, 235, 237, 239, 241-243, 246, 248-249, 253-254, 256-258, 260, 263, 265, 267-268, 270-271, 273, 275-276, 278, 283, 286, 288, 292, 296-297, 300-302, 311, 313, 352, 354-355
seed priming 2-3, 15, 17-18, 21, 23-26, 29-33, 35, 42-43, 46-52, 54, 59-62, 64-66, 70-73, 77-78, 80, 82, 86-87, 91-92, 96-97, 99, 105-106, 116, 129, 135, 144-145, 147, 154, 156, 160, 166, 209, 215, 222, 230-233, 235-236, 238-239, 241-246, 249-258, 260-261, 263, 265-268, 273, 280, 282, 285, 288-289, 293-298, 300-302, 304-308, 310-311, 355
Seed viability 9, 90, 147

seeds metabolism 142
Smart breeding 159
Soil Survey 193, 198, 218
sustainability 3, 16, 18, 31, 38, 44, 70, 82, 105, 111-112, 125, 128, 131, 134, 136, 143, 145-146, 185-186, 211, 222, 239, 253, 264, 269, 282, 285, 288, 316-319, 326-327, 336, 347, 354
sustainable agriculture 1, 10, 13, 15, 17-18, 21, 23, 31, 37-41, 43-44, 76, 80-83, 85, 110, 112-114, 117, 122, 124, 128-129, 131-132, 134-138, 140, 143-145, 151-153, 156-157, 167, 175-177, 190, 207, 215-219, 238-239, 259-260, 263-265, 268, 271, 275, 280, 283, 285-287, 289, 302, 310-312, 316, 319, 327-328, 331, 333-339, 352-353, 357, 359, 361
SWCNTs 90, 93-96, 98, 100

T

technologies 2, 15, 18, 22, 41, 64, 69-71, 90-91, 109, 115, 133, 137, 151-152, 154, 174-178, 180-184, 186, 189-190, 193, 196-197, 210, 217, 222, 241-242, 264, 266, 269, 271, 292, 319, 325, 330, 335-336, 346-347
toxicity 2, 14-16, 28, 30, 39, 66, 69, 72, 77, 83, 95, 97, 99-101, 107, 118-119, 138, 143, 149, 151, 156, 208, 221, 226-227, 235, 237-238, 252, 255-256, 275, 288, 299, 309, 317, 340-341, 344-346, 350
transcriptomic 24, 49, 175, 177, 179, 182-183, 231, 238

V

vigor 2, 7-8, 14, 22, 24-25, 35, 49, 67, 69, 78, 90, 101, 103, 147, 160, 167, 221, 230, 233, 235, 239, 258, 261, 267, 269, 310

Y

Yield 1-2, 9, 11, 13-14, 16, 21, 24-25, 28, 30, 32, 35, 42-45, 49, 64-65, 67, 71, 73, 77-79, 81, 103, 106, 115-118, 123-124, 144-146, 150, 152-153, 171, 173, 175-176, 179, 195-197, 199-204, 208-209, 232-234, 239, 242, 245, 250, 253-254, 256, 259, 263, 269, 272, 275-277, 279-280, 287-288, 290, 293-294, 298, 301-302, 304, 307, 315, 318, 322, 325, 333, 350, 352-357, 359

www.ingramcontent.com/pod-product-compliance
Lightning Source LLC
LaVergne TN
LVHW080348011026
849161LV00009B/2226